Senior High: A Home-Designed Form+U+la

("Form+U+la" is pronounced "Formula."
See page I-13 for an explanation of the title.)

by Barbara Edtl Shelton

(in 1969)

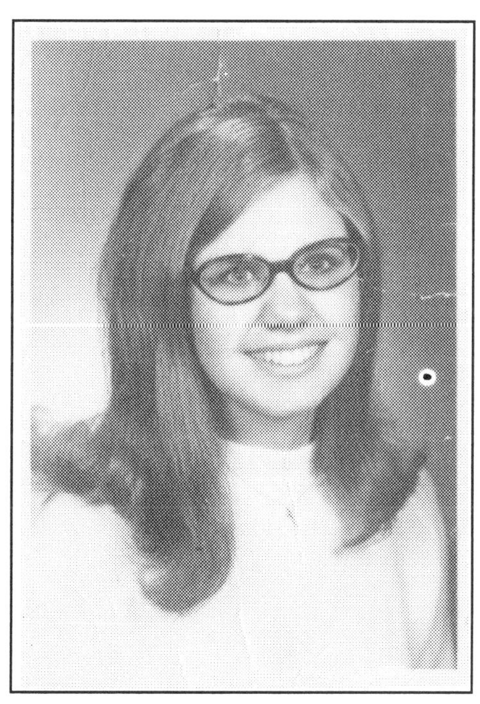

Published by:

Homeschool Seminars and Publications
182 No. Columbia Hts. Road
Longview, WA 98632
(360) 423-4912

SENIOR HIGH: A HOME-DESIGNED FORM+U+LA

Written by: Barbara Edtl Shelton

Published by:
Homeschool Seminars & Publications
182 No. Columbia Heights Road
Longview, WA 98632

Printed in the United States of America

Third Edition (1999 Revision) © 1999 by Barbara Edtl Shelton
Published by Homeschool Seminars & Publications ISBN 1-887639-09-8

Second Edition (1996 Revision) © 1996 by Barbara Edtl Shelton
Published by Homeschool Seminars & Publications ISBN 1-887639-03-9

First Edition / Second Printing © 1993 by Barbara Edtl Shelton
Published by Howshall Home Publications ISBN 1-883288-50-9

First Edition / First Printing © 1993 by Barbara Edtl Shelton
Published by Triune Bible Institute & Seminary Press ISBN 1-55967103-3

For retail or wholesale sales, or speaking inquiries, contact the author at:
Mail: 182 No. Columbia Heights Road • Longview, Washington 98632
Phone: 360-577-1245 *(Homeschool Information Line)*
360-423-4912 *(Home Line)*
Email: BEShelton@aol.com *If no response, please try:* BEShelton@kalama.com
Website: *(As of April 1999)* http://www.ourhomeourschool.com/

COVER DESIGN *(front and back)* by **Barbara Shelton**

ALL FORMS and CHARTS created by **Barbara Shelton,**
using WordPerfect 5.1 for DOS or WordPerfect 7 for Windows, or Word 97

PRINTING & LAY-OUT of cover and pages by **Courier Companies, Inc.**
15 Wellman Avenue • No. Chelmsford, MA 01863-1334

Dedication

This revision of
Senior High: A Home-Designed Form+U+la
is dedicated to my cherished family:

Sharnessa:

My "premier run" on lots of the stuff in here
(Round 1 of homeschooling through high school!);
now at age 21, my precious Daughter and Friend

Tory:

My "polisher" - not just of the stuff in here
("Round 2" of homeschooling through high school!)
but of me personally in my walk with Jesus! ☺

Carlianne:

My fun little companion and sweetheart;
(Sorry, I know you're 12, but you'll always be my "baby"! ☺)
One more chance to do it all again! ~ *(Round 3!)*

Dave:

You are "The wind beneath my wings"
Without you, this book (and our ministry) would not be!
You are my Knight in Shining Armor, my Best Friend,
The Love of my Life!

TABLE OF CONTENTS:

Section Titles and Items of Special Interest

(All individual forms are listed in the Form Index that immediately follows the Table of Contents.)

FRONT STUFF *(Before the Jumpstart Plan and Book actually begin)* . . . . You Are Here

Dedication iii
Table of Contents . . . . iv–vi
Form Index . . . . vii–viii
To the Reviewer (An important note to reviewers or skeptics!) ix
Special Thanks x
With Appreciation xi
Before You Read This Book! xii
To the Undecided xiii
About the Cover xiv
Forludection – by Debra Grace xv
Acknowledgment of My Beloved Parents xvi

JUMPSTART PLAN (SECTION BEFORE THE BOOK STARTS)

Before You Begin (this section) . . . . START
Notes and Tips Before You Begin Your Planning JP-2
Step-by-Step Directions for How to Set up Your Jumpstart Plan JP-3
Sneak Preview of the Bigger Picture JP-4
Jumpstart Class Ideas [*See Form Index for individual listings*] JP-5–10

SECTION ONE: OUT OF FEAR AND INTO FREEDOM

Quote of Marilyn Howshall from *The L. of L. Approach: "A L of L can be a fresh beginning..."* . . . . . START
Quote from *My Utmost for His Highest* by Oswald Chambers *(on disordered, unfocused faith)* I-16
A Few Q's and A's I-17
On Fears, Goals, and Values I-19
Wisdom's 7 Pillars for True Education I-21

SECTION TWO: REQUIREMENTS, COLLEGE, PROMOTION & GRADUATION

Article on "Honors Diploma" . . . . START
What About Requirements? II-1
Several *(awesome and surprising)* Quotes of Helen Keller about College . . . . II-12 and 14
Requirements for Promotion from Grade to Grade II-19
Graduation Stuff II-26
Graduation Ideas and Supplies II-27
Sharnessa's Graduation II-30–31
Tory's Graduation II-32–35

SECTION THREE: RECORD-KEEPING SYSTEM [*Please see Form Index for listing of all forms in this section*]

"Pioneer Girl" by Sharnessa Shelton *(and picture of Sharnessa 'n' Mom)* . . . . START
Record Keeping Stress 'n' Strain Reducers III-11
"Testimony Time!" from Patti in Florida III-15

Section Four: Potpourri of Curriculum Supplements

Q's and A's relating to this section . . . . . START
Intimidometer IV-2
Shelton's Theory of Brain Receptivity IV-3
Learning Guide Explanations IV-5–26
Higher Thinking Skills Chart IV-12
"For Mom to Ponder" by Marilyn Howshall IV-19
Important thoughts on "Education" IV-25
Learning Guides – Masters *[See Form Index for individual listings.]* IV-27

Section Five: Grading Guidelines & Portfolios

One Homeschool Mom Discovers the Heart of *Wisdom's Way of Learning* . . . . . START
Grading Guidelines and Considerations V-1
Q's and A's on Grading V-4
Guidelines for Assembling a Portfolio V-5
Quote of Marilyn H. from *Develop a Lifestyle Routine: "I found that the only schedule..."* V-7
Excerpt from *Uncle Eric Talks About Personal, Career & Financial Security* V-8
The "Grading Criteria & Course Assessment" forms for 19 classes V-9–28

Section Six: Personalizing All this Stuff

Encouraging Letter to Reluctant Homeschool Boy – From Sharnessa Shelton . . . . . START
The Shelton Family Back-to-Basics Schedule ☺ VI-1
One *Actual* Basic Schedule VI-2
What You'll Find in Sections 6-A through 6-D VI-4–6

Section 6-A - Formulating Your Framework

Misunderstanding the Lifestyle of Learning Message - by Luz Marina Lovins . . . . . START
Content-Based Framework Explanation VI-A-2
Goal-Oriented Project-based Framework Explanation *(Mary Schofield's)* VI-A-3
Time-Based Framework Explanation VI-A-3
Time/Competency-Based Framework Explanation *(Diana McAlister's)* VI-A-4
Time/Content-Based Framework Explanation *(Barb Shelton's)* VI-A-4
Unit-of-Life Project-Based Framework Explanation *(Marilyn Howshall's)* VI-A-5
Goal-Oriented Project-based Framework Samples *(Mary Schofield's)* . . . . . VI-A-6 and 7
Time/Competency-Based Framework Samples *(Diana McAlister's)* . . . . . VI-A-8 and 9
Time/Content-Based Framework Samples *(Barb Shelton's)* . . . . . VI-A-10 and 11
Unit-of-Life Project-Based Framework *(Marilyn Howshall's)* . . . . . VI-A-12 and 13
Formulating the Time Aspect of Your Framework; Steps 1–8 . . . . . VI-A-14—29
Excerpts from a Mother's Day letter from Sharnessa to me VI-A-24
But Does It Work? VI-A-30

Section 6-B - Designing Your Classes

"What is Real-Life Learning?" . . . . . START
Quotes of John Taylor Gatto *(woven throughout text on pages...)* VI-B-1, 2,8
Three Class-Creating Approaches Overview *(starting on page...)* VI-B-8
Quote of Marilyn H. from *Get to Know Your Child: "The child must have the privilege of..."* VI-B-9
Collect and Construct Approach VI-B-10
"Blessed are the Flexible" – Quote of Cathy Duffy VI-B-12
Priority Appropriation Approach VI-B-14
"Community Service" – Quote of Diana McAlister VI-B-14
Dissect And Distribute Approach VI-B-18
"Insights On Credits" Quote of Mary Schofield: VI-B-23

"Dual Enrollment: A Homeschool Option?" – by B❤ VI-B–24
Perspective from Homeschool Mom, Betsy Lewis, on Dual Enrollment. VI-B–25
Schofield Family Basic Skills List (*PERFECT* for "Priority Appropriation") VI-B–26
Q's and A's on Designing Your Classes . . . . . VI-B–27—29
Quote of Marilyn Howshall from *The LO L Approach:* ("Many of the principles outlined...") VI-B–29
"Delight-Directed Study" by Gregg Harris VI-B–30
"Delight-Directed" (*One of the 7 Natural Vital Signs of the Learning Process*) Marilyn Howshall VI-B–31
"A Brief Intro to Delight-Directed Learning" by Barb Shelton VI-B–32
Conversations on Figuring Classes, Hours and Subjects VI-B–33

SECTION 6-C - ORGANIZING YOUR MATERIALS

The Wisdom of Noah Webster – Excerpts from *Advice to the Young* . . . . . START
"Sharnessa's Day in Court" VI-C–1
"Shelton Family Memory Page" of Photos VI-C–7
Photo of all five of us ☺ (Just a snapshot I *had* to include!) VI-C–10
List of Class Titles VI-C–13
Quote of Ellyn Davis: "Relationship Skills, Life Skills, Academic Skills VI-C–14

SECTION 6-D - COMPLETED CLASS PLANS

A Panicky Mom's Questions, and Donna Heck's Answers . . . . . START
Q's and A's About These Forms VI-D–2–6
Completed Class Plans and Explanations [See Form Index for individual listings] VI-D–7
Email 101 (for you newbies) VI-D–25
Photo: Tory examining a dissected cow eyeball VI-D–56
Photo: The Shelton Family doing "Homestyle P.E." VI-D–64
Cultural Literacy Resources (A discussion of them) VI-D–82–83
Testimony Tidbits from a Recovering Homeschool Mom (Leslie Gamell) VI-D–84

SECTION SEVEN: LIFESTYLE OF LEARNING APPLICATIONS

"The Heart of This Message" – Marilyn Howshall . . . . . First & Last Pages
Q's and A's about adopting a Lifestyle of Learning at the high school level VII-2–11
"Three Levels of Learning" by B❤ VII-13
"Skeptic Turned Believer" by Pam Heath VII-18
"Finding Him As I Seek" by Pam Heath VII-19

SECTION EIGHT: RESOURCES, REVIEWS, & GLOSSARY

"What Curriculum Do You Use?" - by B❤ . . . . . START
Homeschool Resource Companies VIII-1–2
High School-Related and Legal Resources VIII-3
Handy, Pre-made Catalog Request Notes ☺ VIII-4
A Letter about Form+U+la from Sue Howe VIII-5
A Review of Form+U+la by Julie Lofurno VIII-6
Info and Explanation on the SHF-L (Sr. High Form+U+la) Email List VIII-7
What to Do About Lab Science?!?! VIII-8–9
Homeschool Guide-a-log Write-up and Comments VIII-10
Glossary of Terms Used in This Book VIII-11–12

Form Index

Remember! You will NOT need ALL of these! ☺
This is just a Smorgasbord from which to pick and choose.

Form Title **Page**

Algebra I ACS VI-D-7
Algebra II ACS VI-D-9
Algebra - Grading Criteria V-10
America Online E-mail & Forum Operations TAS VI-D-20
America the Beautiful Vocabulary Worksheet IV-27–28
Armor of God Study Chart IV-29
Art & Crafts ACS VI-D-37
Art of Living TAS VI-D-39
Baseball - Grading Criteria V-11
Basic Cooking Skills TAS . . . . VI-D–47 & 48
Bicycle Maintenance TAS VI-D–51
Book Report IV-30
Book Synopsis IV-31
Building Family Relationships - Grading Criteria V-12
Building Life Relationships ACS VI-D-11
Business & Office Skills ACS VI-D–13
Carpentry Skills TAS VI-D–52
Choreography & Production ACS VI-D-38
Christian Denomination Worksheet IV-32
Christian Religion Types IV-33
Christian Ethics ACS VI-D–15
Class Content Plan III-5
Class and Course Titles VI-C–13
Collect-and-Construct Approach - Mind Map VI-B-13
Collect-and-Construct Approach - Resources VI-B-11
College-Age Weekly Plan III-36
Come and Drink Study (For Mom *or* Student) II-18
Composition (English) - Grading Criteria V-13
Computer Literacy Skills, General - TAS VI-D–21
Computer Skills Survey ACS VI-D–17
Computer Terms For Occupational Ed. IV-34
Computer: Word Processing - Grading Criteria V-14
Concert/Performance Record IV-35
Cooking Terms IV-36
Course Titles/ Class Titles VI-C-13
Creation Science Research Report IV-37
Credit Completion Bar Graph - Mini-Sample III-32
Credit Completion Bar Graph Master - 120 hrs/cred III-34
Credit Completion Bar Graph Master - 150 hrs/cred III-33
Dinner Making Schedule IV-38
Dinner Planner IV-39
Diploma of Graduation - Master II-26

Form Title **Page**

Dissect-and-Distribute Approach - Ideas, Skills, Res's VI-B-19
Dissect-and-Distribute Approach - Mind Map VI-B-21
Dog Report IV-40
Education - Word Study IV-73
Educational Foundations ACS VI-D–27
Elected Officials & Functions IV-41
English I (Comp & Grammar) ACS VI-D–29
English II (Lit & Speech) ACS VI-D–31
English III (Study & Res. Skills / Vocab) ACS VI-D–33
English: Composition - Grading Criteria V-12
English: Spelling/Vocabulary - Grading Criteria V-23
Family Meeting Guide IV-42
Fine Arts Survey ACS VI-D–35
Fort Clatsop Museum Search-it-out Sheet IV-43
General Computer Literacy Skills TAS VI-D–21
Goal-Oriented Project-based Framework Samples . VI-A–6 & 7
God's Actions Study IV-44
Gourmet Cooking - Grading Criteria V-15
Gourmet Cooking TAS VI-D–49
Grade Promotion Requirements II-23
Grading Criteria - Blank V-9
Grading Criteria Examples *(in alphabetical order)* V-10-28
(These are all listed separately here in the Form Index, according to class)
Graduation Announcement II-36
Graduation Invitations II-28, 29, 30, 31,35
Graduation Program (Tory's) II-34
Ground Rules for Studying Other Religions IV-45
Health & Safety ACS VI-D–41
High School Days Memory Booklet IV-46
Home & Life Management Skills ACS VI-D–43
Home Maintenance & Repair TAS VI-D–46
Home Repair (etc.) Job Log IV-47
Home Report IV-48–49
Housework Skills TAS . . . . VI-D–53 & 54
Howshall's Description of Requirements"Bird Book" . . VI-A-13
Howshall's Master Plan for Kathryn *(Unit-of-Life Project-Based)* VI-A-12
Hunting - Grading Criteria V-16
Identifying Prepositional Phrases IV-50
Identifying Prepositional Phrases Answer Key IV-15
Input Box VI-C-11
Internet Literacy Skills TAS VI-D–22
Intimidometer ☺ IV-2

Jumpstart Class Ideas for:
- Biblical & Spiritual Studies JP-5
- Computer Skills Survey JP-6
- Educational Foundations JP-7
- English JP-7
- Home & Life Management Skills JP-8
- Mathematics JP-8
- Social Studies JP-9
- Lab Science Survey JP-10

Junior High Weekly Plan III-35
Lab Science - Grading Criteria V-17
Lab Science Survey / Basic Scientific Literacy ACS .. VI-D–57
Lawnmower, Riding - TAS VI-D–51
Learning Guide Explanations: *These are in alphabetical order on pages* .... IV-5—26
Living Expenses Comparison Study VI-D–55
Location Data Sheet IV-51
Master Framework Plan - Blank Master VI-A-29
Master Framework Plan - Filled-in Sample VI-A-30
Movie Critique & Analysis IV-52
Music Appreciation - Grading Criteria V-18
Music Appreciation Worksheet IV-53
Name Study IV-54
Nutrition - Grading Criteria V-19
Occupational Education ACS VI-D–59
Occupational Skills Interview IV-55
Office & Business Skills Job List IV-56
Open Line & Mind Evaluation IV-57
Organization Skills Job List IV-58
Pace Determiner Worksheet VI-A-27
Parental Educational Input Chart III-22
Performing Arts - Grading Criteria V-20
P.E. / Basketball ACS VI-D–62
P.E. (General - Tory) ACS VI-D–61
P.E. (Soccer) ACS VI-D–63
P.E. (General - Sharnessa) ACS VI-D–65
Personal Healthful Living Planning Sheet IV-59
Personal Organization Checklist (Female) IV-60
Personal Organization TAS (Boys) VI-D–45
Physical Education (P.E.) - Grading Criteria V-21
Play Critique & Analysis IV-61
Plot Wholesomeness Evaluation IV-62
Poet Synopsis IV-63
Priority Appropriation Approach - Ideas VI-B-15
Priority Appropriation Approach - Mind Map VI-B-17
Record of Work Done - 1-column; Content-only type III-9
Record of Work Done - 2-column; Content-only type III-10
Record of Work Done - 1-column; Hour Check-off type .. III-13
Record of Work Done - 2-column; Hour Check-off type .. III-14
Religions of the World - Grading Criteria V-22
Religions of the World Study Sheet IV-64
Requirements for Graduation (WA St. as example) II-5
Requirements for Graduation: Comparison of 6 states II-7
Requirements for Graduation - "Additional Suggested Classes for the Student Who Intends to..." II-13
Requirements for Promotion from Each Grade II-23
Scientific Imaging Techniques IV-65
Scripture Verse Study IV-66
Self Esteem & Understanding - Grading Criteria V-23
Social Studies ACS VI-D–67
Spanish ACS VI-D–69
Spelling/ Vocabulary (English) - Grading Criteria V-24
Spiritual Conferences & Meetings ACS VI-D–71
Spiritual Studies ACS VI-D–73
State History - Grading Criteria V-25
State History & Government (WA) ACS VI-D–79
Staying on Track Record - Master III-19
Staying on Track Record - Sample III-21
Steps 1–8 of Formulating Framework .... VI-15 to end
Tact, Grace & Social Skills - Grading Criteria V-26
Tally Worksheet III-17
Tap Dance - Grading Criteria V-27
Tape Report IV-67
Tel-Med Proof-of-Completion IV-68
Three Levels of Learning Chart VII-15
Time/Competency-Based Framework Samples .... VI-A–8 & 9
Time/Content-Based Framework Samples ..... VI-A–10 & 11
Transcript: Filled-in sample III-25
Transcript: Whole blank master III-27
Transcript Components .... III-29 and 31
Typing - Grading Criteria V-28
Typing / Keyboarding ACS VI-D–75
Unit-of-Life Project-Based Framework Samples . VI-A–12 & 13
U.S. History, Government, & Geography ACS VI-D–77
Vehicle (Car, Truck) Maintenance TAS VI-D–60
Virtual Dissection Record IV-69
Vocabulary Worksheet on America The Beautiful ... IV-27–28
War Report IV-70
Washington State History & Government ACS VI-D–79
Weekly Assignment Plan Filled-in Sample III-6
Weekly Assignment Plan Blank Master III-7
Windows Literacy Skills TAS VI-D–23
Wisdom's 7 Pillars of True Education .... I-21 & 22
Wisdom's 7 Pillars Study IV-71
Word Alive Study IV-72
Word Perfect Operations TAS VI-D–24
World Magazine Report IV-74
World History, Problems, Leaders, & Geog. ACS VI-D--81

To the Reviewer:

If you have taken on the task of reviewing this book, I would like to take this opportunity to share my perspective with you *before* you embark on this possibly-overwhelming, possibly-downright-irritating undertaking. ☺ (Or if you're just skeptical.)

I am a homeschooling mom myself, who, shortly after I started into high school with my own daughter, was deluged with requests from homeschoolers wanting to know *what* I was doing for her for high school, and *how* I was doing it. That is the only reason I wrote it. I am well aware that the book has flaws and will not appeal to or meet the needs of everyone. No book can, and I don't even pretend to do so.

I did not have time to take a course in how to write a book. I was and still am homeschooling. ("Let (s)he who has never tried writing a book *and* homeschooling be the last to cast a stone.") I did the best I could, and wanted to provide a book that was what I myself would have liked to have had access to at a time when I was panicking, feeling alone (because I WAS) with no one to turn to for help in this endeavor!

The homeschool movement needs the freedom to produce methods and materials that appeal to the wide diversity of homeschoolers' needs without fearing their work will be ripped apart. Attempting to review something that you just don't click with is not only pointless, but is like reading someone else's mail – a disservice to the potential reader and to the author whose "labor of love" it is. It's also vital, and only fair, that you actually *read* the whole thing before reviewing.

I chose to write in my conversational style for good reason: I knew there was a need for something "user-friendly" for those who are scared stiff about approaching high school. If a person – reader or reviewer – doesn't like it, and I know not everyone will, they are free to use one of the "brass tacks" books that are out there. Any reviewer who takes a very narrow or critical perspective of another's work (mine or not) and represents it to the world in a bad light needs to be aware that they are potentially closing the door on a book that meets a specific need within the homeschooling community, whether or not it fits their needs or style. Hundreds of people have written to me expressing great delight in the book – even the conversational aspect, which I have chosen to write in for good reason. They say it better than I could...

> **"I LOVE your conversational style of writing!"**
>
> **"This book is like having many conversations with a good friend."**
>
> **"I was SO disappointed to get to the end of your book!"**
>
> **"Your writing style is like visiting with you. And I do love your little notes here and there. It keeps the book interesting!"**

So I just ask you to consider that there is a real person behind this book, written straight from my heart to encourage and equip the homeschooling community. I hope you'll view and review this work with your heart as well as your mind. (I've actually only gotten one bad review, but I'd like the number to stay low. ☺)

From my heart to yours,

❥Barb Shelton❥

To:

Georgia Landis and Luz Marina Luvins

Two sisters in Jesus whom I have only met via the Internet, (Georgia in Japan; Luz in Texas!), but with whom I have a sweet bond in Jesus. Thank you for blessing and serving me so graciously and efficiently during the revision of this book!

And to:

Marilyn Howshall

for profoundly affecting (modeling and remodeling) my thinking about true education and learning, and for graciously allowing me to use excerpts from her writings. (Actually a blessing to you!)

And to: (in alphabetical order by first name ☺)

Anita Treichel, Donna Schmidt, Jean Merritt, Julie Ouellette, Michelle Maynard and Sherrie Johnson,

Six precious sister-friends whom God has blessed me with. Over the years, they have lent me their ears, their shoulders, and their helping hands; opened their homes to me, (especially the Maynards to Carlianne while I did the revision of this book!) Thank you all for blessing me with your love, support, wisdom, grace, and encouragement in Jesus!

And to:

Carolyn "Mom" Shelton

The World's Most Wonderful Mother-in-law, and also my Friend, my Confidant, the "Sunshine" that grows our family, awesome "Gramma" to our three Children, and my Sister in Jesus!

With Appreciation

I want to express my
heart❤felt appreciation to my sister-friend:

Donna Heck

for her ministry of encouragement to:

❥ **Homeschooling Parents** of High Schoolers
She has a gentle, caring way of encouraging those
with fears, frustrations, and concerns with her
"been there, done that" wisdom❣

❥ **Her Family** to whom she is devoted
and loves to brag about! ☺ (Only because it's all true!)

❥ **Me**! – Donna has been a tremendous help to me
in lovingly proof-reading and editing this revision of the book!
She is also my "right hand (wo)man" on the Senior
High Form+U+la email list where she helps me with much
behind-the-scenes and "front-line" legwork,
enabling me to even stay on it!

Thank you so much for all
your devoted hard work and support, Donna!
I don't know what I'd do without you!

❤❤

Before you read this book!

About the quotes that are woven throughout the book... I have selected many quotes that I know you will find interesting, helpful, eye-opening, and possibly even downright irritating. I want to challenge you to read them anyway, and to not consider them "extraneous." Even though they were written by others, they are very much a part of this book and its purpose, not merely "space fillers"; I do not at *all* consider them "extra"! I had a very specific reason for including *every single one of them* and hope and pray you will not skip them, wanting to just get to the "heart of matter." They are *part* of the *heart!* ☺ Some of them actually fit in with the context of the section they have been placed in, but others do not, so for this very reason, I suggest that, even if you are not interested the contents of one section, you flip through that section anyway and read any quotes I have in there. (If you do not *expect* them to fit in, you'll appreciate them that much more.) My guess is you'll probably like most of the quotes more than this book! ☺ That's okay, because part of my intent is to urge you in other directions! Onto bigger and better things!

About the inconsistencies... You will notice – even if you *aren't* a perfectionist – that there are all sorts of inconsistencies in fonts, headers, subtitles, spacings, page number sizes – you name it, I found a way to be inconsistent with it! ☺ It would take another three pages to explain *why* – all the in's and out's behind the lay-out end of self-publishing a book. So if you could just extend grace and assume that either I had good reason for it or was totally oblivious to it, we'll all be the happier for it. ☺

About this one page that you might want to start with... Someone told me that they wish they had seen page VI-A–14 long before they started the book, so feel free to check that out before continuing further with this book! (It's "Step 1: To provide a foundation")

About my humor... I can't leave out the very thing that I personally feel God engifted me with – for a very good purpose. Humor has a way of lightening us up. And as we approach high school – homeschooling through it, to be exact – we are in great need of being "enlightened"! Our perspective made more lighthearted and our burden made lighter as well. So I hope that, if you find my sense of humor a bit "much," you'll ask God for grace to hear my *heart*, and at least understand my intent behind it. ☺

About my conversational style of writing... All my writing is is a written form of me talking. In fact, many people, after hearing me speak at a convention, have come up to me and said "I read your book, and I just have to tell you that before you even started speaking today, I knew exactly what you were going to sound like, and you *did*! You write *just* like you speak!" I wanted this book to be as user-friendly and personal as I could make it. To do so, I had to disregard (not break, just ignore) a few "rules of writing." I use correct grammar as much as possible, but my writing is anything but "formal," so please don't present it to your kids as an example of anything but "writing from the heart!

To The Undecided

In case you have not yet quite decided whether or not this book is for you...

...please refer to the reviews and comments in the "Resources, Review, and Glossary" section in the back of the book. What others have said about it may be helpful to you in deciding if this is what you need or not.

When I was asking one homeschooling mom how she wanted to be signed in a post of hers that I quote in the write-up on the Form+U+la email list (see Sec. 8), she wrote back saying *"If you wanted to put my phone number or e-mail address I would tell all who called to read your stuff and hear you in person and seek Jesus through it all as He truly wants to direct us!!! I suppose I couldn't take the time to answer all the calls but my heart is to share your message with all the homeschoolers I can!"* She is fine with me giving out her email address, and as long as she isn't deluged with emails, she would be delighted to get back to you! Here are some *more* of her thoughts that she wrote to me, and graciously granted me permission to share here in hopes of encouraging any who may need it...

"I just wanted to say thank you for being you once again. For the ministry you have to us moms as homeschoolers. For being in love with Jesus as you so visibly are. For being a pioneer, willing to go through all the difficult places it involves and still coming out smelling like Jesus! Because of God's grace and goodness I went to your session at the Homeschool Conference in Spokane, Wash. this year. Because of your obedience to Jesus to share so very very honestly, I am in a place where I feel like He is able to redeem the mistakes I've made as I confess, repent and seek Him daily. Because of your encouraging e-mail earlier this fall and because of reading *Wisdom's Way of Learning,* and because of your *Form+U+la* book, my hope is that school is going to be much more organized yet flexible this year compared to the rest. We have much redeeming (for God) to do in our family, but we are on a day-by-day path of letting God do His work in us. Just recently a gal at our church asked about homeschooling. She has a daughter the same age as my just turning 5 yr old. I was so thrilled to be able to pass on your *Guide-a-log* and *Make-My-Own Phonics Book* that I had just ordered and received from you. I told her I wished I would have had these books when I started out. I said, "Read the Guide-a-log before you read anything else about homeschooling." It was so nice to help someone just starting out and feel like I led her in the right direction instead of the wrong one. Thank you for producing materials that are understandable and usable, and for just being you. Each time I read your posts on the Form+U+la list, it's like hearing a sermon from my pastor – he has such a real relationship with Jesus, and you do too. It's like when both of you share, you say, "This is how Jesus has worked in me or what he has shown me, now you seek Jesus and see what he has for you."

--Colleen Houger (houger@televar.com)
Homeschooling mom of 5 for 8 years

About the Cover

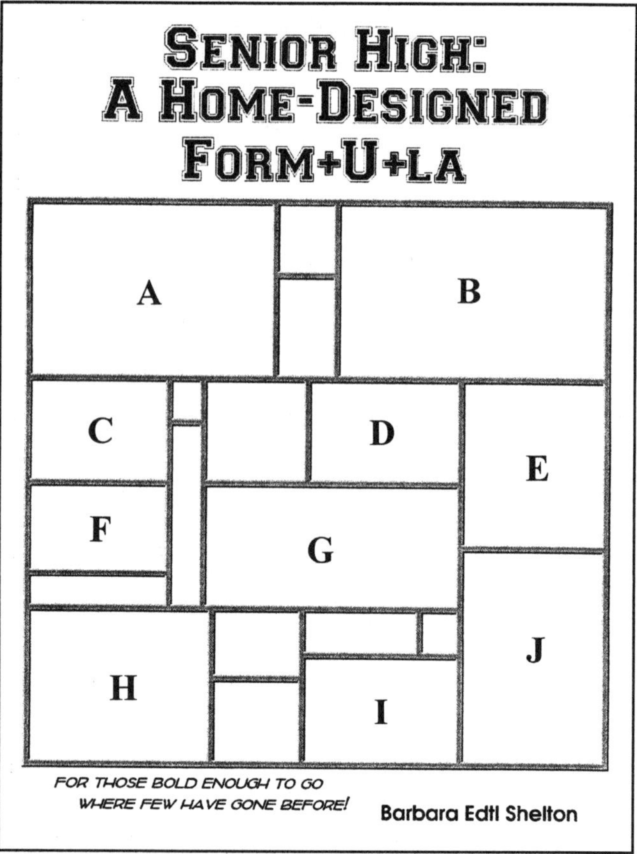

Important Note:
The pumpkin pie is there for the significant reason that it is our family's year-round favorite! And roses are my favorite flower!

A - (Left to right:) Connie Hagen (a mom), and homeschoolers Jamie Cosby, Julie Merritt, Sharnessa, and Katie Lehner work on a craft at our "School Biz" homeschool co-op.

B - Kindra Witt (left), a homeschooled friend of Sharnessa's who had graduated and was attending nursing classes at a community college, helps us with our dissection of a fetal elk (see page VI-D–56 for full story). Next to Kindra is Sharnessa, then Julie Merritt, Carlianne, and Tory. (I'd have to say that Julie's face best reflects *my* original sentiments about this activity! But the more we got into it, and got past our squeamishness, the more *fascinating* it became!)

C - Sharnessa skiing at White Pass in Washington State. (One fun example of "Home-style P.E."! – doesn't have to be at *home*! See page VI–D–64 for yet another example of not-at-home "Home-style P.E."!)

D - Carlianne and Sharnessa had just gotten their hair French braided, and posed sweetly for Mom.

E - Tory works on comb-binding our self-published *Field Trips & Extra-Curricular Activities* book.

F - Carlianne and Tory snuggle up to read *Abe Lincoln* together on the couch.

G - Sharnessa (upper right) and several friends from church pose at the finale of an "Angel Dance."

H - Sharnessa gingerly dissects an earthworm as Mom and Carlianne (gingerly) look on.

I - Dave and Sharnessa (never too old to sit on Dad's lap) share an affectionately fun moment.

J - Tory in action at his beloved game of basketball, "dribbling through the defense trying to create an opening." (a direct quote from his "Basketball Notebook" ☺)

I couldn't decide if this was a
***FOR**eward, a Pre**LUDE**, or an Introdu**CTION**.*
Besides, no one reads any of those anyway.
So here's a...

(Written in 1993, right before the book first came out)

For some strange reason, I feel like I already know Barb, even though we have never met. Reading her material is like crawling into an old, favorite, comfortable chair that has dips and crevices worn into it in the form of my ol' body. Comfy. Relaxing. Renewing. Regenerating. That's how I feel when I read her "stuff" Because of her conversational style of writing, intimidation – which one might expect from a book of this size and nature – is very minimal.

Because of Barb's openness and willingness to share how she and her family do things, the manual has a testimony written into it. Because of her heartfelt sharing of all the wonderful treasures that await us in our relationship with our Lord, she continually rekindles the fire of desire to seek Him more fervently and deeply.

I've counted it a blessing to be given the task of reviewing this manual. I found Senior High: A Home-Designed Form+U+la to be very inspirational and spiritually edifying, and am excited about these high school years now!

And I'm also excited to think of the wonderful things that will happen when this manual hits the hands of homeschoolers across America – life-changing things! May God use it to set people free and draw them closer to Him.

~ *Debra Grace*

Acknowledgment of my Beloved Parents

This book would look quite different if I had come from different parents. (Of course, so would I!) My Mom, in raising seven children, became a master at organization ~ you name it; she can organize it! I used to impress my friends with her laundry room full of labeled baskets and bins, and our family's job charts and color-coded socks! And you should see her office, and all her categorized notebooks, and the Life Notebook she has made and is continuing to compile for each of us kids, which cover our kids too! (I once thought all this was just "normal.") And to keep all these kids' feet warm as they begin life, she has continued a tradition of making booties for all her own babies, and their babies, (and darling sweaters for the girls!)

My Dad, by necessity, became an expert at building shelves to hold all the stuff of this big family! He's actually an expert at carpentry in general, having built many a Barbie house deluxe, tiny-tot table-and-chair set, deck, and fence. He's amazing at landscape gardening! (Their backyard is a haven!), and from before the time I was born, he has kept a daily log of every single day in his yearbooks, all of which stand at attention on a shelf (that he built, of course) in his den, ready to ascertain any date in question. He can tell you anything pertaining to history. and is a great story-teller ~ of history and out of his own rich experiences...

And Mom and Dad do it all with love, grace, and humility.

❦ I love, cherish, and appreciate you both so much! ❧

(Dave laid on his back on the ground to get this picture of us. See the tops of his two knees down in front there? ☺)

The "Edtl Clan" (left to right): Nan, Pat, me, Dad, Mom, Jean, Julie, and Kathy, all of whom I am proud to be associated with as "My Family." John, the sole boy of the seven children *(and please don't feel sorry for him; he usually got his very own room)* is now in Heaven, having been taken by cancer in 1991.

Section Before the Book Actually Starts:

HOW TO SET UP YOUR OWN CUSTOMIZED JUMPSTART PLAN

For those needing to get
their student set up on some sort of
educational plan but haven't had time to read
this book yet, and don't want to spend
a fortune on a bunch of stuff they'll regret
once they know what they're doing! ☺

Before you Begin...

Realizing that some of you
may need to get a student occupied
in some sort of plan before you have time
to read this whole book
and hopefully Wisdom's Way of Learning,
and several others that I recommend for a
Season of Re-education &
Renewing of the Mind
(after which you'll feel much better
about setting up their high school plan),
I have prepared a "Jumpstart Plan"
just especially for you!
Please read everything I have to say
before looking at the charts
that will be coming up shortly;
otherwise you'll very likely become
overwhelmed or confused. ☺

JUMPSTART PLAN FOR HOMESCHOOLING SENIOR AND/OR JUNIOR HIGH SCHOOL STUDENTS

WHAT THIS IS AND ISN'T

The following plan is a smorgasbord of my *ideas* and *suggestions* for you to choose from in getting your teenager set up on a productive, interesting, plan. *Then,* with that set up, *you* will be freer to put yourself through a "Season of Re-education and Renewing of the Mind."

NOTE! It is my personal *strong* recommendation that *every* parent considering homeschooling do this. *And* every parent burning out because of having done "school at home"! As one mom recently wrote to me: "I'm loving your Navigator books. It's taken me thirteen years of homeschooling to find the simplicity I've wanted all along – and *need* now that I have ten children!" Such a season is *vital* to put yourself through *before* trying to come up with a plan for senior high!

Reading this book is only part of this "Season of Re-education and Renewing of the Mind." There is much more that you will find invaluable as you move toward finding out *God's* idea for and purpose of education. This is described in detail in my book, The *Homeschool Guide-a-log.*

Many of the jumpstart ideas here are based on "Learning Guides" in this book, and others on resources you'll find in your home, community, or my book, *the Homeschool Guide-a-log.* The idea is *not* for you to do everything suggested here! This is just a collection of ideas for you to select from. Got it? ☺ *(You really need to get it! Or you'll crash 'n' burn!)*

Whatever plan you come up with for now will *not* be your whole plan for high school! It'll be just enough – all you need for now – to get your student started so that *you* can go through your own re-education season with peace of mind that some good things are happening with your student. With a jumpstart plan, you'll feel less – and hopefully *no* – pressure to get your "Master Framework Plan" set up *right now.* That can wait! You are in a foundation-laying season right now – both you *and* your student – not feeling rushed or forced into the "premature birth" of a larger-scoped plan that you aren't yet ready to create. It would feel very premature at this point, only because it *would* be! ☺

There is nothing "magical" or "official" about my suggestions; such a jumpstart plan could be done in a variety of ways. This just happens to be what *I* came up with.

☺ Once you've gone through your season, you'll have much more vision, ideas, freedom, peace, understanding, and clarity of mind, and you'll feel completely different about setting up your larger plan. *(Really! You will!!)*

TIME TO "DE-BRIEF"

If your student has come out of a difficult situation, or is burned out on school *(even "school at home"),* it's wise to allow a season of "decompression" – some "breathing room" – while still expecting basic help at home, and getting occupied in some "valuably productive" activities, not just TV, phone, and computer games! ☺ No need to rush right into a full-scale jumpstart plan, though it should not be burdensome unless you *make* it so. Read Janie Lavine's article "Bringing Them Home," and ask God for wisdom on the timing for everything!

ABOUT JUNIOR HIGH

There is not a separate set of class content plans for junior high, as just about everything here either would be appropriate for an older junior high student, or could be "geared down" if it's too difficult. I go more by the child's interest and capability than by "grade level."

If you have a "younger" or less mature junior high student, you'll want to use the "olderish" ideas in the Homeschool Jumpstart Navigator for Younger Children rather than this plan.

Please read the remainder of this section in order, not skipping ahead to any other part of the book until I recommend doing so. Just concerned for your blood pressure! ☺

Notes and Tips To Keep In Mind As You Begin Your Planning:

- Whenever a page *(called a "Learning Guide")* is recommended in this book, be sure to read the page *opposite* it in the book for added input, ideas, and sometimes even a chuckle or word of encouragement. *(This book is very "user-friendly"!)*

- In case you go to a form somewhere in this book that has I, P, and M at the top of it, and haven't had a chance to read the directions for that section yet, let me just tell you that the I, P, and M stand for "<u>I</u>ntroduce, <u>P</u>ractice (or <u>P</u>articipate), and <u>M</u>aster." ***Introduce*** means, very simply, that the concept or skill is merely introduced. Mary Schofield, author of *The High School Handbook,* calls it "exposure" to a thing. ***Practice*** means it is done "a few or several times"; and ***Master*** means that it is able to be done alone and fairly well, but not necessarily perfectly. Just FYI! ☺ *(That means "For Your Information" in case you didn't know.)* Oh, and there's an article on this at the end of the Lifestyle of Learning section that thoroughly explains it.

- Do <u>*not*</u> allow yourself to get overwhelmed!!! Anything you don't understand will eventually be explained fully *somewhere* in this book; you're not *supposed* to understand it all now!! *(You've got to have something to look forward to! ☺)* God will give you the grace to get through this part, and I pray that He'll even make it *fun* for you! A *joy!* – not a burden!

- Almost all the resources that I suggest in this plan I am now carrying! If I have inadvertently missed any, please let me know by writing to me at my email address: BEShelton@aol.com or (if that's not working) at BEShelton@kalama.com, or by phone *(via our answering machine)* at: 360-423-4912 or our Homeschool Info Line at 360-577-1245.

- Just keep all the stuff your student does until (at least) you have your Master Framework Plan done. You'll know what to do with it by then.

- Be sure to check out the Task Analysis Sheets (TAS's) in this book as they will offer many practical helps and activities for your student to get going on *now!* They're already all set up for you and offer some very "valuably productive," "real-life learning" ways for students to spend their time, especially as they "decompress" or "debrief" from burn-out! By "ways" I mean activities that will help prepare them for *life* and that will *also* be (make them) a help to the family right now. Remember, these T.A.S.'s are not the *whole* class – only *components!*

STEP-BY-STEP DIRECTIONS
for how to
SET UP YOUR JUMPSTART PLAN!

STEP 1: READ AND DISCUSS: Read through the following suggested subject ideas *and* the "Task Analysis Sheets" that follow the Class Plan (or "A.C.S." = "Assignment Check-off Sheet") for "Home & Life Management Skills." Do this *first* by yourself so that you have an idea of what all is presented here; then with your student, sharing and explaining the things *you're* excited about, and drawing out of him/her what *they're* excited about. Discuss your and your student's preferences and priorities. Together decide on the subjects he/she wants to start out with, marking the specific assignments in each one that he/she will be doing, adding ideas of your own *(that's what the extra line spaces are for)* and highlighting any resources you want to order or hunt down in your home, the library, or from a friend.

STEP 2: SET UP A STUDENT NOTEBOOK Step-by-step directions for how to do this are presented in *Senior High: A Home-Designed Form+U+la,* in section 6-C. You'll want to do a condensed version of this, with just a few subjects, namely the ones in here. It's highly likely that your student can do this himself, though you may want to figure out and purchase supplies together. It's too overwhelming to set up *all* your subjects right now, especially since, at this point, you don't *know* what all you'll want to include in your final Master Framework Plan. So for now, just set up tabbed dividers for the subjects you decide you want to do from this jumpstart plan, which I present on the following two pages. The maximum number of dividers you'll need is 7, unless you decide to start with more ideas out of the book.

STEP 3: GATHER YOUR MATERIALS: Find, order, or borrow all the materials that you have decided you'll be needing for the subjects you have chosen. I have selected several resources that I think would be great choices and included them here in this Guide-a-log for your convenience. However, there will be *many* other resources you'll want and need beyond what I offer here. These are only to get you going and are my personal favorites.

STEP 4: COPYING: Next, make as many copies of the "Class Content Plan" as you have subjects *(see page III-5 of this book; read page III-4 and 6 for directions);* the same number of "Class Completion Records" *(see page III-14 and 15 of this book; directions are on page III-13),* and the Learning Guides you want to use for each of these classes.

STEP 5: FINISH SETTING UP STUDENT NOTEBOOK: Next set up a tabbed divider for each of the subjects you've decided on in the Student Notebook. 3-hole punch each page and insert into corresponding tabbed dividers! *(Are you having FUN yet?!?!? ☺ This is MY favorite part!)*

STEP 6: PLAN CLASSES: Now go to each of the "Class Content Plans" (which should be in their respective tabbed-divider section) and record onto them the assignments you have decided on. The student can easily do this part if the assignments were marked here in this book.

STEP 7: SET UP WEEKLY ASSIGNMENT PLAN: This is where you (or your student) will record all of the student's work and activities for just *this week.* The "Weekly Assignment Plan" is on page III-46, with directions and sample on page 45. *(I take back what I said at the end of Step 5; THIS is my favorite part!)*

Are You Ready For a

A SNEAK PREVIEW OF THE " BIGGER PICTURE"?

READY OR NOT?!

And if you're *not,* then do *not* look! ☺ *Only* if you feel you're ready to look at the "whole picture" for each of the classes should you consider looking at the actual "Class Plans" for them. *(In Sec. VI-D.)* These Class Plans – more accurately called "Assignment Check-off Sheets" *(ACS for short)* – are in alphabetical order in that section. (See the Form Index for page numbers.)

The "Task Analysis Sheets" *(TAS's for short)* that follow some of them *(again, in Sec. VI-D)* contain breakdowns of the specific tasks I wanted my kids to learn as part of those classes, but that (obviously) wouldn't fit onto the ACS. Any *further* questions that will arise due to having seen these sheets will (hopefully) be answered somewhere here in Form+U+la *in due season.* ☺

I know some of you are going to be overwhelmed looking at these. Please try not to be – *if* you insist on looking. If you do, just quickly skip to another section of this book. As you begin to see the whole picture (there's no hurry), everything will make much more sense. I really wanted those who *needed* to see the full line-up of class content elements to be able to do so at this point, before perhaps making some inaccurate assumptions about what all this *might* eventually look like, and how it all *might* fit together.

WHAT EXACTLY ARE THESE SHEETS?

First I'll briefly explain the ACS's for those who truly *want* to know. These are the actual class plans for my son, one per class. The word "plan" is misleading, though, since much of what we did was *actually* recorded "*after* the fact," not *before*, as the word "plan" might have hinted. I *did* plan *some* ahead, depending on the class and the kid. I had a give-me-as-much-latitude-as-possible daughter, and a just-tell-me-what-I-need-to-do son, so I never got in a rut! ☺

However, there will be more planned ahead for our younger daughter, Carlianne (only 12 now), now that I have done high school with the other two. Since they were very different from each other, I now have a broader range of experience and ideas from which to draw!

There will still be much room for Carlianne's unique interests to take root and develop, but I now know more what I want for the "basics" – all of which are included in the ACS's, including the unique-to-each-child particulars as well as the general, It-doesn't-matter-if-it's-a-delight-to-you-or--not-cuz-you-need-it-for-life basics)!

OK, now to briefly introduce the TAS's... As I said, TAS stands for "Task Analysis Sheet." This is what I made up for very specific tasks that I wanted done or learned, like for "Vehicle Maintenance" and "Word Processing Skills." It gave me a very simple, check-off-able way to say "This is exactly what I want learned" instead of just something general like "Learn how to use a word processor," or "Know how to maintain a car," or "Learn what the Internet is and how to use it," or "Learn how to cook." These may be particularly helpful as you start your student onto a path of real-life learning. They give both parents and students the specific parameters and/or lists of things to be learned so that they can both know *what* is expected, *what* the goals are, and *at what point* they are considered "done." I know that I work best in certain areas that way.

These aren't something you will need for every class! Many classes don't even need them at all! So don't feel like you're getting into something heavy. It's just an *optional* tool for those wanting or needing it in some specific areas.

ONCE YOUR STUDENT IS SET UP AND GOING

With your student set up on a plan, you will then be free to get moving on your "Season of Re-education and Renewing of the Mind" – at least I sincerely *hope* that's in your plans for the near future. In fact, that's *really* the whole point of these Jumpstart Plans. ☺

These sheets will be tremendously helpful to you when you get to the serious-planning stages of high school! These are exactly what I wish I'd had to refer to when I was first starting all this six years ago! You won't duplicate these exactly! *No one* will! That's not their purpose. But what they *are* intended to do is give you ideas for the content, and several options for how to organize it on a sheet of paper.

I sincerely hope you're looking forward to it as it will truly be an exciting time, if you are open to God's work in your heart. Yes, you'll feel some "disequilibration" at times, and probably downright uncomfortable, but that's just a good indication that God is truly at work, shaking you up, *not* for the fun of shaking you up(!), but purely for the purpose of loosening and discarding the useless "old" stuff in your thinking and ushering in and establishing the life-giving "new" God has in store for you! He will bring you out of fear and burnout; into freedom and refreshment, which is exactly what the following section (Sec. 1) is designed to do!

Be in prayer *much* as you go through this season; it will be like water to moisten and soften the hard ground, refreshing you in the process, and preparing you for all that God has called you to!

And now for the Jumpstart Class Ideas...

Jumpstart Class Ideas for Senior/Junior High School

Now remember...

These are just <u>ideas</u>!

Not a "To Do" List!

Biblical and Spiritual Studies	
What	Where
Time with God each day: Ideas for what to do during this time are in the "Commune with God" guide.	On VI-D–21
"**Armor of God**" study. *(Also for English)*	Page IV-29
"**Word-Alive" Study** - on the word "education." *(Also for English)* (This was Mary Schofield's idea in her book, *The High School Handbook.* The idea behind this, of course, is to begin renewing the *student's* mind as to what education is!)	Page IV-72
"**Open Mind & Line Evaluation**" - If needed, *(and only you will know if it is or not!),* do some relationship-mending within the family.	Page IV-57
"**Name Study**" on the student's name. *(Also for English)*	Page IV-54
"**Plot Wholesomeness Evaluation**" on a movie or book. *(Also for English)*	Page IV-62
"**Scripture Verse Study**" on a favorite verse (or two, or three...) *(Also for English)*	Page IV-66
"**Experiencing God**" **Study** - This book and study guide is *unusually excellent* study! On the back it says "Learn to hear when God is speaking to you. Experience God doing through you what only God can do." Parent and student could do it *together! (especially good for senior high students and mature junior high students.)*	See the Sr/Jr High Section of the Guide-a-log
"**Mature Christians**" **Study** - The actual title of this book is *Mature Christians Are Boring People (and other myths about maturity in Christ)* and is by Ron Luce of Teen Mania Ministries. *(This study book is especially good for junior high students, but also for senior highers as well!)*	
Study Godly Relationships: Read *I Kissed Dating Good-bye, Romance God's Way, His Perfect Faithfulness,* and *Dating: Is It Worth the Risk?* - *(These make an excellent study to this freeing, heart-healing message that will help usher your teen closer to Jesus and all He has for them!)*	or at our website: www.homeschooloasis.com
Read: *Tortured for Christ, Vanya, Screwtape Letters* - *(Eye-opening, captivating, sobering!)*	

Computer Skills Survey

What	Where
Read ***P.C.'s for Dummies*** - for a basic overview and intro to the wonderful world of personal computers.	See the Sr./Jr High resource section of the Guide-a-log
Read ***Windows for Dummies*** or ***Macs for Dummies*** *(depending on whether you have an "IBM-compatable" computer or a "Macintosh" wanna-be computer. Okay, okay, I'll be nice.)*	
Read ***Internet and the World Wide Web Simplified*** for an intro to the internet that *anyone* can understand. *(even ME!)*	
Learn "***Internet Literacy Skills***" *(See Task Analysis Sheet = T.A.S.)*	Complete T.A.S.'s (lists of specific tasks to learn) are in Section VI-E; See Form Index for pages.
Learn "***General Computer Literacy Skills***" *(See T.A.S.)*	
Learn "***Windows Literacy Skills***" *(See T.A.S.) (Most of this will apply to Mac users as the two do basically the same things, but using different keystrokes and commands.)*	
Learn "***America OnLine Email & Forum Operations***" *(See T.A.S.) (The specific skills listed in this T.A.S. will apply to any email program you may have. They all have their own "idiosyncrisies" but do pretty much the same thing. You'll need to do some editing of mine, if you don't have AOL, but not a lot.)*	
Work on – I mean play around with – a program of interest and use to you (the student). Ideas: a map maker program; an event calendar/schedule; a calendar maker; a genealogy program; a program that makes cards, stationery, labels, business cards, banners, etc. A student could even get good at anything that particularly interests him and make a business of it!	Available at any store that sells computer software
Learn the keyboard and how to type! (There are many programs available, but "Mavis Beacon Teaches Typing" is our family's favorite! And we have heard this from many others as well. The CD-rom version is excellent!)	

Note: If you do not have a computer, I would recommend only minimal reading without simultaneously having access to "the real thing" – even if it is use of one at a friend's or a library. The head knowledge will not have any application, nor any place to "sink its roots." Head knowledge without hands-on experience will be superficial, frustrating, and will not yield much fruit. The idea is not to learn about computers, but to learn computers.

(Hmmm – This same principle applies to God too, doesn't it?!?! We don't want to just learn about Him; we want to get to know Him!!!)

Educational Foundations

What	Where
Read *For the Children's Sake* - by Susan Schaeffer MacAulay to begin learning what true education is, especially for the education of younger children.	See Sr/Jr High resource section in Guide-a-log for descriptions on these
Read *Making a Homeschool Decision* (was "Handbook With An Answer") by Marilyn Howshall to get a good eye-and-heart-opening look at some major issues in education.	
Read *The Right Choice: Homeschooling* by Chris Klicka. *(Sharnessa, our oldest daughter, loved this book as a mid-teenager and would share parts of it with homeschooling friends who were "losing their vision" for homeschooling and wanting to go to "real school." ☺)*	..or at our website: www.homeschooloasis.com
Do a "Book Report" - on any of the above books. *(Also for English)*	Page IV-30

Get your older and younger kids interacting with each other in a productive way during "Homeschool Time with Younger Child" sessions that can be whatever amount of time you feel is appropriate. Twenty minutes is what we set, but it could be more or less depending on how your kids interact. Here are several activity Ideas: *(These are just resources we happen to have on hand. They're not to duplicate, just to give you ideas. I use the pronoun "her" because this is the actual list we used with our younger daughter.)*

- Read a book with or to her.
- Do a gardening/plant project.
- Do a math project.
- Do a science experiment.
- Listen to a history tape.
- Do a craft from one of our craft books.
- Work on a *Booklet Building Book* page ("Around My Town," "Days of Creation," "Tree 'n' Leaf," "Planets"...)
- Bake or cook something out of a children's cookbooks – letting or helping her do most of the stuff.
- Play a board game *(not computer-solitaire or mine-sweeper; something fun and educational that you both agree on.)*
- Work at the computer with her on typing (Mavis Beacon) or "paintbrush."
- Watch an educational video with her *(one time a week maximum.)*
- Listen to a Wee Sing, Psalty, Slim Goodbody, etc. tape with her, and dance, sing, or exercise.

English

What		Where
Literature Survey - For this you'll need any volume of *Compact Classics* also known as *The Great American Bathroom Books* *(which have three volumes that I know of)* each of which contains a *multitude* of two-page synopses of "great literary works," i.e. books, poetry, lives of many leaders, as well as many contemporary works. Parent and student may want to do the choosing together. Use the "Book Synopsis" form that's on...		Form: See Form Index Books: See the Sr/Jr High resource section of Guide-a-log
S.A.T. I for Dummies - part of an English Class - under "Study Skills"		or at our website:
Journal *(Just get a blank journal for this!)*	**Read Biographies** of Famous Leaders	www.homeschooloasis.com
Vocabulary Study - Do the "Vocabulary Worksheet on 'America the Beautiful'" *(The ACS Sheet for this class is not included here, but looks very similar to English II.)*		See Form Index

Note: Several activities can count for more than one subject, so would you count those items for one subject or the other? Or both? The answer is whatever you decide! This book will give you much insight for making such decisions, so for now, just let your student do the work and keep records.

Home & Life Management Skills

What	Where
Carpentry - Let your student to try their hand at woodworking! IF you want a check-off sheet for this, it comes later in this chapter, entitled "Carpentry T.A.S." *(Task Analysis Sheet)*	See Form Index for location of all these
Cooking Dinner - Make dinner __ times a [week / month] *(<u>You</u> decide frequency.)* Use "Dinner Planner" on Page IV-39	
Culinary Arts - Start becoming familiar with basic cooking skills that you'll use throughout life. *(Will also be handy for during Mom's Season of Re-education!)* Use the "Cooking Terms" sheet on Page IV-36	
Also use the "Cooking/Baking Task Analysis Sheet". *(Feel free to change anything that doesn't appeal to you! This list can be completed anytime during the whole Jr./Sr. High experience.*	
Vehicle / Bicycle Maintenance - You'll need to find a dad or a friend who can "apprentice" you in the various tasks. To know where to start, what specifically to do, use the "Vehicle Maintenance Check-off Sheet" And as your teen learns more about the maintenance of your family's bikes, you can record what is being learned on the Bicycle Maintenance T.A.S.	
Housework Skills - Make sure your teen knows the basics of clothing care, kitchen upkeep and cleaning, bathroom, and general household duties. Use the "Housework Skills T.A.S." if desired. *(Again, this is a use-all-through-highschool form, NOT to be done each year!)*	
Home Maintenance & Repair - This includes areas such as plumbing, basic electrical skills, basic flower/plant/garden/yard care, indoors, and outdoors. For specific tasks, see the "Home Maintenance & Repair T.A.S."	

Testing! 1, 2, 3, testing! You're not thinking of these as To Do's are you?!?!?

Mathematics

What	Where
The following may not look like much, but, with everything else there is to do, it's plenty to get your student going on Math. You can also get some kind of a math textbook, but since they have probably had lots of this in school, my suggestion is to just give them a break from the formal stuff and let them do some real-life stuff for a while.	
Calculate "Living Expenses" for a Month - Find out what it *really* costs to live on your own using the "Living Expenses Comparison Study" guide; for 3 "income levels." ☺	In this book; see Form Index
Banking / Accounting - Have student balance your (or *their)* checkbook. (Step-by-step directions are either on the back of your statement or in a flyer at your bank.)	(At the bottom of your purse? On top of the fridge?)
Everyday Math for Dummies or **Investing for Dummies -** part of the "Dummies" series!	(See Sr/Jr High section of Guide-a-log) ↓

...or at our website:
www.homeschooloasis.com

Social Studies

What	Where
Study of the Wars - Find out who fought whom, who won, on what turf the fighting took place in which wars by doing a study on the major wars since the beginning of the United States. These wars are listed on the page opposite the "War Report"	Page IV-70
Worldview Study - This is foundational for every student as they mature into adulthood and for every parent who has never studied this too! *Understanding the Times* was written by David Noebel, President of Summit Ministries. Basically, this course analyzes the popular philosophies and worldviews that drive our culture and objectively compares their consequences to those produced by a Christian worldview.	Available for purchase in the Sr/Jr High resource section in the Guide-a-log or at our website: www.homeschooloasis.com
Constitutional Law Study - study the full texts and Christian interpretations of our nation's historical documents. There's a student book and a teacher guide available.	
Applied Constitutional Study - Learn via actual court cases how the constitution is being correctly and incorrectly applied today! *Where Do I Draw the Line?* by Michael Farris presents many court cases *(some of them blood-boilers!)* including insights and step-by-step guidance.	
Study of Early U.S. History - In the *American Covenant: The Untold Story* (book) and the "American Covenant Tape Series" Marshall Foster reveals the *true* history of our country that is completely left out of public school textbooks! The success of our country does lie in our human ingenuity, as we are lead to believe; it is in the founding principles of this country, which are eroding away by secular humanism and materialism.	
Discover the Faith of Our Past World Leaders - In *America's God and Country* you'll find out how much glory God has received from past leaders in science, politics, medicine – glory and stories that are not mentioned in public schools!	
Current Events, Worldview, and Discipleship "Christian Home Learning Guides" by Marshall Foster. History from the ancient world to today's cultural war between Christianity and humanism; chapters to prepare and disciple in a Christian worldview of literature, art, music, geography, mythology, and science, as well as crime and college success.	
Battle for the 21st Century - to teach students and parents how to be "world changers" for God's cause! Expose deceit and reclaim our culture. *"Those who keep covenant with God win!"*	
Overview of World History - "What in the World's Going on Here?" tape series by Diana Waring. She animatedly covers world history from a Christian perspective (in other words, the already-Christian parts are left IN!), from creation the World War II and the re-birth of Israel.	
Study of Current Educational Trends - Read "Government Nannies" by Cathy Duffy. Be ready for your view of "Goals 2000" to be a bit broadened!	
Study of Homeschool Legality - Depending on the state you live in, parent *and* student may both want to read the booklet "Should Home Education Be Regulated?"! This booklet by Karl Reed will get some interesting discussion and thought going, especially for those who live in a state where home schoolers are faced with burdensome laws.	
Mini-Studies on Various Social Issues - Read the following articles and excerpts: ♦ "But Aren't Schools the Best Place to Learn?" by Michael Pearl ♦ "The True Nature of Public Education" by John Taylor Gatto ♦ "The Great Escape" by Geoffrey Botkin ♦ "The Heart of Testing" by B♥ *(Barb Shelton)* ♦ "A Closer Look at Socialization" by Dr. Fred Worth ♦ "What Is Dating; What Is Courtship?" by Nathan Bailey ♦ "The Problems with Dating" compiled by Nathan Bailey ♦ "Obeying the Law, Testing, & Qualifying to Homeschool" by B♥ ♦ "Excerpts Pertaining to Testing" from *For the Children's Sake* by Susan Schaeffer MacAulay	These are all contained in the Guide-a-log ...or at our website:

Lab Science Survey

What	Where
To do this portion, you will need to purchase "Lab Science: The How, Why, What, Who, 'n' Where Book" by Barb Shelton. Get a 3-ring notebook to keep work pages in. If desired, set up tabbed dividers according to directions given on page IV-1 of the Lab Science book. Have the student do any of the following that sound interesting:	
"Why in the World Would I Want or Need to Study Lab Science?" - a study I designed to help the student gain more of an appreciation of and an awareness of the practicality and value of lab science.	Page IV-26 of LSB
Interview a Scientist - Find a scientist – even just a "science lover" – and do the "Scientist Interview" with him or her *(Could also count for English: "Interview Skills" under "Speech")*	Page IV-27 of LSB
Science Word Study - Do one or more "God's View of Science" Word Studies on "Biology," "Chemistry," etc. *(Could also count toward Spiritual Studies or English.)*	Page IV-24 of LSB
Finding Science in Current Events - Start collecting science clippings in newspapers and magazines using the "Science Clipping Record." Directions for how to use it and what to look for start at the bottom of page IV-2. *(Could also count toward Current Events.)*	Page IV-28 of LSB
To use the following resources you do not need "Lab Science: The How, Why, What, Who, 'n' Where Book," but if you want to know how to structure them into a class or want forms to use portions as part of your class, you <u>will</u> need it for that aspect.	
Unlocking the Mysteries of Creation - tape series and book by Dennis Petersen opens up the world of "Creation Science" (which many are shocked to find out is actually *true* science) in a fascinating, well-documented, understandable manner!	Available to purchase in the Sr/Jr High resource section of the Guide-a-log
Video Microscope Study - Study the amazing world of the microscopic with Warren Hatch as he shows you bugs, micro-organisms, crystals, and everyday objects in a slow-paced, detailed manner that young and old alike will enjoy.	
Evolution of a Creationist - Study the *scientific* evidence (not just "biased-Christian opinion") that it *had* to have taken a Master Designer to have formed the world and that it is literally impossible that it could have formed on its own or by "natural selection."	
Scientific Literacy - Read the various sections in *The Dictionary of Cultural Literacy* on Life Sciences, Earth Sciences, Physical Sciences and Mathematics, Medicine and Health, Technology (a book in themselves totaling 115 pages!) for a good basic background in the sciences. Not a complete study in itself – a good preparation, introduction and reference.	

Section One:

OUT OF FEAR AND INTO FREEDOM

Here's where
we'll get a bit acquainted,
and you'll get released from a lot
of panic, misconceptions, lies, myths,
anxieties, as well as many of your
basic (*and advanced*) fears!

A Lifestyle of Learning Insight...

From Marilyn's Booklet:

The Lifestyle of Learning Approach

"A Lifestyle of Learning can be a fresh beginning for the entire family." [whether you are just now starting to homeschool or have been doing it for years but are dissatisfied with the fruit you are seeing in your family.] "My hope is to challenge you to seek God for a higher level of understanding about education and to encourage you to acquire vision for a wonderful new homeschool adventure, rich in His design for your individual family. Without vision, you will go unrestrained in too many directions. ... This purpose requires getting to know our children intimately if we are to learn to lead them into God's will for their lives. No other learning approach is as accommodating to the Lord than simply using principles along the way of life to wisely apply in whatever fashion He shows you for each of your unique children. ... Learn to trust the leading and teaching of the Holy Spirit. It is God who made us to explore His creation; He knows how we develop and how we learn. He will personally direct and lead anyone who will turn to Him." ❤

OUT OF *FEAR* AND INTO *FREEDOM*

(or "Fortitude, Freedom, and Filet Mignon")

So! Here you are facing high school, all excited and revved up... "Yessiree! Everything's ready to go! We're all set!" You've looked forward to this for a long time now (or maybe even just a few weeks or days), and now (taking deep breath) *this is it!* Full of confidence, joyful anticipation, and vision, you laugh Fear in the face and say "Challenges?! – Eh! Piece of cake!" Like they say, "When the going gets tough, the tough get going!"

Or are you one who's just about ready to *get going* alright – right out the door! And your high schooler *into* the door of the nearest school – where they know how to "do it right"?!? They're the *experts!* What do I think I'm doing even *thinking* about ***homeschooling through high school?!?!?"*** Perhaps the following sounds like you(?):

> ***"I really have "butterflies in the stomach" about all of this. I am feeling very inadequate about being able to offer a quality education to my daughter. Not to mention being able to keep track of her education and write transcripts. I didn't go to college so I have no idea how to prepare for that possibility. ACT's, SAT's, PSAT's; it's all Greek to me. No one in my homeschool group or close circle of friends has children as old or older than mine. I don't have a mentor to guide me through all of this. I am the mentor. I have been so confident during our years of home education. I always said that, the Lord willing, we planned to homeschool all the way through high school. But to be honest, I'm not sure that I ever thought we would finally be here, nor am I sure I realized what an awesome responsibility this is. My confidence is wavering."***

IF I'VE HEARD THIS ONCE...

...I've heard it a thousand times. (And I *have* heard it once.) Actually, there have probably been only a few hundred people – at my seminars and by letter – who have told me that they were doing *juuust* fine with homeschooling *until* – GULP! – they got to ***high school!***

Why *is* this? Why do the fears suddenly come back at this point? Just when you *really* don't need them!?

I believe it's because they were never *completely* dealt with to begin with; only *buried* under a few layers of creativity and a certain level of success and "feeling good" about it all – *at least* feeling *better* than having the kids in school. (*most* days!) That's why it's wise to get these fears – *all* of them – dealt with as early on in your homeschool experience as possible. Unfortunately, I had dealt with *most* but not *all* of mine. So let me back up a few years and share with you a leg of my journey. If you don't identify with me, at least you'll know where I've come from.*

* *"from" (dangling preposition) – See Glossary in Section 8*

HOW IN THE WORLD WE EVER GOT INTO THIS

When we first began homeschooling sixteen years ago *(as of April 1999),* I had no idea we'd ever actually make it to high school. ***HIGH SCHOOL!*** You have *got* to be kidding! Our kids are barely out of night diapers and binkies! They can't be that old yet! *High School* is when your kids are big enough to look you in the eye, take classes that actually require knowledge to teach, and start noticing members of the opposite gender for reasons other than tormenting them. Your daughter could wear the clothing you wore as a teen*, but wouldn't be caught dead in it, and when your son calls on the phone, you have to listen twice to discern if it's him or his dad.

No way – I was not ready for this! So every time the idea of *high school* invaded my thoughts, I made it easy for myself: I simply blanked it out! Worked great! At least for a while...

But, since time has a sneaky way of creeping up on you, ready or not, here it came! Yes, there really were *thirteen* candles on our daughter Sharnessa's cake, and she was now telling me she wanted to start doing "real high school stuff." Ignorance was no longer bliss. I was forced to face the fact that denial would no longer work.

But at the same time, I *knew* that public school wasn't an option for us either. There are *many* excellent books explaining the reasons, though, I think that if you need to read one to be convinced, you must not be fully aware of what's going on in the public school scene today. So, whether or not to homeschool was *not* the question; just *how to do it?!?!?*

* *"teen" - see Glossary at end of book*

SO THIS IS MY QUEST

Having heard somewhere that "not to decide *is* to decide," I knew I had to set out on my quest: *To follow that star, no matter how hopeless, no matter how far.* Which wasn't far from the truth! I followed what meager leads I had, talking with the smattering of people I knew who had gone where few had gone before, reading what I could, listening to tapes, and attending seminar and workshop. Yes, you read that correctly; both of those last two nouns are *supposed* to be singular because that's all that was available back then! *(And, by the way, they were both great – naturally, Diana McAllister taught them.)*

I checked into every option I could find and, to my surprise and dismay, I discovered there were many more options available than I had imagined or dreamed possible! In fact, the whole scene began to snowball! One discovery lead to two more. I'd call this resource person and they'd give me their info and then add at the end of our conversation, "Now you *have* heard of such-and-such a resource or service, haven't you? I'm sure that would be of interest to you." Yes, it certainly was, but it became overwhelming!

There were so many repercussions to all the options I was discovering! Credits! Diplomas! Hours! Requirements! Graduation! And what about *college?* What choices would be wisest for our particular kids? – and for us as a family, since the consequences of the path we'd be choosing would affect all of *us* for several years. Would homeschooling our kids do them and their abilities justice?! Or would we only be jeopardizing their chances for success beyond high school, like in, say for example, *life* to be exact!?

One thing you have to know about me is that I'm terrible when it comes to making a decision. *Terrible!* You should see me at a restaurant! – always the last one to decide on my order, and even then I've changed my mind several times! So every new thought of this *high school* business hit my stomach with a wave of panic! There was nothing easy about this decision, and I did *not* feel equipped to make it. What to do with all these choices?!

I discussed with Sharnessa *her* priorities, current and future. I looked closely at the resources available to us, both within our home and in our area. We looked at her strengths, her interests, and her limitations. And I had to take my *own* strengths and limitations into consideration, as well, knowing that I would be her "direction provider" at least until she was mature enough to take the reins of her own education – one of our ultimate goals.

Now, I need to make a digression here and, for a moment, discuss a current educational philosophy that says kids should be totally in charge of their own education, and that parental or *any* adult direction is not needed, even pretty much disdained. I do not agree! I believe that while "self-directedness" should be a portion of "the ultimate goal," God has a very definite place for parents in the education of their children. I don't believe it should be "parent inflicted," but neither should the parent be out in left field – if even allowed on the field at all! There is one book in particular called *Teen-age Liberation Handbook* that promotes this philosophy. However, I believe it can give a *parent* a broader picture of what "real life education" can be, especially since there is not much available yet in the Christian realm along these lines. The parent can underline parts that would be beneficial for their child, or they can read them together. There is *much* good to be gleaned in it, but I do *not* advise that you just cut your child loose with this book, unless you want to just cut them loose from *you*!

Having studied, consulted, done my homework, and, most importantly, prayed, I eventually got to the place where I was confident enough to feel that I could do this, but only because *"greater is He that is in me than he that is in the world."* And as the years went on, I became *much* more confident, and even excited about the choices we made and walked in*! Sharnessa too! But more on her reaction later in this chapter. . .

* *"in" (dangling preposition) - see Glossary*

WHAT'S ITS NAME?

If forced to give a title to our method of homeschooling through high school, I'd call it something like: "Creatively Structured, Interdependent Free-lance Senior High Homeschooling." Since that's a bit cumbersome, two shorter names could be "Free-lancing" or "Interdependent." Yes, that's In<u>TER</u>dependent, because we really aren't just "independent." We have networked with others in a variety of ways, in resources and in things we did. My favorite title for it, though, is "Spirit-led." However, I can't say that to "just anyone"; I even have to be "Spirit-led" in deciding who I can say that to! ☺

You will most likely get asked about this by those who thought you would finally wise up by the time your kids got to high school age and get them (back) in school where they *belong!* *(I* was even thinking maybe I'd wise up!) However, a soft answer that turns away wrath – or even "just" anxiety – is better than a curt or nervous response like: "It's really none of your business!" – which it may very well *not* be and which you may very well feel like saying at times, *especially* to someone who just insinuated you don't know what you're doing. (Now, of course, this might very well be true, but you are *not* obligated to *tell* them so! ☺)

Most people don't have the time or interest to hear a full-fledged explanation of what all we're doing. They just want assurance that our kids will be adequately educated and socialized, even though, if you asked them what exactly that *means*, or why their choices are *better*, they couldn't tell you! And if you asked them what evidence *(including in their kids)* they have that the public schools are doing a better job, they couldn't give you any.

Let me share with you something that Michael Pearl (author of *To Train Up a Child*) said in a recent "No Greater Joy" newsletter: "It is absolutely true that your view of God and understanding of Scripture are reflected in your approach to parenting. The fact that we get beautiful results, while their approach yields absolute catastrophe, will not persuade them. They will hang on to the 'University mentality" (of child raising) until all their children are in hell and the last Christian has been jailed for producing godly, emotionally balanced, secure, loving, and intelligent children. But we will go on being happy and proud of our children." Mr. Pearl was referring to child raising practices, but I can't help but include homeschooling as a major aspect of child raising.

WHY I HAD TO WRITE THIS BOOK

Whether my mom, dad or my mother-in-law had the time or interest to hear any full-fledged explanations was immaterial. Having indicated a basic interest, (they gave birth to either: A-me, or B-my husband), they each qualified to get my full-blown run-through-the-whole-plan explanation. (This was once I *had* a plan! ☺) They've been very supportive of it since then, but I can't be sure if it's because they agree it's such a wonderful plan, or because they saw how intent I was on it – and they know what I'm like when I make up my mind to do something! *(Some might perhaps choose to call this trait in me "stubborn" or "bull-headed"; I prefer to think of it as "persevering," "unswerving," "resolute," or "tenacious.")*

I'm putting the inner working of our approach into book form primarily because I sensed God was leading me to do so. But as I pondered it further, I saw that there were three really good "sub-reasons" for doing so. Any one of these three could probably have stood alone, but all three together made a triple-braided cord of logic for writing this book:

1) Because I had to get free from the "tyranny of the phone"! Much as I enjoy it, I don't have time to keep explaining the whole picture to everyone individually. Even after describing it for an hour on the phone, I've still not even skimmed the surface, nor begun to do it justice! In fact, I've just opened up a whole new world for them – one they should either see *all* or *nothing* of! The one "quick question" they originally called with quickly multiplies itself exponentially! (Any of you support group leaders know this! There's simply no such thing as a "quick question" for homeschoolers!)

2) Because it's unlike anything else I've seen. I have better things to do than re-invent the wheel that makes homeschooling-the-high schooler go 'round! Part of my stuff is similar in some respects to what others have done, but some (most?) of my focuses are essentially different, covering aspects not covered by any other program, book, or plan that I've ever come across. However, because the communication between homeschoolers nationwide is so disjointed – since there is no "Central Homeschool Agency" to which all, or even *most* homeschoolers connect – a thousand homeschoolers could be doing the same thing, and I'd never know it! But as far as I am aware, the plan you'll see here in this book is unique, and the message is being sought out by many who instinctively know there's got to be "a better way," but just haven't been able to find it or figure it out. (See the quote from the homeschooling mom on page VI-A–14. But come right back! I don't want to lose you to the how-to stuff before you have the heart stuff! ☺)

3) Because Mom said. Very honestly, not only did she give birth to *me*, but also to the original concept of putting all this into a book! One day I just dove in and gave her an in-depth explanation of what all we were doing for high school. Well, that took place over lunch, at a little restaurant table with my notebook and charts and forms sprawled amongst our deli sandwiches, salads, and coffee. When I finished showing her the works, she looked at me with her *"I realize I don't have any jurisdiction over you anymore but I hope you'll be smart and listen to my advice anyway"* look, and said to me, eyes smiling and hopeful, yet very serious: "Now you *are* planning to get all this into some kind of book form and make it available to other homeschoolers who are looking for something like this, aren't you?" She's not a flowery person, so I knew she wasn't just buttering me up. *(Besides, she was already paying for lunch.)* I answered, "Hmmm. I just don't really know if anyone would be interested." We finished our coffee and then relocated out to her van where we continued brainstorming for another hour on all sorts of organizational and practical technicalities of putting such a plan into action. *(Her specialty if you read my Acknowledgment of my Beloved Parents.)* I took it from there and ran with it! . . . What you are reading right now was "beget" back then! – the fruit of *(violin, please)* one mom believing in her daughter, and *(sniff, pause for effect)* caring enough to verbalize it!

I DID NOT WRITE IT BECAUSE . . .

. . . I think I know it all! I'm keenly aware that I *don't*, and *you'll* realize it before long, too, if you haven't already. ☺ I have learned much in the past six years since I wrote the first edition of this book, having now done it with two very different-from-each-other kids, and I'm sure that by the time we graduate Carlianne, our now-twelve-year-old, my ideas will have changed even more! However, as I am doing this 1999 revision, I am realizing that I'm not really *changing* a lot, but am mostly just *adding to* it! As with anything, we have to start *somewhere*. Where God has put me is the best place I can think of to be. And the same is true for *you!* Therefore, my intent is not to give you a totally completed, step-by-step blueprint of every single thing you should do for high school, and how to do it. And I wouldn't if I could, no matter how much knowledge and experience I gain, for two reasons: 1) There is no "one best model" to follow! God has a unique plan for each child! And 2) I would be stepping out of what God called me to do. I have received much assurance from God and from others who are using this book that what I'm presenting here is as complete a picture as it *needs* to be for what God wants to do with it in homeschoolers' lives of. I'd be taking the place that God and God *alone* has in your life – that of Counselor, Lord, and Provider! Yes, *Provider!* We need provision for more than just money and material goods! He's got *all sorts* of good things, like resources, knowledge, and experiences that He wants to provide!

WHAT THIS IS

Rather, what I *do* present here is a few basic "tools" for you to use as you see fit. The framework and design are presented as "unfinished furniture" *(or as Dad used to call it, "unfurnished finature"* ☺) – now *you* get to finish making it a creation after your own "vision"!

With my input in section six, you will custom-design a flexible framework. for – and probably *with* – your own child. You might call this the "skeleton" on which you'll be hanging the "meat" of your own choosing. Whether your meat ends up being more like "hamburger" or "filet mignon" will be determined totally by where you go to get your groceries and what you do with them! If you go to the One who is the "Personal Private Education Specialist" with all knowledge about not only children in general, but about *your* child in particular, you'll end up with a "Top Quality, Grade-A" plan.

God has also graciously shown me some pretty good stuff that's been the "muscle" and "fat" *(the good kind)* of our plan. These will both strengthen and add flavor to your program of study – and play! But consider them "ideas" – not "to do's"!

In the last section of this book I give you a list of several "grocery stores" you can shop at for resources, including but not limited to us, since this can be a major frustration for those who don't live near a homeschool resource center, or have access to a convention when they need something.

Throughout the pages of this book, you'll find "just enough" structure and suggestions to assist you in designing a learning "program" that is creative, natural, preparatory for real life, *"parent-directed-but-hopefully-and-eventually-turning-into-student-motivated-and-directed, a perfectly-suited-for-your-student-and-your-lifestyle* plan! *(Whew!)*

This manual won't and *can't* be everything you'll need – *no* resource could – but it will be more than sufficient to get you going and off to a good start! There's enough structure to give the "free-lance" approach some roots, at the same time giving you wings to take off on your own!

WHY WE WENT THIS ROUTE

The plan contained in these pages satisfies the needs, hopes, and requirements that *we* had for "whatever" would fill and fashion these senior high years of our kid's lives. We have now graduated two high schoolers! And Carlianne, now 12, will be starting high school before we know it. The "plan" or "approach" I came up with:

⏲ - allows flexibility in scheduling and in utilizing resources in the community throughout all the high school years.

✂ - facilitates weaving the student's strengths, weaknesses, talents, and future aspirations into the fiber of the curriculum, not merely taking them into consideration or giving them lip service.

☞- encourages and directs students to learn for themselves; how to become self-educated for life. This is not the same as "self-centered" learning, *(more on this in section 7)* but rather is self motivated, and leads one into a closer relationship with, awareness of, and dependence upon God as the *sole* Source of all knowledge and truth – the <u>only</u> *sure foundation. (A bit narrow-minded, aren't I?)*

❀ - is rich in meaningful, life-preparatory experiences.

✍- provides a way for a student to earn and record hours for credit in a variety of subjects in one year, rather than being confined to just five or six, to the exclusion of other areas of interest.

☺ - helps alleviate the fears and anxieties of grandparents and family friends: "Oh good, at least they're doing *something* with that poor child!"

☞ - gives the student more control over his time and more input into what he'll learn. (If you had hoped this would help you figure out how to cram pile of curriculum into your student's head and call it "homeschool," you're in the wrong place. It's one thing to impart our values and present our "program" to our children, especially during these last "older" years, but it's another thing to *cram.*)

☺☺ - promotes a healthy blend of student input and parent direction.

↔ - gives the student a sense of being "on the same educational level" – *credit-wise*, but not *content-wise* – as his or her peers. (Though the truth is that he will very likely surpass them.) I personally believe that this "need to be on the same level" can be a valid need or desire, and is not the same as "negative peer pressure." Our daughter, Sharnessa, had several Christian friends in public school throughout high school, and it was important to her early on (not later, when she figured out "reality") to "really" be at their level academically. As a homeschooler, she is already "different," which she is generally fine with. *(See "Pioneer Girl" at the very end of this section!)* But if I could find a way to satisfy her need to feel "the same" as her friends while not sinning or going against what I believe God is saying, and still filling those credit-hours with the content I desire, then I had no problem with it. We're actually getting "the best of both worlds." ... And we *did!*

$ - is about as financially inexpensive as you can get for high school. Cheaper is not always better, but for us it was a major consideration. Now that we're down the road quite a ways, I can't imagine being more satisfied with any prepared curriculum or program that we would have paid a steep price for! In more ways than mere money!

☻ - has the potential for learning to become more of a joy for the student.

IS YOUR DIPLOMA "VALID" OR "MEANINGFUL"?

When I explain to someone the educational path that we have chosen for the high school years, I'm often asked: *"Will they be getting a diploma?"* When I answer "Yes," the next thing I'm typically asked is either: *"Will it be valid?"* or *"Will it mean anything?"*

I figure this question must have *really* been burning in their mind because I think it would take a lot of nerve to ask a homeschooling parent him-or-herself such an audacious question! Just to be contrary, just *one* time, I'd love to answer: *"No, I'm glad you asked that... It actually means absolutely nothing. We are homeschooling only because we were just too lazy to get the kids up and out of the house in the mornings! Does that answer your question?"*

LET'S FIRST DEFINE "MEANINGFUL"

You'll be relieved to know I keep my cantankerous comments to myself, ask Jesus to nail them firmly to the Cross and cover them with His blood, and try to realize that these people are (probably) only concerned for our kids. And I'm sure they're a bit curious, as well. So instead I might say: "I guess we have to determine what you mean by *Does it mean anything?..."*

If you're asking whether it means anything to an employer, the answer is *"Most likely; as much as any other diploma means."* Have you ever, even once, been asked to show your diploma for a job? But mostly they just want the name of the high school – we named ours "Tree of Life Learning Center" and that solved that! – and "yes" next to "Did you Graduate?" From there, once hired, it doesn't *matter* what it said on your application about *anything;* only who you really are and your work performance will matter.

If you're asking whether a college, the armed services, or a financial aid board would consider it valid, the answer is *"Sometimes yes; sometimes no; but you don't necessarily need it."* You just need to find out ahead from wherever you're hoping to attend what *they* require, and gear your program accordingly. Another option is to take some classes at a community college. (More on that in just a second...)

There are actually *many* options in this wide-open arena, far too extensive to cover them all here where our purpose is primarily to determine the validity of this type of diploma. At any rate, you can either have an official looking diploma made up professionally, or make one yourself using a computer or press-type lettering (available in stationery stores) which is as professional of an appearance as you need. I do include one in this manual to use as a last resort, but it's not as spiffy as some you can purchase, or make yourself on a computer. My friend had a computer that was better than mine, so I used that! And it turned out *beautifully!* We purchased a spiffy diploma cover from Josten's (1-888-479-3336) and had our school name engraved in gold on the cover.

But back to the meaning of such a diploma. First, if it doesn't mean what you *want* it to mean to the party you want it to mean something to – like the Dean of Admissions at a college, perhaps – the student can always take a few classes (a year is even better) at a community college. This automatically waives or overrides the need for a diploma. It's kind of like when someone has taken Algebra II, they never ask if you've taken Algebra I, it's simply assumed that you know the material for the simpler if you have attained the more difficult.

And, you know, you just *have* to wonder if *their* diploma means anything to God.

ABOUT THE ARMED SERVICES

If there is one organization that has more than its share of red tape and inflexibility, it's the Armed Services. In speaking with a Navy "Ed Spec" (Education Specialist) about these requirements six years ago, I found that a traditional diploma was required in most instances. But attending college could waive this requirement. So what if a homeschooler wanted to go into the service right after graduation? Not after "some college"? They *couldn't.*

Well, this generated a little steam under my collar, so I pursued it with this "Ed Spec" and ended up having some enlightening-to-him, frustrating-to-me conversations with him. I suggested to him *(respectfully, and sitting down, trying to stop my shaking knees)* that they might want to change their unfair and restrictive criteria. I told him – congenially but directly – that if the policies are not changed, *they* will be the ones who lose out! I told him that homeschoolers would make some of their "top quality" men and women, because, generally speaking...

✯ homeschoolers aren't burned out on school

✯ they have higher ideals, finer social skills (the kind that count), greater integrity, better attitudes, and a greater love of learning – which increases one's morale and capacity to learn adapt to any new job or situation.

✯ they are generally more "wholesome," an image the Armed Services pay millions of dollars to put forth.

Well, he agreed, but said he was still bound by regulations. *(Grrrrr!)* I told him *(kindly)* that their regulations have probably been changed many times over the decades, and they need to catch up with times.

Since then they *have* changed their entry requirements several times. (I'm sure it was *entirely* because of his and my chat that day), but I haven't kept up on it. No one except someone who's applying right *now* needs the current rules anyway, but a quick call to your local recruiting agency will bring you up to date, as policies are changing all the time. Even if you heard something three months ago, you'd still want to know the *very latest!*

PLAN AHEAD!

If your student intends to enter the service, or another certain institution upon graduation, check into their requirements now, then work toward them - or work them in to your program -through these years. Some or all of those college credits could even be taken during the high school years. Not my favored choice, but if your student *really* knows where they're headed, it's certainly a possibility.

But I suggest even trying to work around their rules and policies by seeing if it would be possible to work with one of their counselors in finding "equivalencies" in meeting requirements. This is breaking new ground; scary, but it *has* been done before, and needs to continue to be done if homeschoolers are to forge ahead and (gently) take their rightful place in society. This is usually done one brick at a time at a grass- roots level – where you find a lot of dirt, bugs, and rocky ground!

SO BACK TO THE QUESTION

But I still haven't totally answered "Does your diploma mean anything?" in the way I *want* to answer it. (And in case you couldn't tell, I'm rather excited to do so!)

Warning: It's not an answer I would actually give someone who was honestly asking the question. Rather, it is for your benefit. (or maybe "entertainment"?) The reply I'd *actually* give a questioner would contain *elements* of the following response, but the whole thing would blow them away (as in "Ka-BOOM!") and would probably never be the heart of God. (Bummer!)

Actually, my answer to *"What does your homeschool diploma mean?"* starts out in the form of a question...

> *"Hmmm. Well, let me turn that around and ask you: What does your diploma mean? How has it prepared <u>your</u> high schooler for the "real world" where offices don't usually have 25 people all the same age, and work days aren't usually divided by bells into six segments? Has your diploma prepared your student to be a loving, considerate spouse? Or a wise, nurturing dad or mom? – the most important, impactful, and life-long-lasting roles they'll probably ever hold...*
>
> *Did what your student learn in science reveal to them the awesomeness of God and give them more of a sense of His hand upon the making and history of - the universe? Did your diploma help your student acquire wisdom and true knowledge? Or mere head knowledge? Do they now view all knowledge as being founded by God and planted in faith, and do they have a grasp of "truth" only as it aligns with the truth – God's Truth? Or do they hesitate to make such a "biased" statement?*
>
> *Has your curriculum lead your student closer to God, more assured of His love for him?* Does your diploma mean he is better equipped now to accomplish the work God has 'prepared beforehand' for him to walk in throughout his life?... Just thought I'd ask..."* (Though, again, I really never *would*!)

*"him" - see Glossary

BACK TO REALITY

While it's true that high ideals usually precede high attainment, the two do not necessarily balance out on the scale in real life. So I can assure you that the diplomas Dave (my husband) and I awarded our kids didn't carry with them a guarantee that the recipients had attained all these high ideals either. We can do all the planting we want; only God can do the growing. But our diplomas *do* signify that we went the direction we believed God was taking us, with His values as our "Student Learning Objectives," with His vision our vision. And as Dave and I look at our two graduates, we both, with great joy and pride (*parental* pride; the *good* kind) say "This is our beloved daughter *(who graduated in '96)* and our beloved son *('98)*... in whom we are well pleased"! *(And I assure you I'm not just being melo-dramatic!)*

ONE MORE OPTION FOR THE "DIE-HARDS"

However, recognizing that some will still feel a need for a "valid" diploma, I have made arrangements with Triune Biblical University to offer an "official" diploma. Full details are given in Section 2 entitled "Getting It Validated – If You Wish." But please don't get stuck in that section yet! If curiosity compels you to take a peek, please hold your thumb in this spot and then come right back, okay?!. . .

OK, SOUNDS WONDERFUL - BUT IS THIS METHOD *LEGAL?*

This is not an easy or simple question to answer. First, each state is different, having its own law(s) regulating home-based education. Some are very lenient to homeschoolers – and wisely so – while others are so restrictive that it's impossible for anyone but a certified teacher to homeschool legally.

But I would be remiss in my presentation of this issue if I did not mention the other side of the "legal coin." It's such a broad topic that I couldn't begin to do it justice here. So I will confuse and confound you, and then let you take it to the Lord from there and find your answers and refuge in Him (and Him alone!)

ON OBEYING THE GOVERNING AUTHORITIES

There is much more to the "obey the law" issue regarding homeschooling than meets the law-abiding eye. Some believe that if a homeschool law is not restrictive to homeschoolers, meaning if it doesn't place unreasonable requirements upon them, and that if it does not impose upon one's religious beliefs, it should be obeyed. I'm not going to say you *shouldn't* obey it, but neither am I going to say you automatically *should.* There is a "bigger picture" to consider here...

Are you aware that the highest governing authority of this country, the U.S. Constitution, guarantees your right to homeschool? Note, neither the state nor our country *give* you this right, because it is inalienable, given to you by your *Creator* – not by any agency. *(Though the state tends to get a bit confused about this.)* The Constitution of the United States *(at least as of May 1999)* merely *acknowledges, guarantees,* and *protects* our right to educate our children in a manner we as parents see fit. And states are not permitted to make laws that prohibit us from doing so, only *define* how that will look and be applied in their state. Yet laws have been enacted in many states that *do* at least greatly hinder a common citizen's ability to comply with said law. Like having to be certified as a teacher, or having to submit an application and lesson plans for the approval of a school board. Many homeschoolers around the country have chosen not to comply, not out of rebellion, generally speaking, but out of a sincere desire to obey God rather than man; to "render to Cæsar that which is Cæsar's, and to God that which is God's" *(Read the story in Dan.1– 3 for further insight into this matter.)* They do not see it as being possible to do *both.*

Our children are not owned by the state. Nor are they even owned by *us,* their parents! *God* owns them, and has charged us, their parents, with raising them to walk in His ways and for His glory, leading them into knowledge and *truth.* I do not see these two, knowledge and truth, as being inseparable. Yet in public schools only *man's* knowledge – not *truth* – is taught. Teaching the *truth* – as only God can define because He *is* truth – is now punishable to teachers by being fired. Even Christian *students* in many cases have had a very hard time finding a place on their school's campus where they are allowed to have a Bible study *off hours!!!* "Separation of church and state" has become grossly misinterpreted – because people "whose minds are darkened" for lack of walking in *truth* (see Romans 1) are in charge.

Home Spun Schools by Dr. Raymond Moore, tells of a few who believed to the core that they should be able to raise their children in the nurture and admonition of the Lord without interference, direction, or regulation from the state even though their state's law stated otherwise. It's hard to believe, but right here in the "Land of the Free," jail cells have been the "home of the brave" for a season for more than just a few God-fearing homeschoolers. And I hate to be a pessimist, but as the end times get closer, the heat of the "frying pan" is not going to lessen. Brothers and sisters, *we have to know what we believe and why. We must become people of conviction.* Are we homeschooling out of mere "preference"? – because it's a "nice" or "better" idea? Or are we doing it out of conviction? More on this in a moment.

LEGAL DEFENSE

The Home School Legal Defense Association, if you have joined before litigation is brought against you, will defend you on either side of the coin. This organization, which is made up of attorneys who specialize in homeschool litigation all over the U.S., has been very effective in many such situations. They have been able to keep nine out of ten cases that arise against homeschoolers out of court by intervention immediately after a charge has been brought against a "truant" homeschooler. They are also able to help you find out about the homeschool law in your state, if you want to know.

The more "reasonable" a law is, the more difficult it is to defend a family whose conviction it is to not "sign up" with the school district or abide by the state's law, but the right remains ours, according to the Constitution. And H.S.L.D.A. will defend such families as long as it is their honestly-held conviction. ... The Rutherford Institute is another excellent legal option. See the Resource Section for more information.

CONVICTION OR PREFERENCE?

How to know which is which? ... A *conviction* is an honestly and firmly held belief that you *will* hold on to no matter what the cost. If you would go to prison or even die for a *conviction.* But if the threat of jail or death could sway you from your belief, then it's only a preference, a "nice idea."

If you hadn't realized it before, I hope you are seeing that this is one matter that merits a well-researched, thoroughly prayed-through decision. Why? Because we are living in uncertain times; the Word calls them "evil" and "perilous" – anything could change in a moment. Your state may have a good homeschool law right now, but come the next legislative session, that law could be overturned, or more likely, a new regulation tacked on (slipped in) that would make homeschooling either totally impossible for you to carry out or "just" very cumbersome.

At that point you would need to know *exactly* what you believe and *why* you believe it. My intent is not to infuse fear *or* rebellion into you, but to raise your awareness of the fact that your decision of whether or not you will comply with the state requirements is something you need to *think all the way through, start to finish.* I may get stoned for this (by *Christians!*), but I do *not* believe this question is as simple as "Christian, obey the governing authorities."

HEAR GOD, NOT FEAR

I realize I have just opened Pandora's Box here, but this is too critical a matter to be treated in a superficial manner or with across-the-line generalities. If you are undecided in this matter, before you make a decision either way, I recommend you do some *praying* to prepare your heart for God's way for you, some *reading* to familiarize yourself with the issues (we carry several), before making your decision, and then some serious *listening* to God! What might seem to be the "easier" or "scriptural" choice may or may not be in the long-run. Please don't just assume I'm right *or* wrong. Do your homework! A conviction must be rooted in a solid foundation.

AREN'T THE HOURS CUMBERSOME?

A few have asked if having to fill so many hours isn't cumbersome. "I thought one of the advantages of homeschooling was fewer hours. If they have to spend so many hours doing school at home, why not just put them in regular school?!"

Well, let's think this through... A student in "traditional" school (either public or private) spends about seven hours at school a day. But then how much more time must be spent to complete the added homework for those classes? This varies from student to student, of course, but it's safe to count on at least one or two hours, more likely three additional hours a day. As I was told by one public school girl who got straight A's, three to four hours is the minimum. Obviously much more time is required for not-so-gifted students who want good grades. Even if the student has time for other interests, the energy needed to do them has already been *spent!*

And speaking of "homework," let me step up onto my soapbox and say something about *that...* How this practice got to be such a "thing" – I'd even call it a "high place" – in our culture I do not know. Think of how divisive it is to families! The children – and often parents – are gone all day from each other. They come home, fix dinner, do dishes, and then what do the kids have to do? *Homework!!!* I know one mom whose kids were in school, coming home with homework – of course, it's the "American way" – and she got tired of it! So she went to one of the children's teachers and told her that she did not feel her child needed to be doing so much, if *any,* homework. She said "You've got my child for six hours a day! That ought to be plenty of time to get everything done. When they come home, they are *mine.* They will not be doing more work for you." The teacher told her that if she insisted on that, her child would fail. The mom then told her that she would then just have to take her child home to be homeschooled. And that's exactly what she did!

I know another mom who, when her child was in *second* grade, was very upset (and was lamenting about it to me) that her child didn't have *enough* homework! You can be sure she got *no* sympathy from me! In fact, I was utterly disgusted and speechless. *(Believe it or not! Remember the old adage: "If you can't say anything nice, then don't say anything at all...")*

Homework

Next, consider how those class hours are being *filled.* For one thing, time spent in a classroom with twenty or more other students is much less efficiently used than with the low teacher-student ratio at home. (Unless you have thirteen children! But even then, older ones typically help teach the younger, and the ratio again lowers).

We as homeschoolers, in a tutorial setting with a low teacher-student ratio, are able to accomplish much more minute-for-minute than a classroom teacher with a room full of kids and their accompanying distractions and discipline challenges to deal with. This fact *alone* instantly raises the quality and quantity of a home-based education.

So what about the question "Aren't the hours cumbersome?" It depends on a few things: *who* you ask, *how* you fill those hours, *whether or not* you enjoy record-keeping (or how open you are to letting God define what that ought to be for *you* and/or give you an affinity for it), and *what* you're comparing it with.

If you really want to know how ***I*** feel about whether or not the hours were cumbersome, I'd have to say "No, not really." *Documenting* the hours was more cumbersome at *first* than what we *did* with the hours, but it gets to be second-nature before too long. *(Really, it does!)*

But you're probably more interested in how my *kids* felt about it. So I asked Sharnessa if she felt the hours are cumbersome and she said: "Definitely NO. I'm learning in much more fun and interesting ways than kids in regular school. There are always things coming up that interest me, so I go to the library and get books on them."

I asked Tory, and his short-but-sweet answer was "No." So I prodded: "But do you know why?" "Nope; they just don't." ... But *I* know why. . .

For a couple of reasons: for one thing, with him we're going the "take five years" route. (More on that later in the book, both in Sections 3 and 6.) Also, and more importantly, the hours don't feel cumbersome to him because I've laid down (as in given up) *most* traditional, cumbersome-feeling forms of education. The bulk of his learning was hooked into his real life. However, in his senior year, when he was more "sober-minded," he did do more in-depth studies; but by that time he was *ready*. There are a few "schoolish" aspects, but those are not the main bulk of what we do. More on that later, too!

OVERWHELMING NUMBER OF HOURS?!

What about that almost overwhelming number of hours we need to fill, intimidating to those who have been "more experimental and less structured" in our homeschooling approaches (as stated in Washington state's homeschool law)? When formal studies have been kept to a minimum, 120 to 150 hours for one measly credit -- and they need at least five times that in one year! – can feel overwhelming to put it mildly!

Let me start with a question. Have you ever figured up how many hours you spend a year doing dishes? How about eating? Or reading or watching television? It would add up to quite a sum, wouldn't it? *(More on some specifics later in the book!)*

How many waking hours do teenagers have in a day? After the nine or ten hours for sleep, *(allow for more as much as needed especially in the puberty years)* that leaves them with 14 to 15 waking hours. Is five or six hours of "academics" – especially if you've "gotten the picture" and are including real-life stuff too – *really* overloading them? Even after the hours spent at eating, grooming, and chores, there's still *plenty* of time to pursue other interests – although, with the system I present, many enjoyable "outside" activities will count toward high school credit. And if you move into the Lifestyle of Learning realm (Section 7), much of that study will eventually transform into delight-directed study!

TIME OF TRANSITION

These high school years are the time to *prepare* for adulthood. Not *enter* it yet! I want my kids to be *kids* as long as they should be and can, and to *enjoy* being kids! And that's part of why I don't feel they should be "heavy-laden" with a burdensome pile of academic studies and homework that would fry even an adult's brain!

But this is a season to begin preparing for whatever field God is taking them into – be it missions, motherhood, money management, ministry, the military, or McDonald's. High school is a season of transition from childhood to adulthood; a time of learning a variety of skills that will carry the student – the person – through life.

OVER-PROTECTIVE? COMPARED TO WHAT?

A major objection to homeschooling through high school is that the kids will be "over-protected." I have even heard Christians say that such sheltering is "not trusting God with our kids." With all due respect – especially since some of these conscientious objectors are dear to me, and may even be reading this! – I believe that such a view is rooted in a world-tainted and even *world-based* idea of our role as parents. I believe it is also a misconception of what Jesus meant by *"Be in the world but not of the world."*

I don't understand how we can expect kids to be distanced from the world, especially in philosophy and "world view," when we send them into a system that's saturated with humanism ("self as God") where they drink it in daily! While a few classes might be considered "neutral," like maybe Math and Wood Shop, most *must* have an underlying philosophy. Personally, and just for the record, I believe that not even the "safe" classes are immune.

Some might argue; "But our kids are old enough to discern when non-Christian material is being presented and, with all their Christian training, they'll be able to take in the good and reject the bad." I wish I could believe that. Satan hasn't been labeled "the deceiver" for just a few kindergarten-level capers; I'm certain he has a Masters Degree in it! His deceptions are so subtle and brilliantly clever.

I myself, age 46, *(I don't care if you know my age; either I look it or I don't),* having been a Christian for most of my life, regular church goer, fairly mature Christian, *(according to a recent survey),* attender of numerous seminars; I, even *I,* was just recently *(earlier this week)* deceived *once again* by the dirty rotten scoundrel! I didn't *stay* there, but my point is that, even with my age and Christian foundation, I'm not immune!

High schoolers who are growing in the Lord have not yet had the time or experience to have their "faculties trained" and therefore do not possess enough spiritual proficiency to consistently discern which "voice" they are hearing.

A question I have is: Who knows what the public schools are putting into our children's formative minds? Donna Heck, who is graciously editing this manuscript for me, just inserted the following that I want to include: "Barb– Regarding this... Did you see the news last night about the survey that is being given to jr. high kids in some school in Oregon? It was very sexually explicit, given by a teacher – they say, by mistake – however, at least one parent has come forward saying that her daughter also took this survey test back in the 1980's! It's only coming to light NOW!"

And this is just one of many cases! So this question of *Who knows what the public schools are putting into our children's formative minds* is one that demands the attention of any parent who has submitted their children to *any* institution for their minds to be filled. The answer may very well be that much of the knowledge taught within those walls is "good stuff." But we know that much is questionable as well. When our children are out from the scope of our influence, we as parents have no oversight of the good and bad that goes in. There are no long-distance remote-control switches we can turn off when subtle "anti-truths" and outright lies are fed into our children's minds. So are we then absolved from accountability? Just because we can't be there to evaluate what's going into their minds? No! ***God still holds us as parents responsible!***

So one of my main goals as a parent *educator* is to cultivate the needed tools in our (my) high schoolers before they leave our nest. And I think that the *main* tool is just a deeply-imbedded knowledge of the *truth.* And what does Jesus say He is? *"I am the way, and the truth, and the life."* If this is not foundational to home educators, nothing is!

While the world looks with disdain upon a parent's desire to regulate negative influences, or to "over protect" our children from them, I happen believe (to the contrary) that this is our *job* and our *high calling!*

When parents are able to participate in choosing the content and direction of their child's education according to their own priorities, the many hours a child has available each day become much more valuable. The child's mind is much more protected, and the child is "trained up in the way he should go." Now if I'm still trying to shield my kids when they've moved out of the house, then you are free to call me "over-protective," even to my face! *(That number to dial is: 1-800-LET-EM-GO! ☺)*

THE "REAL WORLD" CONTROVERSY
(OK, let's get into it!)

So the high school years are a time of transition, a season of preparation – as they prepare to enter their next "real world." The world they have been living in at home has been the "real world" for them for these years since birth. Just because they aren't exposed to the gangs, assaults and rapes of big cities, or to the drugs and hideous language at the public schools doesn't mean they aren't in the "real world." Thousands of people educated in public schools are assaulted or murdered every year; how did public school "prepare" *them* for the "real world"? Very simply, it can't!

What people are probably *most* concerned about when it comes to the "real world" issue is that kids need to learn how to take pressure, unfairness, competition, the "hard knocks of life." These well-meaning people have somehow come to the conclusion that such knocks prepare a person for life and make him well-rounded.

I agree that trials build character. Adversity definitely has a way of "toughening us up." But I also believe that adversity needs to be God-orchestrated for it to produce the fruit He wants to produce in our lives. Many people are so "toughened up" because they've been over-exposed to the harsh elements of the world that they are either hardened or are "walking pain tanks." Obviously, some end up in sad mental conditions, but the funny thing is that most of these people are the ones we work with and live next door to. Perfectly nice, normal people! Normal-*looking*, that is. And in many ways they *are,* but they aren't experiencing the fullness of life and joy and love that God designed them to experience because they "short-circuited" emotionally somewhere in life just to be able to survive. They are emotionally unable to receive true love, God's or people's.

Yes, God will cover our blunders. He has an abundance of grace and mercy! But if we will submit our children's education to Him, for Him to Master-Mind, there's a whole lot less He's going to have to heal and fix. I don't want to put my kids through unnecessary wounding just because I wouldn't take the time to hear Him in this area of my life!

Speaking of which, I am one of those people who seems perfectly normal *(well, at least somewhat normal)* to the casual observer, but was full of hurt – inflicted by school kids – on the inside for years. And I attended some of the better private schools, and these were perfectly "normal" kids! Since I was heavier and wore glasses, I was the recipient of many a gradeschool-boy-concocted nick name. Of course I got the usual "Four-eyes," but "Bubbles," their term to describe my rotund figure, lingered as a painful memory for a long, long time. It also distorted my perception of people.

Kids will be kids, and they have a keen ability to be cruel. Even the "nice" ones! Those, like me, who are objects of frequent ridicule *have* to do something to *survive*. Of that, I am only *too* aware! ☺ Not being the type to retreat into silence, I developed quite a sharp tongue. I cringe at the thought of how many kids must still bear scars from the damaging flames that shot from my mouth! *(Oh, God, forgive me - and heal them!)*

I was clueless as to just *how* deeply it had affected my self-esteem and confidence until, when He felt I was ready, God started taking me through a season of healing. That's another entire story *(book)* in itself, but my reason for bringing it up is simply to encourage you not to fear it! The pain and discomfort I went through in getting healed and restored was slight compared to the glory of knowing Him as I do now! It's a blast! And it wouldn't have happened without the season of healing! There just ain't no shortcuts to wholeness! *(By the way, Jesus has redeemed the name "Bubbles"; He changed it to describe my personality rather than my body!)*

So homeschooled kids are just in a *different* "real world" – one that I believe God intended kids to be in during these "greenhouse" years. Their "stems" need to be strong, their "roots" deep before being plunged into the harsh elements and storms of life.

WHICH REAL WORLD?

When you think of it, what is a typical "real world" anyway? Think about this: Do you know any two people whose "real worlds" are the same? The "real world" of a farmer is nothing like the "real world" of a business executive, which is unlike the "real world" of a nurse, which is not even close to the "real world" of a truck driver. We each have our own place, our own mission in life, our own personality and style. God wants to use these high school years to prepare our kids for the particular and unique"real world" He has in store for them.

Of course, what people might mean by the "real world" is having to deal with competition, work with "difficult people," having to work for a living, etc. But how can anyone think that growing up in a classroom of twenty five peers parallels the "real world" or prepares a child for it? How many of your teen's friends possess the maturity to teach your child, either by example or precept, how to deal with difficult circumstances? *But that's what they are doing!* Dr. Raymond Moore stresses that whomever the child spends the bulk of his time with, *that* is whom he will become like and become dependent upon. Do you want your children to be unhealthily "peer dependent" or healthily dependent upon parents and God? – ultimately to be dependent upon God *alone.*

INTIMIDATED BY PEOPLE'S REACTIONS?

If the thought of telling your parents, in-laws, or a close friend about your plans to homeschool through high school turns your knees to jello, I encourage you to start by praying. You can't just talk yourself into a "no fear" state.

First ask God to establish you firmly in the security of His will. In order to feel secure, you have to *know* that He called you to this. So *ask* *("You have not because you ask not")* and *listen* *("My sheep hear my voice"),* though chances are that if you've made it this far in this book, you have already been hearing Him all along. I just want you to be *sure.* You will need this anchor when the going gets tough or the vision gets dim. *(especially on a P.M.S. day!)* I don't mean to be negative, but you can pretty much expect the flesh to grow weary in well doing and the enemy to do some fiery-dart shooting *sometime* in the course of these high school years, especially if your child is yielding to what God has in mind for him. You'll need this whether you have challenging people to tell about your decision anyway. ☺

Next pray for God to soften your parents' or friend's hearts, relieve their anxieties, break any pride that might be operating, and open their eyes to His work. If they aren't Christians, this will be less likely to happen, but nothing is too difficult for God! He will *at least* help you deal with it.

Finally, ask Him to give you "a soft answer that turns away wrath" when faced with those challenging – in more than one way – questions or comments.

You'll also need to accept the possibility that it may be a few years before they're persuaded. And that is *if* they ever are! It took my dad nine years! Sixteen years ago (as of 1999) when we first started homeschooling, I couldn't even *begin* to visualize him approving our choice to homeschool. He never voiced his disapproval, nor tried to make it obvious, but I could tell.. ☺ He had a high regard for "education" and didn't want his grandchildren missing out on an adequate one.

That might explain why tears welled up in my eyes in the radio recording studio of KGNW as I listened to Dad speak into the microphone next to mine. We were guest speakers on a program called "Homeschool Connections" this particular episode on "Homeschool Grandparents." Craig Kunnard had just asked Dad if he had any words of encouragement for the listeners. "On air," before who-knows-how-many skeptical grandparents-of-homeschoolers who were tuned into the radio talk show, my dad answered him: "Yes. Yes, I do... Be open. Don't take as long as I did to come around." *(Teary pause before going on to the next paragraph...)*

If it gives you any hope, his sentiments had first started changing when I finally decided to ask him – holding my breath – to teach our daughter and four of her friends Washington State History, his beloved area of expertise. (More about this in the "Potpourri" and "Completed Class Plans" sections.) It also helped that he observed that our children were actually getting an education, which he values highly. He told me he had noticed that Sharnessa possessed poise and the ability to carry on an intelligent conversation. Not that our other two kids didn't, but she, being the oldest – the "first fruit" of our homeschool – was just further along. So he had been watching her! ☺

I can't tell you how precious to me his observations were, in particular the fact that he verbalized them to *me*. I had waited a long time to hear them. But the interesting thing is that by the time I received his approval, I didn't *desperately need* it. That doesn't mean I didn't enjoy or savor it! But God had been doing a deep work in me during the "drought" years. Not having received "the Blessing" early on, when I needed it (in an insecure way), I had to get it from *somewhere*. So I had learned to go to my Heavenly Daddy - for a lot of other needs and healing, too. And He was so faithful to give me not only ample affirmation, but also ample grace to do the job! So when my earthly Dad's approval came, I received it as a blessing, with joy, not as a desperately needed affirmation.

It may be that your earthly dad and/or mom will never like what they see in your kids; they may never bestow the blessing of approval. Or perhaps you no longer – or never did – have a dad or mom to give it. But you *do* have the same "Daddy God" I have – and He will give you everything you need to feel the confidence you will need to enter the high school years as a homeschooler "on your own," although you're not *really* alone. He will be there for you and with you, and will give you the deep assurance that the decision you have made was based in *His* wisdom and love. He will personally take you through these years. David, in Psalms 92 (Living Bible) must have had the heart of one who is leaving traditional ways of education and entering this new territory in mind when he wrote: ". . .for they are transplanted into the Lord's own garden and are under His personal care."

IN A DIFFERENT PASTURE?

What if you're not a Christian? Or you think this "hearing God" stuff is for the birds – the *charismaniac* birds! If you're not *(a Christian)* or do *(feel hearing God is impossible or crazy)*, then I'm utterly amazed that you're still with me! But as the song goes, His grace really *is* pretty amazing. You do realize, don't you, that it's only by His grace that you're still reading this book? No self-respecting, God-fearing *(woops! sorry; scratch that last adjective)* atheist or agnostic would be caught *dead* reading such a thing as this if they knew what was really in here. *(And there are four really blatant hints on the back cover that indicate that God is at the heart of this book!)*

It may be that this whole subject of high school is the very thing God is using to get your attention, and then capture your heart forever! He's so good at that! He does it when we aren't even looking! There are only four reasons you haven't chosen God. *(And He'll pick 'em off one by one!)*

1) Your life is going just fine without God, thank you very much. You don't need Him.

2) You simply don't know Who He *really* is, or how good and loving He really is. All you've seen is Christians who are a lot more forgiven than they are perfect. Some are even hypocrites. But that's like trying to judge Thomas Edison by the vast majority of his inventions which were not nearly as successful as the light bulb. *He* was a whole lot smarter than some of his creations! *(How's that saying go? "Only those who dare to fail greatly can ever achieve greatly." Robert F Kennedy said that! I love it!)* Anyway, look at the *Creator*, not the *created*.

3) You're angry at Him or feel wounded by Him. The two usually go hand in hand. You feel He let you down or purposely hurt or abandoned you when you really needed Him.

4) You are disobeying Him, know it, and have no intention or desire to leave or repent from your sin. It feels good and you can't imagine living without it.

I'm not here to preach or counsel. *(Okay, maybe I am a little)* But my intent is to crack open Pandora's Box a bit, or the proverbial can of worms, and get you thinking about some matters that I know God has already been speaking to you about. I could say *a lot (no surprise),* but there is just one thing I feel He has put on my heart to say here:

I (the author of this book) love you as a person and as a (potential) sister or brother in the Lord. God *(the Author of you!-)* loves you infinitely more! Obviously I can only love you from a distance; He is as close as your heart and breath. Wherever you're at, whatever your past or present, it's not too late. Seek Him with all your heart because He has already sought *you,* and is waiting and anxious to receive you to Himself, personally. Not "just" for eternity, which is certainly a good deal, but He wants you *now!* "Seek first the Kingdom of God, and His righteousness, and all these things – a whole lotta really good stuff – shall be added unto you."

Please don't let it drop here, and just move on with the book. You'll get SO much more out of this if you get your heart and life right with God before going another step! Get away with God and deal with anything that comes to mind, and let Him meet you where you're at. It will make *all* the difference in the world – and with your homeschooling efforts as well because how we educate our children is really an extension and application of our walk with God! ... Okay, I will not bug you anymore about this for the remainder of the book.

WHAT ABOUT "FEAR OF THE UNSTRUCTURED" FOR YOU NEWCOMERS?

Even aside from receiving support or approval from family or friends, it's natural to feel intimidated by change of any kind, especially if you're changing from a secure "known" form of education, such as traditional high school or a prepared high school curriculum, to a scary "unknown" form. This is just as true even if the "known" has not been fruitful or effective! There are a few things I can say that may help, (and of course I will!) but ultimately your source of boldness and confidence is going to have to come directly and freshly from the Lord. His hand is there, ready to lift you out of the frustration and fruitlessness you've been in, ready to lead you into something better, especially for these last few years with your budding adult.

The first thing that needs to take place is that the truth must be exposed. In reality, you are not going from "structured" to "non-structured," which is perceived on the emotional level as going from the "safe" to the "unsafe." Rather, you are going from an old, possibly unprofitable, *familiar* structure to a new, flexible, more fruitful and exciting, though as-of-yet unlearned and *unfamiliar* structure. As you will see in reading through *all* of the manual, there is plenty of structure, but it's not the "typical schoolish" kind of structure with which we have felt most comfortable. The Lifestyle of Learning section presents what I consider to be the "Cadillac of Education"! Don't skip it! I hate to boss people around, but I just *have* to get firm on this *one* point! ☺

SO HOW DO I PREPARE FOR THE "NEW"?

The first step is to make room for it by allowing the "old" to *die.* "Remember not the former things, [*ways of education*] nor consider the things of old. Behold I am doing a new thing [*in your homeschool*]. Now it springs forth, do you not perceive it? [*You will if you take time to listen.*] I will make a way in the wilderness and rivers in the desert. . . for I give water in the wilderness, rivers in the desert, to give drink to my chosen people [*who are thirsty for My Way for their High School Plan*], the people whom I formed for myself that they might declare my praise!" (Isaiah 43: 18-21.)

As you, *both* parent *and* teen, enter a season of trying to figure out what to do for these high school years, I encourage you to draw closer to the Lord and seek *His* will for you. We all, having been raised in our society, possess many "old" preconceived ideas of what education is, how it should take place, and what structure it should have. We've probably never even given much thought as to how valid our attitudes and assumptions are, but we find out just how strong they truly are when we start trying to leave them!

To help give you a more complete picture of what *true* education is I devised "Wisdom's 7 Pillars of True Education," which you'll find at the end of this section.

I'm not saying to throw out every traditional method or philosophy of education, but I do believe there is much *new* life that God wants to breathe into it, as well as some *old* that He just plain wants to get rid of. But the "old" we are talking about here seldom dies a quick, easy death! It needs time. You will do well to let God take you, starting with the parents, through a season of drawing close to Him for the purpose of letting Him do the needed work in your heart. Your mind needs to be renewed, a feat that takes more than a quick prayer. This season will indeed feel like and be for you a wilderness. But know that it will bring about refreshment in the Lord and a new outlook on what education really is.

GETTING FREED UP

Before we go on, I want to encourage you with the words of Sharnessa back when she was fifteen. God had been taking *her* through a process of change and "renewing of her mind" as He was mine. This didn't happen by osmosis; she spent a good chunk of time in the Word every day. And she and I spent much time talking! She said:

> *"It feels good to be getting freed up from the idea that education always has to look like 'school.' I used to look at what my friends in public school were learning and feel intimidated because I wasn't doing the same workbook type stuff. But now I've discovered that I can learn the same things – and more – in much more creative ways. There are many interesting and exciting ways to learn other than sitting at my desk... The more I've gotten into it, the more the fear of not learning enough goes away."*

WHAT TO SEEK DURING THIS SEASON

There are two things *at least* that you'll want to be seeking in this season of re-education and renewing of the mind. First of all, ask God to help you to, very personally, fall (more) in love with Him. You'll probably even need Him to show you what that means. Not on an intellectual level, but I mean take you into the Holy of Holies! He just wants to be intimate with us! It's not something we can make a doctrine out of, put into a religious formula, or concoct in our minds. Who feels awkward or fanatical screaming their head off for their team at a football game?! Full-grown men and women, otherwise cool, calm and collected, think nothing of hollering, jumping up and down, waving their arms wildly, letting uncontrolled emotion burst from their innermost depths when a touchdown or a bad call is made!

Yet most of these very same people will cringe at the thought of experiencing just 1% of that passion with God! "It's too emotional." "Doesn't look reverent or religious." That's exactly right! It's *not* religious at all! And I struggled with it until God finally got me to lay down my concept of what was "religiously acceptable." It was a long, slow "death process," but the more free I became the more "holy fun" I started to have. I have a blast as a Christian; in a holy, righteous way, but in a way that God intended us to live. It's not the world's idea of a "blast," but it is God's. And since He drew up the game plans, He wins.

Proverbs 14:14 says "The godly man's life is exciting, but the backslider gets bored with himself." I could go on and on, but my point is that He invented our emotions. He made our bodies. Passion was *His* idea. We need to be freed up from the world's perverted idea of this word to recognize the wonderful, full *life* He died for us to experience – not just "eternal life" but starting now! "To the holy, all things are holy." I don't go flaunting my body in an unholy manner, or letting my emotions or my passions run my life *(though they make a commendable effort on my P.M.S. days)*, but I have allowed Jesus to sanctify them and give them the place in my life that He designed them to have!

Which brings me to the best part: I believe that the thing of greatest value that I have to give my kids is a passionate love for God. A pretty radical statement, but I firmly believe that this intimate love is what constitutes the only soil healthy enough for all our "head knowledge" to be planted in and grow from. Planted in any other soil, knowledge will only turn rancid and be barren and fruitless. Without a passionate love for the One from whom all knowledge came in the first place, "education" is like clanging cymbals without a song; noisy and pointless.

LET GOD ALLURE YOU

Perhaps the thought of entering such a season actually sounds mysteriously intriguing to you. If so, know that you're hearing the voice of God and feeling *His* drawing because *no* other "voices" will be leading you in that direction! You can be sure that neither Satan nor the flesh, the two main "forces" in our lives, are the least bit interested in letting you get away alone with God – nor will either make it sound attractive. Because even if you did nothing but just sit there with God, doing absolutely nothing, you'd still get more out of that spiritually than not having gotten away with Him at all. But that's another whole subject – another book for that matter....

Marilyn Howshall wrote *Secrets of Solitude* which is the story of her life-changing experience in this very area, and I guarantee that if you read it, let it soak in, and then do it yourself, as God leads *you,* it will be life-changing for you. Homeschooling is primarily an extension of our walk with the Lord! We must first receive from and walk with Him if we truly want a "pearl of great price" to give our children.

"Therefore, behold, I will allure her, and bring her into the wilderness, and speak tenderly to her. And there I will give her vineyards." A vineyard is often thought of as a ministry or work, the second thing you'll be seeking in this season. He's not bringing you "out there" to leave you exposed to the harmful elements or to the blinding, burning sandstorms of the wilderness! He just wants to get you alone with Himself, where He will have your complete attention; with no other voices alluring you or pulling you away from His plan for you and your family. There you'll be free to hear his heart on *your* vineyard.

What does "allure" really mean anyway? Its definitions include "to tantalize, entice, tempt, attract" – in other words, to make something look so good you won't be able to turn it down! We typically think of *Satan* doing this and therefore associate it with sin. If what we're being allured to is sinful, then we're certainly wise to flee in the opposite direction because where Satan leads, only death and destruction follow. Being the father of lies, he *tells* you one thing, but *means* another. Oh – but when God does the enticing, it leads to *life!*

No one but God knows what we really need. And He alone has the ability to deliver the goods, completely and perfectly, and without sinning or maxing us out. He is all-knowing (of what's best for us); all-loving (even when we are yet unperfected); and all-powerful (able to follow through on what He promises), so we have nothing to lose and everything to gain!

BROKEN CISTERNS

In Jeremiah 2:11-14, we see that God has good reason to want His people to forget the former things, as I previously mentioned. I'm going to insert a few words which clarify how God spoke these words to me:

> *"But my people have changed their [educational] glory for that which does not profit. Be appalled, O heavens, at this, be shocked, be utterly desolate, says the Lord, for my people have committed two evils: they have forsaken me [and my concept and purpose of education], the fountain of living waters, and hewed out cisterns [curriculums] for themselves, broken cisterns, that can hold no [living] water."*

This should produce more anxiety than any fear of the unknown or less-traditional!!! Still, it's understandable that anxiety attacks would spring from trying to figure out what to do for the high school years – at home. But as I have come to know a loving God who knows my weaknesses and strengths better than I can imagine *(or would have liked)*, and as I have experienced the reality *(not the concept or the doctrine)* of His grace, my anxieties have quieted. He has shown me that I need only take one day at a time, not the whole picture at once!

Grace is a minute-by-minute provision. I see it as a "Spiritual I.V." *(I.V. = "Intravenous feeding.")* It was never intended to be doled out in large doses and therefore doesn't come by the gallon; only by the I.V. method, as we abide in Him. It's kind of like the difference between having a water boy give a ballplayer a gallon of water all at once at the beginning of the game before he really needs it, and having him be available all through the game to give it exactly when needed, as he gets thirsty.

He will give you what you need to step from the structured to the new-structured. He's so thoroughly trustworthy. That means, if we want to stick with the ballgame analogy for a bit longer, He covers every base! Can you imagine Him saying; "Well, I've got just about everything covered - well, except this one detail! I'm simply at a loss for what to do about that, so I'll just watch her squirm for a while until I can figure out something." His plan may include some waiting, and I'm sure He has some things to teach us along the way, even aside from patience, but His heart is only to love us, and to bring us to greater levels of experiencing that love for ourselves! – not just *hearing* about it!

Jesus said, "Let not you hearts be troubled, neither let them be afraid. . ." Why did He say this? Because there was nothing to be afraid of? No. because the peace that He gives in the midst of the stuff we're afraid of isn't fragile like the peace the world offers. (John 14:27) In fact, the "peace" the world offers is so unlike the peace God offers that you can't really call the world's version peace at all!

My prayer for you is that you would take it step by step, with each step lead and ordered by the Lord.

YOUR ONE-STOP SHOPPING CENTER?

I'd like to be able to say that this manual will be your "one-stop shopping center" for high school. But I know I can't because I know it isn't. To be honest, I wouldn't even *want* it to be!

Since no one has all knowledge except God, you'd be missing out on a lot if you tried to get all your groceries from just me, or just any *one* other author or place. Missing out on what? On the wisdom and insight God has given to a diversity of wonderful people whom I've gotten to know during the past several years.

I'd hate to think of not having the expertise and creativity of Diana McAllister, the understanding of the learning process of Marilyn Howshall, the knowledge and research of Mary Schofield, and Cathy Duffy, just to name a few. These women have all, in varying ways and to various extents, shared with me their wonderful insights, ideas, and how-I-am-doing-it's – *all* of which were instrumental in the formulation of my own ideas and focus. They are all pioneers in their own arenas in this rapidly blossoming arena of high school homeschooling.

BEFORE YOU GO OUT AND BUY CURRICULUM

While the system and plan I present in these pages is about as inexpensive as you can get for high school, you do need to plan on spending some money, depending on what you want to get out of it and how elaborate you like your supplies and resources. But I *highly* recommend that you hold off on buying a bunch of stuff until after you've read and processed this book and a few other books as well that I recommend:

- ***Wisdom's Way of Learning*** by Marilyn Howshall
- ***Homeschooling the Highschooler*** by Diana McAllister and Candice Oneschak
- ***The High School Handbook*** by Mary Schofield
- ***Christian Home Educators' Curriculum Manual for Junior/Senior High*** by Cathy Duffy

These resources are listed in the Resource Section, but I didn't want you to miss them just because they're way back there! In a variety of styles and emphases, they will more thoroughly acquaint you with the heart and soul of homeschooling through senior high, with considerations for the college-bound, and a variety of resources available for home-based education. They will give you a broad foundation from which to branch out and devise your own personal plan. *(More specific help on that is coming up in Section Six, "Personalizing All This Stuff" which includes up through Sec. 6-D)*

As you read what different homeschoolers have done and are doing, it will help you to clarify your own vision. And remember, *your* vision will not look like anyone else's. God will give you something *fresh* and totally *right for your student,* your student's style, and *God's* purpose for his or her life! *(It's so exciting I can hardly stand it!)*

RECIPE FOR SUCCESS

What you'll find in these pages is a "form+U+la" *(pronounced just as "formula")* that was designed to make it easier for you to take on this exciting challenge and meet it with confidence, and without fear – or at least with a greatly reduced amount of fear *(less than 1%)* than you would have had otherwise! ☺ But "NO FEAR" *is* definitely the ideal goal! And it's not unreachable!

One definition of the word *formula* is "a recipe or prescription." The system I present in this book gives you the opportunity to create your own "recipe" and implement it according to your own views. The end product will truly be a "Home-Designed" plan, having started as a "recipe" in my home and ended as a savory, personalized, spiritually and academically nutritious "smorgasbord" in yours!

The meaning of a similar word, "formulate," is perhaps even more appropriate for our use here: "to devise or develop, as a method." This, too, is exactly what you'll be doing – with help and input from me! ☺

In the title, Senior High: A Home-Designed Form+U+la, I capitalized the letter "U" in Form+U+la to indicate that, while ***I*** offer "form" and "format" that will hopefully make the task less formidable, ***U*** (as in ***you***) get to choose, add, and mix your own ingredients! ☺ So the end "Form+U+la" will be of your very own making!

We all have differing recipes for what we see as producing "success." If you read this book in its entirety for the "meat" of it, not just the forms, you'll eventually see what *my* recipe is. However, I believe that success is from the hand of the *Lord!* It comes only as we seek and get *His* heart and *His* goals for our *own* lives. We apply ourselves and we study to show ourselves approved of Him; He furnishes the direction and the grace to accomplish and personalize it for *our* family.

ABOUT MY WRITING STYLE

As you Grammar Experts *must* have noticed by now, I write conversationally. Since much of what I present in this book is on mildly to extremely intimidating topics, and since my system and forms take some explaining to get across clearly, I wanted my writing to be as personal and easy as possible.

You've found, I'm sure, that I am liberal in my use of italics, "quotation marks," paren(theses), and c,o,m,m,a,s. Believe it or not, I used them as sparingly as I could, but I wanted you to be able to grasp this stuff as quickly and effortlessly as possible. Italics facilitate that by helping you, the reader – who is totally dependent on the written word – to "hear" the inflections in my writing "voice." A "just-the-facts, ma'am" approach is not my style. I *speak* with animation; I *write* with animation. Many people have told me that when they came to hear me speak at a workshop after having read my writings, they knew exactly what to expect because they could already "hear" me in my book because of the way I write. So my overuse of certain punctuation marks and font styles keeps you from having to go back and re-read paragraphs over again and helps you quickly "hear" what I'm saying. At least that's my intent and hope. Besides feeling that God is leading me that way, enough people have told me this is the very thing about my writing that they most appreciate, that I feel good about it. In fact, here's a quote:

> "I just purchased your Form+-U+La and LOVE it. I have been dreading high school, feeling inadequate, unprepared and totally ignorant of what to do. Through word of mouth I found your book. After reading it for only a couple hours I had complete peace about the next four years...and actually I can't wait to get started. I am working on digesting all the material. I just bought your Lab Science book, also. I am telling everyone I know about these books. They are so easy to read and I love the way you write. I felt like you were in the room with me and we were laughing together is some parts, crying in others."
>
> *– Dawna Renne, homeschooling mom of 2 for 8 yrs; NJ*

Have you read *DOS for Dummies?* Or any of the *"[Whatever] for Dummies"* books? If not, let me just tell you that the author of some of them, Dan Gookin, has a wonderful lighthearted way of explaining technical computer stuff and helping hysterical people – who just lost everything on their hard drive, or who just translated an entire document into an ASCII file without even trying – regain their composure and start breathing again in the midst of a computer-originated panic attack.

Homeschooling at the high school level is similar in that it elicits panic in the hearts of most normal, red-blooded homeschooling parents. My system, at first glance, might *appear* to be exceedingly technical to anyone with an aversion to forms and charts. I had already decided to stretch a few rules of language crafting and was well into my style and the original manuscript for this book when I ran across one of these books by Dan Gookin. His use of humor confirmed to me that I was on the right track – at least the track I want to be on for *your* sake.

Another of my hopes is that this book will be a simulation of you and me getting together in person for an afternoon and hashing over all this stuff. If you don't like my style, there are other more "brass tacks" type books available. And if you are a reviewer, did you happen to see my note to you at the beginning of the book? *Please* take a moment to read that! Or even if you're not a reviewer, but are maybe just having a hard time with my writing style, it may help! *(But don't count on it! ☺)*

FORMS WITH A FLAIR

Many of the forms presented in this manual can be used in conjunction with a variety of programs, resources, and educational styles. If you have a distaste for forms, try thinking of these instead as being "doors to learning how to learn." That's why I call them "Learning Guides." *(However, I have a hunch that if you truly dislike forms, or aren't at least open to a novel way of regarding them, you probably wouldn't have made it this far in this book!)*

You'll find the typical forms that one would expect to find in a manual of this nature: forms for organizing your goals, planning course content, and recording completed work and fulfilled credits. You'll also find forms that I'm sure you've probably never seen anything similar to! ☺ I have taken great care to explain each one well enough for you to be able to easily understand and utilize them. Please don't be insulted if the explanations seem too elementary. I've had the opportunity to "interpret" and apply these charts and forms to/for several people personally. Some have figured it out immediately with just a few words of explanation; others were totally overwhelmed and needed step-by-step instructions. Since I don't have the privilege, luxury or advantage of having personal contact with each one of you, and can't see the look on your face, I have to go ahead and *write* to accommodate the totally overwhelmed. ☺ If you're not one of those, be thankful, pray for those who *are,* and just move on to the next section!

The reason it's so crucial to me that you get it, even if it takes a lot of effort to figure it out, is that within this system there is much freedom and flexibility! It will enable you to have the time – and the education – of your life! When you take the time to study this manual, you will be set free to homeschool through the high school years with confidence to implement the vision God has given you for your children's education. So the forms are unto *freedom,* not *bondage*! (More on that later in the book!) Many of the "Learning Guides" I have created require the student to actively participate in the learning process. For instance:

- interviewing church leaders about their denomination's beliefs
- interviewing people in various occupations
- reading with Mom and Dad and discussing together what's read, even if it means disagreeing! – respectfully of course!
- discussing "the state of our (own) union" with family
- doing research on a topic of interest to the student
- leading a family meeting
- reporting on material learned in some creative ways
- noticing various aspects of a concert or performance and how they "intertwine"
- discussing "cultural literacy" issues *(see last 2 pages in Section VI-D)*

All of these are part of the "requirements" for various classes and units of classes such as "Tact, Grace, and Social Skills," "Marriage and Family Life," "Building Family Relationships," and "Religions of the World." Each of the above ideas has a form of some sort to help direct the student to learn needed skills or knowledge in a delight-directed, search- it-out non-to-semi-traditional manner. So in one type of form, the kids either experience, do, or live the material, and then use these forms to document what they learn or do. I've found this to be the most difficult aspect of my stuff to explain. You'll just have to take a meander down "Potpourri Lane" in Section Four to see what I mean. And don't just skim! They look traditional, but they're not! They're another type of form I designed to help the student extract from the material what I see as being valuable. I want to develop in my kids the skill of seeking out and gleaning information from "real life books," rather than it being handed to them on a platter in the form of prepared curriculum. These Learning Guides require the student to go and seek out knowledge, and learn for themselves. I call it a "search-it-out" method. And it's just one style of or approach to "real-life learning."

This is also *productive* learning as compared to *consumptive* learning. Here the emphasis is on the student being productive with his energy and study, although this is not to be confused with emphasis being on the product itself. (Kind of like we worship the *Creator*, not His *creation*.) Learning is a *process* that develops the mind, the spirit, the entire being of a person. In consumptive learning, it's all pre-digested. Someone else, even a wise, educated "someone else" has already done the knowledge-acquiring for the student – and probably had a wonderful time doing so *and* got a fascinating education in the process. All the student has to do is consume it.

There's nothing inherently wrong with this form of learning, but if this is the *main* avenue of learning – it will only produce boredom and "dullness of spirit." It will not – because it *can* not – produce a love of learning or an ability to learn on one's own once one is out of "school." I liken prepared curriculum to TV dinners; handy and fast, but not advisable for a steady diet, especially for one who wants to *learn* to cook!

This is what *Lifestyle of Learning* stuff is all about. If what I've been talking about strikes a chord in your heart, do *not* miss Section 7 of this book! And if it *doesn't* strike that chord, then you *really* need to read it! ☺ If you're *already* ready to get right into it, then *Wisdom's Way of Learning* by Marilyn Howshall is the place to start! She has also come out with many other smaller books that will be tremendously helpful to you at this point! Books on developing vision, a lifestyle routine, "coming home" from homeschool, getting to know your children, walking by the Spirit... which will all help you create your own lifestyle of learning. Please do not wait until you have everything planned to read these! They're *vital* for where you're at right now, whether you've been homeschooling for ten days or ten years!

GRADING GUIDELINES

The "Grading Guidelines" are forms that you might want to use in two ways. The obvious way is for help in determining ways to assign a grade. Actually, the parent doesn't necessarily *assign* a grade, but rather sets up certain standards – *attainable* ones – from which the student decides what he wants to "go for." Before you decide if you even want to do grading at all, be sure to read the first part of that section.

The other way you could use these grading guidelines is as further ideas in planning the content of your classes. These are not complete plans for a class, as you'll see (those are in the "Completed Class Plans" section!), but they'll definitely help get your creative juices flowing as you plan your stuff!!

Hopefully they will inspire you to utilize the resources all around you, in your home, church, relationships, and community, as an integral part of your "curriculum" – not just an add-on or elective! I am convinced we learn better and get more out of the "hunt" when we have the opportunity to go and search for the "treasures" ourselves!

GOALS - YOURS, MINE AND OURS

My goal is not to present you with a fully laden shopping cart, but rather a starting point – shelves on which you'll be able to "assemble" your own plan and approach. As you seek God on all this, ask Him to help you set up the priorities, goals, and grading systems *(if you even decide to grade)* that will be the best plan for *your* child. Even if your sister who has homeschooled "from scratch" has some wonderful ideas that are working great with her kids, if you don't feel comfortable with them, express your appreciation, and then do as *you* feel God is leading *you* for *your* kids. If she pushes, have her read this part of the book! ☺ *(Hi, Sis! ☛ Good to meet you! ☺)*

Then, once you start sensing some direction, and as you continue in an attitude of letting Him direct you, the best thing you can do is *relax*. Rest in Him! Let Him breathe His life into all of this and show you His will one step at a time. You'll be amazed at how He will reveal things and bring opportunities your way. It probably won't come all at once, so don't let yourself get all flustered and panicky trying to get one big revelation *now*. Instead, as you start walking in the glimmers and rays and bursts of light that He shows you, His purpose will be unfolded to you. If you've never had the joy of walking in faith, I can't think of a better time than now to start experiencing it! *This, my dear sister, (or brother,) is what homeschooling through high school is all about! It is a faith walk in every sense of the term!* That doesn't mean you don't do some planning. There's probably some good stuff God has already been putting in your mind and heart. But it does mean you do not lean on your own understanding of the world's ways. Let God unfold His plan to you as you go. I can't tell you what it will be, but I do know it'll be good!

THE WONDERS OF WITE-OUT ™

I've found that most people are more comfortable beginning with a solid idea and re-doing it than they are with starting from scratch. So feel free to take the ideas presented here and change them. If alterations are few, a little Wite-out™ will do the trick. If you have bigger plans, start from scratch using my concepts as a point of reference.

INTIMIDATION SIMPLY NOT ALLOWED!

Please do *not* take all this so seriously or get so intimidated by educational buzz words or focused on "requirements" that you miss having the time of you life! These years can be some of the best, most memory-filled, God-illumined, years of not only your high schooler's life, but of *yours* as well! You'll get to participate in helping God grow your child "in wisdom and stature and in the knowledge of the Lord." I encourage you to *not* settle for mere "hamburger" – cheap, "pre-digested," fast 'n' easy to fix – but rather take some *time* and make the investment into "filet mignon"! And for you vegetarians, what I really mean is "prime meat." Prime means "choice, of high caliber, superior, top-quality, surpassing all others in workmanship and worth!" Having graduated two, and now enjoying the "fruit of my labor," I can tell you with all my heart that it is worth every ounce, every moment, every penny! I wouldn't trade homeschooling my kids all the way through to graduation for all the pounds or dollars in the world!

FORTITUDE – IN JESUS

The original title of this section was "Fortitude, Freedom, and Filet Mignon." Thus far you have heard about Freedom and Filet Mignon, but what about "Fortitude"? Well, in order to obtain all the wonderful things I've been talking about, you're going to need a little fortitude. And even though the title of this section has changed, your need for fortitude hasn't. I looked up *fortitude* and was in for quite an eye-opening blessing! It means: "nerve, endurance, forbearance, tenacity, stamina, strength, grit, and vitality!" Woah! That's just about everything you'll need in your bag for this journey! But you're not going to be able to just muster up all this of your own accord. "Your strength must come from the Lord's mighty power within you." (Eph. 6:10, TLB)

And amplifying the word *vitality*, this verse comes to mind: *"Let your roots grow down deep into the soil of God's marvelous love, and draw up nourishment from Him. Live in vital union with Him."* (Col.2:7 TLB)

God has got plenty of nourishing groceries available! *"My God shall supply every need of yours according to His riches in glory in Christ Jesus."* (Phil 4:19) There's no shortage in *God's* supply house, which *does* include "Aisle 14: Homeschool Needs"!

Before we go on to some Questions and Answers, I'd like to share this excerpt from My Utmost for His Highest by Oswald Chambers. This is from "The Way to Permanent Faith" for April 4. **(*Comments in brackets are mine.*)**

"Indeed the hour is coming...that you will be scattered..." (Jn 16:32) Their faith was real, but it was disordered and unfocused, and was not at work in the important realities of life. The disciples were scattered to their own concerns and they had interests apart from Jesus Christ. After we have the perfect relationship with God, through the sanctifying work of the Holy Spirit, our faith must be exercised in the realities of everyday life [including homeschooling!] ... "...you will leave Me alone." Have we been scattered and have we left Jesus alone by not seeing His providential care for us? Do we not see God at work in our circumstances? Dark [confusing, frustrating, I-don't-know-what-to-do] times come to us through the sovereignty of God. Are we prepared to let God do what he wants with us? Are we prepared to be separated from the outward, evident blessings of God? Until Jesus Christ is truly our Lord, we each have [educational] goals of our own which we serve. Our faith is real, but it is not yet permanent. And God is never in a hurry. If we are willing to wait, we will see God pointing out that we have been interested only in His blessing, instead of in God Himself. The sense of God's blessings is fundamental. "...But be of good cheer, I have overcome the world." (Jn.16:33) Unyielding spiritual fortitude is what we need."

I looked up "fortitude" and it means:
"courage, determination, nerve, persistence, resolve, stamina, guts, strength, backbone, boldness, fearlessness, resolution, steadfastness, tenacity, valor, vitality."
Woah! Do you need any of these?! Jesus says:

"My grace is sufficient for you; my power is made perfect in weakness."

Another translation words it this way (which I like even better):

"My power shows up best in weak people." (II Cor. 12:9)

God will strengthen you where you are weak if you let Him lead and teach you.

"For you are in the Lord's own garden and are under His personal care." (Ps. 92: 13)

A Few Q's and A's

Q: I feel like maybe my reasons for homeschooling are not the right reasons. I'm not really sure the Lord lead me to this, or if I just jumped in with both feet, and a blindfold on. I'm already burned out...

A: My own guess is that God lead you to the "mansion" of homeschooling, but that once you got in the front door, you went into the most easily accessed, largest, most attractive double-door labeled "School at Home." – where He was probably not leading you, which I'm gathering only from the fact that you are burned out. ☺ The door to *God's* way of doing it – for *your* family – is not the most glamorous-looking door out there in the front lobby of the mansion. It's very unassuming, "narrow," with a Vine trellis adorning the doorway, and just a humble sign of rugged wood, blood-stained, over the door saying "I am the Way. Enter here, by Me." Only the Holy Spirit is authorized to usher you in; you can't do it on your own. You've come this far, now *yield*, and let Him bring you the rest of the way.

~~~

Q: (...I am already burned out...) Yes, we are basically doing "school at home" which *none* of us are enjoying.

A: I'm just glad – and so should *you* be! – that you are figuring this out early on! Some people waste *years and years* before realizing this or being willing to admit it! They bought all that curriculum and "by gum, we're just gonna USE it!" So they just trudge, trudge, trudge, thinking they are doing this great thing for God, sacrificing for His cause! I call this the "Onward Misdirected Christian Soldier Syndrome." Well, no doubt they *are* trudging and sacrificing), but if they would just *listen*, God has something much better!!! "Obedience is better than sacrifice," right?! But we have to *hear* His voice before we can *obey* it! *(And, in case that intimidates you, helping people learn to discern His voice is at the very heart of Marilyn's and my ministries!)*

~~~

Q: I truly want to enjoy being with my kids again, but I feel the pressure everywhere for doing grades, following scope-and sequences, etc. And even at that, don't feel I'm doing a good job, I guess that's where the burn-out comes in, right?

A: Partly. Burn-out comes when we are listening to the wrong voices and therefore *doing* the wrong things that we thought all along were the *right* things because that's what everything and everyone around us told us was right! Or we may be doing it for the wrong reasons – or for *no* reasons, other than "Well, I thought that's what I was *supposed* to be doing!" The vast majority of home-schoolers, IF they seek God at all, just ask Him which curriculum company they should go with! I'm not saying God *never* leads people in that direction, but what I'm saying is that most people don't even *ask!* They take on a burden God never gave them, and then figure it must be either because we or our kids just aren't doing it right – we just aren't disciplined enough! – or we chose the wrong curriculum company! NO!!! We burn out when we try to carry a burden God never designed for us to carry, or designed *us* to carry! Jesus says "Come to Me all you who labor and are heavy laden and I will give you rest." Man – not God – designed the public school system, and then he made laws to assure that God could have NO part of it! The educational system we have today kicked God out of its core many years ago, and therefore even the *system* itself – not just the *content* being taught in it – is void of God's wisdom, design, and blessing. Yet how many of us spend hundreds of dollars and all our heart, soul and mind trying to imitate it within the private walls of our homes where God's ways should have free reign! If this has been the case, no need to bang our heads against a wall or cover ourselves with sackcloth and ashes. ☺ Just repent, and "let there be tears for the wrong things you have done." Stay there only as long as you sense God keeps you there. And then *move on!* Into the land flowing with milk and honey that He has awaiting you! You've got a *life* to live!

~~~

Q: What's the first thing I should do?

A: My suggestion is that parents – yes, *every* parent planning to homeschool, and even ones who have been at it for a long time – put themselves through a "season of re-education and renewing of the mind." This is what I lay out for you in the *Homeschool Guide-a-log.* There are several different resources I recommend to cover every fear and concern that I have ever faced! *(And, yes, my suggestions for*
~~~

such a course include a hefty dose of Marilyn Howshall's Lifestyle of Learning books! ☺) Mom needs to homeschool *herself* before she can expect to effectively and peacefully homeschool her children! This will not be *separate* from your "real life" – what you learn in this course will be like the "stake" set firmly and deeply "in the soil of God's marvelous love" upon which the vine of your new and ever-changing life will cling and grow straight and strong. If you have a husband who would like to see some evidence of a "plan" in the midst of this less-structured approach, that's why I wrote what you'll find directly before this chapter, the "Jumpstart Plan." It helps you "jumpstart" your kids into a temporary, non-burdensome plan of real-life learning, and translate it into a "Homeschool Lesson Plan" on paper – while you educate yourself and become more equipped and prepared to come up with your "ultimate plan." – which is exactly what this "season" is designed to accomplish! Form+U+la is *part* of it, but there's much more – for which you'll actually be *thankful* the further you travel on this road!

~~~

Q: How concerned do I need to be with "catching up" or "keeping up" with kids their age in school?

A: I have long since shed my concerns about "catching up" or "keeping up" with the public schools. They simply aren't my standard in any way, shape or form! They're not God's standard, why should they be ours? And, amazingly enough, for those who feel they have to submit to testing, it doesn't take much to get kids "up to par" with where the schools are at.

~~~

Q: I guess if it doesn't work out, we can always just put them back in school; no harm done – right?

A: Wrong! ☺ I don't believe quitting should *ever* even be an option! If you're not satisfied after a time, all it means is that you need to keep seeking God. The schools are not going to solve one thing or be one bit more satisfactory than homeschooling! I *(and I hope you)* know too much to believe that!

Q: My homeschooling friends have told me that it is very stressful beginning to homeschool.

A: Actually, the main things that are stressful are beginning (and continuing!) to do the "school at home" approach, and being unwilling to let go of it! This is part of *dying to Self* – who really and truly believed that doing "school at home" was the *only* and the *best* way to go! Once we catch *God's* vision for what we're supposed to be doing, that portion of the source of the stress can be gone in a flash!

There are other sources, though, like kids who are rude to each other and you and who won't obey, too much "coming and going" for both Mom and kids (always getting ready to go somewhere or trying to get settled back in), a disorderly home and/or schedule, sick children, elderly parents that need a lot of time and care, too many commitments outside the home, just to name a few. God will be dealing with these, one at a time. Stress does not come from Him dealing with the junk in our lives. Coming to Him will only relieve us of our burden! Jesus said "Come and learn from me, for my yoke is easy, my burden is light." (Matt.11:28) Rather, stress comes from our *resisting* what He is trying to do.

Allow God to settle you. "Settle" means "to establish; to make *stable* or *solid.*" This is not something you actively seek, but that will come as a natural result of seeking to hear God's voice, and allowing the "voices" of the world – which are *many* and *loud* – to subside under the lordship of Jesus! ☺ His voice must become the *main* one you hear if you want to homeschool through high school with all the peace, joy, and freedom God intends you to have. "My sheep know me, and hear my voice." This is our birthright, not something reserved for "leaders only" or the "spiritually elite"! ☺ So we need to *go for it!* Or as the Nike ad says:

"JUST DO IT!"

♥♥♥

Coming Up!

On the following two pages is a "conversation" between myself and a homeschool mom. She had written an email with some insights and concerns, and I answered it. It's broken down into portions that I responded to one by one. What she had written is inside the **<<double arrows and in bold>>**. *I wish I had saved her name so I could give her credit! Whoever you are, if you read this, THANK YOU for sharing your heart so that others could be encouraged and blessed by sharing in your process of growth and change!*

Then on the two pages after that is an article on "Wisdom's 7 Pillars for True Education." (that I wrote.)

On Fears, Goals, and Values

<<Sometimes I panic that my children won't hold their own in the world when they leave the nest>>..........*I firmly believe that parents with children in the public schools have more reason to panic over this than we do, my dear sister(s)!*

<<When I feel this, I remind myself that my goals are not the world's goals.>>..........This is *so* good! And if you can *keep* this at the forefront of your mind all the way through high school, you'll be sparing yourself all sorts of grief and panic! ☺ We need to keep in mind that, even though the public schools have a lot of good goals, their *means* of accomplishing them is very inefficient. Just because they have wonderful scope-and-sequence approved, full- color, 3-D, college-prep goals and curriculum doesn't mean that very much of it is actually *sticking* in the kids' hearts and minds! And we must also consider that almost *none* of it is forming their character – at least the way God would have it formed – which is a *huge* part of what true education is supposed to do. "If I speak with the tongues of men (seemingly well-educated men!) and of angels, but have not love, I am nothing."

<<I want to raise children to love and serve the Lord! I don't care what their income is or their status compared to everyone else. I remind myself that the big picture is eternity and I often ask myself, "how important is this or that in the scope of eternity?">>..........You have touched the heart of my message right here! (and Marilyn's!) If each one who deals with fear can grasp this, and let everything else go, they will never feel intimidated by the traditional school system again! "Perfect love casts out fear"! And what's that verse about gaining the whole world, but losing your soul?

<<We are walking to the beat of a different drummer, often different than our brothers and sisters in Christ, and that is not easy.>>..........So true! But you know what's going to be even *less* easy? When parents who have had their children in public (and even Christian) schools for the whole thirteen years do not see in their graduated children what they had *hoped* and *thought* they were going to see, and yet *do* see what they (and God) do not like!

Yes, many kids who graduate from public schools have wonderful qualities and are great people, but how much was "lost" in the system? Only God knows – and I'm not experimenting on my kids to find out! I'm talking about the products of having been in and formed by the system and ultimately assimilated its way- off-(God's-)base thinking. This is not easy to see! Hopefully those parents who have needed, for whatever reason, to have their children in the public school system have become aware of the possible pitfalls and are taking the time to talk to their kids and give them input and resources that counter the humanistic, evolutionistic, Godless, self-centered thinking that has permeated the school system. Not that every teacher in the system thinks this way, but even if the teacher *doesn't*, the kids they are spending the bulk of their life with *do*. At best there is a tremendous void there *(which Satan has a hey day with)*, and the textbooks, more often than not, are based on this thinking as well. How can they *not* be? Martin Luther in the 16th century said:

> *"I am much afraid that the schools will prove the Gates of Hell unless they diligently labor in explaining the Holy Scriptures and engraving them in the hearts of youth. I advise no one to place his child where the Scriptures do not reign paramount. Every institution in which men are not unceasingly occupied with the Word of God must be corrupt."*

Pretty hard words, aren't they? I used to think they were *too* harsh. But as I understand more and more of what *true* education is, I see that there is no form of learning that is not unto *something* – and if it is not unto God, it *must* be unto something else; and if God is not only intentionally left out, but is now also scorned and ridiculed, then the end of such "education" is ultimately folly and destruction. Period. I don't care how powerful the NEA is; God wins in the end. That's the team I want to be on! The battles before the final victory are not going to be pleasant, and the battle for our children's *minds* is the most dangerous, deceiving, and insidious of all! We have much to lose if we do not "wake up and smell the coffee"! We must *be watchful!*

<<I have a close friend who started homeschooling this year and is very concerned with grades and achievement. When I see what they're doing I feel very inadequate and yet I know this is not for us.>>..........Again, you have hit a key point! ☺ We have to look at *why* we feel inadequate. I saw a high school girl who was sitting in a waiting room with her Trigonometry book. I asked her if she was taking that class and she said yes. I then asked her how much math she plans on taking and she said "Oh, all four years!" (as if that were kind of a stupid question). Undaunted, I asked her if it was for an area she was going into and enjoyed, and she said "No, I just need it for college." ... OK, let's pause for a moment and take a closer look at this picture. This girl is taking advanced math classes that are not only taking up one hour of her day in class, for every day of the four school years of high school, but since it's *advanced* math, you can be sure she has *at least* an hour of homework in that class every night. Let's do some quick math here (I only went up to Algebra I, but I can DO this!!): There are 180 school days, so that's probably 180 hours of homework at one hour per day (and this isn't counting preparing for tests) plus about 150 clock hours of class (which are usually 50 minutes X 180 days = 150). So with 150 hours of class and 180 hours of homework, that's 330 hours for the year.

OK, let's just assume that all this girl really needed for where God is taking her in life is two credits of math. (I personally think this is generous and that it's really more like *one* credit, but we'll give her the benefit of the doubt.) If she is going for two *more* credits of math than she really needs – 660 hours – then this girl has approximately 660 *fewer* hours to pursue her delights, who and what God made her to be, and/or knowledge of God himself. She's getting ready for college, and is qualifying dandily for the Honors Program at her high school, but I have to say "SO WHAT?!?!" I repeat and continue:

"So *what* does this really mean? Where does this *really* get her?!?!" It means she knows a lot about math (which will probably fall right out of her brain after the tests; facts she will probably never use.) It means she will do well in her college math classes. If she were going into a field that was math-based, great! But our example gal here is *not!* (as many other girls are *not!)* She's just doing it so she can "do well in the system"! *THAT IS IT!!!* She's not getting prepared for the future God has for her, she's *not* learning diddlysquat about God or how to serve in His Kingdom, her relationship with neither God nor people is deepening, she's not developing her giftings – the ones God wants her to use while on earth, and she's *not* getting better prepared for *life* in general! She's just "doing well in the system" – and she's spending at least 660 unnecessary hours of her precious *life* on doing so!!! 41 *waking days* of her life, not even counting sleep! I'm sorry, I know I'm getting heated about this, but this is pathetic. It *grieves* me! The system is robbing kids of *who* they are!

Read John Taylor Gatto's *Dumbing Us Down* to see and understand more of this travesty. He was voted the New York State Teacher of the Year back in 1991, and he has some *very* interesting things to say about the school system! And he's not just talking about the occasional abuses and misuses that occur in the dark corners of the system, but the system itself! Here's a quote from *Dumbing Us Down:*

"I've come to believe that genius is an exceedingly common human quality, probably natural to most of us... I began to wonder, reluctantly, whether it was possible that being in school itself was what was dumbing them down. Was it possible I had been hired not to enlarge children's power, but to diminish it? That seemed crazy on the face of it, but slowly I began to realize that the bells and the confinement, the crazy sequences, the age-segregation, the lack of privacy, the constant surveillance, and all the rest of the national curriculum of schooling were designed exactly as if someone had set out to prevent children from learning how to think and act, to coax them into addiction and dependent behavior."

This man is not a Christian, so if any of you have non-Christian spouses who are not interested in your "radical Christian" views of education, this would be an excellent book for them to read!

<<The thing though that really lets me know I'm on the right track is my kids hearts!>>..........I cannot amen this enough! Good for you! If a person has all the knowledge in the world, but has not love (which will result in a soft heart toward God and others), then their knowledge is nothing more than a clanging cymbal, noisy, irritating, and of no good use. I don't understand why we invest so much money and esteem in "clanging cymbals" when we can get pure *gold vessels* by simply allowing God to do His work in us. Yes, yes, yes; part of His work will include academics, but let's pull them *down* off the pedestal – the *high place* – that our culture has them on – and give *God* and *His* priorities alone that place! He knows how to lead us perfectly; we just get to take our rightful place as His beloved (but dumb) sheep and *hear* Him. ☺ This is not something that is reserved for the "spiritually elite"! It's one of the most basic privileges and necessities, really, of being a believer! How can we follow a leader whose voice we can't even hear? If we lack in this area, no better time or place or practice ground to start on than in hearing Him on how to educate our children! (Sometimes I think that my learning to hear Him is one of *the* most important lessons I am learning in homeschooling!) And if it's any comfort, I believe that He covers our inability to hear Him well early on! I am *still* learning! ☺ Notice that the verse (I can't remember the reference) says "Without faith it is impossible to please God." Look closely at that: It doesn't say "Without *a highly-honed ability to hear God*" or "without *the best curriculum* it is impossible to please God." It's FAITH that will please Him! Just trusting in *Him* – nothing else! "Seek first the Kingdom of God and His righteousness, and *all these things shall be added unto you."*

YEE-HAH!!!

WISDOM'S 7 PILLARS OF TRUE EDUCATION

The choices we make regarding homeschooling, whether at the beginning or throughout it as well, are going to be based completely upon our view of education. We will choose materials and resources that we think will help us fulfil or create that view. So it is vital that our view be correct, or we will waste a lot of time, money, and effort, not to mention create problems that will take a lot of God's redeeming power to correct.

Part of that is understanding what it *isn't*. And it *isn't* just "schooling them at home." Yes, many homeschoolers do this, but while they are following God's *directive* to teach their children, are they really following God's *directions* – His *ways* and *methods* – in so doing? Are they really accomplishing God's purposes?Webster's 1828 Dictionary defines "education" as:

> *"...the bringing up, as of a child; instruction; formation. Education comprehends all that series of instruction and discipline which is intended to enlighten the understanding, correct the temper, and form the manners and habits of youth, and fit them for usefulness in their future stations. To give children a good education in manners, arts and science ins important; to give them a religious education is indispensable; and an immense responsibility rests on parents and guardians who neglect these duties."*

What is the main focus of this definition? Academics? I don't think so; academics are alluded to, but could hardly be considered the focus. No, *character* is the focus, including preparation for "future stations." But even this preparation has more to do with the *heart* of the person than with his head knowledge.

As I have studied "education" over my past 16 years of homeschooling, from a variety of perspectives including the thoughts of many authors, mentors, leaders, and friends that I highly respect, I have seen a very different picture of what education *truly* is than what is typically thought of in our culture today. I saw that there were actually *several* components that make up this larger picture. What you see below is nothing official, just what I came up with after looking at and compiling what were, to me, the most important aspects of education that I wanted to be part of my own, and my children's, *true education.*

Look at all seven of these "pillars" below... Which of these are present in today's educational system? Which is or are the main focus? I think it's pretty obvious that it's a "Basic Foundation of Knowledge to Function in Society." It's certainly not to give a revelation of the Knowledge of God and His will – in fact, that is actually completely and effectively *squelched*, as is developing a "Closer Relationship with God." How about "Preparation for Roles in Life?" Very little, and more and more this is to promote tolerance of worldly and ungodly "lifestyle preferences." Even if there is a focus on this, it can't be very specific for unique students with 20 of them in the same class. What about "Preparation for Specific Life Calling/Purpose"? It can't even touch that because this has to do with God's purpose for the student's life. "Character Development"?

I don't know what you've seen of the typical high school environment, but this is about the last place godly values are present let alone taught or caught. "Equip With Tools to Learn for Life"? Hardly – students are told what, not *how* to think. Few students graduating from high school know how – or have the desire – to learn beyond graduation. Lacking 6 out of the 7, their "Basic Foundation of Knowledge to Function in Society" is sadly lacking. "All the above" are the total picture that's needed if we are to function as *God* would have us function in society!

On the following page is a collection of scripture verses that illustrate more of what each one means.

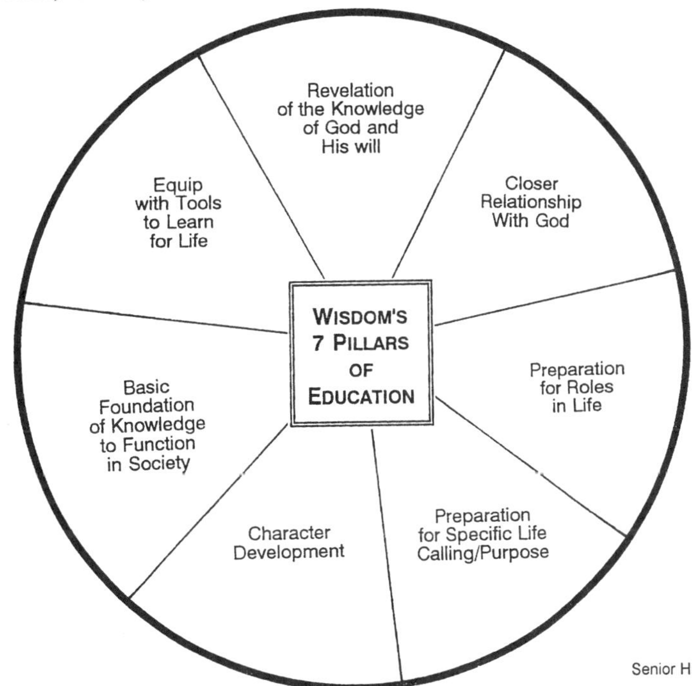

Scriptures for Wisdom's 7 Pillars of True Education

Revelation of the Knowledge of God and His Will - "For these things are yours (see previous verse, given in "Character Development") and abound, you will be neither barren nor unfruitfl in the knowledge of our Lord Jesus Christ." (II Peter 1:8 NKJV) "And this I pray, that your love may abound still more and more in knowledge and all discernment, that you may approve the things that are excellent..." (Philippians 1:9–10 NKJV) "We... do not cease to pray for you, and to ask that you may be filled with the knowledge of His will in all wisdom and spiritual understanding; that you may walk worthy of the Lord, fully pleasing Him, being fruitful in every good work and increasing in the knowledge of God..." (Colossians 1:9–11 NKJV)

Intimate, Personal Relationship & Fellowship With God - "The one thing I want from God, the thing I seek most of all, is the privilege of meditating in His temple, living in His presence every day of my life, delighting in His incomparable perfections and glory. There I'll *be* when troubles come..." (Ps 27:4–5) "My heart has heard You say, 'Come and talk with Me, oh my people,' and my heart responds, 'Lord, I am coming." (Ps 27:8) "And now, just as you trusted Christ to save you, trust Him, too, for each day's problems; live in vital union with Him. Let your roots grow down into Him and draw up nourishment from Him." (Col. 2:6–7) "For they are transplanted into the Lord's own garden and are under His personal care." (Ps 92:13) *(All from TLB)*

Developing Godly Family Relationships - "Let the older women teach the younger women to live quietly, to love their husbands and their children..." (Titus 2: 4–5 TLB) "Husbands, likewise, dwell with your wives with understanding, giving honor to the wife as the weaker vessel..." (I Peter 3:7 NKJV) "Children, obey your parents..." (Eph. 6:1 TLB) "Fathers, do not provoke your children to wrath, but bring them up in the training and admonition of the Lord." (Eph. 6:4 NKJV) "If you can find a good wife, her worth is far above rubies..." (Prov. 31:10–31 TLB)

Preparation for Specific Life Calling and Purpose, and for Roles in Life - "If anyone purifies himself form what is ignoble" *(= dishonorable, evil, common = average, standard, traditional, conventional, familiar, customary"* "then he will be a vessel for noble use, consecrated and useful to the master of the house, ready for any good work." (II Timothy 2:21) "Jesus said 'My food is to do the will of Him who sent Me and to accomplish His work.'" (John 4:34) Also: "Anyone who lets himself be distracted from the work I plan for him is not fit for the Kingdom of God." (Luke 9:62 TLB) "Therefore, brethren, be even more diligent to make your call and election sure, for if you do these things you will never stumble..." (II Peter 1:10 NKJV) "We are His workmanship, created in Christ Jesus for good works, which God prepared beforehand, that we should walk in them." (Eph. 2:10)

Character Development - "But also for this very reason, with all diligence, add to your faith virtue, to virtue knowledge, to knowledge self-control, to self-control perseverance, to perseverance godliness, to godliness brotherly kindness..." (II Peter 1:5–7 NKJV) "For the moment all discipline seems painful rather than pleasant, but later it yields the peaceful fruit of righteousness to those who have been trained by it." (Hebrews 12:11 NKJV)

Basic Foundation of Knowledge to Function in Society - *(I couldn't find this anywhere in the Word, yet it seems to be universally accepted wisdom and common sense. However I did find the following references that give specific guidance for gaining knowledge:)* "Do not let others spoil your faith and joy with their philosophies, their wrong and shallow answers built on men's thoughts and ideas (empty deceit), instead of on what Christ has said. For in Christ there is all of God in a human body; so you have everything (you are complete) when you have Christ, and you are filled with God through your union with Christ. He is the highest Ruler, with authority over every other power." (Col. 2:8–10 TLB) "Therefore, if you died with Christ from the basic principles" *(seems like this would include "the basic principles of education")* "of the world, why, as though living in the world, do you subject yourselves to regulations..." *(requirements, criteria, etc.)* "...which all concern things which perish with the using – according to the commandments and doctrines of men? These things indeed havew an appearance of wisdom in self-imposed religion, ... but are of no value against the indulgence of the flesh." (Col. 2:20–23 NKJV) "For the Lord spoke thus to me... and warned me not to walk in the way of this people" *(and the public school system is one of the "Top 10 Ways of This People")* "saying: 'Do not... fear what they fear, nor be in dread. But the Lord of Hosts, HIM you shall regard as holy; (= "to be revered") "let HIM be your fear, and let HIM be your dread, *(which boils down to: Let HIS be the Voice you heed!")* and He will become a sanctuary." (Is. 8:11–13)

Equip with Tools to Learn for Life - *(Again there isn't really a verse for this, but it is a biblical principal. See Wisdom's Way of Learning for specific, in-depth teaching in this area.)* "But everyone who lives on milk" *("milk-like curriculum" = bottle-fed: for less mature)* "is unskilled in the word of righteousness, for he is a child. But solid food is for the mature, for those who have their faculties trained by practice to distinguish good from evil." (Heb 5:13) *The tools we need most:* "The days will come when I will establish a new covenant with the house of Israel... I will put my laws into their minds, and write them on their hearts, and I will be their God, and they shall be my people... and they shall all know me, from the least of them to the greatest... In speaking of a new covenant, He treats the old as obsolete. And what is becoming obsolete and growing old is ready to vanish away."

"If any of you lacks wisdom, *(yes, even, if not especially, about how to homeschool)* ***then let him ask God, who gives to all men generously and without reproaching, and it will be given him."*** (James 1:5)

Jesus said, "My sheep hear My voice, and I know them, and they follow Me." (John 10:27)

Section Two:

REQUIREMENTS, COLLEGE, PROMOTION AND GRADUATION

Insights, options, guidelines, suggestions, considerations, and new perspectives on "all the above" that will help free you up, as well as equip you to determine your own family's "requirements." Also photos of Sharnessa and Tory's homeschool graduations, and program and invitation samples!

Re: "Honors Diploma"

Barb Shelton

Someone on the Form+U+la email list asked for input (from anyone) on what's involved in earning an "Honors Diploma" through the school system. Someone else posted an answer (about a mile long) with all the requirements, adding that it probably wouldn't be too hard for a homeschooler to fulfill most of them. Without even looking at the requirements, I had a "gut reaction" that I felt compelled to share on the list. Here it is...

I'd just like to interject a comment here that I hope won't feel like a "wet blanket," but when this topic first came up, something inside me kind of cringed, and I wasn't sure what it was. As I ponder it, I realize that I don't have a problem with this topic itself, but I guess I just want to interject a little "heart check" here. I think what bothers me about the very idea of an honors diploma boils down to this: What I talk about so much in my stuff is the idea of pulling out of – unplugging from – "The System." The idea of an "honors diploma," to me, infers that not only is there *not* an "unplugging" but that there is the desire to be at the *top* of this system, what *it* considers to be of value. The whole foundation of my message is striving after what is "honorable" to God, not to the world.

Perhaps your student may be wanting to get a scholarship within the system, and that's why they want an "honors diploma," but I guess I just want to challenge us all to make sure our hearts are right in the midst of doing so. I'm not saying anyone's heart *isn't* right – just wanting to make sure they *are*.

After looking at the requirements, much of what I saw listed throughout them is *very* binding, *very* worldly-minded, and *very* "system-ish"; there isn't any way you could "please both God and man" in fulfilling most of them. (With *some*, you could; many others, *no way.*) These requirements could far too easily become "Lord" of your child's education without you even knowing it. So again... what do you think I'm going to say? ☺ Yup...

Hear God!

(*To this, one mom responded:*

"Thank you Barb for being a "wet blanket" and helping me to be able to breathe again! ☺ ")

WHAT ABOUT "REQUIREMENTS"?

WHY THE INTIMIDATION?

There are probably many reasons that requirements intimidate us, but I want to focus on what I see as being the main one, which is primarily a matter of the *heart*. The foremost intimidator, responsible for keeping many families from homeschooling let alone from being free to homeschool how they feel deep in their hearts they'd *like* to, is that we just don't think we're *capable* of making such crucial decisions. We're afraid we'll do it all wrong and probably even wreck our kids. What makes it worse is that we imagine that when our failure does finally get exposed, it will be much to the delight of that human cloud of witnesses we fear are lurking behind the nearest bush, ready and waiting to say "I told you so!"

Even though you are leaving one system (traditional schooling) and going to another (homeschooling), your mind may still have some renewing that needs to take place before the *system* will leave *you!* Otherwise you'll just bring that system into your home and try to proliferate it in a place it was never intended to be proliferated! If you want to be part of the pure, *new* thing God desires to do in your family, you must first be willing to let Him dismantle the old system so that He can build with *new* materials. If you don't, you may get to the end and find that your labor was in vain. If you have brand new floorboards, studs, and walls, but never replaced the electrical system, you won't have a nice new house for very long. At some point, one of those frayed *(or should we spell that "fraid"?)* wires will short out and damage, possibly even *destroy* the house.

THE TANGLED, INVISIBLE WEB

The amazing thing about systems is that they have the ability to govern our lives without us ever being aware of them. Whether we know it or not, we all have a web of them at work in our lives. This network is so broad, it covers every area of our lives without our even knowing it. But when a system is challenged, you know it! It's earthshaking!

How, then, do you know when a system exists? Just try challenging it! One example: You say to your parents one evening at the dinner table: "Well, we're thinking about taking Sarah out of school and homeschooling her." Unless you have very progressive parents who have taken some time to enlighten themselves in the area of education, you are about to *hear* the tremors emanating from a system that has just been shken. It will help *(I hope)* to remember that these reverberations are not necessarily anything you should take personally; they are mostly from the quaking, as a system *firmly established* within them is defied.

Defied?!? All you said was that you want to homeschool your child! You aren't contesting anything!

Oh *yes* you are!!! ☺ By wanting to provide an alternative form of education – alternative from what the person you are speaking with used or is using – for your child you are making a silent but *loud* statement that the system you're leaving *(or not entering)* has some sort of deficiency. You're saying you can do it better.

Furthermore, you are also saying that all the wonderful, cherished memories of "high school" are to be denied your child! The funny thing is that time has a way of distorting our memories. Most people's very *best* memories were of *leaving* high school, I've found! *(Take a poll! Before they find out what you're doing!)* And the favorable memories they *do* have usually have little to do with *education,* which I was under the impression "school" is supposed to involve. The more I talk with people and see the fruit in their children's lives, the more I'm convinced *solid* of this!

So for you to have departed from – or never entered – the traditional school scene, you have already challenged a system that's firmly set in place in our society. That doesn't mean anyone who wants nothing to do with "the system" is acting out of rebellion (which we want nothing to do with since it's counted as witchcraft); it just means we are choosing to not take advantage of an *option* that has been offered to us. Do you realize that, because of "compulsory school attendance laws," our schools are no longer really "public schools" but are "government schools"? It's quite true. To be truly "public" they have to be *free* to take or leave, as we choose. So we are still required to educate our children, but we get to choose *how.*

IF THE SHOE FITS...

Exactly *how* difficult it will be for you to depart from the traditional school scene depends entirely upon how convinced you are of its value – deep down. You can say one thing, even think it, but *really* believe another. Some of you have older children who went through the traditional system whose self esteems and love of learning were shredded bit by bit as they fell through the cracks. It's likely they're *still* recovering in some ways. For you it was *easy* to look for something else – just about *anything* would be better. ☺

Others of you have a deep conviction that God is calling you to homeschool your high schooler, but not seeing any major deficiencies in traditional forms of education, you subconsciously see it as being the standard to which you should conform *if you want your child to "get what he needs."* You may even be from a long line of public school loyalists that includes teachers and administrators, or even just Honor Roll students. For you, it's harder. You are trying to step into shoes (that you think are) much larger than your feet. ☺

Would it help if I told you they were the wrong style of shoes to begin with? Trying to duplicate the public school system is like trying to climb into size 45 army boots when God is calling you to step into *just-your-size* sandals. God isn't calling you to teach an "army" of kids all sorts of head knowledge; He's calling you to teach *your* kids what *He* knows they need to be ready to walk in the good works that *He* prepared beforehand for them to walk in. He's got a plan! You just get to cooperate with Him!

Analogies go only so far. I'm not saying that God's plan for homeschooling through high school is like wearing sandals that don't give you any support or protection. My point is just that the shoes He has for us are *light* and airy: *"My yoke is easy and my burden is light."* (Matt. 11: 28) Plus you can see through them: *"Where there is no vision, the people perish."* (Psalms 28:14) That may be stretching the analogy a bit, but I think you get my point. ☺

"ONE NATION, UNDER GOD, INDEPENDENT, WITH LIBERTY AND JUSTICE FOR ALL"

This book is about *independent* homeschooling. And by "independent" I don't mean "alone." We are not "an island unto ourselves"; we have contact with other people: homeschoolers, family, church family and friends. So that's not it.

Rather, I mean "*non*-dependent" in that we are not dependent upon, affiliated with, or "under" the state or a curriculum company. We have used all sorts of resources from all sorts of curriculum companies, so don't get the idea that I create every piece of our kids' "school stuff." And I've read a lot, gone to seminars, and listened to tapes.

So why do I even bring up "requirements" in the book, with any thought of *utilizing* them? I am using the state's requirements not as "requirements" but as mere *guidelines.* Why? Because I feel they constitute a valuable line-up of courses. Course *titles,* actually, because the content of our classes is quite different from that of these same course titles in public schools, which can be questionable. So as long as we are free to fill or "slant" them as we deem appropriate, these guidelines are helpful.

The diplomas we have issued our (two older) children, therefore, were in alignment with my and my husband's priorities rather than the state's." *(...unless we choose to get it "validated" through an institution of some sort; even then, it would not be "state-approved" but approved by whatever institution that we chose to put ourselves under.)* Therefore, please understand that *our diploma will not be state-issued or state-approved.* We are a private, home-based school; not state-funded or even state-affiliated.

LOOKING AT REQUIREMENTS

The graduation requirements that you'll see on page 5 embody the *minimum* education the state feels a student should receive. I know that many if not most public schools require even more to graduate from their institution. Two to four additional credits is common for public schools in our area. And many Christian schools, to allow for Biblical studies, require even more.

My own feeling is that a state's list of requirements for graduation is a good basic place to begin, but not complete until I have added what I feel God would have us add. This is assuming I have the freedom to *fill* these subjects as I feel led by God. That doesn't mean I will "invent" or create everything they learn! Just that I will *choose* and *direct* their education.

I like to view the "requirements" – which we will adapt and form into our framework – as the "shelves" upon which we place the "groceries" of our learning.. I want our children to have a solid foundation in a broad spectrum of subjects. We especially want them to be proficient in the skills of writing and speaking. A good working knowledge of the practical application of mathematics is important too; maybe even some advanced math. But what good is mastery of Calculus if you can't balance a checkbook or understand what "18% Interest" *really* means and how it works on a practical level?

Primarily, we want to provide an environment in our home and give our kids the tools to be able to *learn for themselves* the most significant knowledge they can ever acquire: knowledge of God Himself, and then it follows that they'll learn what He wants to do *in* them, *for* them, *through* them, and *with* them; how much He loves them; how mighty He is; and how able He is to heal and lead His people! This is *our* heart for our kids as they walk through the high school years! Perhaps it is yours too. If so, you'll find some encouragement and practical help for doing so within these pages. See my "Wisdom's 7 Pillars for Education" charticle (article and chart) at the end of Section 1, if you didn't already. ☺

FOLLOWING IN FREEDOM

One thing I want to make unquestionably clear is that, even though we (our two high schoolers) followed the *general guidelines* for graduation requirements for our state, we are not bound *by* them or *to* them. We are enjoying the freedom Washington state *protects.* Not the freedom the state *gives us.* Huge difference! I may start out thinking that a certain line-up of credits and resources is a good idea, but if I feel the need to change mid-stream – and I *did!* – then I do not feel I am letting anyone (including myself) down, nor am I "fudging." I am free to grow and change and be led by God from day to day and year to year. I cannot submit to a system that takes away that – that *what?* Is hearing God on a day to day basis a *right?* Or is it a *privilege?* No!!! These are just the very basics of walking in relationship with God!

God has charged us to raise up our children in a godly way. In Deut. 7 *(one of the "Homeschoolers Hall of Fame" verse)* it says,

> *"And these words which I command you this day shall be upon your heart; and you shall teach them diligently to your children, and shall talk of them when you sit in your house, and when you walk by the way, and when you lie down, and when you rise."*

Now I fully admit that I'm a little biased, but I'm not sure *how in the world* I'd be able to do that if my kids were gone all day?! I'm not saying everyone should homeschool – though I think many more could and would if they weren't so tangled up in "the system" and knew what homeschooling *could* and *should* be – but I *am* saying that I for one am taking God up on this, and that ***my freedom comes from God, not the state.***

WHY HAVE SUCH A "FRAMEWORK"?

I had several reasons for wanting to devise and implement guidelines, credits, and hours into our kids' plan for high school, all of which constitute a flexible framework. Marilyn Howshall and I discuss some of these in greater depth in Section 7 on "Lifestyle of Learning," but here are the ones I want to emphasize for now:

- to provide a set of definable, reachable goals that, once attained, give the student a feeling of accomplishment
- to indicate that some program of study has been completed thus indicating the student's diligence and "stick-to-it-ness"
- to motivate an undisciplined student – to give him a tangible, realistic goal
- to help in-laws (and maybe even *you?)* feel better about the plan – since they tend more toward the structured and logical
- to give a student *who wants it* the feeling that they are on *somewhat* the same as their friends.

So let me state again, just for the record, that we are not "complying" with any line-up of requirements for graduation; we are merely *using* them as guidelines and are filling up the courses with our choice of content.

WASHINGTON STATE AS AN EXAMPLE

To present the concept as simply as possible *and* to keep from overwhelming you, I'm going to start with just one state's list of credit requirements for graduation. However, I'm also going to present a parallel line-up of several states' requirements a bit later in this section. *(No fair peeking ahead – you'll regret it!)*

Having looked at these other states' requirements *(I had to peek to be able to write this)* I can tell you that Washington State's requirements are very typical. So whether or not you live in Wash., *if* you want a standard set of guidelines to go by, this is a good place to start.

If you're *not* a Wash. resident, and would feel more comfortable following your own state's guidelines, contact either Home School Legal Defense Association or your state's homeschool association and request a list of your state's guidelines for graduation. Contact your state's Dept. of Education *only* if you *know* your state is "homeschool friendly."

Each of the fifty states has its own set of requirements for graduation from a public school in that state, not to mention their own homeschooling laws. So there are about – no, *exactly* – 50 different answers to that question. ☺ Because there ought to be *some* advantage for those who live in the same state as the author, I present Washington State's credit requirements in just a moment, on page 5. If your state is in the process of establishing or revising a homeschool law, Washington's is a good model to follow. It is known as being one of the better homeschool laws in the country *(as homeschool laws go)*. I present the highlights here for Wash. State residents who are interested *and* for those of you who want ideas if working on formulating a homeschool law for your state.

NOT HEAVY ENOUGH?

Someone had just found out about a more "weighty" list of requirements in order to be prepared for college, and noted that it was much more than was listed in my book. I just want to mention a few things... First, the examples of requirements I list on page 5 are not "college prep" requirements, but are just for graduation from high school. On a later page in this section I list additional requirements for the college-bound student. It may or may not be closer to the list that she saw, but it really doesn't matter, because...

There is no *one* universally accepted list of college-prep requirements. Each college has its *own*, so you need to ask what the college your student wants to attend requires and, if you feel so led, follow those.

But more importantly than this is the fact that I do not simply present requirements and suggest that you follow them. This is far from the focus and purpose of my message! In this section my main reason for presenting the high school graduation requirements for several states is *primarily* to show you all the holes left by any one high school program! What I have seen in doing seminars all over the country – at which I always ask for input from the attendees when I do my "Intimidometer Poll"(coming up in Section Four) – is that most homeschooling parents of high schoolers suffer from severe "Gap-a-phobia." ☺ This is only natural, but who wants to be *natural* when we serve a *super*natural God?!?!? Not me!

PRESCRIPTION FOR GAP-A-PHOBIA

We *must* get a bigger picture of education – *God's* view of it – if we are to be free from this fear of leaving gaps. There is so much more to true education than merely academics. *(Again, make sure you don't miss my "Wisdom's & Pillars of Education" charticle at the end of Section 1!)* Look at the gaps being left by any public or even *private* school! I am *much* more concerned about *those* gaps than about any I would leave as a homeschool parent! But my priorities probably differ – and *that* is the key! We can be intimidated only by that which threatens that which we *really* value. What do you *really* value?

Most kids graduating from high school have mere head knowledge, which is knowledge that is totally "out of context." So what's the correct context? The context of II Peter 1:5-7 which gives a very clear picture of the exact position knowledge must have if we are not to be (or produce) mere fools who are puffed up with pride – which is exactly what the Word says we will be if our knowledge is not built on the right foundation!

> *"For although they knew God, they did not honor Him as God or give thanks to Him, but they became futile in their thinking and their senseless minds were darkened. Claiming to be wise, they became fools..." (Rom. 1:21-22)*

Please *consider*, with respectful heart but open eyes, that people who do not honor God are primarily the ones setting the public school system's requirements – that *we* are so diligently trying to fulfill! There is a steep price our children pay to *fulfill* those requirements! If we can "in a sense" (God's sense) do it in a manner that still gets them where God wants them, and bears the fruit in their lives and in our families that He wants to see, then great. But four credits of math that a student will never need in life is taking up an awful lot of their precious, extremely limited time! *Just* to fulfill a *requirement?!?!* Yet *everyone's doing it!!*

REDEEMING THE TIME

Your high schooler has only so many hours left with you at home. Then they are *gone!* If they're wise, they will still come to you for counsel and friendship. But your years to influence them while they are still in your home and are still "wet clay" will soon come to an end. Now is your only chance to "redeem the time." *Redeem* means "to take something of little or no value and, in exchange, get something of great value." Who is the Only One who can make such an amazing transaction with us? The Master Educational Consultant, of course!

Our oldest daughter, now 21, is gone – no longer living in our nest. She doesn't want us to do anything to her room, but she hasn't lived in it since the fall of 1997! We homeschooled her from scratch and graduated her two years ago. Our main job in her life – raising and training her – is now *over*. We still have a wonderful relationship, but my (our) years of being the primary influence in her life are past.

Hear God *now* on what to do with the precious time you have. Bow down to HIM, not to requirements. Let Him use requirements as a tool in His hand; do not let requirements use, control, or intimidate you. Just filling requirements is not necessarily – or even *probably* – going to prepare your child for *life* – and for the *"future works God prepared beforehand that they may walk in them."* A "one-size-fits-all," "cookie-cutter" education will not prepare our children for all that God has for them. I see too many people who are far more concerned about their children's high school experience being "college prep" than they are about it being "life prep"! This is not to put anyone down for wanting to prepare their children for college! I just want to re-direct the focus of any who may be overly concerned about this to the exclusion of even seeking God on what His heart is for these vital years. I ask again: *What do you really value? (If the shoe doesn't fit, don't try to put it on!* ☺) The bottom line of my message, as *usual*, is...

Hear God for your family!

REQUIREMENTS FOR GRADUATION:

Washington State's Graduation Requirements – a typical set of requirements:

English	3 Credits
Mathematics	2 Credits
History/Social Studies: *Distributed as follows:*	2.5 credits
• *U.S. History & Government* ♦	1 Credit
• *WA State History and Government* ♥	.5 Credit
• *Contemp. World Hist., Geography & Problems* ■	1 Credit
Science	1 Credit
Lab Science	1 Credit
Occupational Education	1 Credit
Physical Education (and Health) ☼	2 Credits
Restricted Elective ✫	1 Credit
Electives	+ 5.5 Credits
	TOTAL = 19 Credits

The law does not specify what the content of these subjects shall be, except for the following:

♦ U.S. History and Government shall include study of the Constitution of the United States.

♥ WA State History shall include study of the Constitution of Washington State. The Wash.State History & Government requirement may be fulfilled in grades 7 or 8, but (if so) the credit earned may not apply toward the minimum number of credits required for graduation.

■ Courses in Economics, Sociology, Civics, Political Science, International Relations or related courses with emphasis on current problems may be accepted as equivalencies.

☼ The Physical Education requirement shall be met by course work in the areas of personal fitness development, leisure activities, health education life skills management, and healthful living program design. P.E. and Health are listed together in the law; it does not state one year of each; only that the above areas be covered.

✫ This 1 Credit requirement must be selected from fine, visual, or performing arts, or any of the subject areas listed.

THE GOOD SIDE OF REQUIREMENTS

When considering "requirements," something to keep in mind is that life is full of them. ☺ At home, on the job, in our cities, on the freeway, within a family, even in our walk with God, but in a grace-based way. "He has shown thee, oh man, what is good, and what does the Lord require of thee." It's a matter of "which ones" we submit ourselves to, and "who" is determining what they are, for what purposes, and how they are interpreted and carried out in the student's life, not to mention how much time is left after filling them for the child to develop in their personal, God-given areas. And what *are* God's requirements? – the rest of the above verse? They are "...to do justice, and to love mercy, and to walk humbly with thy God." What doth "humbly" mean? Nothing more than to be *teachable.*

In the school system, there would be many *more* requirements that were set by people who know nothing of my children, their temperaments, their futures, or (worst) their or *our* values. And then, to top it off, they would be required to use whatever time they might have at home to do *more* of the system's requirements! Our (my) requirements were not burdensome for my kids; they were *work* and required *effort,* but they still had plenty of time for their own interests.

WASHINGTON STATE'S REQUIREMENTS

In this state, the only requirements relating to *curriculum (for* ***all*** *grades, not just high school)* are to teach the eleven required subjects: Mathematics, Science, Occupational Education, Language, Social Studies, History, Health, Reading, Writing, Spelling, and Appreciation of Art and Music. The law does not mandate specific levels of proficiency other than "making reasonable progress consistent with his or her age or stage of development." Nor does it state that homeschoolers – or *any* school, (unless they want to be accredited, which requires even **more** credits) *must* comply with Wash.'s graduation requirements; just that the number of hours spent in school should be equivalent to those required for public schools. But it includes the following qualifying statement:

> ***"The legislature recognizes that home-based instruction is less structured than the instruction normally provided in a classroom setting. Therefore, the provisions relating to the nature and quality of instructional and related educational activities shall be liberally construed."*** *(which basically just means that it will be interpreted or understood with a lot of grace, freedom and latitude)*

And in another section, the law also goes on to say:

> ***"...all decisions relating to philosophy or doctrine, selection of books, teaching materials and curriculum, and methods, timing and place in the provision or evaluation of home-based instruction shall be the responsibility of the parent except for matters specifically referred to in this law."***

I understand that the state has some "compelling interests," but I believe they far over-step their boundaries and take much more control than the Constitution allows. In fact, this has been proven in many court cases. So whether I agree with those last nine underlined words would fill up another book, and many good ones have already been written; one by John Whitehead and Wendell Bird, entitled *Home Education and Constitutional Liberties*. Another by Michael Farris (head of H.S.L.D.A.) is *Homeschooling and the Law.* Both books discuss the volatile issue of whether or not, *according to the Constitution of the United States of America,* homeschooling should be regulated by state laws. Since they do a good job of it and it's obviously a subject that much has been written on, you can see why I don't want to get into the subject here! I couldn't begin to do it justice by just dabbling in it. I have written a five-page article in my *Homeschool Guide-a-log* entitled "Obeying the Law, Testing, and Qualifying to Homeschool" in which I deal with these matters in more detail. I hope you'll read it.

There are also requirements in Washington's law relating to number of hours, filling out a "Declaration of Intent" – not "application" or "request" – *big* difference!) to homeschool, testing or assessing, how parents can "qualify" (under the *law*; that is; God has already qualified them) and what to keep in the child's "permanent record" which is kept by the parents until such time as the student is placed back in a public school. But we don't need to get into that aspect any deeper. For now we'll stick with high school issues.

COMPARING WITH OTHER STATES

Next let's take a look at the requirements for six different states put side by side. These may very well be outdated as I compiled this in 1993. However, the point of this is *not* to give you up-to-the-minute guidelines, but to make a point, which I'll do on page 8, *after* the chart.

How did I choose my sampling of states? These are the home states of the first homeschoolers who requested information on this manual when it was still....

Text continued after chart ☞

COURSES	STATE REQUIREMENTS					
Some state requirements use semester credits. For ease in comparison, I've translated all requirements into the same base.	**IN**	**KS**	**NH**	**OR**	**TX**	**WA**
English (Lang. Arts)	4	4	4	3	4	3
Mathematics	2	2	2	2	3	2
Science	Life Sci - 1 Phys Sci - 1	2 (incl 1 credit of Lab Science)	Phys Sci - 1 Biol. Sci - 1	2	2	2 (including 1 Cr. Lab Science)
Social Studies	⇓	(3) ⇓	2½ ⇓	⇓	(2½) ⇓	(2½) ⇓
U.S. History & Geogr.	1	1	"included"	1	1	1
World, (or Global) History, Culture, &/or Geography	—	—	—	1	1	1
U.S. Government	½	½	"included"	½	½	(w/ U.S. Hist)
State History	—	"A Course"	"included"	—	—	½
Economics	½	—	(½) (part of Soc Studies)	1 + Pers. Finance	½	—
Fine Arts	—	—	(Elec Choice)	1 OR or Applied Arts	—	(1 option for Restricted Elective)
Foreign Language	—	—	(Elec Choice)	- OR - 1	—	—
Physical Education	½	1 ½ cr may consist of: ⇓	1	1	1½	2
Health (& Safety)	½	(½)	¼	1	½	
Occupational/Voc. Ed. or Career Devel.	—	—	(Elec Choice)	½	—	1
Electives	8 (w/ certain restrictions)	9	3 - Restricted Electives ------- 4 - Open Electives	—	7	1 - Restricted Elective ------- 5½ - Open Electives
Total # of req'd credits:	**19**	**21**	**19¾**	**22**	**21**	**19**

PLEASE NOTE: These are as of 1993 and, especially with the introduction of "Goals 2000," many requirements may have changed by now. But the intent of this chart is to show the differences in states' priorities, not to give an up-to-date list of what you should follow!

forthcoming! ... There are two things I want you to take note of here: First, how *similar* the requirements are; and second, how *different* they are. I'll take those one at a time.

FIRST, THE SIMILARITIES

As you can see, the requirements in the subjects of Math, English, Social Studies (History, Government and Geography), and Science are very close from state to state. So for these subjects it doesn't make a lot of difference which state's guidelines you follow. You can either choose the line-up of your own state, go with that of any state whose line-up you like, or look at any of these options as "nice ideas" and use them to come up with your own completely unique and personalized list.

TOTAL NUMBER OF REQUIRED CREDITS

Now look at the differences in the total number of credits! The states on this chart range between 19 and 22. Let's translate these differences in totals into *hours.* First you might wonder *can we even do that? Are the credits and hours for them even the same?* While some states didn't specify the number of hours required for one credit, most did. And it was that "one credit" equals 150 *clock hours* of instruction. *(In Washington state it's actually 180 50-minute hours because the school year is divided into that many days, but it still comes out to the same amount of time. 180 50-minute classes comes out to 150 hours for the year.)*

If you want to know the specifics for your area, you could just call up your local school district and find out. *(You may want to disguise your voice or pay a friend to do it – from a pay phone.)* At 150 clock hours per credit, a difference of 450 hours is quite a chunk of time! At 6 hours of study per day, that's a 25-day difference!

OTHER VARIANCES

What about the other subjects? If you live in Indiana, Kansas, or New Hampshire, you can technically graduate without ever having studied World History or Geography. (The local high schools probably tack on those requirements anyway, but my point is that *technically* you *could* graduate without it.) If you live in Indiana, Oregon, or Texas, you can graduate without knowing one teensy fact about the state you live in. Residents of Washington or Kansas don't have to know a thing about economics. In Texas, Kansas, and Indiana you run the risk of being "uncultured" since those states require no fine arts. In Oregon you're only half as likely to know anything about getting a job as someone in Washington state because the latter requires 1 full credit of occupational skills whereas in Oregon, only ½ credit is required. But they're much better off than the other five states that don't require it at all! In New Hampshire you'll only be half as health-wise as you would be if you graduated from a high school in Indiana or Texas, ¼ as health-wise as a Washington State graduate, but a whole lot better off than your health-illiterate cousin in many other states that don't require that subject. Which may not be so bad after all because at least you'll be better prepared for *college* if you fulfill the foreign language option – which you can do only at the risk of being uncultured and unhealthy. *(Please note; this was all said "tongue in cheek.")*

SO WHAT'S MY POINT?

Freedom. With all these variances between states, specifically, with all the holes in both *time* requirements and *content* requirements between *any* given states, *we* should certainly not have any feelings of anxiety or inferiority about leaving gaps. *Especially* since we can choose which subjects are most important to us so that the particular holes we leave – and we *will* leave holes, may as well accept and get over that now – are the holes *we* feel best about leaving. You have a lot more leeway in making your choices than you might have thought.

As one homeschooling mom said to me: "My job isn't to cover *everything* and give my kids *all* knowledge. I couldn't do that anyway. It's just to teach them basic skills and give them a good, solid *foundation for life.* I choose things that I feel are important to the context of their life and include as much of the Christian world view as possible."

No state will cover it all. And even if they did, *(which is humanly impossible)* the manner in which they cover it wouldn't include a Christian world view *or* an approach that makes the students able to *retain it all.* No approach *can;* but some can come much closer, with a different *purpose* to begin with than just "filling the heads with knowledge."

And you can be sure that not one word in the entire array of courses will have been uttered pertaining to God's hand in history or in our lives, nor of His design of the universe including its creation and the laws of science that regulate it. I don't know of anything more important! The gap they are leaving is foundational to the person's entire "paradigm"! Their way of seeing the world! Any gaps *we* might leave can be easily filled; you can't just "fill" an entire understanding of the world, the way it got here (which was by divine *design,* not purpose-less *chance)* and how it functions.

Frankly, this is *so* important that I wouldn't want a secular institution teaching it because you have to have God's perspective to be able to present it accurately. So why am I bringing this up? Because *supposedly* everything that "should" have been taught *was* taught. And yet it *wasn't*, and without God *cannot* be. That means it's up to you and me; and that we need not fear gaps.

OTHER "RED TAPE" REQUIREMENTS

Because I'm presenting requirements of particular states, I feel that I must also tell you that just because you teach a particular group of courses, it doesn't necessarily mean you're fulfilling *all* of the state's requirements. I'll explain what I mean...

What you *don't* see on the six-state Requirements Chart is all the jargon regulating *how* these credits shall be earned. I'll give a few examples *only* to demonstrate what I mean, and will add my own hopefully-not-*too*-snide comments after each. But please don't get caught up in or intimidated by the specifics! You do not need to follow these! So I want you to *read* these next few statements only to get my point, but then *totally disregard* them, okay?

- Alternative graduation requirements established by any local board of education shall comply with statewide educational goals adopted by the state board and shall have broad-base community involvement in their formulation. *(Broad-based community involvement? Hmmm, nice idea, but what does the community know about my child's future and what God is going to be doing with their life?)*

- Each teacher shall hold a valid certificate with the appropriate endorsements for the subject and level of assignment. *(Are you aware that research has shown that there is absolutely NO relationship between student performance on standardized achievement tests and teacher certification?! None. Zippo. Nada. In fact, homeschoolers score an average of 20 points higher on achievement tests and for the most part are not taught by teachers with certificates! That doesn't mean students automatically score higher if not taught by someone with a certificate; just that you don't have to have a certified teacher to score higher or learn!)*

- The district school board may adopt policies for alternative learning experiences, such as credit by examination and credit for off-campus experiences; and with any modification of the attendance requirements for graduation, school district staff shall consider age and maturity of students, access to alternative learning experiences, performance levels, school district guidelines and the wishes of parents or guardians. *(Know what? You can do this just as easily and validly. You know your student and the extent of their experiences better than any teacher with a classroom full of students, (5 or 6 classes a day) ever could! And you can help him get much more out of it too! And even do it with him!)*

- A school district may grant high school credit for experiences prior to the ninth grade which are equivalent to high school courses within the school district; however, the granting of such credit cannot reduce below 22 the units of credit to be completed in grades 9 through 12. *(Why not? Think about this. Control? Funding requirements?)*

- The [Unnamed] *(don't want to make enemies)* State History and Government course requirement may be fulfilled by students in grades 7 or 8 or both. [However] credits earned in grades 7 or 8 shall not be applied toward the minimum number of credits required for high school graduation. *(Now what good reasoning could there be for that guideline?! Learning is obviously not the concern because the content of the class would be exactly the same, wouldn't it? They're just wanting to make sure the student stays under their jurisdiction for the full amount of time in high school. My guess is that it has something to do with full state-funding for attendance. Could be different, but whatever it is, being in high school longer is not in the student's best interest, no matter what the motivation!)*

- Any student who has completed high school courses while in grades 7 or 8 shall, upon the request of the student and his or her parent or guardian, be given high school credit which shall apply toward fulfilling high school graduation requirements if the academic level of the course is equivalent to or exceeds the requirements for courses in grades 9 through 12; and the student takes the course with two or more students enrolled in grades 9 through 12 who are receiving high school credit for the course and the student successfully completes the same course requirements and examinations as the other grade 9 through 12 students..." *(Okay, think about this. Yes, the state and school districts* ***have*** *to have requirements to govern their programs. But for a homeschooler to take on these last restrictions is utterly pointless. It makes as much sense as going to Michelangelo and saying "Okay, Mike, your art work will qualify as good art as long as you do it with at least two others who are considered artists and are successfully completing the same projects. Also, you must do the work when you are 35 or older. If you are 34 or under, it doesn't qualify."... Since when does good learning take place only with two or more people or at a certain age?! Most of the learning I've done in my life has been totally self-directed and alone! This is what I mean by the "run-em-through-the-same-mill" mentality. It has absolutely* ***nothing*** *to do with* ***real*** *life or* ***real*** *learning. So your student is no Michelangelo, you say? That makes no difference in the principle of the thing. Think about it!)*

Again, I am *not* including all of this to pile more regulations on you, nor to imply that I think you ought to be following them. If you're not a certified teacher, you aren't even "eligible" *according to state guidelines* to teach *their* requirements in their institution in the first place! So part of why I'm including this is to keep myself out of trouble with you and/or your state *only* in that I don't want you to have followed *what you thought* were your states' guidelines only to find out the state actually required more than just a list of required subjects.

But more than anything, I want you to see how unrelated regulations and guidelines are to real learning, the kind that's enjoyable, relevant, and sticks for life.

WHAT ABOUT GETTING A G.E.D.?

The GED (General Education Diploma) is an option that has worked well for some homeschoolers. Since this is not the focus of my book and it is discussed thoroughly in Cathy Duffy's Jr./Sr. High Curriculum Manual, I refer you there for an in-depth discussion of the pro's and con's and how-to's of getting a G.E.D.. But since you asked, here's my *personal* view about it.

First, in case you weren't aware, some employers regard it as "second-rate" or inferior to a high school diploma. There's a certain "stigma" attached to it. But there's also a certain stigma attached to homeschooling that's probably even worse at this point in history, although it *is* improving as homeschoolers across the country are proving their competency – academically, socially, and morally. Spiritually, too, but the world doesn't care about *that* aspect generally speaking and so doesn't respect it, and the world's approval is really what this is all about. We already know we have God's approval and are not in need of a certificate or diploma to prove that.

However, back to the stigma... As *homeschoolers* we have the chance to "redeem" *(see glossary)* this stigma. The homeschooling movement is now at the "coming of age" and "proving itself" stage. Sixteen years ago, when I was beginning, homeschooling was at the "you are *crazy and stupid* to even be *thinking* about this" stage. So we've come a long ways! ☺ I believe that eventually, because the proof is *always* in the pudding, home schooling will be the hands-down, overall best choice, but we're not there yet except in "pockets" where this is obvious to the ones involved in individual cases.

I'd rather go entirely on the merit of our own homeschooling experience and bag the G.E.D., which, to me, makes it appear that we couldn't quite "make the grade" on getting a diploma, so we tried for the "next best thing." Don't get me wrong, that's not what it really *is*; it's how it might appear. I think homeschooling produces people with far better employee potential, so I'd rather just go on the merit of our homeschool diploma. You may not want to, and that's okay; this is just where I'm at.

But my other consideration in not wanting to go that route was that I did not want it to be a way for our kids to "get done" early by passing the G.E.D. test, even though it *is* partially life skills oriented. I've said it somewhere before, I'll say it again: I want these high school years used to not merely amass a certain amount of *testable* knowledge, but to *spiritually* and *really* and *truly* prepare our kids for *life* and to become more established in our *relationship with them.* That doesn't mean that those who get their G.E.D., or get it early, don't have a good plan; it just doesn't fit with *our* vision.

ACCOMMODATING TO ALL HIGH SCHOOL STYLES

This manual (Form+U+la) accommodates *all* types of programs: college-bound, minimum requirements only, no requirements, getting a G.E.D., the *Lifestyle of Learning* approach, and any combination between – and not being limited to or stuck on any one exclusively.

I do not provide a lot of information on college requirements simply because, again, books are already written that have not only invented that "wheel," but that have also thoroughly greased and fortified it! However, since college is a concern for many, I will add my (and Helen Keller's) two cents worth... (Actually, *my* thoughts are worth about 2 cents; Ms. Keller's are worth *much* more, as you'll soon see.)

SO WHAT ABOUT COLLEGE?

Okay, so maybe we can choose subjects that are comparable to a typical high school's line-up of subjects, but what about college? How can we know if we are choosing subject content that is "college-preparatory"?

First, don't (as in *do not*) let this issue of "college" intimidate you. I can sense, even as I write, that blood pressures are escalating a few points with each mention of the words "college" and "requirements." I bring tidings of great joy! – not nearly as life changing as the tidings brought by that renowned band of angels 2,000 years ago, but I guarantee they will profoundly affect your high school world *and* lower your blood pressure! Hark!...

First, you must be willing to ask yourself if college is *truly* where God is taking your student. You will save yourself much wasted time and money if you start out with the assumption that God actually has a plan for your student that may or *may not* include college rather than starting with the assumption that college is automatically God's plan for your student!

Next, some states give a suggested course of study for the college-bound or "advanced" student. You may – or may *not* – want to check into that.

Next, from yet another perspective, Cafi Cohen wrote a book for homeschoolers called *And What About College* that many have found very helpful. *(We do carry it.)* Also, there are ample resources available to help you "college-prep-ize" your classes, coming right up...

COLLEGE PREP RESOURCES

Diana McAlister and Candice Oneschak's book ***Homeschooling the High Schooler*** presents a wealth of information especially to help you choose college-prep course content. It's presented in a way that's easy to understand and apply to your own homeschool. They present much helpful material on charting your course as well as "core and non-core content requirements" for many high school courses.

Cathy Duffy's ***Junior/Senior High Christian Home Educators Manual*** is full of information that will enlighten you and give you confidence as you get behind the wheel and start heading down the curvy road before you. In fact, her book has a way of taking quite a few curves (that might throw you for a loop) *out* of that road! She has reviews of many products available to homeschoolers which will help you make choices that are appropriate for your student.

Mary Schofield's book, ***The High School Handbook*** gives resources and ideas for doing college at home, earning college credit by examination, and books to help prepare for college. She discusses, as do the other two, college admission, college entrance examinations, and financial aid. This is on top of many class format and goal ideas for high school classes.

This last one has nothing to do with "college prep," but I wanted to mention *somewhere* that I have written a book called ***A Christian Homeschooler's Guide to Field Trips and Extra-Curricular Activities.*** It covers topics mentioned in the title as well as how to put on a Home School Presentation Night, how to organize a support group and co-op, which can be very helpful for people wanting to work together – or *co-op* – on doing high school classes, college-prep or not. The first chapter, called "It's About Time," is a practical, spiritually uplifting challenge to hear God on what all is wise to include – and *not* include – in our extra-curricular program – which for us was actually *intra*-curricular. It's so easy to live by a set of misleading and unhealthy standards that result in our family life and closeness getting swept away in a sea of activities.

SUPPLEMENTAL IDEAS FOR GRADUATION REQUIREMENTS

As we all know, there's more to succeeding in life than doing well academically, whether this is at the high school or college level. Preparing our children to enter "the real world" is a challenge we have taken on, having joyfully accepted it from God.

In this section *(page II-13)* I offer suggestions for you to consider as you *and* your high schooler *and* Jesus – who is *the Way* – determine the *"way"* you will go.

Much of what you will find in the "Potpourri" section was birthed out of my desire to equip our children for *life,* not merely for a job or college. In that section (4) you will find ideas for the non-traditional classes that are listed in the starred (*) sections of this chart.

Note: the plus (+) symbol means "in addition to" the usual graduation requirements listed previously.

OR NO COLLEGE AT ALL?

There are some excellent non-college programs available that you may want to check into when your teen starts wondering "is there life after graduation?"

Youth With A Mission (Y.W.A.M.) has an excellent Christian discipleship training program with several outreach bases they call "D.T.S.'s" or "Discipleship Training Schools." Their teachings are sound and life-changing.

A.T.I.A., the homeschool program created by "Institute in Basic Life Principles" (Bill Gothard), has an Apprenticeship Program. *(Some just call it "ATI")*

Christ for the Nations is another post-highschool option, located in Texas. And I think there might be one in Canada just north of Washington state.

(I'm sorry I don't have up-to-date information on these as far as addresses and phone numbers. I have found that it's almost pointless to try to keep up with information as it changes constantly – especially area codes! So if you do try a number that I've given anywhere in this book and it doesn't work, try finding out if the area code has changed before assuming they're out of business!! This happened to us! Our area code <u>*was*</u> *206 but a few years ago it changed to 360.)*

At any rate, no matter what you do or don't decide to do after high school, if you already know that college is definitely **not** in the picture, you have no reason to be one bit concerned with college-prep courses or with any *particular* graduation requirements, other than your own. You should not feel you *have* to be bound to any requirements set forth by man's law, unless you believe God is leading you that way. God entrusted the education of children to the children's own *parents,* not to the state. It is not the state's calling. True, it is now the state's *business,* but only because parents, wanting more time for their own pursuits, or *needing* to work outside the home (which I personally think is not a true "need" in many instances) gladly handed it over to the state, one chunk, one law at a time. But that still doesn't make it the state's *calling.*

BEING ON THE "SAFE SIDE"?

If your teen is unsure about their future (as it relates to college), and you think there's a good chance that they *will* go to college, it might be a good idea to be on the "safe side" and plan, at least for now, as if he *is,* unless you don't mind having to do some cramming to get caught up in the later years of high school, or at least before or upon entering college.

But then again, it might *not* be so "safe"! There's a trade-off! To fill up those precious hours with college-prep work leaves less time for studies and activities that might be more fulfilling and preparatory for *life* for your student. You have the freedom to pick and choose only what you consider to be valuable and enhancing to your child's education, not to mention the most interesting. *(However, Diana, Candice and Mary do a good job of incorporating plenty of the "interesting" aspect.)* How much of that freedom are you willing to relinquish for the sake of being "maybe safer"? It's up to you.

Consider the Colfax family in California. Not one of their four sons ever took a single "college prep" course, yet all four were admitted to Harvard! And this was a direct result of the education their parents had provided – mostly *allowed* and *encouraged.* Their acceptance was based on portfolios they had assembled that highlighted accomplishments in areas they had time to pursue and become experts in because they weren't preoccupied with "college prep" courses. And if you suspect "superior genes" as being a prime factor in the boys' intelligence, guess what – two of their four sons were *adopted!*

If you do choose college-preparatory work for your courses, you can be more confident that your student will do as well in college as any public or private school student, but you also have to ask yourself if it's worth the time which will necessarily have to be taken from *somewhere else.* Only you can answer that. *(Again, oh God, we need to hear your voice!)*

I personally am not overly concerned about all this. Maybe I *should* be, *(I certainly used to be!),* but I'd rather send my budding adult out into the world equipped with a character-fortifying, real-life-preparing education than one that merely prepared him for college.

I haven't come to this place of "rest" naturally. I'm so glad that I – ***we*** – have a Master Curriculum and Career Counselor who is able to guide us into what will be best for *our* student and *our* situation! I firmly believe He has a specific and *good* plan for each one who earnestly seeks Him. If we ask our Father for a loaf of curriculum, will He give us a stony reply instead?

OTHER OPTIONS

Just for the record, I want to establish that I neither recommend nor "dis-recommend" college. However, I do admit that in many cases I would lean *away* from it. I believe there are many other more valuable paths to choose from these days. College was once viewed as being the "ultimate" direction to take after graduation. No other choice was considered nearly as prestigious, wise, or ambitious. This attitude still exists, but to a lesser degree, at least from what I have seen and heard. As our economy is shifting and college graduates no longer always get the higher-paying jobs they once did, the trend is changing.

And I am glad because I believe that, while college is the best option for some, it is not for others; and is turning many unprepared students away from God. *How* does that verse go? "For what profit is it to a man to gain the world['s credentials], but lose his soul?"

More people are becoming aware of the wide range of options including vocational schools and apprenticeships. The tremendously high costs, for which many remain bound to paying back for years after graduating, also make one think twice about whether this is really what God has for them.

Also, as the world's values and morals continue to spiral (plummet) downward, the chasm between the values of Christians and those of the world has widened dramatically. God has an abundance of work for His people in these latter days and perilous times! I believe we need to be more concerned with preparing ourselves and our children for *His* work – and finding out what that is for *us* – than with preparing ourselves to merely attain a high-paying career or gain entry into a prestigious college.

THE *ULTIMATE* OPTION – FOR EVERYONE

I'm not saying that God can't use college to prepare us for His work. Nor am I saying He doesn't desire to prosper us. What I *am* saying is that if we are clinging to "old wineskins" when God is trying to pour "new wine" into our lives, we need to stop and *listen* to His voice, and let Him have those useless, cracked old wineskins! Are we considering college for perhaps the wrong reasons? "Let us study (through these high school years as well) to show ourselves approved by *God,*" – not man! *That* is the "ultimate option" for every single person who calls themselves a follower of God.

CONSIDERING COLLEGE

If any of you are finding it difficult to consider anything but college as being a valid option, I gently encourage you to take this matter before the Lord. There is so much more to consider than merely a degree. *Especially* if "just" getting further education is your goal, consider that college is not necessarily the best or the only way to get it. As I read chapter 20 of *Helen Keller: The Story of My Life* I found that she expressed my very thoughts about college. She wrote:

> *"I began my studies with eagerness. Before me I saw a new world opening in beauty and light, and I felt within me the capacity to know all things. In the wonderland of Mind I should be as free as another* [with sight and hearing]*. Its people, scenery, manners, joys, and tragedies should be living tangible interpreters of the real world. The lecture halls seemed filled with the spirit of the great and wise, and I thought the professors were the embodiment of wisdom... But I soon discovered that college was not quite the romantic lyceum I had imagined. Many of the dreams that had delighted my young inexperience became beautifully less and "faded into the light of common day." Gradually I began to find that there were disadvan-*

ADDITIONAL SUGGESTED CLASSES FOR THE STUDENT WHO INTENDS TO...

...ATTEND COLLEGE AFTER HIGH SCHOOL:

Community colleges do not have "requirements" per se, but students wanting to be well prepared for and most successful in the college setting should consider spending elective credits on the following classes. These are only suggested guidelines; consult the institution you want to attend for specific recommendations. Some states give a list of college-prep recommendations with their high school requirements. (I know that Indiana and Texas do so, out of the 6 I have checked into.)

English	1 Credit +
Math *(Higher level Math)*	1-2 Credits +
Occupational Skills	1 Credit +
Science	½ - 1½ +
Fine Arts	1 Credit
Foreign Lang. *(For 4-year college, credits should preferably be in 1 language)*	2-3 Credits

...SUCCEED IN THE "REAL WORLD":

You can either include these subjects in with other courses (such as Occupational Skills or Biblical Studies), or just have them stand as their own subjects. These are just my own personal recommendations; vary your inclusion and quantities according to your own priorities and "hearing" (from God)!

Life Skills	1 Credit
Home Economics	1 Credit
Tact, Grace and Social Skills	½ Credit
Applied or Consumer Math *(As part of the Math credits.)*	½ - 1 Credit

...SUCCEED IN MARRIAGE AND FAMILY LIFE:

Building Family Relationships	½ Credit
Marriage Preparation	½ Credit

...BE "EMPLOYABLE IF ALL ELSE FAILS":

(as Dad frequently said as we 6 ***girls*** *and 1 boy were growing up! I think he wanted to make sure he didn't end up with 6 "old maids" all back at home!*

Typing	½ - 1 Credit
Computer Skills	½ - 1 Credit

...SPEND A LIFETIME AND ETERNITY WITH GOD:

Bible and Spiritual Studies	1 - 4 credits
Self Esteem and Understanding	½- 1 credit

tages in going to college. The one I felt and still feel most is lack of time. I used to have time to think, to reflect, my mind and I. We would sit together of an evening and listen to the inner melodies of the spirit, which one hears only in leisure moments when the words of some loved poet touch a deep, sweet chord in the soul that until then had been silent. But in college there is no time to commune with one's thoughts. One goes to college to learn, it seems, not to think. When one enters the portals of learning, one leaves the dearest pleasures – solitude, books and imagination – outside with the whispering pines. I suppose I ought to find some comfort in the thought that I am laying up treasures for future enjoyment, but I am improvident enough to prefer present joy to hoarding riches against a rainy day."

I (Barb now) have *always* felt that college was "too much in too compacted a time." I also feel that the learning in that environment is disjointed and too non-applicable to anything of the person's real life to be of any lasting value – even in the area being studied! And, yes, I went to college for three+ years. Sure, you get your degree to which there *are* advantages career-wise. But if *learning, higher* education, and *truly* mastering your "area" are your goals, I think that taking classes *while* you are also working at least to some degree in your field is a much better, more effective way to *really* learn and *retain* it, because you are applying all your head knowledge to something it can stick to!

My dad did not go to college before marrying, and so took night classes at community colleges and independent study courses for many years. Throughout the years, even though he had as much or more skill and knowledge in his field of engineering than many of his co-workers, he was often bypassed for promotions and job opportunities or disregarded for the lack of a degree. Eventually he had gained enough "education" through various course work and experience that he decided to take the State of Washington P.E. (Professional Engineer) examination. This is a license granted by the State required of any engineer – graduate or non-graduate – to practice engineering in the public sector in the State of Washington.

At that examination, he sat next to a man who had a *doctorate* in engineering who was there taking the examination for the *fifth* time – he had failed his first four attempts! *(And now, here's my favorite part...)* My dad, who did not have such lofty academic credentials, only an associate (two-year) degree, but instead, had years of "learning as he went" both on the job and "one class at a time," passed the exam the first time"!

So does this totally negate the value of attending college all at once? No. It was very difficult for my dad to try to attend college *and* support a family of (eventually) seven children. It was hard on my mom as well. So I'm not saying "don't" or "never." I *am* saying that I hope you rethink your ideas about college as being the best or the only way to acquire "higher education." Be creative. Consider apprenticeship, or taking a college class as you work in the job of your field. This may not always be practical, but I just want you to consider your views and priorities, and where you are really going. consider Helen Keller's continued thoughts on this:

> *"Many scholars forget, it seems to me, that our enjoyment of the great works of literature depends more upon the depth of our sympathy than upon our understanding. The trouble is that very few of their laborious explanations stick in the memory. The mind drops them as a branch drops its overripe fruit. ... Again and again I ask impatiently, "Why concern myself with these explanations and hypotheses?" They fly hither and thither in my thought like blind birds beating the air with ineffectual wings. I do not mean to object to a thorough knowledge of the famous works we read. I object only to the interminable comments and bewildering criticisms that teach but one thing: there are as many opinions as there are men.*

In Section 4 of this book (on page 3) I discuss "Shelton's Theory of Brain Receptivity." This is a theory I came up with myself *(so it's very unofficial)* that basically just states that the human mind has an optimal "functionability," and the schools push you way beyond what's "optimum." I quote a Far Side cartoon in which one student in a classroom has his hand raised and is asking the teacher, "Mr. Osborne, may I be excused? My brain is full now." I think this is *sooo* true! Even aside from the "full brain" issue is the fact that so much of the knowledge is unrelated to anything in the student's life! Heads are being filled, but minds wasted! Miss Keller said it much more eloquently...

> *"There are times when I long to sweep away half the things I am expected to learn; for the overtaxed mind cannot enjoy the treasure it has secured at the greatest cost. ... When one reads hurriedly and nervously, having in mind written tests and examinations, one's brain becomes encumbered with a lot of bric-a-brac for which there seems to be little use. At the present time my mind is so full of heterogeneous matter that I almost despair of ever being able to put it in order.*
>
> *Whenever I enter the region of my mind I feel like the proverbial bull in the china shop. A thousand odds and ends of knowledge come crashing about my head like hailstones, and when I try to escape them, theme goblins and college nixies of all sorts pursue me, until I wish – oh, may I be forgiven the wicked wish! – that I might smash the idols I came to worship."*

Miss Keller's thoughts astounded me! Not so much

her actual views – because she put into words *exactly* what I have felt – but because she spoke them *so* eloquently *and* dared to say them to the world! College is such a high place in our culture; in fact, I believe her term "idol" is very accurate – and no exaggeration! Allow the Lord to examine your heart about this matter and let Him breathe *truth* and life into your perceptions, values and goals. I can guarantee you, only because the Word of God guarantees it, that you will thoroughly enjoy the peace and secure confidence you'll receive as a result of letting God have His way in your and your child's views and life.

> *"Remember not the former things, nor consider the things of old. Behold, I am doing a new thing. Now it springs forth, do you not perceive it? ... I will make a way in the wilderness and rivers in the desert,...for I give water in the wilderness, drink in the desert, to give drink to my chosen people, the people whom I formed for myself, that they might declare my praise."* (Isaiah 43:18–21)

I don't know about you, but I don't want to miss out, because of dull hearing or hazy vision, on perceiving what it is that God is bringing forth! I want a new wineskin so I can contain all that God has for me and my family! And I *don't* want to go thirsty in the harsh desert of this life, passing up the refreshing drink He offers. More than that, I want to be eligible to do my Daddy-God's "stuff"; to be used for his highest purposes, including being equipped and qualified to "declare His praise" to whomever He crosses onto my path.

If we are faithful to *"cleanse ourselves of what is evil"* – and our curriculums of what is value-less (of less value) – then we will be *"like vessels of gold, for noble use, consecrated and useful to the Master, prepared for any good work."* (Isaiah 43:18–21)

HOMESCHOOLING FOR EXCELLENCE

I say that these requirements are a place to *start* because I'm assuming you are homeschooling for *excellence.* I don't say that in a competitive way, but in a spirit of "doing all things well for the glory of God," in readiness for doing the will of God. It is *nourishment* to us! Jesus said, "My *food* is to do the will of Him who sent me and to accomplish His work." The first and foremost prerequisite one must possess to be eligible for such work is – Calculus? No. ... Two hundred hours in Speech? Nope. *(Look at Moses!)* How about a good Bible class on "The Words of Jesus"? No!

Very simply, all God needs is a heart completely yielded to Him. A "good education" placed in the correct perspective of being the result of yielding ourselves to God for His purposes now and for our whole lives, will only enhance and equip us *to that end.* *(See "Wisdom's 7 Pillars for True Education at the end of Section 1, if you missed that.)* Nothing "educational" should *ever* take pre-eminence over our call to become *educated* to serve and know God – or cause the fire of our love for Him to wane, or challenge or wear down our faith in Him.

WHAT ABOUT RE-ENTRY INTO SCHOOL?

You have a couple of options here. But first I want to say that there is one important thing to keep in mind about the system presented in this manual. You've probably even figured it out already: Going totally independent of "the system" as I suggest in this book is not going to be conducive to a student re-entering the regular school system somewhere down the road, prior to graduation from high school. I'm sure it's better, *record-wise,* than a totally undocumented experience, but if or since you are not filling your credits with public-school-equivalent material (which is half the point and advantage of homeschooling) you might have some difficulty transferring credits to a public school. Of course, in the school district we live in, only six homeschool credits will even transfer, even if they were taken through a full-fledged traditional curriculum program! (I call this out-and-out discrimination.) But how *any* homeschool credit would be accepted is never predictable. It will really depend completely upon how accepting of – or biased against – homeschooling your school district is – or rather I should say *the person dealing with this in your school district.*

A NOTE ABOUT TRIUNE

If you are keeping good records of your course planning and completion of work *(see the Record-Keeping Section)* you have the option of having your courses made "official" through Triune Biblical University, a small Bible college here in our area. They offer both a diploma and transcript service. If you have your child's documented work "authenticated" (record-wise) through them, they can forward an official transcript to the high school to which your child transfers.

By the way, Triune is not a fully-accredited college, and doesn't pretend to be, but their validation of your high school program is all you really need in most cases. The point is not the accreditation of the college, especially since many public high schools are not accredited either; but generally they just want to know *that* you graduated, not how accredited it is. Several homeschool moms have told me it worked just fine for them. If it is a concern to you, check it out ahead, or go with a different program than theirs. I arranged it with them only for your convenience since most people don't have access to such a program.

THIS ISSUE OF "ACCREDITATION"

Since this comes up so frequently, I felt I should address this issue here. As with this whole book, all I can give you is my own perspective. You are free to take it or leave it – just as you are free to do with the rest of this book. ☺

The possibility of our homeschool credits not being deemed valid was not at all distressing or intimidating to me. We planned from the beginning of our high school experience to homeschool all the way through to graduation. Therefore we were fully free to enjoy the flexibility and variety that my system offers. My aim was never to get what we did approved by the schools or to get a diploma from them, because I didn't approve of what *they* did (in large part), so why would I want to be approved by them? Just so I'd be approved by another institution? For me, that would have been doing it under "false pretenses." I felt I would have been deceitful to ask an institution I would never place my child in to approve what *we* had done! I had seen other homeschoolers do this and felt that it wasn't fair to the school either. I have translated what we did into a format that is concise (one class per page, see the Completed Class Plans), looks good, and is understandable by the school system. But I am not doing it to be approved by them. I never have and never *will* put myself in such a position where I *need* to be approved by them. And if they were to look closely at what we did, there would be much they couldn't approve of – which is precisely my point.

So the public school system's "accreditation" means, well, pretty close to nothing to me. I truly do *not* mean to put anyone down in saying this, but I truly believe it. So it just doesn't matter to me if our diploma is "accredited" by any institution or not. And neither Tory nor Sharnessa's diplomas *are* accredited by any institution. If we needed to, we could have Triune do it, but we haven't needed to. And believe me, the fact that I write homeschooling books is not going to count for *anything* or impress *anyone* in the public school system one bit!

Not wanting to sound like I just think I'm *sooo* spiritual, I have to say that my reason for staying non-accredited is two-fold:

1) I wanted to retain my freedom to plan high school as I saw fit *over* striving to be "college prep," and...

2) I chose to trust God that, if we followed Him all through high school, and obeyed Him on how to educate our children, and concerned ourselves more with seeking and following *His* requirements, then *He* would open the doors that *He* wanted to open for my kids when they needed to open. And, if you think about it, all that needs to open next is just *one* door – we don't need to try to make our homeschool approved by forty one colleges! ... And so, you ask, what *are* God's requirements?

> *"He has shown thee, oh man, what is good, and what does the Lord require of thee, but to fear the Lord thy God, to walk in His ways, to love Him, to serve the Lord thy God with all thy heart and soul, and to keep the commandments and statutes of the Lord which I command you this day for your good?"* (Deut. 10: 12–13)

This doesn't negate fulfilling some requirements that one might need for entering a particular college, but that's where this section and the other resources I've mentioned come in – lest you think I'm "so heavenly minded that I'm of no earthly good"! I just want to help you get a heathy view of and feeling about requirements, as well as help you find ways to fulfill them without compromising your desire to follow God's plan for high school, which can *easily* get obscured or "drowned" in our attempt to satisfy requirements!

And just to let you know what has transpired"down the road" for our two "Grad's"... Sharnessa didn't even need her diploma to get into the School of Creative Ministries in London! *(Phooey! I really wanted to show it to someone!)* And Tory, having graduated from our homeschool last summer (of 98), plans to attend the local community college this coming fall (of 99), and is doing a year of "independent in-depth and discipleship studies" this year to prepare for that *and* for life. If he decides to go to a four-year college after that, he'll have all the "credentials" he'll need from the community college.

SHOPPING IN THE RIGHT DEPARTMENT?

Please bear in mind that this manual was designed for those planning to homeschool ***all*** the way through the high school years. It was not intended for those wanting to just stick only one toe in to just "try it out." Nor is it for those planning to put their kids back in school as soon as the air clears in a difficult situation they're in at school or whenever their child catches up to grade level on their achievement tests.

I designed this book for those who are in it "for the long haul," which we were – and still *are* with Carlianne, our youngest. It did make it easier for me that, over our previous eleven years of homeschooling prior to starting high school, God had been slowly but surely prying my security away from the traditional school system. Comments like the following quote from *Dumbing Us Down* by John Taylor Gatto helped tremendously in this process: *(By the way, this is one of the most important books you can read, in case you haven't gotten that yet from it being oft-quoted in here!)*

> *"Independent study, community service, adventures and experience, large doses of privacy and solitude, a thousand different apprenticeships... – these are all powerful, cheap, and effective ways to start a real reform of schooling. But no large-scale reform is ever going to work to repair our damaged children and our damaged society until we force open the idea of 'school' to include family as the main engine of education. If we use schooling to break children away from parents – and make no mistake, that has been the central function of schools since John Cotton announced it as the purpose of the Bay Colony schools in 1650, and Horace Mann announced it as the purpose of Massa-*

chusetts schools in 1850 – we're going to continue to have the horror show we have right now. The 'Curriculum of Family' is at the heart of any good life."

If you are not yet convinced that you can do *at least* as good a job of educating your children as the public school system, then you need to read *all* of this book!

I'm not in rebellion against the system, but neither do I trust it to truly educate my children. God is dissolving in me the notion that professionals can do a better job of educating my child than can a parent who is called by God to do the job. Believe me, this is a "notion" that they would prefer no one took on; it's insulting *(they paid good money and went to school for many years to do something that a regular person like me can do without their training?!?)*, and it's expensive – as in it is already depriving school districts of much income! The tax money is "there," but it's not appropriated to *them* unless the student is going to school there. This is why many school districts are now pushing to entice homeschoolers *(back)* by means of classes and services.

This might seem innocuous on the surface, perhaps even well meaning, but we have to look deeper than the immediate benefits to us. First, parents who opt for such services are really not out of the public school system. I spoke with a school district official a few months ago who was wanting me to join a committee to help set up such a program in our school district. First he had sent me (and others) a letter, but I hadn't responded. I just didn't quite know how to tell him I didn't agree with the "heart and soul" of the program. But then he called, so I finally had to talk with him about it. He *was* well-meaning, and, even if he did have dollar signs in his eyes (which his secretary came right out and *admitted to me!),* he did have a heart for kids. He told me that, while parents like myself are able to do a good job, at least half of those wanting to homeschool really *aren't* capable. This is where he and I come to a fork in the road. *He* wants to pull (or keep) them into the system and keep them dependent on it, and keeping the system as their standard! *I* believe that once parents get the system out of their system, and get new (God's) ideas about education, they can do a wonderful job with their own kids! So *that's* the heart and soul of *our* ministry!

I no longer regard the traditional school system as any kind of "standard" that I am to live by or up to, or as being one smidgen *better*, or more *valid*, or more *effective* than what God can lead us into.

While there is *some* value in the "groceries" offered by regular school curriculum, they nonetheless lack what I see as being the "nutritional value" necessary to grow men and women the way God's Word tells us He desires for them to be grown. Not so much because the school system has failed at what they do, but because that system *itself* does not – *cannot* – accommodate what I see as being "real learning." Why, then, would I want to imitate right in our home a system that isn't accomplishing what God (or I) want it to accomplish?!

Public and private school educators do the best job they can on their own turf, but I have no reason to feel that traditional school methods, systems, and goals will produce a better product on *my* turf.

FINDING "QUALITY MERCHANDISE"

People are hungry for good "groceries"! They are intensely thirsty for knowledge – ***true*** knowledge! And I don't mean just Christians! *We all* want a fulfilling life! Every billboard, commercial, and magazine ad is an attempt to help show – usually *sell* – someone's idea of "the way" to gain true happiness! Every human being wants that – but unless we're seeking the Source of true and lasting happiness, we look in vain.

It's not hidden, but God has entrusted us, His people, with the privilege of getting to be part of helping others find the Way. Jesus Christ, the Way, Truth, and Life, is standing right outside the front door of every human heart's door, knocking! We need to be – and we need to be raising our *children* to be – mature enough in Christ to "accurately handle (divide, dispense) the Word of Truth," equipped to give His hope away to the thirsty. When we know the One who is ***the*** Way, ***the*** Truth, and ***the*** Life – not just ***a*** Way, or ***a*** truth – then and *only* then do we have access to all the blessings of ***true*** life, ***true*** happiness, and ***true*** knowledge!

The book *Wisdom's Way of Learning* will help open your eyes, mind and heart to an intensity of yearning for what God wants in your life and homeschool. If this is what you want, you will find this message most encouraging and refreshing! – and I do mean *most!* *(See "Lifestyle of Learning Applications" – Section 7.)* Of all homeschool messages, *none* will affect your entire life and *thinking* like this will! You will probably get pretty tired of hearing me say this throughout this book, but so be it. ☺ If it gets you to reading these resources whereas you otherwise wouldn't have (which *many* have told me was the case for them), then it's worth it to me. ☺

RUN WITH HIM!

God spoke to me in the first three verses of Isaiah 55 *(TLB)*. I don't know His plan for you, but I encourage you to neither let fear direct you, nor to run ahead of Him. Lay down at Jesus' feet all the anxieties you might be dealing with right now, and wait to hear His voice deep in your heart of hearts concerning all this.

"Say there, is anyone thirsty? Come and drink, even if you have no money! Come, take your choice of wine and milk – it's all free! Why spend your money on foodstuffs that don't make you strong? Why pay for groceries that don't do you any good? Listen, and I will tell you where to get good food that fattens up the soul. Come to Me with your ears wide open. Listen, for the life of your soul is at stake...

...This plan of mine is not what you would work out, neither are my thoughts the same as yours! For just as the heavens are higher than the earth, so are my ways higher than yours, and my thoughts than yours." *(Isaiah 55:1-3; 8-9)*

STUDY TIME!

I've prepared a short study-exercise based on the preceding verses for you to do. If you're interested, it's below. So take a moment to do that if you like!

This will give you a bit of sneak preview of what some of the forms are like that you'll find in the "Potpourri of Curriculum Supplements" section *(Sec. 4)*, although I've used a little more humor here than I do there. ☺

"Come and Drink" Study

on Isaiah 55:1-3 and 8-9

NAME ________________ **AGE** ________ (Just kidding!) **DATE** ________________

Find the one word that is mentioned three times: ________________

In the first mention of this word, what *verb* are we to do as well? ________________

In the second mention, what verb are we to do? ________________

In the third mention, whom are we to come **to**? ________________

How? ________________

Now find the word that's mentioned two times: ________________

After the first mention, what does God say He'll do? ________________

____

After the second mention, He tells *why:* ________________

____

This plan of His is [remarkably similar to / sort of like / not like] the plan you would have thought up yourself.

God's thoughts are [pretty much the same as / not even close to being the same as] our thoughts.

His ways and thoughts are [at right about the same level as / lower than / higher than] our ways.

What is He saying to you? ________________

____

(If you haven't heard anything, then you didn't do the study. ☺ "Just browsing" is a handy line to give salesmen in stores, but it's not going to reap you any blessings when it comes to hearing God.)

REQUIREMENTS FOR PROMOTION FROM GRADE TO GRADE

(AND ASSOCIATED ISSUES)

What originally got me going on this was that Sharnessa wanted to feel she was "validly" in her grade, not just "theoretically," nor by mere merit of her age. She wanted to be called a Sophomore, for example, *only* if she had fulfilled requirements for her Freshman year. She didn't care *what* those requirements were; she just didn't want to be "faking it." ☺ Your student may not care one iota about this, and if so, there's no need to even bring it up with them, *or* to read this section. But since I know many will deal with this, I share what we are doing about it.

Since we had never, in our entire eleven years of homeschooling up to that point, abided by a traditional curriculum, this was a little tricky! I had always enjoyed our freedom. As Sharnessa entered the senior high years that I felt had the potential to be "an experience of a lifetime," I didn't want her to suddenly now be restricted in her opportunities for a uniquely, creatively *excellent* education by being confined to five subjects a year, to the exclusion of opportunities in other subjects that would be arising. It seemed like a waste of resources to cram all 150 (*or* 120 *or* 180) hours of 1 credit into one year.

LET'S BREAK IT DOWN

Credits first: 20 credits is an average number of required credits. So, for ease in calculating, let's use that figure. Divide 20 by 4, the number of years our theoretical student wants to take to earn these credits, and we get 5 credits to be earned each year.

Hours next: It's typical for 180 50-minute hours to make up 1 high school credit. Since we, as homeschoolers don't have bells ringing 50 minutes into a class, and are not going to six classes a day, 5 days a week, 180 days a year, we can convert this time, 9,000 minutes for 1 credit, into clock hours. This figures out to 150 clock hours per credit. Just for an example, let's take a look at a typical one-year course of study that a traditional-schooled child might take to fill up those five credits:

English	1 credit (at 150 hours)
Math	1 credit (at 150 hours)
U.S. History	1 credit (at 150 hours)
Typing	1 credit (at 150 hours)
Health	1 credit (at 150 hours)

For obvious reasons, this particular line-up of classes means that during *this* particular year there will be no Science, Occupational Skills, Home Ec, P.E., State History, Life Skills, Bible or Fine Arts. Or at least credit will not be earned in these areas. I have two reasons for not being excited about this. The first centers on what the student *isn't* getting. While being immersed in these five subjects exclusively, the student is *not* getting any involvement in others. Now if the subjects he's bypassing aren't too captivating, this poses no problem or loss. But what if the subjects *are* personal interests? They might be just what God wants to use to provide spiritual, mental, physical and emotional balance *this* school year, not just one, two or three years down the road.

My other reason centers on what the student *is* getting. I feel that 150 hours is *a lot* in one year, which is unavoidable if you need to complete four credits of any one subject (like English) by the end of the four years, but rather overdosing for the other subjects.

I don't know about you, but I love pancakes; really thin; slightly burnt, as big as the skillet. I like 'em for breakfast, but I also like to eat them at different meals, for lunch, or even for dinner! Much as I like pancakes, (*hang on; this really is going somewhere!*), I can guarantee you that if I had to eat them at the *same* time, *every* single morning, *five* days a week, *all* nine months of the school year, I'd get sick of them pretty fast. Oh, but what if the cook tried new ways of serving them? Different toppings, like powdered sugar one day, boysenberry syrup the next. Or different shapes and sizes! A Mickey Mouse head, or my name, or cute little dinky pancakes! Or green or pink, just for variety. The fervor of such creativity would undoubtedly tack on several more days of interest, maybe even a few more weeks. However, guess what! Bless the cook's heart, but *they'd still be pancakes!* Oh sure, I'd still get them eaten, I'd have a full stomach, but my love of pancakes would disappear long before the year ended.

Same with school and the typical five-credits-in-one-year course of study. Kids *still* get educated, they *still* get their brains full, but how many *enjoy* the meal? How many have a desire to keep gaining knowledge once they've tossed their graduation caps? I doubt that many of those jubilantly-flung confetti specks are being flung out of joy – the joy of having acquired a useful, valuable foundation of knowledge. More likely the joy stems from finally being free to abandon the drudgerious (*if that wasn't a word, it is now*) process at long last! And most kids who go on to college aren't doing so because they're excited to *learn;* they generally just want to get a good paying job! I'm not condemning that; I just feel they're missing out on so much of the joy God intended learning to bring us!

"COME ON, BARB! GET WITH IT!..."

"This is the System! Accept it! It's just the way it is!" you say. *"The experts know what they're doing! It's their profession! They were* <u>*trained*</u> *to do this and have been doing it for years!"* Yes they have, and look at the product! Again, the *system* is the problem, not the educators! Some of my dearest friends and relatives are teachers and administrators in the public school system, and they are very caring and conscientious people with a sincere desire to teach kids, with love.

So what's wrong with the system? Basically, it just isn't working. According to Jonathan Kozol who wrote *Illiterate America*, there are sixty (yes, *60*) million functional illiterates in the United States! According to one study, it's *one third* of the U.S. population; another more conservative estimate puts it at one in five adults. At any rate, a great many Americans cannot read the poison warning on a can of pesticide, a letter from their child's teacher, the front page of a daily newspaper, or function effectively in society! And they're not all in minority groups or children. 16% of these people are *Caucasian adults!*

Educators are trained in systematically educating the masses, an indispensable job since most parents are quite satisfied to hand over this job to the "experts." It's an insurmountable job that I would certainly not want. I am allergic to teaching large numbers of kids! You couldn't pay me enough to do it.

I tried doing it one semester in our School Biz Homeschool Chapter. I taught Creative Writing to 25 kids. Well, sort of. This isn't a subject where you just spout information and they take notes. They have to *process* it. It has to *connect,* and *then* it has to be applied. I can accomplish this one-on-one because I can see exactly what and how they're doing, and go at their pace. They don't get it? Then I'll explain it again, maybe a different way. They understand? Then let's move on! But 25 kids?! How can 25 kids all get it *all at the same time?* Very simply, they *can't,* they *don't!* – and they *didn't!* I spent a huge portion of our class time trying to re-explain it to those who didn't understand it, and disciplining those who really did not want to be there, which was at least 90% of the class! (Yes, sadly, this was homeschoolers!)

On the other hand, I've *tutored* (one-on-one) primary through college-level English, and I *love* it! I've seen a 13-year-old boy catch on to three *totally new,* complicated concepts in *one hour:* "prepositional adverbial phrases," "transitive and intransitive verbs," "predicate nominatives and adjectives"! *(Full story elsewhere in this book.)* Now, I know you grammar aficionados are chuckling right now because you know that these concepts I've mentioned are *really* not that difficult. With a little explaining, a little patience, and more or less practice depending on how many Grammar-receptor cells you have in your brain, it comes. It really does. ☺ But kids – and most adults – don't know that! The grammar terms alone intimidate them. So I get a real kick out of turning lights on in the brains of people who *think* they have "bad" brains! All they need is someone to explain it to them *at their pace!* This, by the way, points out one of the chief advantages of homeschooling, and is the reason that we can spend fewer hours and obtain better results!

PREPARING FOR THE "REAL WORLD"

Maybe I'm just more sensitive to this whole issue of mass teaching because of some negative experiences I had in school. I can still vividly remember the utter panic, the desperate frustration of sitting in a Math class and not understanding the stuff! It always took *me* two explanations more than everyone else (or so it seemed to *me*) before I was able to grasp a new math concept. Whenever I put myself back in that particular classroom, my stomach experiences the nauseating sensation of *drowning!* Not only did this repeated experience deprive me of the joy of learning for many years to come, but it shredded my self image!

Someone might be saying right about now, "Well, sweetheart, it's time to wake up and smell the coffee. Welcome to the *real world!* This is how we get *toughened* up, prepared for *reality!*"

While I agree that this is *definitely* the "real world," I cannot believe that God ever intended us to partake of this aspect of it to the extent that we become entrapped and wounded by it.

"BUBBLES" AND "SLUG BRAIN"

[DIGRESSION:] I'm going to digress a bit here and open myself up. ☺ If you are excited to get on with the meat of this section, feel free to skip ahead a couple pages to *"Grades aren't everything."* What I share between here and there is "supportive," but not crucial, especially if you're in a hurry. *(This is just a personal "aside" that may minister to you,* <u>*or*</u> *to your student.)*

Until just a few years ago, I was convinced that I was *really and truly* inferior! "Defective merchandise" says it well. Possessing a body shape that drew such sing-songs from gradeschool boys as: "Bubbles. Bubbles. Bubbles. Pop! Pop! Pop!" (referring to my body shape, not my personality!), and this "review" from a fellow male college-mate: "You know, Barb, from the waist up you look great; but from the waist down..." and he trailed off.

So I never had the looks to be accepted in society on the basis of being physically attractive. It's just as well, because it compelled me to pursue *inner* beauty rather than external. But, unfortunately, the ruthless comments had done already much damage to my self-esteem before I was mature enough to *realize* it was "just as well" to pursue inner beauty, or that I really should not be putting any weight *(no pun intended)* in such comments. And as I saw my 40th Birthday roll in at 12:01am of July 26, 1992, seated at the computer, clicking out *this* page of the original edition of this book, I said to myself: "Self, you know what? Your body is not going to get much better!" But it has served me well; and is continuing to help me serve the Lord. I work at keeping it healthy and functioning well, but I will never be sought after as a model, ☺ so it's good that "body beautiful" is not my life's ambition. The inside of me is getting healed, but it's taken a *loooong* time. Because I couldn't catch on to the material as fast as the others, I was labeled, by others and in my mind, as being "not too sharp," "slow," "slug brain." *And it stuck for years.* ☹

I said that I'm getting healed, and I *have* come a long way. But, you know, I just now remembered a situation that came up *(allow me to digress from my digression for just a moment)* when my sister Kathy was teaching me the computer program that I needed to know to write this book. There was an overwhelming mound of knowledge to learn and assimilate in a short amount of time, so that alone added pressure. But what hit and disturbed me was how difficult it was for me to believe I actually had the *ability* to learn it! Those sickening old feelings of panic shot through me many times throughout those first two intensive days of learning WordPerfect 5.1 for DOS! (Really advanced in the *olden* days!) In fact, we came to this one word processing operation that looked particularly difficult to me, and I actually asked Kathy *(who has a computer for a brain)*, "Kathy, do you *honestly* think I'm actually *smart* enough to learn this?" She smiled, looked me straight in the eye and said, "Yes, Barb. I think you are." I can't tell you what that meant to me. (Even if she was lying; which she says she wasn't! ☺) Her words reminded me of the verse that says "the power of life and death lie in the tongue." Our words have the ability to minister *life! – or death!*

SNUGGLE UP; IT'S STORY TIME

(If you're trying to get past my digressions, keep going; this is still part of the digression. Keep going to "Grades Really Aren't Everything" to the right there.)

I wish I could say my difficulty in learning was in just that one math class, but it also happened in eighth grade English *(with "Sister Laura")*, in History, and in a college microbiology class. And it happened "big-time" when I went to train for a new job!

You say, *"Hey – no biggie! Not everyone is an "A student."* Well, if I didn't have this one particular experience to share with you, I'd say, "Yeah, you're right. Grades aren't everything anyway." And they aren't. But the ability to learn *is.*

So snuggle up
with a pillow** (and maybe a kid) **and
listen to this little story...

Once upon a time, there was this one math class I took in ninth grade. "Algebra I," to be exact. We had this funny older man for a teacher who was awkward and unsure of himself. It didn't increase his confidence level that the kids always made fun of him and cruelly mocked him, even to his face. But, fortunately for him, he didn't usually catch on. Once in a while, when he did figure out their mean intentions, he would get flaming angry, and even stormed out of the room a few times, the kids chalking up another conquest.

Granted, this teacher was the perfect type for such mocking! But I'm sure his awkwardness is what endeared me to the guy. I identified with him and even defended him to the other kids.

Well, to add insult to his injury, this guy wasn't a very good teacher. Quite a poor one, actually. He could never explain things right! It was like he was doing a problem in his head, mumbling parts of what he was thinking as he worked the problem on the chalkboard. And this was his oral presentation of the material! Even the smart kids didn't get it. Being vocal, and having a soft spot in my heart for this guy, I somehow became the "official class spokesman." I wasn't afraid to raise my hand – in this class only – and say "I don't get it! Would you explain it again?" And if I still didn't get it after that, I'd ask again.

Because the other kids thought it was "uncool" to ask so many questions, they put me up to asking their questions for them! Time after time, kids would relay messages in whispers across the room: "Tell her to have him explain it again." But I was usually ahead of them in thinking up the questions because I wasn't getting it myself! And this one guy, Kim, who sat next to me (a neighborhood friend I walked to school with) would sometimes ask questions too. But being the quiet type, he preferred to have me do it for him, too.

Well, this teacher wasn't real wild about these "interruptions." In fact he'd get pretty exasperated with us. Not with me quite as much, because he could tell I liked him and was nicer to him, and even stuck up for him in front of the kids. I'd tell ones who were mocking him to knock it off. *(Now you can see at least one reason I had so few friends: I didn't go along with the crowd, even if it was to my own advantage to do so!)* But he still felt he had explained it well enough for us to have gotten it the first time.

But I didn't, so I kept my pesky interruptions coming, and he did keep answering them until I got it. As I continued doing this, and then as I started catching on to one concept, and another, and another, I started discovering that this Algebra stuff was actually fun! Like solving puzzles! And the accompanying feeling of success – very unfamiliar to me – was exhilarating! Well, I left that class with a love of Algebra and without my previous fear of this subject. Guess which two students were the only two students in that class to get A's! **[END OF DIGRESSION]**

GRADES REALLY *AREN'T* EVERYTHING!

In fact, they're very low on my educational totem pole. But they can indicate how much of the material the student is assimilating. If it's worth learning at all, *(and of course, this is frequently the question)* then it's worth learning, period. The usual side effect of low grades, on top of not knowing the material, is that they put the student into one of two modes:

> **perpetual frustration:** "I keep *trying* and I just *can't* get it! I'm just not smart enough!" – which results in a damaged self esteem, *or...*

> **apathy:** "I don't care about this junk anyway" resulting in *no* growth; a non-education even after being in the system for 13 years! Often the student just "drops out" – if not physically, emotionally and mentally, essentially the same.

But look at even those students who *do* earn good grades, how many of them graduate with a love of learning *and* a desire to *continue* learning after high school? Some do, I'm sure, but the majority do *not.*

BUT WAIT! THERE'S MORE!

Even aside from the "overdose" aspect of cramming so much material into one year, there's another matter to consider. Let's say a great ten-hour "Creation Science" seminar by Dr. Don Chittick comes to your church, or a one-time, prime-opportunity-for-fine-arts concert to your community. What will you do with these? Pile them onto your student's already-full load of credits? Pass them up? That would be a shame! But there are only so many hours in a day!

So another disadvantage to taking the narrow line-up of 5 or 6 credits in one year is this: *The hours spent at pursuits other than the 5 credits being worked on will either be lost or will have to be added to the many hours your student is already putting in – at least if you want them to "count" toward his graduation goals.*

SO WHAT'S A PARENT TO DO?!

There *is* a more creative way to arrange and distribute credits that won't lay added time or study pressure on your busy student! I know I want(ed) my kids to thoroughly enjoy these high school years as much as possible! Yes, I want them to learn. Yes, I want them to be disciplined. But I also want them to have fun! To *me*, learning is out-and-out, pure, unadulterated *fun.*

Referring back to that five-credit "typical course of study," we see that it figures out to, *(let's see, that's 5 credits × 150 hours = 750 hours)* a total of 750 hours a year.

Okay, let's do something different with these 750 hours. Let's divide them up a different way and then rearrange them. ... If a student were to accumulate 750 hours of *any* combination of classes, not just these 5 classes, he would have completed as much *work,* and accumulated the same amount of *time* at the end of one year as a student who completed 5 credits in 5 subjects, right? Same number of hours; same amount of work.

MAKE SURE YOU GET IT!

I don't mean to belittle you, but if I have lost you please go back and read that again. Don't go on without understanding it because this is *crucial* to the rest of what I'll be explaining. If it sounds confusing, ask God to shed His light on it. I have to tell you, I can still remember sitting on the couch, buried in *lists* of requirements and scratch paper, trying to figure all this out! If I hadn't been doing it for my beloved daughter, I probably would have given up! If *I* can get it, believe me, *anyone* can get it! It's certainly an opportunity to sharpen those math skills! (On the other hand, you math whizzes are probably chuckling at my strained efforts, wondering why I'm belaboring the point. I'll tell you what: I'll cut you some slack in English; you cut me come slack in Math! ☺ Deal? ... Just my luck; you're probably good at both Math *and* English!)

LOOKING AT A THEORETICAL EXAMPLE

Now let's see how this works out on a practical level with a theoretical line-up of subjects. There are *many* other combinations of subjects, numbers of credits, etc., besides what I'm about to present her, but we want to just see how all this works out, so we had to start somewhere.

ENGLISH: 4 credits - Since no one wants to take 2 credits of this subject in one year, you need to just plan on taking one credit of English each of the four years if you're requiring 4 credits of English. (I ended up requiring only 3. It was plenty for our purposes.) So looking over at the opposite list of Requirements, you see one credit of English for each of the 4 years. Nice 'n' simple. *(Appreciate it; it's the simplest subject to explain.)*

MATH: 2 credits - First, this is the *one* subject that I didn't have our kids count up their hours for because high school math programs are generally designed to be completed by the *average* student in one year, according to one administrator I questioned. Ila Edwards, of Hewitt Prep School, said that if the student is understanding the Saxon Math material fairly easily, without too much stress or struggle, then he's at the appropriate level and should be able to complete the work in close to 150 hours. It was designed that way, she told me.

So we were done with math when the Saxon Math book was completed. However we did include the number of hours a Math class *should* take when figuring hours for the whole plan, because that time needs to be accounted for in the whole picture. Otherwise Math hours would have to be piled on top of the rest of the

REQUIREMENTS
FOR
PROMOTION FROM EACH GRADE

BASED ON: English: 4 credits / Math: 2 / Social Studies: 2½ / Science:- 2 / Bible: 2 Occ. Ed.: 1 / P.E.: 1 / Fine Arts: 1 / Health: 1 / Home Ec: 1 / Typing: 1 / Life Skills: 1 / Computer: ½

FRESHMAN:

- ☐ **English** - 1 Credit (150 *(or 120)* hours)
- ☐ **Math** - ½ Credit (75 *(or 60)* hours)
- ☐ **Social Studies** - 94 *(or 75)* hours
- ☐ **Science** (Gen. + Lab combined) - 75 *(or 60)* hours
- ☐ **Bible/Spir.** - ½ Credit (75 *(or 60)* hours)
- ☐ **Any combination of other subjects** - 281 *(or 225)* hours

SOPHOMORE:

- ☐ **English** - 1 Credit (150 *(or 120)* hours)
- ☐ **Math** - ½ Credit (75 *(or 60)* hours)
- ☐ **Social Studies** - 94 *(or 75)* hours
- ☐ **Science** (Gen. + Lab combined) - 75 *(or 60)* hours
- ☐ **Bible/Spir.** - ½ Credit (75 *(or 60)* hours)
- ☐ **Any combination of other subjects** - 281 *(or 225)* hours

JUNIOR:

- ☐ **English** - 1 Credit (150 *(or 120)* hours)
- ☐ **Math** - ½ Credit (75 *(or 60)* hours)
- ☐ **Social Studies** - 94 *(or 75)* hours
- ☐ **Science** (Gen. + Lab combined) - 75 *(or 60)* hours
- ☐ **Bible/Spir.** - ½ Credit (75 *(or 60)* hours)
- ☐ **Any combination of other subjects** - 281 *(or 225)* hours

SENIOR:

- ☐ **English** - 1 Credit (150 *(or 120)* hours)
- ☐ **Math** - ½ Credit (75 *(or 60)* hours)
- ☐ **Social Studies** - 94 *(or 75)* hours
- ☐ **Science** (Gen. + Lab combined) - 75 *(or 60)* hours
- ☐ **Bible/Spir.** - ½ Credit (75 *(or 60)* hours)
- ☐ **Any combination of other subjects** - 281 *(or 225)* hours

hours. We wouldn't want that! It needs to be counted in with the rest of the hours. Whether it actually took longer or shorter is immaterial at this point. When it's done, we're done. *(By the way, I call this a "content-based credit." Classes based on number of hours spent I call "time-based credit." With both aspects, "time/content-based.")*

Whether you prefer your credits to be "time-based," "content-based," or *both,* you still need to somehow figure the hours into the total for graduation. So even though we went with the "content-based" system for this class, we still *counted* it as 120 hours, even if it ends up taking more or less time than that. *(We opted to have one credit be 120 hours rather than 150 hours; more on that in just a moment.)* So how do you figure this in to the whole picture of math requirements for graduation? You can do one of two things:

Option 1: Take one whole credit per year, which leaves one or two years "Math-less," and which means you have to re-figure your "Remaining Subjects" total. This is really no big deal; just a little more math computation. *(You know, a person could probably earn ½ credit of Math just figuring this system out!)* If you *do* choose this route, know that college-bound students are advised to take Math in their *last* year of high school so that the brain is "primed" for college math.

Option 2: You might prefer to spread these 2 credits of Math out over the 4 years, taking more time to complete the work. *(A good way to relieve pressure for those for whom Math does not come as easily.)* If spreading it out like this, simply divide the total number of hours for the 2 Math credits. For our example here, let's say 2 *(2 × 150 = 300 hours)* which you then divide by 4 *(for each of the 4 years)*, and you get 75 hours per year.

Obviously "Option 2" is the *simplest* way to compute your Math hours, and it's what I have illustrated on this theoretical chart. However, once again, simple is not always going to be better *for you.* It may work better for your student to complete a whole year of Math in one year, and do that for two of the years, leaving two years "Math-free." So 2 credits of Math could be taken, for instance, in grades 10 and 12.

Our preference was to figure it out according to the "spread-two-credits-over-four-years" method, so this is how the Math credits appear in this list of requirements.

SOCIAL STUDIES: 2½ credits - This divides out to 94 hours per year. (More details later on the breakdown of that into the various *classes;* we're just dealing with overall subjects right now.) One advantage to being free to cover this subject all the way through the four years is that it frees you to do Current Events (like with "World Magazine") on a continuing basis and to also take advantage of special events at various museums.

SCIENCE: 2 CREDITS - I've listed this in the yearly sections not as *½ credits* but as the equivalent in *hours* because you'll probably be combining hours for two different *areas* of science, lab and general science. So 2 credits equals a total of 300 clock hours. Divide that by 4 and it comes out to 75 hours a year.

BIBLE/SPIRITUAL STUDIES: 2 CREDITS - This is nice 'n' simple: ½ credit, or 75 hours, a year.

ALL OTHER SUBJECTS: Now we get to play with all the "leftovers"! After the above hours are added up, we deduct that total *(sub*total, really) from the original 750 hours needed for one year. Now this is for if you're doing 5 credits *worth* each year. My "Requirements For Promotion" list is figured out for 20 credits.)

FILLING ALL THOSE HOURS

For ideas on how to fill these precious hours, see the "Potpourri" in section 4, the "Grading Guidelines" in section 5, and the "Completed Class Plans" in section 6-D. This is where it gets fun, so hang in there!

REALITY TIMING FOR CREDITS

Rather than counting 180 ***50-minute*** hours as one credit, we count 180 ***40-minute*** hours. In other words, we count one credit as being 120 *clock* hours instead of 150 clock hours. *And I don't feel in the least bit that we are cheating! Why?...*

The original revelation that spurred me into considering this to begin with was that, as students in the one-on-one tutorial setting, homeschooled kids are getting much more out of each minute than are kids in the classroom setting. Once a concept is mastered we move on. If a concept isn't yet understood, we stick with it until it is. So our time is spent much more efficiently.

But upon visiting a *real* live classroom, we got liberated even further. On several occasions (about once each year), Sharnessa was invited by friends to spend a day with them at a public high school as a "visitor." What she saw and experienced there freed us up enough to count 40 minutes as one "class hour" rather than 50 minutes *without feeling we were cheating!*

Think about it. There is *no way* that every one of those classroom 50 minutes is being spent on "education." I've even had several *teachers* tell me how much time is wasted in the classroom! He or she has to deal with discipline problems, getting the unruly students settled down, handing papers in and passing them out. There are directions to be explained, and explanations to be repeated to make sure everyone gets it. Homework is given out, which takes time away from "education."

So I could even whittle it down to a 50/30 ratio: 30 minutes of homeschool = 50 minutes of regular school, and still retain my integrity. Let me reword that: I would not feel in the least that we were cheating by allowing our kids to count 30 minutes of homeschool time as being *"equivalent"* to 50 minutes of regular school time. But having discussed it with our kids, we all felt comfortable with a 50/40 ratio, or the 120 hours per credit route. Of course, that meant they *really* needed to be making full use of minutes that are counted. They time most of their activities, which got to be second nature, believe it or not. So if they get a phone call, have to go to the bathroom, or if Carlianne interrupts them, the "time stands still."

Now this totally changes all the figures I've just given you in this chapter. I didn't do it this way to begin with because I needed to present just one concept at a time. "Precept upon precept." It would have been overwhelming and confusing to have presented this variance right at the start. If it's any encouragement, I know of one support group that uses this same measure for their graduation requirements. But some of you may feel more comfortable sticking with the 50/50 ratio, or 150 clock hours for one credit. That's fine. Again, you get to make the choice. What serves *your* purposes the best? What does your student need?

REFIGURING THE TOTAL HOURS/CREDIT

So let's take a quick romp through the numbers and re-do them with the 50/40 ratio (120 hours = 1 credit) instead. It's really not all that confusing; just takes a little figuring and a little Wite-out ™. *(Though I've already added in the figures on the previous chart in small italics.)*

Okay, one credit now equals 180 **40**-minute hours. That's 7,200 minutes. To figure the clock hours, divide 7,200 by 60 *(60 minutes in one hour)* and you get 120. So 120 actual *clock* hours of work equals one credit, in this system. And ½ credit is 60 hours.

When you go to do your tallying for the year, you'll need to figure how many credits you are requiring for graduation (just like you would have done anyway in the 50/50 system.) and multiply the number of credits per year by **120** instead of 150.

So if you're requiring a total of 20 credits to graduate from high school, that's 5 credits per year which equals 600 hours for the year. ... One more example: A total of 24 credits comes out to 6 credits a year, or 720 hours per year.

What if you have an "odd" number of credits that won't come out to a whole number per year? Like 22. You can either divide the total number by four and come out with a number like 5½ credits per year, or 660 hours, or you can have the student earn 5 credits in two of the years and 6 credits the other two years.

You can work it to your advantage either way. Since the student is more likely to be driving and working at a job after turning 16, it might be wise to have the two 6-credit years be the Freshman and Sophomore years leaving the last 2 years freer for non-academic pursuits. It *really* varies according to your own style, motivations and goals.

FATIGUED BRAIN CELLS?

It took me many weeks to write what you just read. I figure I both sharpened and lost many brain cells in the process. ☺ You're probably a bit tired, too. You may need to go through the math stuff again. I wish there was another way to explain it more simply, but that's as good as it's gonna get! I got a kick out of a comment from Kathy Hutton, one of my original "Homeschool Parent Testers" for this book: *"I did all the math stuff, so it's OK! And I can't do Math. I think I have dyscalculia – and that's a **real** learning disability."*

REMEMBER THE RIGHT STUFF

Once you understand the whole picture, you can be as flexible and creative as your situation warrants. Remember the elephant that was tethered to a pole for many years and, even when finally cut loose, still would not go any farther than the rope had allowed? So we tend to be with our new-found freedom in homeschooling! Don't waste this precious independence by staying leashed to old, unproductive, ineffective frameworks that need to be changed.

Remember the Scripture verse I referred to earlier in Isaiah 43:18–21? It's the one that says to *"Remember not the things of old, nor consider the things of the past. For behold I am doing a new thing... I will make a way in the wilderness and rivers in the desert... for I give water in the wilderness, rivers in the desert, to give drink to my chosen people, the people whom I formed for myself that they might declare my praise..."*

This is my heart's desire and the place I *rest* regarding dealing with all this requirement stuff. I know He has an answer for not only me but for *all* who will call upon Him and ask for His wisdom regarding their individual children. In fact, verse 22 which immediately follows, goes on to say: ***"...Yet you did not call upon me, O Jacob."***

God had a wealth of blessings prepared for him, and yet he missed them *(at least at **this** point)* because he—well, let me just give you a multiple choice quick-quiz: Jacob missed God's blessings because he:

- ☐ A. wasn't the firstborn
- ☐ B. didn't have a high school diploma
- ☐ C. didn't really want them; *(already had lots of blessings)*
- ☐ D. wasn't good enough or handsome enough
- ☐ E. just plain didn't *ask*

ANSWER KEY: I have a feeling you don't need an answer key to know the correct response! If you know the Lord, you *have* the Key – to not only the answer above, but also to the best Answer for *you and your family!*

In this section I'm excited to present a variety of ideas and experiences for a most memorable homeschool graduation! As a homeschooling family, you need miss out on *none* of the fun, specialness, and memories of "Graduation"! In fact, I think a homeschool graduation can be *better!* And *many* homeschoolers who have done it share my sentiment!

GENERIC DIPLOMA

This book simply wouldn't be complete without at least *one* option for a diploma, humble as it may be. (You can certainly find spiffier diplomas, complete with gold lettering, elsewhere – like at Josten's, for one place.) Copying it onto parchment is a nice option. I did not provide the border lines because they are too hard to avoid when cutting, and you *don't* want any tell-tale lines around the edges. For your school name, press-type lettering makes as polished an appearance as possible without use of a computer. It's available at stationery/office stores in a variety of fonts (lettering styles) Adding your homeschool logos or a notarized seal of some sort are also nice touches. *(The page number is at top right to avoid diploma.)*

Diploma of Graduation

From:

This certifies that

Has satisfactorily completed our Course Requirements prescribed for Graduation
and is therefore awarded this

Diploma

Given this ____________ day of ____________________________, in the year ___________.

____

Principal

GRADUATION IDEAS

❧ Invitations designed by the students or parents

❧ Parents and even siblings and grandparents come up and present the diploma to the graduate

❧ Have songs, speeches, performance, presentations be done by the graduates themselves in their area(s) of expertise.

❧ All graduates could wear the same color caps and gowns as each other – in their "class colors"; or each student could wear their own chosen color(s) or those (that) of their own home's homeschool.

❧ Mom and Dad share personal reflections on the homeschooling years with their Grad(s).

❧ Make a program; include: Grad, time, date, place *(front)*; Order of Events *(inside)*; Special Thanks *(back)*

❧ Have an "open mic" time at the graduation for guests to share about the grad(s); humorous memories, stories, qualities, etc.

❧ Do a "Grandparents' Legacy" *(See "Tory's Graduation" for details on what we did.)*

❧ Ask friends to share tributes, "personal impartations," or a few words of encouragement.

❧ Ask the pastor of one of the students to give a sermon on a topic of relevance to the homeschool nature of the graduation. But ask any of your speakers to be sensitive *(as in non-offensive)* to those with other educational preferences or convictions.

❧ Have a "Prayer over the Grad" time. If people will do it over the microphone and thereby get it on tape *(if you are video or audio taping the event)*, have them do that as it is very special to have it forever.

❧ Have a "candle ceremony" and/or Communion... Honor "special input" people in the student's life.

❧ Consider putting on a "Baccalaureate Breakfast" if you have more than just one or two grads.

The Reception:

❧ Each student could have his own display table featuring the student's "star work" and projects, awards, memorabilia, notebooks, and photo albums. Tory had his favorite lunch box, his basketball hoop that went over a wastebasket, and his baseball lamp. ☺ He also had awards from basketball and soccer that he had played at the Christian school that allowed homeschoolers. *(a door that opened the year we hoped it would, and closed right after Tory graduated! Prayer!)*

❧ Make a photo collage and/or slide show with pictures from childhood through present (humorous shots are fun to include!)

❧ Have a sheet cake done for each student in colors and decor of own choosing – at his table.

❧ If the student is renowned for baking a particular item, that would be fun to have on their table, too.

❧ Have napkins, plates, balloons, and streamers in the grad's "homeschool colors," and decorations *(which don't all have to match the homeschool colors.)*

❦ GRADUATION SUPPLIES ❦

Josten's is "the place" for graduation supplies! They carry everything you could possibly think of for graduation! No one would need *all* of them (unless they're obsessed with graduation), but it's handy to know they're available. Their Central Office number is: 1-888-479-3336 from here they'll direct you to an area representative. They carry:

- Caps and Gowns
- Binders and Portfolios
- Announcements
- Yearbooks
- Attractive Diploma Covers *(in 2 sizes)*
- Tassels *(in your own customized colors!)*
- Awards: acrylic and wood
- Banners
- Class Rings
- Diplomas *(nice)*
- Writing Instruments; Pens
- Imprinted Apparel *(Jackets, T-shirts)*
- Lapel Pins and Charms

"Pomp and Circumstance" is available on a "Sound Choice" accompaniment track audiotape entitled "Graduation Songs" which also has on it "Friends," "Do You Know Where You're Going?" and "Forever Young" *(not recorded by the artists.)* The ISBN number is 29913-46134.

GRADUATION INVITATIONS

I thought you'd enjoy having a few samples to get some ideas from... First, here's the one that Todd Misclevitz' family designed themselves. It's in a tri-fold format, and the house – on the cover – is their actual home, drawn by a friend. On the inside top third of the invitation (*to the right* ☞) is a scroll that contains the meanings of the names of each of the graduates, combined to form an amplified version of Proverbs 22:1! *(How clever!)*

The "You have been a significant influence..." note (directly below) was on a separate sheet of white paper in place of the traditional tissue. The "Class Scripture" on the back of the lower tri-fold (which I didn't have room to include) reads: ***"A good name is rather to be chosen than great riches, and loving favor rather than silver and gold."*** Todd's pastor, Roger Lucas, spoke on "A Good Name" as part of his sermon at the Graduation Ceremony.

You have been a significant influence
in the life of
TODD MISCLEVITZ
You are cordially invited to the
reception immediately following
graduation to acknowledge the work
that God has accomplished through
instruments such as you.

Insert ⇑

Cover ⇒

Inside of Todd's Graduation Invitation ⇨

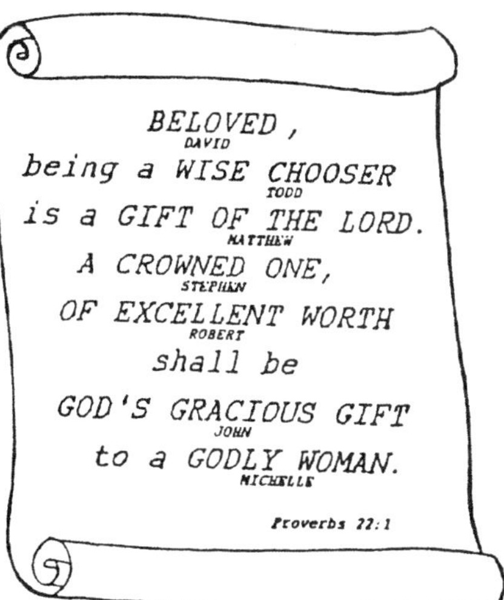

In conjunction with

the Lower Columbia Christian Education Association

I am pleased to announce my graduation

from my home education program.

Because I have reached this goal in my life,

I wish to share my joy and achievement with you.

Thursday, the evening of the first day of June,

the Nineteen hundred eighty-ninth year

of my Lord and Saviour, Jesus Christ

at five minutes past the hour of seven o'clock

Kelso First Assembly of God

GRADUATION INVITATION

Here is Sharnessa's Graduation invitation, which we did up on the computer, cut-and-pasted (literally) a corner border from a border booklet I had. What you see below is the opened-up invitation. We printed it on a sky blue parchment cardstock, getting two out of each sheet, and got matching envelopes. I have reduced the cover here so that it would fit on this page, (directly below and on the right) so in case you can't read the verse, it says "For we are His workmanship, created in Christ Jesus for good works which God prepared before hand, that we should walk in them." On the back of the invitation it said:

"Dance in the rain."

Sharnessa made this up, and it's very fitting for her, partly because of her love of dance, and partly because this is so true of her in the spirit. She's always full of joy and has a wonderful way of seeing the bright side of life, even in the midst of difficult or painful times. ☺

As you can see, *(I hope it's not too small)* her actual graduation was with W.H.O., (Washington (State) Homeschool Organization) up in Tacoma, 1½ hours north of us. This meant only a few people could go, which was five of her friends and all four grandparents. The whole event was very well organized and put together; a good choice in many respects. Sharnessa didn't know a soul, but, being Miss Social Butterfly, this didn't daunt her in the least. She just drove up by herself and made a bunch of new friends! *(As if she needed more!)* They put on a "Grads' Night Out" the night before, with a nice dinner and program at the Seattle Space Needle and a horse-drawn carriage ride in Downtown Seattle. At the Graduation, they wore caps and gowns, the guys in red and the girls in white, and the ceremony itself was nice. Our whole family, including grand

It is with
a sense of satisfaction for her accomplishments
thus far, and great anticipation for
the path she has yet to walk
that

Sharnessa Lee Shelton

announces her
Graduation from Tree of Life Learning Center,
the Shelton Family's Home School.

The two following events
will mark and celebrate her achievement:

❦ ***A Homeschool Graduation Ceremony:***
sponsored by Washington Homeschool Organization
to be held at University of Puget Sound
in Tacoma, Washington at 7:30 p.m.
on Saturday, June 8th, 1996

❦ ***A Private Reception:***
to be held at the Shelton Home,
182 N. Columbia Heights Road, Longview, Washington,
on Sunday June 9th, 1996
from 1:00 p.m. to 4:00 p.m.

You are warmly invited
to attend either or both events.

A cherished saying of Sharnessa's:

"When you were born,
you cried, and the world rejoiced.
Live your life in such a way that when you die,
the world cries, and you rejoice."
– Old Indian Saying

parents, went up onto the stage to present her diploma to her. And afterwards there was a nice reception. But I have to say that I *personally*, just from a homeschool *Mom's* perspective, felt a keen sense of disappointment in it not being smaller and more personal. There was almost nothing of our own choosing, nothing that represented *Sharnessa*. Because there were 92 graduates, each one got a whole twenty seconds allotted to them to share! *Thirteen* years of homeschooling compressed into just *twenty* seconds!?! With so many Grad's, it couldn't have been any other way. It *was* fun for Sharnessa, as it should have been, but it's not what I would generally recommend if you want something more personal and meaningful. I think that something more intimate would have been even *more* fun for Sharnessa too.

On this page are the front of her Graduation Invitation, a picture of us all at her Graduation on the stage, and her senior picture.

TORY'S GRADUATION

You can be sure that as the time of Tory's Graduation time approached, I did a lot of praying and talking with him! *(The praying was to God not Tory!)* The only homeschooled friend he had who was graduating that year did not want to wait to graduate until August, when Sharnessa would be home *(from being a nanny in NY)*; he wanted to do it in June and get on with *life!* ☺ But Tory didn't want to be the center of attention as the "Lone Graduate." Even sharing the stage with just *one* other person would have diverted the focus from being on just *him*. But that wasn't an option. So I prayed, asking God what to do with and in this situation. He eventually changed Tory's heart to be content with being the only Grad, and thus we ended up with a late August Graduation, complete with Sharnessa! Tory was *very* pleased with how it turned out! All the honor(s) given him throughout it by those who shared made it worth all the nervousness. ☺) ... On the following page is a write-up on Tory's Graduation, then a page of photos; then his Graduation program; and then his invitation *(which was folded into fourths; I have them all right side up here for viewing ease)*.

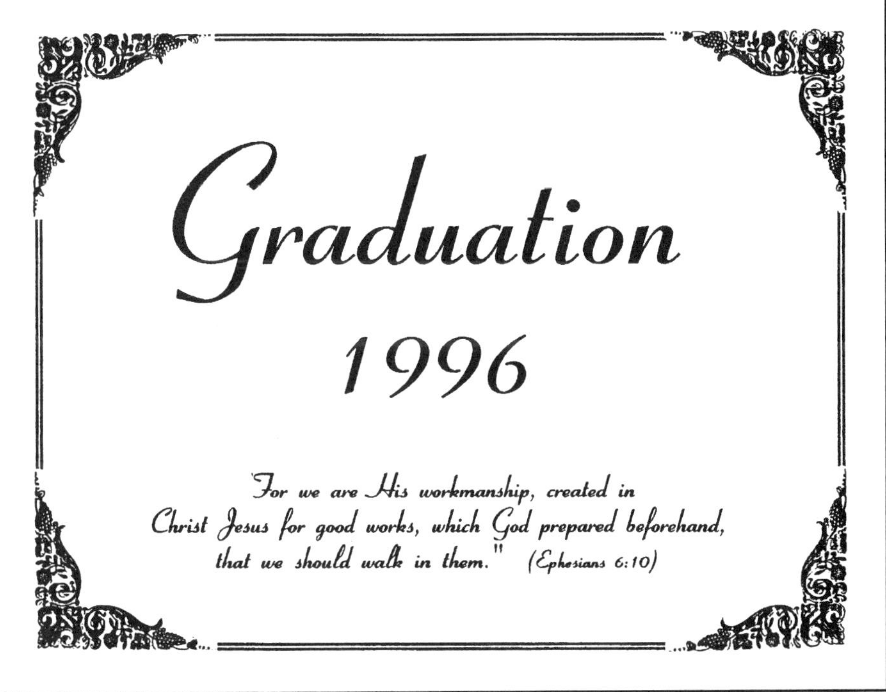

Graduation

1996

"For we are His workmanship, created in Christ Jesus for good works, which God prepared beforehand, that we should walk in them." (Ephesians 6:10)

Photo by Mr. C's in Longview, WA

Sharnessa Lee Shelton
1996 (Our lovely Sweetheart!)

Tory's Graduation

Our son, Tory William Shelton, graduated on August 31, 1998. Yes, a rather odd time of year to be graduating, but we wanted our older daughter, Sharnessa, who was just arriving home from having worked as a nanny in New York the past year, to be able to participate. Starting time? 6:05 p.m. – Seattle Mariners time – a little early to accommodate the need of a few "key guests" to get to other meetings. *(Our choice of days was limited.)*

The number of Graduates was low – only one student, Tory, but it lacked nothing in tradition, "specialness," and elegance! Of course, it *would* be on the hottest day of the year in a non-air-conditioned room! I opened the event with: "You probably think that those gold "98's" positioned around the room represent the year Tory is graduating, right? Well, they are actually *thermostats*, and I see they are working quite well!"

Purple, grey, and black balloons and streamers bedecked the room and helped welcome the Graduate as he walked up the aisle to "Pomp and Circumstance" wearing a black cap and gown, with a tassel of purple, grey and black – the "homeschool colors" we chose fifteen years ago! Tory sat in a chair on one side of the stage facing the audience and whomever was at the microphone. Dave ("Dad") opened the ceremony with comments about our years of homeschooling with the Grad, followed by Mom's comments of a similar nature.

We then had a candle ceremony called "Grandparents' Legacy" where Dave and I shared what each of our parents had given us that we want to pass along to our son. All four parents came up on the stage, and we gave them each a candle, in a color that represented unique "legacies" that each one had given to Dave and me that we wanted passed on to Tory. As I gave my Dad a *green* candle, I told him it reminded me not only of his love of the woods, but also of his lifelong love of learning and his continual growth. I gave my Mom a *pink* candle not only because it's her *(and my!)* favorite color, but also because pink represents the warm unconditional love that she has so given her family all these years, as well as her organizational skills. Dave gave his Dad a *blue* candle representing his loyalty and faithfulness to his family – something Dave could always count on; and a *yellow* candle to his mom representing the sunshine of her love for her family, and the beautiful yellow roses she grows in her garden. Then Dave and I, together, lit Tory's *orange* candle – for his favorite orange shorts *and* his "golden" heart – as we shared with him the "legacy" we've received from *our* parents. *(sniff)*

Sharnessa's part was to sing two songs that Tory had chosen: "Refiner's Fire" and "Make Me Like You Lord" which she sang with Chuck Tilton, Tory's youth minister *(and Sharnessa's choir and ensemble teacher – the one class she took at a Christian school)*, with whom Tory is now working in the youth ministry.

I had asked a few people to share just a bit about Tory – anything they wanted to commend him on, encourage him for his future, "personally impart" to him, what they saw in him that they wanted to bless... These included his former basketball coach, Grandma Shelton, a woman who has been a "prayer warrior" for Tory for many years, and two "men friends" – one of whom introduced himself as a "Tory Wanna-be" saying "I want to be just like Tory when I grow up." (He's 40. ☺) We then opened it up for people to come and share, *(which we had "fore-warned" about in the invitation)* and several came forward with words of advice and encouragement, humorous anecdotes and memories, and words of praise – some of which were sweet blessings to Mom and Dad as well. After this, Tory responded from his heart, and then presented Mom with roses and Dad with a gift. A friend commented to me: "It was so wonderful and encouraging to see Tory thank his parents for homeschooling him and for raising him the way you did. The tears he displayed throughout the graduation also added to the specialness of it all."

And then... the Awarding of the Diploma, the traditional handshake, and hugs. And after that Tory's three pastors made a fun but meaningful presentation of several "Tools for the Graduate to carry Into Life" – the "prop basis" for which was a set of toy tools, like a hammer – for securing his foundation; like a level – to help set and keep his heart aligned straight with the "level" of Jesus; all presented with love, humor, and sound advice. We then had Tory come down in front and invited any guests who wanted to, to come and gather around Tory and pray (into the microphone so all could hear) for him. And it's all on video!

A display table in the back of the room held several notebooks, awards and mementos of Tory's years of homeschooling "from scratch." The notebooks included his High School Notebook *(containing the class plans for all of his classes)*, his "Basketball Notebook," his Washington State History & Government Notebook, and his "Lab Science Survey Notebook..." His senior portrait portfolio sat on the table along with his baseball lamp, ☺ his favorite lunch box, his baby book, and all his trophies and awards for basketball and soccer. It was a sentimental representation of his life and homeschooling years – which happen to be one and the same!

A "Congratulations Graduate 98" cake *(in our school colors, of course)* was bordered by napkins and plates *(in our school colors, of course)* on the cake table; and an array of hors d'oeuvres, salads and fresh fruit were beautifully arranged on a long table by a dear friend of the family, having also been *made* by her. A savory ending to an evening of sweet memories that we will savor forever.

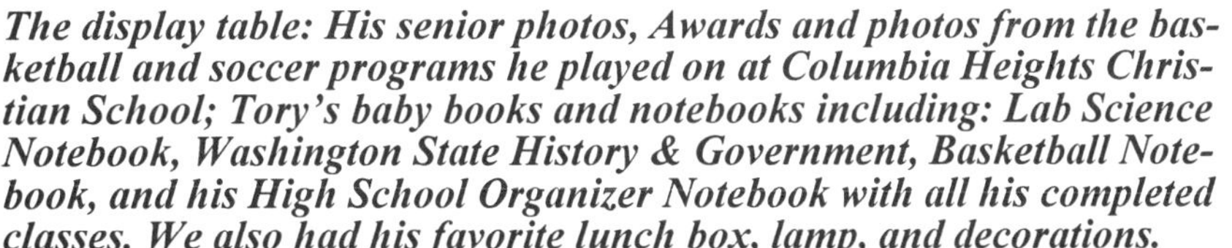

The display table: His senior photos, Awards and photos from the basketball and soccer programs he played on at Columbia Heights Christian School; Tory's baby books and notebooks including: Lab Science Notebook, Washington State History & Government, Basketball Notebook, and his High School Organizer Notebook with all his completed classes. We also had his favorite lunch box, lamp, and decorations.

Photo by Mr. C's in Longview, WA   Used with permission

Tory William Shelton
1998

(What a handsome dude!)

The whole Shelton Family with the Grad ⇨ holding his diploma (with "Tree of Life High School" imprinted in gold on the cover – by his left thumb). He had already tossed his cap, but his hair looks like it's still there! ☺

⇦ Here I** (Mom / Barb) **am sharing at the microphone with Dave** (Dad) **right by my side. Tory's on the left sitting in a chair to on the stage.

(I look a little mis-shapen because the sun is casting an odd glare across my frontside! ... 98's are on quite a few of the balloons but they all refused to face the camera for the picture ... The candles are down in front of the stage on a table, ready for the Grandparents' Legacy.)

Order of Events:

Processional: "Pomp and Circumstance"

Welcome & Prayer: "Dad" / Dave Shelton

Reflections of a Homeschool Dad and Mom

Song: "Make Me Like You, Lord"
by Chuck Tilton (who wrote the song) & Sharnessa

Grandparents' Legacy: by Tory's Parents, Presented to:
Frank and Carolyn Shelton –and- Lee and Mary Edtl

Tributes to the Graduate: Carolyn (Grandma) Shelton
Clint Van Fleet - Former Basketball Coach
Dave Coons - Friend of Tory's (and of the Sheltons)
Ruth Nelson – Friend & Prayer Warrior

Personal Impartation from Pete Ouellette

Open Mic – for Our Guests to Share!

Words from the Graduate

Awarding of the Diploma

Tools for the Graduate to Carry Into Life:
Tory's Pastors: Jim Bass, Larry Russell, and Paul Treichel

Prayer Over Tory – Any Guests Who Would like to Come Up and Do S

Song of Blessing: "Refiner's Fire" - Sharnessa & Chuck

Closing Prayer: Chuck (Tory's Youth Pastor)

Inside - right ⇧

The Program

On back of program ⇨

August 31, 1998
6:05p.m.

at Father's House
Longview, Washington

Special Thanks to:

Jean Merritt – *Refreshments (tonight's lovely spread)*
And Encouragement All Along the Winding Road ☺

Keith Merritt – *Videotaping the Event*

Father's House & Pastors – *Use of Facility*

Chuck Tilton, Karen Smith & Sharnessa – *Music Par Excellance*

Stuart Schmidt – *Sound Man*

The Ouellette & Johnson Families – *Set-up, Serving, Friendship*

All Our Guests and Each "Sharer" This Evening!

Marilyn Howshall – *Wisdom, Vision, and Insight into*
Learning and "True Education"

Dr. Raymond & Dorothy Moore – *Original Impetus*
and Practical Advice to Homeschool 16 years ago

Diana McAlister – *Vision and Guts to even consider the crazy idea*
of attempting to Homeschool through High School

God – *The Wind Beneath Our Wings*

"Character
is what you
are in
the dark."

Dwight L. Moody

It's not enough to have a dream
Unless I'm willing to pursue it.
It's not enough to know what's right,
Unless I'm strong enough to do it.

It's not enough to join the crowd,
To be acknowledged and accepted,
I must be true to my ideals,
Even if I'm left out and rejected.

It's not enough to learn the truth,
Unless I also learn to live it.
It's not enough to reach for love,
Unless I care enough to give it.

Author Unknown

Tory William Shelton
invites you to
his Home School Graduation
on
Monday evening, August Thirty-first
nineteen hundred and ninety-eight
at 6:05 o'clock in the evening
at Father's House
1315 Commerce
Longview, Washington

Please R.S.V.P. 423-4912

GRADUATION ANNOUNCEMENT

This came from the Latham family when the first of their four sons graduated from their family's homeschool with a G.E.D. A very simple announcement could be made as follows for *whatever* route you take. *(I include this sample as an option because our invitations were anything but simple!* ☺)

King's Academy of Longview, Washington
is honored to announce that
R. David Latham
has successfully completed all requirements
for his high school graduation
and on April 20th Nineteen hundred ninety-three
received his General Education Diploma
from Lower Columbia College.

I invited Norma, the homeschooling mom of David, to be on a panel of homeschool moms a few years ago. During the question and answer time, this question was asked her: "What would you say has been your greatest challenge in homeschooling four boys?" to which she thoughtfully smiled and then answered, "To be honest, I'd have to say *feeding* them." ☺

Section Three:

Record Keeping System

Many options and ideas for record-keeping forms to help you create your own system – one that will be a right fit for YOU!

(You will not use them all! Keep in mind that this is just a smorgasbord from which you can pick and choose!)

"Pioneer Girl"

This is an excerpt from a letter Sharnessa wrote when she was "nannying" in New York

(to earn money to attend the School of Creative Ministries in London, which is what she's doing now, May of 1999, at the time of this writing.)

"... I was on the phone with my mom, and we were talking about Eric and Leslie's "Romance God's Way" seminar. And for some reason, I started, just in my mind, getting upset. Not about the seminar, or anything my Mom was saying! But I started thinking to myself, "Why do I always have to be such a loner in the major areas of my life? In starting a "weird" church *(Shekinah Christian Center)*, we were pioneers. In homeschooling, we were pioneers. Then the whole "courtship" thing, *(I don't really like that word! It's just "purity and God's timing in relationships")* we were pioneers. So for all of my life, I've dealt with people constantly questioning me, and thinking I was "not the norm." (which actually doesn't bug me! I know I'm weird!) All of this was going on in my head! And then all of a sudden I was overwhelmed by this feeling of total exhilaration, total joy! This thought then popped in my head, "But God has given me the right character, the right attitude, the right spirit to be a pioneer! I'm my mom's 'Little Joan of Arc'! I love to be different! I love to be challenged! (You could say, without sounding too weird, that I "get-off" on challenges!) I'm a "fighter!" In fact, a quote from "My Utmost For His Highest" that I highlighted *and* double-underlined was: "Huge waves that would frighten an ordinary swimmer produce a tremendous thrill for the surfer who has ridden them." Isn't that incredible! I just freaked out when I read that, because that is SO me! Anyway, back to the point. Well, later that day, I read my little daily saying calendar thing, and guess what was that days quote? "It is not easy to be a pioneer, but oh, it is fascinating! I would not trade one moment, not even the worst, for all the riches in the world." Can you believe that! I just started bawling! I called my mom and read it to her, and of course, she started crying! That was one of those "Whoa, God! You are so awesome!!!" things!..."

– Sharnessa Shelton at age 19

See also some excerpts from a Mother's Day card she wrote to me (on page 6-A–24) for a bit more insight into Sharnessa's heart. ☺

INTRODUCTION to the RECORD-KEEPING SYSTEM

"In the beginning...the earth (and the whole prospect of homeschooling through high school) was without form and void..." *(Gen.1:1-2)* Is this where you're at right now? As God's "vision" for your family starts coming into focus – and it *will* – you'll need a place to put your plans. The forms in this section are what I offer to you as "the places"!

It will be easier for you to come face to face with these forms if you are prepared beforehand for what you'll be seeing. At first glance, a form can seem overwhelming and binding. People have gotten the impression from looking at them that our home is a place an Army officer (in active duty, not retired) would feel very comfortable with all the regimentation. Nothing could be farther *(or further; either is acceptable according to Webster* ☺) from the truth! (See the first part of Section 6 for some eye-opening *truth!)*

The forms contained herein actually *free* us up and enable us to be very flexible and relaxed in our approach to high school because it gives a clear picture of what our goals are and how we are coming along in completing them. If you embrace this method of home schooling through high school, they will become your "good friends" because they'll help make and keep the record-keeping end of this plan as simple as it can be – which, by the way, cannot be any simpler than what you're doing. ☺

DOCUMENT!

In all the research and interviewing I've done, the common-thread message I've heard over and over has been: "DOCUMENT." Even if you do not yet know your goals, if you *document,* you'll have something *concrete* to gather up and put into *some* "form" or "format" later, or to be put together by someone else doing the evaluation of your student. So it will very likely prove helpful no matter what direction you go. Documenting will ensure that all your hard work will *not* be for naught!

Allow me to introduce the friendly faces to you that will be assisting you in your record-keeping task! ☺ I'll just briefly describe each of them here in order to present an overall picture, and then I'll get into each one more specifically as I explain and present the separate forms.

CLASS PLANNING SHEET

All the resources you'll be using and all the assignments you'll be giving, no matter whether the source is traditional or more non-traditional – *everything* you or your student plans to do goes on a Class Planning Sheet. One per class or subject. Think of each of these planners like a grocery store aisle – all of which comprise your own *personally-stocked* grocery store! You will have the Science aisle, the English aisle, the Fine Arts aisle, and on. I can just hear Jesus, the Grocery Store Owner, now: "Good morning, SafeWay-Truth-'n'-Life shoppers! I have a special in the Science Department today! Dr. Chittick is coming to town next week. Don't pass up this shopping opportunity!"

WEEKLY ASSIGNMENT PLAN

This is the "shopping cart" that the parent, the grocery store branch *manager*, loads with just a week's worth of groceries at a time for the student. In our home(school), the kids were (are) encouraged to help choose and gather the groceries.

(SLIGHT DIGRESSION: I've got to say this *somewhere,* may as well be here... This present and past tense stuff is rather awkward; you'll find I switch back and forth because, at the time I'm writing this revision, I have already graduated our two oldest, but still have our youngest, age 12, coming up in about 3 years. When I'm in the middle of thinking through all this stuff, I often use the present tense because this is what I *did* for the past six years! Even though the last one graduated six months ago, I'm still not out of that mindset; plus with Carlianne coming up, I'm not totally out of that mindset! All my beliefs and "practices," whether I'm currently using them or not, are still in effect, if you know what I mean. ☺ And if you don't, just resign yourself to being confused.)

RECORD OF WORK DONE

This form, the "Record of Work Done," is for completed assignments to be recorded on *if* you want a final record of what was actually done. Your "Class Content Plan" will work just fine if you want to use that; you'll just have to adjust it to what you actually *did.* (Does *anyone* ever stick to a *plan* perfectly? I never have! I thought that the definition of "plan" is "something you intend to do because it sounds good at the time, but that doesn't mean it will actually get *done!"* I think it's in "Shelton's Revised-for-me Dictionary.)

It is from these that you can make up whatever "Completed Class Plans" you might want to make, although I found that, for Tory, I really didn't need to do these because everything I needed was on his "Weekly Assignment Plans" and it would only have been doubling up on recording to have to do them by hand here too. Actually, I guess you could just say that my "Completed Class Plans" (that you'll see a ton of in the section by that title – 6-D) *are* my "Class Completion Records" all done up nice 'n' fancy 'n' formal. If you do those too, (what you see in 6-D), then these would just be your "rough drafts."

Only *you* can determine if you actually *need* them. And this brings up a *major* point for this section: I do ***not*** want anyone thinking they need to do *all* that I have done just because *I* did it! Not everyone has a *reason* to do as much as I did. If you aren't going to show this to anyone, then you probably *don't* need to do it. So don't take on any burdens that Jesus isn't placing on you!

TALLY WORKSHEET

If you are into figuring hours, this form will enable you to tally up the hours you are accruing, how ever frequently you want to check. *(I highly suggest you not wait until your child's senior year!!! Foolish, foolish, foolish!!!)* You could think of the Tally Worksheet as being the set of "inventory shelves" where we gather the hours before distributing them. However, you *do* have another option for doing this, if you prefer, just because different styles work best for different people, and I want you to have several options. *(Remember, this is a smorgasbord!)* So this other option is the Credit Completion Bar Graph...

CREDIT COMPLETION BAR GRAPH

This is definitely *my* favorite way of figuring up the hours! (If you have the original version of Form+U+la, it's not there. I hadn't invented it yet!) To me, it's much neater and more concise! You don't need *both.* What's fun about this is that it's all in one place, it's easy to just fill in the hours in the bar graph (in 5-hour increments), and you can SEE the progress being made! I'd say it's utterly brilliant, but I don't want to brag. ;-) (So can I say "thank you, Jesus" without being misunderstood?)

STAYING-ON-TRACK RECORD

This form will help those of you who want something very specific to help you make sure you stay on track, hours-wise, and have everything done by Graduation time! I never used it myself, the Credit Completion Bar Graph sufficed for us, so I asked on the SHF-L email list *(see the last section of this book for what that is)* if anyone was actually *using* it, because if no one did or had or was, I may as well take it out! But I got a few responses, including this one: *"I used it quite a bit actually. If you looked back at all my records, it is actually a timeline of the changes that took place in "me" during my "re-education"! It had to be re-done almost every year!* ☺ *I've pretty much got it memorized now..."* And another mom answered: *"ME!! I'm using it right now for my son 'as is.'"* I heard from a few others as well; enough that I felt it warranted keeping it in. So, it stays. ☺

TRANSCRIPT

Your transcript is where all your classes are listed as well as grades and basic "school" information. The only reason we have needed one so far is for car insurance. And guess what! It *worked!* ☺ The International Bible Institute of London that Sharnessa is currently attending *(yes, it's really IN London!* ☹ *way far away)* did not require a transcript. And Tory hasn't applied at a college yet. So we'll see! ☺ *(Will our story ever be "complete"?!)*

(No more forms, let's *<u>talk</u> about this now...)*

NO-DOZE WARNING!

I suggest you be fully awake before entering the rest of this section. Of all sections in the book, this one requires the greatest amount of concentration. Attempting to tackle it when you're tired will only result in major frustration and an exaggerated perception of its difficulty. ☺ It will probably be challenging for most people to understand *at first*. It certainly was for *me* when I was first figuring it out! But an alert brain will be able to digest it fairly easily, one bite at a time. (You'll be getting other warnings from me. ☺)

You may wonder why my system seems a little more complicated than others you may have seen. Good question. The answer is *not* that it's because I enjoy making something simple seem more complicated. What you'll find is that this system offers a tremendous amount of freedom in utilizing resources all around you *all through* the high school years. As I have mentioned elsewhere, but will say again here, if you are locked into five or six subjects in one year, you will either have to pass up some wonderful opportunities that come up in your community *or* load extra work and time onto an already-busy-with-study-and-life student.

THE BEST OF SEVERAL WORLDS

One of my favorite aspects of this system is that it enables us to "have it all." The idea is to free up the student to enjoy the *process* of learning. For one thing, it will keep dull, mundane, tedious schoolwork to a minimum, though there will still probably be a certain amount that's unavoidable. The flexible structure also accommodates a healthy variety of subjects, rather than an overdose of just a few.

RECORD-KEEPING MADE SIMPLE(R)

Keeping records can burn out even the most avid writer, not to mention freak out those afflicted with recordaphobia! The objective is to make record keeping serve us; not end up with us enslaved to *whatever system* of record keeping. We don't want it to become cumbersome, or any *more* cumbersome than it has to be.

The ***art*** of keeping records lies in making a given form attractive to the eye as well as being logically laid out on paper. The ***science*** of record keeping involves making the form as efficient as the task demands, equal in complexity or simplicity to the task.

For example, if you make (or use) a form that is simpler than the task (like dishing out a student's work for the day or week), then the recording will be inefficient. "Holes" will be left, and areas uncovered, which will, in the long-run, make your task *more* complicated.

Likewise, if you use a form that's too complicated for the task, you make your task more burdensome, recording things that aren't necessary or important to *you.* So the key is *not* simplicity or complexity, but ***compatibility*** of the form to the task it is serving.

FOR EXAMPLE

Take a look at the Weekly Assignment Plan later in this section. This is exactly what our family needs for what *we* do on a daily basis. I could have made up a form with just one empty box for a day. Simpler? At first glance. And for some it may be. But not for us. That would mean I'd have to think up, every day, what subjects we'll be covering, not to mention write them out every single day. (I also have a pre-reader level Daily Report, as well as one for the elementary years, which we used as the kids grew up, so they have grown up with a system that has grown along with them! ☺)

Let's say your son wants to go for a bike ride. You say "Hey! Sounds like a wonderful idea! Got your WAP *(Tory's nickname for the Weekly Assignment Plan)* stuff done up till now?" "Uh, not exactly, Mom. *But this is P.E.!*" "Yes, Son, I realize the importance of P.E. and I can't tell you how excited I am for you to get your exercise but you are as free as the wind to go and do that *as soon as you have your stuff done.*"

If you need to, you can mark, as with a highlighter those items *(jobs, assignments)* that you expect to be done before he is, indeed, free to go do whatever. I do this a lot with my younger child, but it also works great with older kids who are still getting used to a new system.

Back to "perfect fit" record keeping... Let's say that the thought of building a car engine totally intimidates you. (Not hypothetical for me!) So to try to make the task simpler, you say, "OK. Ten parts is my limit. This engine I'm going to build can have NO more than ten parts!" The car engine you'd end up with wouldn't take your vehicle very far, would it? Simple may be easier, but not for long. Eventually you'll find out it isn't necessarily one bit easier *or* better *or* more efficient.

Also please remember as you mosey through this section that, as with *all* my stuff, this is a *smorgasbord!* Lots of choices. Try on the ones that look like they might fit, and if they don't, either alter them, or try on a new one. Won't cost you a thing *(you already <u>have</u> the book.)* So maybe it'll a cost you a little time, but better to spend a little time finding the "perfect fit" than wasting a *lot* of time spinning your wheels with a wrong-fit.

Explanation of ☞ "CLASS CONTENT PLANNING SHEET"

This is where all the student's class work and assignments will be recorded. It's not intended to be for completed work, though you can use it that way if you want to keep things to a minimum. There's no "best way" to do this! Just a "best way for ***you***"! But rather this is a "lesson plan" for your student to refer to find out – or refresh his memory on if he helped with the planning – exactly what needs to be done to earn the credit for a class.

And speaking of "planning," I need to make it *perfectly clear* (and you're going to read this elsewhere!), but I/we do much of the "planning" as we go! So I use that term (planning) very loosely! ☺ Often our "plan" is written down *after* the thing was done!

First, minor point, but it may seem silly or elementary-schoolish to have the student's name on this sheet, but since I have two students in high school at the same time, we want to be able to see, at a glance when the notebook is open, whose page we're looking at. Or if, for some reason, one of them needs to take a planning sheet out of the notebook, and I find it later, I may not be able to figure out whose it was. (Of course, *my* kids never leave things lying around.)

It's not actually *necessary* to fill in the number of credits required for graduation on this sheet since this is listed on your Master Framework Planner. But it's convenient to have it right here and not have to flip back and forth when planning lessons or trying to figure how much more time we have left to fill.

To the right of this it says "No. Of Credits Planned for now:____" This is where you put how many hours, credits, or fractions of a credit you are planning *on this page.* For instance, you might require a total of 4 credits of English for graduation, but on this particular page you're planning for only one *class* of English, like Literature or Grammar. *(More about this in the Completed Class Plans section, 6-D!)* Let's take Sharnessa's "Fine Arts" class, for example. We divided this one-credit course into several smaller units or classes: "Tap Dance Class," "Christmas Angel Dance Presentation," "Concerts and Performances" and "Choir." Part of why each of these classes has its own planning sheet *(and its own Class Completion Record – more on that coming up)* is to keep the instructions separate and simple for the student. If Sharnessa wants to know what's required for "Community Concerts," she doesn't have to go sifting through the assignments and resources for all the other subjects. It's all on one sheet – *nice 'n' simple!*

One other important point I need to mention here is that I don't plan *ahead* on how long all the activities or lessons will take, or how interested or "into it" the student will end up being. I want my kids to be free to turn down different "lanes" on the main road! So I write up only a bit of the content at a time. As they get close to completing them and need more, they either let me know they're running out, or tell me an idea of their own – which I love, and which is almost *always* fine with me!

Let your child know they need to tell you in *advance,* or they may have a waiting period since you may not have time right away to do more planning. I do like to do enough at one time *(I usually get "on a roll")* to last a while so I'm not having to continually come back to the planning stage. Of course the "ideal" is for your student to do more and more of the planning, especially in areas he's interested in. Hopefully the "delight level" will rise as time goes on and he is given the freedom to pursue areas he finds particularly interesting. We may "negotiate" on the maximum number of hours a certain thing can count for, which I call a "ceiling." Since it's possible that you'll have more ideas for resources than the student has time to do, you might want to put an "**R**" next to the ones you require and definitely want your student to read or do, and an "**O**" next to the ones you consider optional.

CLASS CONTENT PLANNING SHEET

STUDENT: Tory **Class or Unit** Computer **Course** Occ. Educ.
Total number of Credits required for Grad.: 1 **Being planned for NOW:** 1

ASSIGNMENTS

Learn how to operate WordPerfect word processing
Learn how to use Print Master graphics program.
– Make: • Banner • Card • Poster • Calendar
Learn Ed Tech Educational Games:
• Facts 50 (U.S. Geography) • Spanish • Jeopardy
Learn and take quiz on Computer Terms (get from Mom)

CLASS CONTENT PLAN

STUDENT ______________________ CLASS OR UNIT __________________ COURSE ________________

Total number of Credits required for this class to Graduate: _____ No. of Credits Planned for now: ____

ASSIGNMENTS / RESOURCES / PLANS / IDEAS / GOALS:	DONE	DATE

Explanation of

☞

"WEEKLY ASSIGNMENT PLAN"

This is the place where daily assignments and activities – whatever *you* call *learning* – are written, 5– 7 days' worth on one sheet. We would generally have four "school days" – or days on which "schoolish" stuff was done – but I ended up finding out that there were things we wanted to record on the weekend too, thus the master you have here has room for the weekend.

I recommend that this sheet be kept, along with other assigned written materials that will be needed for the week, (like work booklets, papers, etc.) in a pee-chee type folder. Put "Assignments To Do" on the left side, "Finished Assignments" on the right.

To protect the WAP, as we shall affectionately call it for short, slip it into a plastic report cover, the type that opens on one side and has a slip-on plastic spine. Holding the cover in your left hand with the cover's fold on the left, insert the left side of the sheet into the fold of the cover. This way the cover can be lifted up on the right side and entries easily made on the sheet... don't use the spine that comes with the plastic cover; it would just pop right off. Thanks to static electricity, the sheet actually stays put!

Name and date go at the top. As for the "entries," I do as much "ditto'ing" (quotes beneath a written entry) so that as little writing is done as possible, taking as little time as possible. The time goes in the space at the bottom of each box. Some of these are "known," like for our reading-together times *(yes, I read aloud to them all the way through to graduation!)*, we knew each of the three of these we did in a day were 20 minutes. Other things we wouldn't know the amount of time for until it was done.

You may notice that all the times come out pretty even here (below); that's because we round off to the nearest increment of 5, usually 15 or 20 minutes. Some prefer to do it to the nearest half-hour, not always rounding up, but rounding to the *closest,* which may mean *not* counting some time! That will it really will "all come out in the wash."

WEEKLY ASSIGNMENT PLAN for *Tony Shelton* Time Period: *March 2-8, 1993*

DAY ↓	SPIRITUAL	ENGLISH (Lit., Vocab., Spelling, Grammar, Composition)		ED. FOUND'S	P.E.	EVERYTHING ELSE Computer / Fine Arts / Occ. Skills / Home Ec & Maint. / Life Skills Math / Educational Foundations / Typing / Bldg Life Relationships / Science / Soc. Studies					
MONDAY	"Armor of God" Study 30min	VOCAB Wordly Wise Ex. 1-A 20min	LIT. Book Synopsis: "Of Mice & Men" 30min	Read on own: "The Right Choice: Home Schooling" 20min	Push-ups, sit-ups, weights 30min	TYPING Mavis Beacon 30min	SCIENCE Read w/ Mom Visual Dict. of Hum Body "Cells" 20min	HOME SKILLS Assemble end table	LIFE SKILLS Deer Hunting	SOC. STUD. Read w/ Mom Where Do I Draw the Line? 20min	
TUESDAY	Read w/ Mom: Foxe's Book of Martyrs 20min	" " EX. 2-A 20min.	" "Diary of Anne Frank"	" 20min.	" 30min.	" " 30min.	" " "Hand, Skin" 20min.	" Plan dinner (on form) for Friday	Bldg Life Rel's	" " 20min	
WEDNESDAY	" 20min	" " EX. 3A & C 20min.	" "Lion, Witch & Wardrobe"	" 20min	Basketball practice 1 hr.	" " 30min.	" "Brain" 20min.	" Help re-org. Storage shed	" Deer Hunt • gut out • hang • skin	" " (finished) 20min	
THURSDAY	" 20min	" " Ex. 4-A	" "Grapes of Wrath"	" 20min	Push-ups, sit-ups, weights 30min	" " 30min	" "Bones, Skeleton 20min	" Make dinner "stir fry"	" Help dad prepare "Swiss (Deer) Steak"	" Listen to tape: American Covenant." 20min	
FRIDAY	" 20 min	" Do review crossword puzzle for the chapter	" "Malcolm X"	" 20min.	" 30min.	" " 30min	" "Spinal Cord" 20min.	" Help Dad repair broken sink	FINE ARTS Concert: "Carman" 2 hrs.	" " 20min	
WEEKEND		FINE ARTS Watch play: "Fiddler on the Roof"				Home Skills Work on Warrantees Notebook 30min.		" Help Mom mount wall paper border			

WEEKLY ASSIGNMENT PLAN for ____________ Time Period: ____________

DAY ↓	SPIRITUAL	ENGLISH (Lit., Vocab., Spelling, Grammar, Comp)		EDUC. FOUND'S	P.E.	EVERYTHING ELSE Computer / Fine Arts / Occ. Skills / Home Ec & Maint. / Life Skills Math / Educational Foundations / Typing / Bldg Life Relationships / Science / Soc. Studies					
	○	○	○	○	○	○	○	○	○	○	○
	○	○	○	○	○	○	○	○	○	○	○
	○	○	○	○	○	○	○	○	○	○	○
	○	○	○	○	○	○	○	○	○	○	○
	○	○	○	○	○	○	○	○	○	○	○
	○	○	○	○	○	○	○	○	○	○	○

Explanation of

☞

"RECORD OF WORK DONE"

The "Log-in Method"

Next I've got a few different possibilities for keeping a record of work done, and titled the same. ☺ These first two sheets are essentially the same method, what I call the "Log-in Method," but are for different length entries. By "log-in," I mean that the activities are "logged in," and the time spent at them "logged in" at the same time. Every time something is done, it is recorded here, and a running total of this is kept in that far left column there, and then forwarded to the next sheet. (Just put the forwarded subtotal on the top line of the next sheet, and put dashes in the "Date," "Work Done," and "Initial" areas on that top line.) If you do not need or want as much detail, you may prefer the *next* method, which I describe directly after the two sheets for this method.

This 1-column "Class Completion Record" is for entries that require several words to describe, such as the ones shown below:

RECORD OF WORK DONE

STUDENT Sharnessa Shelton CLASS OR UNIT Crafts COURSE Fine Arts

TIME	DATE	WORK DONE	Initial
2hrs.	3-20-92	Make porcelain roses out of white bread & white glue	J.M.
2½hrs.	4-7-92	Make paper sack vases — spray paint and shellack	A.J.
2hrs.	4-15-92	Make hand-blown eggs—decorated as babies (faces, hats)	BSS
3hrs.	5-1-92	Make tole painted earring holder (just paint—not cut out)	JW
3hrs.	5-15-92	Make string of 3 fabric sewn (stuffed) bunny rabbits	J.M.

The 2-column version for shorter entries, doubles the amount you can get on one page.

RECORD OF WORK DONE

STUDENT ______ CLASS OR UNIT ______ COURSE ______

TIME	DATE	WORK DONE	INI	TIME	DATE	WORK DONE	INI
1½hrs.	9-6-92	Floorboarding supports	DS	67hrs.	☞ SUBTOTAL OF HOURS FROM PREVIOUS COLUMN		
2hrs.	9-8-92	Lay insulation in floor	DS	2hrs.	9-30-92	Help Mom wallpaper	BS.
3hrs.	9-11-92	Help lay subflooring	DS	1½hrs.	10-1-92	Help D. install light fixture	DS
2hrs.	9-16-92	Help frame in rooms	DS	1hr.	10-2-92	Help D. mount outlets.	DS
1½hrs.	9-17-92	" " "	DS				
1hr.	9-18-92	Help mount drywall	DS				

RECORD OF WORK DONE

STUDENT ____________________ CLASS OR UNIT ________________ COURSE ______________

TIME	DATE	WORK DONE	INITIAL
	☞ TOTAL OF HOURS THIS SHEET *(CARRY OVER TO NEXT SHEET)*		

RECORD OF WORK DONE

STUDENT ____________________ CLASS OR UNIT __________________ COURSE _________________

TIME	DATE	WORK DONE	INI	TIME	DATE	WORK DONE	INI
					☞ SUBTOTAL OF HOURS FROM PREVIOUS COLUMN		
	☞ SUBTOTAL OF HRS; CARRY TO NEXT COL				☞ TOTAL OF HOURS THIS SHEET		

WARNING: Please do not continue if you are getting bleary-eyed or flustered! ☺ If this ends up leaving a bad taste in your mouth, you won't want to come back. And if you feel tired and overwhelmed when you go *into* this, what you're reading now *isn't* going to help! Get some rest, or grab a cup of coffee, herbal tea, or a triple-strength espresso, *(juuust kidding)* and I'll see you after your snack! – or tomorrow, whichever comes first!

Take a break from the forms and read my...

RECORD-KEEPING STRESS 'N' STRAIN REDUCERS

In response (I'm sure) to the prayers of all of you who suffer from recordaphobia, here are some suggestions for easing the anxiety of keeping track of work completed,

RECORDING BOOKS: If you were to record every single reading session individually on the Record of Work Done, the volume of entries would fill up a book themselves! Instead, when you first start reading a book, time how long it takes to read one page aloud, and write that on the front page of the book, (in pencil, if the book is borrowed or you want to erase it later), and then when the book is finished, multiply that minutes-per-page figure (1 to 2 minutes/page is usual) by the number of pages in the book. That *one* number would then go onto the Record of Work Don, once the entire book is completed.

ONE SUBJECT AT A TIME: If you are using the Record of Work Done, I highly recommend making up one per class, unit, or activity. I mention this, not to overwhelm you with one more detail, but to prevent you from making the same mistake I did... Sharnessa was participating in several activities that, at the time, seemed would fall under "Fine Arts." So we entered all of these projects and activities on one page entitled just that: "Fine Arts." All entries for her Tap Dance class, Choir, Community Concerts, and a Christmas Angel Drama she was in all went on the one page. The problem arose when I realized that hours for Fine Arts were adding up quickly and that, since P.E. was low on hours, it make more sense to put the Tap hours into P.E. But all of her hours were mixed in together! We realized then that each of the classes should have their very own sheets so they could be more flexible. So unless you are absolutely certain beyond the shadow of any curricular doubt that the activity couldn't *possibly ever* count for *any* other subject, it'll be more efficient in the long-run if you to keep track of just *one* subject or activity on *one* Class Completion Record. Sure, it uses a little more paper, but remember, you want the *system* – and in this case, the paper too – to serve *you.*

SHORT-TERM ACTIVITY: If a short-term activity takes the same amount of time each day, for instance, the kids watch a Creation Science video series for ½ hour every day for two weeks, instead of making 10 identical entries 10 days in a row, at the end of the 2 weeks simply write one lump-sum on the Record of Work Done.

Explanation of

☞

"RECORD OF WORK DONE"

The "Record & Check-off" Method

You will probably notice two things about this; one, that it looks a lot like the previous two pages; and two, that it has a nice chart full of boxes down at the bottom. This form is similar to the previous "Log-in Method," but is actually less record keeping. On the other form, every time something was done, it was recorded, the time logged in and a running total of it kept. For this one, you'd log in the basic activity – but only once. However, you keep track of the time -- *not* in the "Time" column (that's gone) but down at the bottom, in one, ongoing, bit-by-bit increasing "lump" – in graph form! ☺ You make your entries at the top; time goes below. I'll explain...

Let's look at the Hour Check-off Box at the bottom... In it you see there are 180 boxes. This is to make it versatile for *whatever* number of hours per credit that you decide upon. (See section VI-A, Formulating Your Framework, for more on this). The most you *might* have is 180. If you decide on 150, use a dark wide-tip felt pen to black out the first thirty boxes, from 180 down to 151; and you're left with 150. Or if you're going the 120-hours-per-credit route, then block from 180 down to 121; 120 hours are left! (See sample below.) Custom-made! *(Now, I know perfectly well that this is going to drive you perfectionists crazy, so all I can say is I hope you have a computer on which you can start from scratch and make your own, or that you're good with a ruler and pen! ☺)*

As for "checking-off" time, you can either do it with actual check marks, one in each box, or you can use a light colored highlighter (wide-tipped) and fill in the hours as they are done with that. And finally, to answer the "do I go top to bottom or bottom to top?" question... Whether you begin your highlighting with "1" or at the top with your number of hours per credit depends on whether you want to know how many hours you've *done* or how many hours you have *left.* ☺ It's actually *that* simple! I'm going to simulate a 120 hour class for which the student has done 27 hours of work, and they're recording from "top down" cuz we like to know how many hours we have *left.* ☺

RECORD OF WORK DONE

STUDENT ________________ CLASS OR UNIT ______________ COURSE ______________

DATE	WORK DONE	INI	DATE	WORK DONE	INI

HOUR CHECK-OFF CHART

180	179	178	177	176	175	174	173	172	171	170	169	168	167	166	165	164	163	162	161
160	159	158	157	156	155	154	153	152	151	150	149	148	147	146	145	144	143	142	141
140	139	138	137	136	135	134	133	132	131	130	129	128	127	126	125	124	123	122	121
120	119	118	117	116	115	114	113	112	111	110	109	108	107	106	105	104	103	102	101
100	99	98	97	96	95	94	93	92	91	90	89	88	87	86	85	84	83	82	81
80	79	78	77	76	75	74	73	72	71	70	69	68	67	66	65	64	63	62	61
60	59	58	57	56	55	54	53	52	51	50	49	48	47	46	45	44	43	42	41
40	39	38	37	36	35	34	33	32	31	30	29	28	27	26	25	24	23	22	21
20	19	18	17	16	15	14	13	12	11	10	9	8	7	6	5	4	3	2	1

RECORD OF WORK DONE

STUDENT ______________________ CLASS OR UNIT ________________ COURSE ______________

DATE	WORK DONE	INITIAL

HOUR CHECK-OFF CHART

180	179	178	177	176	175	174	173	172	171	170	169	168	167	166	165	164	163	162	161
160	159	158	157	156	155	154	153	152	151	150	149	148	147	146	145	144	143	142	141
140	139	138	137	136	135	134	133	132	131	130	129	128	127	126	125	124	123	122	121
120	119	118	117	116	115	114	113	112	111	110	109	108	107	106	105	104	103	102	101
100	99	98	97	96	95	94	93	92	91	90	89	88	87	86	85	84	83	82	81
80	79	78	77	76	75	74	73	72	71	70	69	68	67	66	65	64	63	62	61
60	59	58	57	56	55	54	53	52	51	50	49	48	47	46	45	44	43	42	41
40	39	38	37	36	35	34	33	32	31	30	29	28	27	26	25	24	23	22	21
20	19	18	17	16	15	14	13	12	11	10	9	8	7	6	5	4	3	2	1

RECORD OF WORK DONE

STUDENT ____________________ CLASS OR UNIT ________________ COURSE __________________

DATE	WORK DONE	INI	DATE	WORK DONE	INI

HOUR CHECK-OFF CHART

180	179	178	177	176	175	174	173	172	171	170	169	168	167	166	165	164	163	162	161
160	159	158	157	156	155	154	153	152	151	150	149	148	147	146	145	144	143	142	141
140	139	138	137	136	135	134	133	132	131	130	129	128	127	126	125	124	123	122	121
120	119	118	117	116	115	114	113	112	111	110	109	108	107	106	105	104	103	102	101
100	99	98	97	96	95	94	93	92	91	90	89	88	87	86	85	84	83	82	81
80	79	78	77	76	75	74	73	72	71	70	69	68	67	66	65	64	63	62	61
60	59	58	57	56	55	54	53	52	51	50	49	48	47	46	45	44	43	42	41
40	39	38	37	36	35	34	33	32	31	30	29	28	27	26	25	24	23	22	21
20	19	18	17	16	15	14	13	12	11	10	9	8	7	6	5	4	3	2	1

This stuff is just toooo dry! I think it's...

Testimony Time!

When I began reading the Form+U+La, I, at first, looked at it with eyes used to curriculums. That can make it seem a bit "difficult." I sat back, prayed, and kept reading; the more I read, the more I began to "get it". When I was finally free of the "curriculum mentality" I found Form+U+La to be one of the most freeing and easiest concepts I had ever encountered. You make it fit what you ARE DOING – or rather, it allows you to fit what you are doing into a framework called "high school." Here are two examples of what WE did (remember each individual will have a different focus and different priorities):

☆ Seth did Ecology (with a lab) by participating in one year of 4-H Soil and Land Judging competition, and two years of participating in the Envirothon competition. We never opened an ecology book – it was all real-life, hands-on learning! His team took many field trips to places such as nature preserves with Rangers and scientists as teachers! They studied the materials provided and met in study groups. I just kept track (in general) of these trips and competitions. And at the end we had over 150 hours for a full credit! (We originally planned it as a 1/2 credit) By the way, his teams all placed first in regional and quite high in state competitions. ☺ Those experiences are what "finally" convinced my parents we weren't ruining our children! ☺

☆ We have a farm. Seth gets credit as he repairs fences, builds rabbit pens, repairs water lines, feeds stock, and so on – all the regular chores that need to be done are part of his "school." I shuffle them into appropriate credits. Some go to industrial arts, some to agriculture electives, some to woodworking.

Every time you go to a play (even the free church plays), listen to music, do a craft, you are doing "school." After a few years you have 1/2 or even a whole fine arts credit without even trying!

I rarely use texts. Most of our books are from the library, government offices (such as Fish and Game, FL AG in the Classroom), and 4-H (your tax dollars at work. ☺)

In our state students can also go to community college with free tuition at age 16 for dual enrollment. We are doing high school in 5 years (with Seth's permission) to take full advantage of that option. He will have two years of high school dual enrollment credit and almost an AA by the time he graduates.

Patti; FL

Explanation of

☞

"TALLY WORKSHEET"

This is the place where, at various points through the year, you can tally up hours logged thus far to find out how far you've come and how many more hours are needed to complete the year's requirements.

List all *courses* you've decided to include in your line-up of classes in the left column under the "Class" heading. You will not need to have separate entries on this sheet for all the *"units or classes"* because all you really need to know is how you're coming along on your hours for the *course.* You can just add up all of the individual classes on a separate sheet of scratch paper and enter the total for the course on the Tally Sheet. For instance, we'd add up all the Tap, Crafts, Angel Dance and Choir hours before entering them under the Fine Arts.

Now just to mess you up, because you may prefer to be more detailed, you certainly *may* want to add up all the *classes* on this sheet as well, just to keep your running totals of all of them. But be sure to put parentheses around the *classes* to differentiate them from your *courses (which you could possibly circle),* because the courses are the only ones you actually *need* to keep track of for our purposes here. *(Unless, of course, you do it all differently.)*

The time figures we'll be adding up here are taken directly from the "Record of Work Done" on which the student has been logging hours. First add up all the minutes and hours logged for one class on the "Record of Work Done." To make sure you don't count it again at the time you do your *next* tally, you need to indicate on your sheet where you left off. I simply put a little right angle corner, using a colored pen so it's quickly viewable, below and to one side of the last item added this count. *(See example below.)* Then I write the date out in the column , and below that the number of hours added on this time, which I circle. Write these subtotals in a different color from the work entries to make it easy to spot where you left off for the next time you do a tally. ... That's it! You can do this how ever often you like!

A portion of a tallied "Record of Work Done" for a "Crafts" class:

2 hrs. 3/20/92 Make porcelain roses out of white bread & glue J.M.
2½ hrs. 4-7-92 Make paper sack vases – spray paint & shellack A.J.
2 hrs. 4-15-92 Make hand-blown eggs, decorated as babies B.S.
3 hrs. 5-1-92 Make tole painted earring holder P.W.
3 hrs. 5-15-92 Make string of 3 fabric sewn bunnies J.M.

1-18-92 (6½ hrs)

A few lines of the Tally Sheet, to show how to make entries: *(We're on the second check here.)*

entered here

CLASS	1ST CHECK	2nd CHECK	3rd CHECK	4th CHECK	5th CHECK
Date of check ➪	10-30-91	1-18-92			
Bible/Spiritual	24 +	20 = 44			
Crafts	16½ +	(6½) = 23			
English Total→	28 +	27½ = 55½			
(Speech)	(4) +	(6 = 10)			
(Lit.)	(16) +	(10 = 26)			
(Grammar)	(8) +	(11½ = 19½)			
Home Ec	14 +	9 = 23			
Life Skills	8½ +	16 = 24½			

TALLY WORKSHEET FOR THE _____ SCHOOL YEAR

CLASS	1st Check	2nd Check	3rd Check	4th Check	5th Check
Date of Check⇒					

Explanation of ☞ "STAYING-ON-TRACK RECORD"

How does one know if or how well one is staying on track in completing credits and hours with such a flexible system? You have two options. One is to use a "Staying-On-Track Record"! (Your transcript is where you'll record the list (titles) of *finished* courses and classes, but that won't help you stay on track, if, indeed, you'd *like* to stay on track!) Another option is the "Credit Completion Bar Graph," which we'll get to right after this, but for now I'll explain the "Staying-on-Track Record." I'm giving you two versions of this form: one blank and one filled out. *(See next page.)*

HOW IT WORKS

You're probably wondering: *Why all the blank lines between the subjects?* These are for you to write in your own individual *classes* under each of the *courses.* For instance, your English course might have a class of "English Literature," "Spelling, Vocabulary and Grammar," "Creative Writing" and "Speech." Use Wite-out™ to make any deletions or substitutions, fill in your required number of credits to graduate, and then at the end of each year, fill in how many hours or credits were earned – transfer from the Tally Sheet if you used that – and how many hours or credits you have left to go.

For instance, you might have 4 credits required for English. So at the end of the Freshman year you'd (hopefully and smartly) have 1 credit earned; with 3 left to go to graduate.

You might need 2 credits of Math to graduate. At the end of your Freshman year you might want to have ½ credit earned. Or you might prefer to just do one whole credit in the Sophomore year and skip it the Freshman year. If you need 1 credit of P.E., you could put 30 hours (on a 120 hrs./credit system) completed with 90 hours to go. (Or you could put ¼ credit earned with ¾ credit to go.) Use whatever measure will make the most sense and be the most useful *to you.*

MORE SPECIFICS

If you're just starting out, you may feel rather bound to imitate what you see here. But as you start doing it yourself, especially a year or two into it, you'll see how flexible the whole system really is. Once you understand how this set-up works, you'll feel free to make all the adjustments your heart desires!

You'll get the running totals for all the classes (that you've filled in down the left column) from the Tally Sheet where they've been kept track of. Once a *credit* requirement for the year (such as 1 credit of English) or an *hour* requirement (like 113 hours of Social Studies) has been fulfilled, it is entered under the "Credits or Hours Earned" column on this sheet. If you're at the end of your year and your credit or hour goal has not been met, I would simply not put anything down in this column until it *has* been earned – *whenever* it's earned.

(Examples coming up on the next page...)

STAYING-ON-TRACK RECORD

COURSE/CLASS TITLES ⇓	Req'd Number of Credits ⇓	FRESHMAN		SOPHOMORE		JUNIOR		SENIOR	
		Credits or Hours Earned	TO GO	Credits or Hours Earned	TO GO	Credits or Hours Earned	TO GO	Credits or Hours Earned	TO GO
ENGLISH									
MATH									
SOCIAL STUDIES:									
SCIENCE									
BIBLICAL/SPIRITUAL									
OCCUPATIONAL EDUC.									
P.E.									
HEALTH									
FINE ARTS									
LIFE SKILLS									

A FEW EXPLANATIONS AND EXAMPLES

Let's start at the top of the "COURSE/CLASS TITLES" column with **English.** Since I've divided this *course* into the different *classes* or *units,* the number of hours earned in each of these (taken from the Tally Sheet) is first recorded in the spaces provided. After you get all your classes and times entered, add up all the classes that make up one course to get your total number of hours for each of the courses.

In case you're wondering, the reason I've divided up the courses into individual classes is that I want to make sure we spend adequate time on each. For instance, I don't want just 3 hours out of 120 spent on Grammar and 117 on Literature. However, each subsection will not be allotted "equal time" either. The number of hours spent in each area of study will depend *completely* on the student's proficiency, need, and interest. This won't tell me how much to allot where, it just gives me a way and a place to keep tabs on it.

ABOUT ENGLISH

Let's say that 168 hours of English were earned one year, 18 hours *above* the required 150; or 48 over the required 120. What do you do with those extra 18 or 48 hours? You would still write "1 credit" (eating up 150 or 120 of the hours) in the "Earned" column, but then enter the extra "18 or 48 hours" into the *following* year's Tally Sheet under the appropriate English class(es). *(See, this isn't so hard! A little* ***dry*** *maybe, but not hard!)*

ABOUT MATH

Math can seem to present a problem at first, but actually it is simple. Since there are two systems for figuring credits, I'll deal with them separately to keep the two distinct:

If you are going the **"content-based"** route, then *completion of the class material* is the point at which a credit is awarded. In our case, as soon as this year's Saxon book (or you might have A.C.E. paces, etc.) is completed, a credit is earned and can be entered onto this sheet and the appropriate amount of time deducted.

On the other hand, if the **"time-based"** system is used, and let's say 2 credits of Math are being spread out over the 4 years, then as soon as the 75 hours (half of 150) are logged, the *hour* requirement for the year is satisfied and can be recorded here. (This is one of the reasons I went to a "Time AND Content-based frame work; I felt that, to truly earn a full credit in Math, *all* the book should be completed (even though I don't think we *ever* completed a full math (or *any)* textbook when I was in school. ☺)

HISTORY/SOCIAL STUDIES

If these classes have been kept track of as separate classes on your Tally Sheet, then the separate totals would be recorded here once the required hours (for all three of them put together) are logged. These three subtotals are added, and that total put in the "Earned" column. Again, if there are leftover hours, transfer them to *next* year's Tally Sheet.

Another way to do this is to keep track of *each* of your Social Studies classes totally separately *right from the start.* (Note: You'll probably need to keep track of them separately during at least the Senior Year to make sure each of these classes is covered individually. But it's not all that necessary until then.) The only reason I would not have been able to do that, even though, *starting out,* I *thought* I would have, is that as it turned out, I didn't know which classes we'd have enough hours in to actually make a *whole* class of it. There are *lots* of possibilities for Social Studies classes! Economics, Educational Foundations, Contemporary World Problems, World Religions, Western Civ, and of course your various history and government classes, state, country, and world.

AND SUMMING IT UP...

After you have totaled up all your classes that have definite requirements (like probably English, Math, Science, Social Studies) the *classes that remain* on the page should, *all together,* total how ever many *more* hours are needed to be promoted to the next grade. Refer to your own Master Framework Plan for your specific total. *(You'll learn how to do that in Section 6-A.)*

Some of these classes will have only a few hours accumulated, *not* totalling a full credit. Those *hours* are still recorded in the "Earned" column. For instance, let's say that you are requiring 1 credit of Fine Arts and you're on the 120-hour system. Your student has accumulated a total of 83 hours this year. Subtract *that* number of hours from the *total* number of hours required for that subject. 120 minus 83 equals 67. So put "67" in the "TO GO" column. (This is *why* you can enter either credits *or* hours in this column.)

Now wasn't that FUN?!?! ☺

STAYING-ON-TRACK RECORD

COURSE/CLASS TITLES ⇓	Req'd Number of Credits ⇓	FRESHMAN		SOPHOMORE		JUNIOR		SENIOR	
		Credits or Hours Earned	TO GO	Credits or Hours Earned	TO GO	Credits or Hours Earned	TO GO	Credits or Hours Earned	TO GO
ENGLISH	4	1	3	1	2				
(Speech)									
(Literature)									
(Spelling/Grammar)									
(Composition)									
MATH	2	½	1½	½	1				
(Algebra)	(1)								
(Consumer & Appl'd)	(1)								
SOCIAL STUDIES: 300 hrs = (75/yr.)	2.5	75	225	75	150				
(World Hist & Curr. Evts)	(1)								
(U.S. Hist & Govt.)	(1)								
(WA St. Hist.)	(½)								
SCIENCE	2	½	1½	½	1				
(Lab)	(1)								
BIBLICAL/SPIRITUAL	2	½	1½	½	1				
OCCUPATIONAL EDUC.	1								
Typing 120 hrs. = (30 hrs/yr.)	1	30	90	30	60				
Computer	1	30	90	30	60				
P.E.	2	½	1½	½	1				
HEALTH	1	30	90	30	60				
FINE ARTS	1	30	90	30	60				
LIFE SKILLS	1	30	90	30	60				

not kept track of separately

Parental Educational Input Chart

Assignment/Input/Idea	Subject	Do w/ Dad	w/M	w/T	w/C	On Own	Frequency; or Do It On	Due By	Done
"Today's Christian Teen" articles	Occ. Ed. Spiritual		✓				1 article per day	4-21-95	✓
"Sr. High: Out of Fear" audiotape	Educ. Found's			✓			4-23-95	4-23-95	✓
"Court Terms" Search-It-Out Sheet	WA St. Hist.					✓	one SIO per day	5-8-95	✓
"Two Types of Court Cases" S.I.O. "	"					✓			✓
"Municipal & District Court" SIO	"					✓			✓
"Superior Court" SIO	"					✓			
"Supreme Court" SIO	"					✓			
"Educating the ADD/L.D. Child" tape	Educ. Found's		✓	✓			4-30-95	4-30-95	✓
"Dealing w/ the Spec. Needs of the ADD Child	"		✓	✓			5-2-95	5-2-95	
Change the oil in van	Vehicle Maint. (Home Econ)	✓					5-3-95	5-3-95	
Watch "Hidden Keys to Lasting Relationships" video series	Spiritual	✓	✓	✓	✓		—	5-12-95	
Tap Dance Lessons w/ Carlianne	Occ. Ed.				✓		Fridays at 2-3pm	—	
"American Covenant" seminar with Marshall Foster	U.S. History	✓	✓	✓			Oct 28-29	—	

Explanation of Parental Educational Input Chart

Sharnessa got to the point of eventually wanting to make up her *own* Weekly Assignment Plan (Daily Report), and didn't want my handwriting on it! *I* had thought our handwriting was pretty similar, but she said, no, it's *very* different, and she preferred to have just the *one* style of handwriting on there. ☺ Really, I just think she just wanted to be independent, but that was fine because she was not being rebellious in *having* that desire, nor was she disrespectful in how she *expressed* it to me.

The only problem was that I still had stuff I wanted to go on her plan, but with this new independence thing, didn't have a way of *getting* it there — *until* I came up with this! I didn't have space to include a blank master here, but it's pretty simple to make up, and yours would be a little different anyway.

Why such an ostentatious-sounding title? Well, I thought it just added a little "dignity" and "officialness" to it. ☺ Besides, if you were her, would *you* want a form with the cumbersome title of "Stuff My Mom Wants Me to Include on My Weekly Assignment Plan"? ☺

The Art & Science of Making A ☞ "TRANSCRIPT"

If the very thought of doing a Transcript intimidates you, you're not alone. It does the same to many of even the most experienced, successful, capable homeschoolers. And it did so to me, *until* I started getting set free from the things that were at the *root* of my intimidation. (which is what section one, "Out of Fear and Into Freedom" is for; so I hope you didn't skip that!)

But suffice it to say that as you allow God to deal with your fears in this area, you'll find, to your joy and surprise, that The Transcript begins shrinking down to "normal size." (Hey! That could be a poem!) Besides, my personal feeling is that the more you come to realize that the *main* thing you need for the right door to open (and only *one* door needs to open *next)* is to be seeking and following *God. "Seek first the Kingdom of God and His righteousness, and all these things shall be added unto you."* I'm not saying to do nothing and leave it all to God (or I wouldn't have written a book of this nature!), but I do think that God has to come first or all our other efforts will be in vain.

WHY SO INTIMIDATING?

They're intimidating because they hold within their thin grasp the power to open or close doors of access to our hopes and dreams. That's all. Think about it... If you needed a transcript just to be admitted to a dressing room at your favorite clothing store, you wouldn't feel nearly the same intimidation. Yeah, you might get a little irritated, but you'd either say "Well excuse me, but you must not need my business very badly; I can buy my clothes elsewhere." Or, Plan B: you'd just go ahead and buy the item without trying it on, save your receipts, try the thing on at home, and if it didn't fit, just return it. My point is that *there would be a way around it!*

But if Ms. Sternface, the Dean of Trepidation at I.W.U. (Impressive, Wanna-go-there-bad University) asks you, lowly Bud or Betty Justahomeschooler, for your transcript before they'll accept you, you probably wouldn't feel like you were in much of a position to say, "Well excuse me, Ms. Sternface, you must not need my business very badly! I'll just go to college elsewhere!" Of course, you'd never be so irreverent, (although Ms. Sternface is surely nowhere near deity, even though sometimes we – and *she* – seem to think so,) but ideally, *really,* saying something to that effect, but more graciously, isn't a bad idea! Because there really *are* plenty of good colleges.

GOD-ORDAINED COLLEGE?

But let's just say *(because some of you are, I'm sure)* that perhaps you have a good, perfectly logical reason to assume that *this* really is *the* one and only college God wants you at for one of the following reasons:

- The scholarship you received is good at this school *only.*
- This is your parents' (*and* your grandparents', *and* your grandparents' parents') Alma Mater. *Tradition!!!*
- God's purpose for your life will be totally thwarted if you can't go there and you'll never *ever* be able to walk in His presence or His will again because *the* training He wanted you to obtain can be obtained *only* at I.W.U. and no other.
- And besides, your best friend is going there and the two of you made a pact in third grade that, no matter what, you'd room together in college. *(Or death would part you instantaneously, no chance for explanations.)*
- And besides *that*, you saw I.W.U.'s full-color brochure (or their travelling vocal ensemble) and are just *sure that* God's-will-for-my-future-spouse looked like he/she could be in amongst those happy, wholesome, God-radiating faces.

At any rate, let's say you have adequate reason, (and it really is a good one, despite all my jesting) to trudge through the pains of making up a transcript and just get in and DO IT. Fine. But *how?...* Comin' right up!

NOT CREATED EQUAL

I should preface my upcoming directions with a confession: I am *not* your friendly neighborhood Transcript Expert. Friendly, yes; expert, no. I looked at several of my own transcripts, requested a blank one from a couple of local school districts, and got another elsewhere. My brilliant deduction? Even here in the Land of the Free and the Home of the Brave, all transcripts are NOT created equal! *Every one of them is different!* So at least we know there's *no set-in-stone way* of tackling this task. I would have loved to have been able to just call up the National Transcript Formatting Authority Service, but, alas, no such place exists!

Since transcripts each have their own entirely unique styles, fonts *(type of writing),* and forms of validation, you can do it pretty much any way you want. But here are some generic points to consider as you ponder how you want to design yours:

- Seals seem to be widely accepted as being "official." Or at least they have a way of making something *appear* to be official – and I don't just mean our homeschool transcripts!I heard of one homeschooling family whose graduated son wanted to go into the Armed Services, but because he didn't have an "official transcript" was not going to be admitted. They told him his diploma didn't have an "official seal." So the homeschool leader said, "You mean if I go get, like for instance, a notary public's seal on it, it'll be considered 'valid'?!" He said, "Yes." They did it, and the homeschooler was accepted! Now I realize that this might not happen again, but it does tell you the weight a seal *can* carry! (I added the bit at the bottom of the transcript about the "blue ink and seal" from what was on one of my own transcripts.)
- Typing or computer-generated lettering seems to be the most professional looking; the least "suspicion-evoking." If you know someone with a good computer, it'd be well-worth a bribe-berry pie to get them to let you use it, or, if you're totally computer-illiterate (and they don't even want your fingers touching their precious computer!), to get them to do it *for* you.
- Keep in mind that hand calligraphy, while it may look great, can be a dead give-away that this *isn't* "traditionally" generated! How many traditional schools have secretaries with time to hand-calligraphy the school name onto every transcript? *(Only the small ones, is my guess.)* You want your transcript to just sorta blend in with the crowd and *not* be outstanding among the others, for better or for worse! But then again, Cafi Cohen (author of *And What About College?)* says this can actually be an *advantage* to homeschoolers! They can "stand out in the crowd"! So, again, *you* decide. ☺

As you can see by looking at the transcript over on the next page, I've filled in the blanks of our hypothetical homeschooler's transcript. There's nothing difficult about it, but I've been told that examples are appreciated, so let's take it one aspect at a time. *(By the way, the homeschool laison officer at Oral Roberts University told me that my transcript is EXACTLY what she likes to see come in!)*

PORTFOLIO INSTEAD???

Something I want to interject here, one thing you need to consider, in the midst of giving you all this info on creating a transcript, is that it's *possible* that your college may very well accept a *portfolio* in lieu of a transcript. Of course the two serve different purposes. A portfolio does not offer a grade point average, nor does it state which classes were taken. But in many cases it can give enough information to an admitting institution to convince them that this student merits their consideration. I personally think giving *both* is best.

(I attended a workshop by Elaine Brouwer, an expert in the arena of transcripts. I then compiled the lists she gave as handouts and re-wrote them up into a 4-page article on putting together an effective, impressive portfolio. This appears in this book toward the beginning of Section 5 on "Grading.")

Sample Filled-in Transcript ☞

THE SCIENCE OF MAKING A TRANSCRIPT

Beginning at the top of the transcript, let's go through each of the components. What I'm giving here, though, is not the "last word" on the subject. It's just what I've found in looking at several transcripts.

- **HOMESCHOOL NAME:** This will be discussed on the next page.
- **STUDENT DATA:** Just enter data as requested. Circle gender of student or put an "X" over what they're *not.*
- **INSTITUTION DATA:** Just fill in as shown.
- **ACADEMIC RECORD:** Enter all the classes taken, the dates they were completed, the number of credits earned, the letter grade (or "Pass" or "Fail"), and the equivalent grade point. *(Refer to "Grade Points Equivalents" Chart at bottom.)* Typing these seems to be pretty standard, which makes sense because the class and grade entries are added year by year, *unless* they have it on computer. What you see here is a computer font that *looks* like typing! ☺ *(Mine is called "Courier New.")*
- **CUMULATIVE SUMMARY (Credits + G.P.A.):** Here you enter the total number of credits earned (in left box) and the grade point average (right box). For those who are rusty in math, just add up all the G.P.A. numerical (intelligent-sounding word for *numbers)* equivalents in the far right column. Divide that sum by the total number of *credits* - **not** just the number of figures you added, since some of those numbers *may* represent *two* credits.
- **THE "THIS TRANSCRIPT IS OFFICIAL..." BOX:** (lower left corner) Feel free to use "Wite-out" and change or omit this part. Sounded good to me. One of my college transcripts has something similar.

OFFICIAL TRANSCRIPT

§ **Our Very Own Senior High** §

« STUDENT DATA »	« INSTITUTION DATA »
NAME: Last, First, Middle: Cated, Ed U.	NAME: Our Very Own Senior High
ADDRESS: 321 N.E. Homestead Rd.	ADDRESS: 321 N.E. Homestead Rd
CITY/ST/ZIP: Hometown, WA 00000	CITY/ST/ZIP: Hometown, WA 00000
PARENTS: John and Mary Cated	TELEPHONE: 000-000-0000
BIRTHDATE: 11-7-76 SEX [(M)/ F]	

ACADEMIC RECORD

COURSE/CLASS TITLE	Date Completed	Credits	Grade	G.P.A.
English - 1st Year	5-18-92	1	B+	3.3
English - 2nd Year	6-1-93	1	A-	3.7
English - 3rd Year	4-29-94	1	A	4.0
English - 4th Year	5-2-95	1	B-	2.7
Math - "Algebra"	3-1-93	1	A	4.0
Math - "Consumer & Applied Math"	5-3-95	1	A	4.0
Science - including 1 credit of Lab Science	11-7-93	2	A	4.0
Social Studies:				
U.S. History & Government	11-13-93	1	A	4.0
World History & Current Events	1-4-94	1	A	4.0
Washington State History & Government	2-6-94	1/2	A-	3.7
Occupational Skills	4-15-94	1	B+	3.3
P.E.	5-27-94	2	A	4.0
Health	5-30-94	1	A	4.0
Fine Arts	4-30-95	2	A	4.0
Biblical & Spiritual Studies	5-23-95	2	A	4.0
ELECTIVES:				
Typing	8-26-95	1	A-	3.7
Computer	5-1-95	1	B-	2.7
Life Skills	3-4-94	1	A	4.0

GRADE POINT EQUIVALENTS							
A = 4.0	A- = 3.7	B+ = 3.3	B = 3.0	B- = 2.7	C+ = 2.3	C = 2.0	C- = 1.7

(Grades D through F not awarded)

CUMULATIVE SUMMARY	
TOTAL CREDITS:	CUMUL. G.P.A.:
20.5	3.27

This transcript is official when the principal's signature and the date are in blue and a portion of each is sealed by a notary public embossing.

Signature of Principal: John Cated

Date: 8-30-94

WHAT GRADES DO I PUT ON MY STUDENT'S TRANSCRIPT?

This can be a tricky question for homeschoolers who have never graded up until now. The Colfax's solved it by giving their son, who was confronted with this very question, a blanket 4.0. It's not really that hard to figure out. If your child has done well in a given class, give him an "A." If he did "good," give a "B." If he did "OK" have him do it over or better. (I don't believe homeschoolers should do less than "B" work. If they get less than 85% on a test, *if* you test, then they don't know the material yet and you just need to keep working at it until they *do!)*

If you want specific ideas for 20 traditional and more creative, experiential classes (like "Hunting," "Gourmet Cooking," "Building Family Relationships,") see Section 5 of this book, on "Grading Guidelines."

Blank Transcript ☞

AND NOW, THE <u>ART</u> OF TRANSCRIPT MAKING

If you like and are going to use the one you see here, the only thing left for you to do is to get your school name on there and copy this onto nicer paper. First, about the school name...

You *could* just type your school name on, but if you want it to look really sharp, you might want to use press-type lettering which, in case you've never used this method before, is actually very simple to use. Tons of letters, all in nice neat rows, *(<u>unless</u> your 2-year-old, bless his/her heart, gets hold of it,)* are affixed to the bottom(back)side of a small sheet of transparent, waxy paper. You lay this sheet down directly over the position you want it to end up at on the paper. Then, holding both the letter and the paper you're transferring the letters onto firmly in place, just rub over the letter with something hard, like a popsicle stick. Wah-lah! Your letter is now on the paper! All you have to do *now* is the *rest* of the letters! *Evenly!* (If this doesn't make sense, don't worry, complete *illustrated* instructions come with the lettering packet. I just wanted you to know now how simple it is so you wouldn't be intimidated into not doing it before you even try it.) This press-type lettering, which you can find at most supply stores, comes in all sorts of gorgeous fonts.

Or, if you know someone with a fancy computer with some fancy fonts and a laser printer, you might coax them into printing out your school name in a few different sizes and fonts. *(Hopefully you can get this with the same bribe-berry pie.)* You'd then cut-and-paste your favorite one onto the transcript, taping the edges so they don't copy as lines.

Before you can *do* any of the preceding, you need something paperish to *put* it *on!* So at some point in this process you're going to need to get some paper! Because you've never made a transcript before (I'm assuming this, otherwise why would you still be reading this section?), I will give you some possibilities for color combinations and paper choices...

- a grey, mottled-looking paper with red and black print
- a yellow or creamy colored, "linen-look" paper with green and black print
- a mauve, grainy-looking paper with blue and black print
- a smooth, solid gold paper with blue and black printing

Where to get said paper? Find a printer that sells paper by the piece. You don't need a ream of this! Ask to see their higher-grade paper with interesting textures and grains. Ask if they have a booklet with samples in it. It might be overwhelming, but just take a deep breath, look at what they have available, and let your creative right brain (or is it left?) have some fun. Avoid highly textured paper (with ridges you can feel) as it doesn't copy well. The toner has a hard time getting down into the little valleys. And parchment, pretty as it is, has a couple of drawbacks: the print tends to rub off of it and besides, it's *terribly* formal looking. Kind of like showing up at McDonald's in a formal gown. You don't want to stick out like a decorated thumb *(sore or not!)*.

I suggest getting a few samples of the types of paper you like and running them through the copy machine(s) that you'll actually be using to make sure it'll "work." You need to find a copy machine that has a hand-feeder (not difficult to find). *Only* when your practice sheets come out exactly how you want them do you use your expensive, spiffy paper.

Remember, you'll only do this (make up a transcript) once, so if you want something professional-looking, you'll need to put a little time and effort into it. But if spiffiness is not a high priority, don't feel any pressure to go to all the trouble! I've seen transcripts from good schools that have tacky-looking transcripts.

I recommend making up several copies of your finished transcript so you'll have some "insurance" for ones you mess up when entering classes and grades. *(It happens to the best of us, consistently, so just plan for it and prevent a few months from being shaved off your life.)*

Coming up on the next four pages you'll find info on making 2-color transcripts. Even if you have no plans to make one, you might want to read pg. 30 anyway for other insights and more attempts at humor, especially under "If You End Up Using This Transcript."

§ OFFICIAL TRANSCRIPT §

« STUDENT DATA »

NAME: Last, First, Middle
ADDRESS
CITY/ST/ZIP
PARENTS

BIRTHDATE SEX [M / F]

« INSTITUTION DATA »

NAME:
ADDRESS:
CITY/ST/ZIP:
TELEPHONE

ACADEMIC RECORD

COURSE/CLASS TITLE	Date Completed	Credits	Grade	G.P.A.

GRADE POINT EQUIVALENTS

A - 4.0 A- - 3.7 B+ - 3.3 B - 3.0 B- - 2.7 C+ - 2.3 C - 2.0 C- - 1.7

(Grades D through F not awarded)

CUMULATIVE SUMMARY

TOTAL CREDITS:	CUMUL. G.P.A.:
____.____	____.____

This transcript is official when the principal's signature and the date are in blue and a portion of each is sealed by a notary public embossing.

Signature of Principal: ______________________

Date ______________________

One Piece of a Two-Color Transcript ☞

ADDED VALIDITY

To add a bit more "validity" to your transcript and to make it *really* "official," you might want to print it in two colors. I have been told that this is how they're able to tell whether or not the transcript is a photocopy. Now it may be that not all schools know or go by this. So it's not a guarantee. Plus, with computer technology so highly developed now, I'm sure this is not as important anymore.

"*What's the big deal about copies?*" you understandably ask. The institution wants assurance that the transcript has not been altered from its original state. This is why a student cannot obtain a transcript from a school office and take it to the next school that's requesting it; it *has* to be sent directly from the previous school to the next school with no intermediary.

The reason a just-black-and-white transcript may not be accepted as "official" is that it can be retouched (in particular, by a student who doesn't like a particular grade he got), and then easily copied again without the change being noticed. So if the transcript is done in two colors, it *couldn't* be a photocopy. (Actually, with today's technology, it *could.* I imagine that's why an embossed seal and a *blue* signature was required on my college transcript. But neither of those was required on the transcript I obtained from a Washington State high school.)

IT'S NOT "ALL THERE"!?!

No, the sheet you see on the previous and opposite page *isn't* a printing malfunction; what I'm giving here is just separate *portions* of the transcript. See the *following* page for the *rest* of it. This page and the next are like the two puzzle pieces of a 2-piece puzzle. In fact, if you place one over the other and hold them up to the light, they'll look like one. Be sure to use scratch paper for your trial-and-errors! *(We're still in the drafting stage.)* Don't run your good paper through the copy machine until you have closely examined the "final copy"!

Being aware that every copy machine is different will help you not go crazy. One machine prints on the underside; another on the top side. When your copy comes through, you'll have to check it out to decide if you put your first one through with the correct *side* and the correct *end* first. Keep running 'em through till you figure it out. Or if this is just too much for you, ask the clerk help you figure out which way to feed it in so that it all comes out right. But often they don't know the machine either, so be prepared to experiment.

DECIDING ON COLOR

The next step is to decide which part of the transcript you want to have in which color. Again, for those who prefer to not have to make such a weighty decision, I do have a suggestion: print the section with the heading in **black** because black is more prominent, and then (obviously) print the next page in your other color. But, if you're like me, you'll do whatever you want.

Once you're ready to copy it, place one of the two originals face down on the glass, taking note of which direction you are placing it (so you can place the *other* original in the same direction when its turn comes.) Run a few copies on *plain* paper. *(We're still in the drafting stage.)* Yes, a *few,* like maybe 6 or 7, even though you need only *one* in the end, because you'll need several extras for trials. You never know till you *try* which direction they'll come out, unless you're fortunate enough to have a clerk who really knows the machine.

OK, so your (let's say) black copy comes through. Now you put the *other* original face down on the glass in the same direction as the first one was laying, and run that (let's say) red copy through. This is where the trial-and-error part comes in. Every machine is different. When your copy comes through, you'll have to check it out to decide if you put your first one through with the correct side up and the correct end first. Keep running 'em through till you figure it out.

How will you know when it comes out "right"! Because everything will match up! Both originals will be on one page and will *look* like they were printed as one – providing the guides on the hand-feeder were tight and the paper went through straight. Again, *only* when you have it exactly how you want it do you use your expensive paper.

Since I have some space here, thought I'd include an excerpt from a note I received from a homeschool mom:

> "While reading the articles in the Guide-a-log, I found myself nodding my head in agreement and applauding you for speaking out on issues you feel so strongly about. I also share many of the same convictions and want you to know that you have touched my heart with your message and given me the courage to seek God's <u>best</u> for our family."

« STUDENT DATA »	« INSTITUTION DATA »
NAME: Last, First, Middle ADDRESS CITY/ST/ZIP PARENTS BIRTHDATE SEX [**M** / **F**]	NAME: ADDRESS: CITY/ST/ZIP: TELEPHONE

ACADEMIC RECORD

COURSE/CLASS TITLE	Date Completed	Credits	Grade	G.P.A.

GRADE POINT EQUIVALENTS	CUMULATIVE SUMMARY	
A = 4.0 A- = 3.7 B+ = 3.3 B = 3.0 B- = 2.7 C+ = 2.3 C = 2.0 C- = 1.7 *(Grades D through F not awarded)*	TOTAL CREDITS: _____._____	CUMUL. G.P.A.: _____._______

<u>Other</u> Piece of a Two-Color Transcript ☞

MAKING A LASTING *IMPRESSION*

One other "officializer" is to purchase a rubber stamp that says "OFFICIAL TRANSCRIPT." These cost only about $10 to $20 and could even be shared by several homeschooling families or a whole support group.

You might also want to purchase an embosser, the round type, with your school's name that you impress onto your transcript. These cost about $20 to $25. I got ours from "Artistic Greetings" / P.O. Box 1623 / Elmira, NY 14902. Write and ask for their catalog called "The Personal Touch" or for a flyer with the embosser.

WHY GO TO ALL THIS TROUBLE?

So why would any intelligent homeschooler, parent or student, knowingly put themselves through all the previously described trauma? Because it might someday prevent an even *worse* trauma – like not being accepted for a scholarship, or at a college that one had one's heart set on, certain that God was directing them in that college's direction.

THE VALIDITY QUESTION

I know exactly what you're thinking. That this *still* isn't a "valid" transcript. Right? That we're just "faking it" and really don't have any business trying to produce a "valid" looking transcript for an "invalid" school – like your *homeschool*! Right? ... *All right, enough is enough...*

STOP RIGHT THERE!

We need to *talk!!* ... Why on earth would you think your homeschool is *invalid!?!* (Sorry, I tend to get a little agitated on this subject.) Listen to these definitions of the word ***valid:*** *"founded on truth or fact; sound; fit; healthy; genuine; capable of being justified or defended."* If *God* gave you that child and called you to educate him/her, what *earthly* defense or explanation would or *could* make it any more *valid* than *that!?* – assuming, of course, that you are trying to follow Him in doing so. If you link up with the Almighty Teacher, Counselor, and Helper, your kids are going to get an education that is *at least* as "valid" as anything a traditional school could provide, if not a whole lot *more* "valid" in the *true* sense of the word. So why should we feel guilty making it *appear* valid? What I'm doing here is simply enabling you to make your transcript "do justice" to your child's *excellent* education *in form and appearance!* We're just learning how to speak their language, "Acedemese," and translate our 'stuff" into that language.

IF YOU END UP USING THIS TRANSCRIPT

If anyone encounters any successes or failures with this, I'd love to hear from you. Well, I'd *love* to hear about the *successes;* any *problems* I could live happily without, but probably *ought* to be made aware of. *(Just try to not hit me up with any <u>major</u> blunders or omissions on "Day 27," OK? I might crack. For an update of my current monthly cycle dates, send an S.A.S.E. to— juuuust kidding!!!* ☺ *This book <u>was</u> written for women, you know!)*

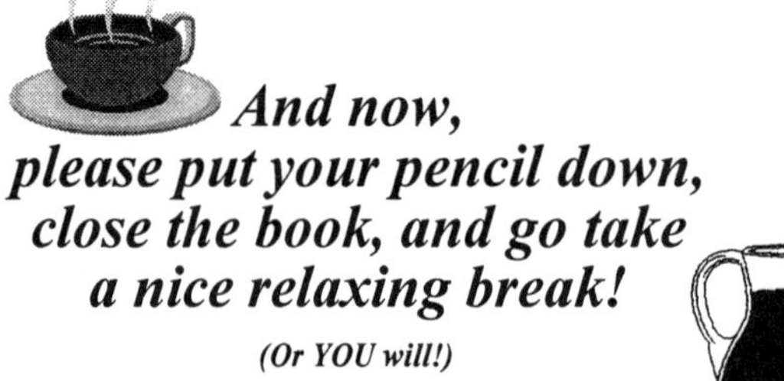

And now,
please put your pencil down,
close the book, and go take
a nice relaxing break!

(Or YOU will!)

Just one more form to go,
and I want you to be fresh 'n' ready for it!

Official Transcript

§ §

This transcript is official when the principal's signature and the date are in blue and a portion of each is sealed by a notary public embossing.

Signature of Principal: ______________________

Date ____________________

Directions for

☞

"Credit Completion Bar Graph"

1) COUNT COLUMNS: First you need to get out your Master Framework Plan (which you need to *do* first!), and count the number of classes and credits you have. If a class, like English, will be 4 credits, then you'll need 4 columns for all that. If a smaller class, like (for us) Health or Current Events, has only ¼ or ½ credit, it still will need one whole column. Be sure to use the right one; I have them here in 120 and 150 hours/credit.

2) FIGURE NUMBER OF PAGES: Now figure how many pages of the Credit Completion Bar Graph you're going to need, considering there are 10 columns per page. You'll be connecting them together – unless you prefer to keep them separate in a notebook. We liked having *one* "Total Picture" to look at.

3) COPY THE PAGES that you'll need. You may want to cover up the title on the second (and third?) pages if you want just the one title to be at the far left. Doesn't matter. You may also want to cut off the excess columns.

4) CUT THE FIRST ADD-ON PAGE (one with no title) along the left edge-line of the chart, cutting *right* next to the line. You may or may not want to include the "Accumulated Hours" Column; it's a whole page away from that first "Accumulated Hours column on the main sheet, so you may want it there so your finger doesn't have so far to travel to refer back to it, but if you prefer continuous columns, just cut it off, cutting along the left single line (between the double line) of the first "__ **Credit** ○" column. I find it easiest to turn the chart up-side-down to cut; then the line I'm cutting along is to the left of my scissors. (I'm right-handed.)

5) Lay this just-cut add-on sheet over the right end of the main sheet, lining up the lines evenly. Place 3 small post-a-notes at the top, middle, and bottom of the "seam" to hold both sheets in place. Do *NOT* tape the front!!!

6) Now turn the whole double sheet over and tape along the whole back seam, allowing just enough extra tape, about ¼" at the top and bottom ends, to bring the tape around from the back and fold over to the front, which will keep it secured better.

7) Write all your subjects in the boxes at the bottom, and below that, the courses along the courses line, adding vertical lines between courses to divide them from each other.

8) Across the top, write the number of required credits for each subject (¼ to 1 credit). Put a checkmark in the little circle (to the right of "Credit") once it's completed.

9) Draw a thick, black line along the dotted line that corresponds with the number of hours required for each class. I like to make a little archy type thing as shown.

10) As accumulating hours are computed with each tally you do, fill (color) in the columns from the bottom up, with a highligher. Mine are in 5-hour increments, but you can estimate between.

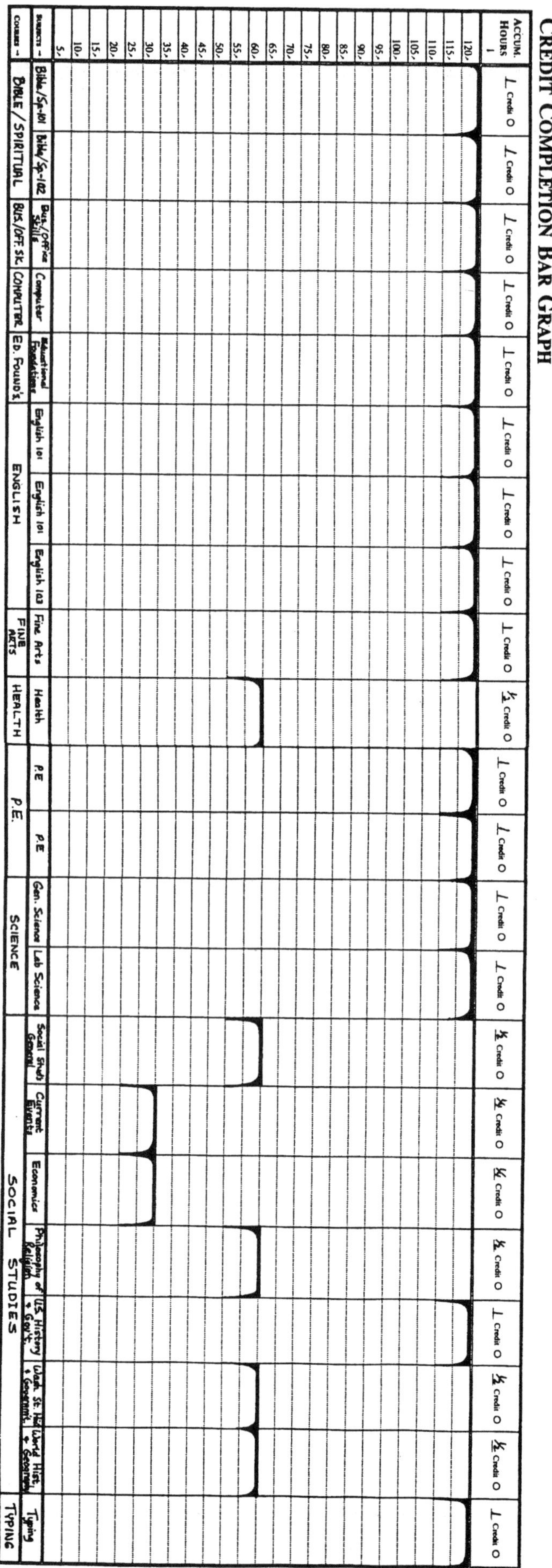

Credit Completion Bar Graph

Accum. Hours ↓	___ Credit ○	___ Credit ○	___ Credit ○	___ Credit ○	___ Credit ○	___ Credit ○	___ Credit ○	___ Credit ○	___ Credit ○	___ Credit ○
150										
145										
140										
135										
130										
125										
120										
115										
110										
105										
100										
95										
90										
85										
80										
75										
70										
65										
60										
55										
50										
45										
40										
35										
30										
25										
20										
15										
10										
5										
Subjects →										
Courses →										

CREDIT COMPLETION BAR GRAPH

ACCUM. HOURS↓	___ Credit ○	___ Credit ○	___ Credit ○	___ Credit ○	___ Credit ○	___ Credit ○	___ Credit ○	___ Credit ○	___ Credit ○	___ Credit ○
120↗										
115↗										
110↗										
105↗										
100↗										
95↗										
90↗										
85↗										
80↗										
75↗										
70↗										
65↗										
60↗										
55↗										
50↗										
45↗										
40↗										
35↗										
30↗										
25↗										
20↗										
15↗										
10↗										
5↗										
SUBJECTS →										
COURSES →										

TWO BONUS FORMS!

JUNIOR HIGH WEEKLY PLAN

Here are two more "Daily Plans" (a.k.a. "WAP's") that I'm adding to the "form smorgasbord" for you to choose from. I came up with these "after the fact" – *after* the rest of the system in this book was completed. They are what I created for Junior High (for Carlianne) and Post-Highschool (for Tory), but they would actually work just fine for High School, so you might want to carefully consider these as you decide what to actually use; maybe even a combination.

I wanted something different for Carlianne for a weekly plan when she reached junior high age. Having just recently completed Tory's plan on the opposite page, I just used components from that *(the top chart)*as well as from the regular high school WAP *(earlier in this section)* and came up with this! Since you would use neither of them as masters, but rather only for ideas, I am giving only reduced samples here.

"Spiritual" is her quiet time and anything she and I read or discuss of a spiritual nature; "Math" is her workbook; "Writing" is any writing she does on any of her academics or "productive time"; "Typing" is the "Mavis Beacon Teaches Typing" program; "History" right now is the Little House books and a biography we're reading on Mother Teresa; "Reading on Own" is any books she's reading, which right now are the Chronicles of Narnia, *I Kissed Dating Good-bye*, *Romance God's Way,* and the "Christy" Books. Her options for "Productive Time" (for the afternoon) are listed there.

CARLIANNE'S WEEKLY PLAN

♦ DAILY DISCIPLINES/ACTIVITIES ♦

TASKS / ACTIVITIES	MON	TUE	WED	THU	FRI	SAT
Regular Home Jobs *(on Job List)*						
Extra Job(s) *(as/if needed)*						
Homeschool with Mom *(See below)*						
Homeschool on Own *(See below)*						
Reading on Own *(1 hour per day)*						
P.E. *(Bike, Trampoline, Walk, Exercise, etc.)*						

STUDIES ↓	MONDAY	TUESDAY	WEDNESDAY	THURSDAY	FRIDAY	WEEKEND
SPIRITUAL						
MATH						
WRITING						
TYPING						
SCIENCE						
HISTORY						
READING ON OWN						
"PRODUCTIVE TIME" OPTIONS: Baking, Cooking, Tole Painting, Sewing, Choreograph, Art or Craft Project, Work with Dad or Mom on a project, Gardening, Work on Booklet(s), Piano, Write a letter, card, or note, Housework, Babysitting, Reading (approved), Educational Videos (limited)						
YOUR CHOICE FROM ABOVE OPTIONS						

College-Age Weekly Plan

I needed a new form for Tory's "season of in-depth study and discipleship" since he would be operating completely differently from how we had done things in high school. I wanted him to have more control over his educational choices and his use of time than he did before, so I needed a form that would reflect and accommodate that.

There are two main areas: his "Daily Disciplines / Activities" which are just the things he regularly does, and that don't "count" toward his 25 hours a week of study, which is reflected in the chart directly beneath that called: "Study / Academic Work." There are 25 boxes (for the 25 hours) which Tory is free to arrange how ever he sees fit, according to his interest(s) and how he thinks he can best accomplish his goals. A few of the resources are ones that he and I together agreed he needs to have done by the end of the year; others are ones he himself chose for his own reasons. (We later went down to 20 hours of study a week because of his work schedule – he's also working to save up for college next year. 25 hours just felt too heavy to him, so he made a "godly appeal" to me to lower it, to which I acquiesced. ☺) At first I made up each week's sheet *with* Tory: me at the computer, us *together* deciding how much of what he would do for the coming week. But I wanted to eventually get him to the place where he did it all, which is how it is now, and has been for some time. Tory sets this up each week, deciding new studies to start when he has completed one, and how much of each..

Tory's Weekly Plan

♦ Daily Disciplines/Activities ♦						
Tasks / Activities	**Mon**	**Tue**	**Wed**	**Thu**	**Fri**	**Sat**
Q.T.						
Home Jobs *(on Job List)*						
Extra Job(s) *(as/if needed)*						
Study/Academic Work *(See below)*						
P.E.						
Youth Group / Church / Prayer Time						
KPD job						
Parks Dept. job						

♦ Study/Academic Work ♦	
Specific Resources / Studies	**Hours**
Y2K Research: (Book, Internet, etc.)	☐ ☐
Read *True for You, But Not for Me* By Paul Copan	☐ ☐ ☐
Read *A Ready Defense*	☐ ☐ ☐
Watch "Battle for the 21st Century" by Marshall Foster	☐ ☐ ☐
Attend "Strategies for Spiritual Warfare & Y2K" by Tom White	☐ ☐
Do "Mavis Beacon Teaches Typing"	☐ ☐
Prayer Meeting with Fathers' House	☐ ☐
Read *Understanding the Times*	☐ ☐ ☐
Do *Experiencing God* study	☐ ☐ ☐
Work on Word Study on "Ministry"	☐ ☐
OTHER CHOICES THAT MIGHT COME UP:	☐ ☐
	☐ ☐

Possibilities to be thinking about for next week or later:

- *Dedication and Leadership* by Douglas Hyde (Ex-Communist)
- "World Changers" Video-seminar by Dr. Jeff Myers
- "American Covenant" Audio-seminar by Marshall Foster
- "Battleground for the 21st Century" Video-seminar
- *Constitutional Law for Christian Students* by Michael Farris
- "Countdown to the Year 2000" Video-seminar

Section Four:

Potpourri of Curriculum Supplements

A smorgasbord of
non-traditional "Learning Guides";
to help your student glean more from a
variety of experiences, and for you to
give specific guidelines for
learning assignments.

Let's start with a couple Q's and A's relating to this section...

Q: My daughter is an avid reader...2+ books per day. She reads a lot of Nancy Drew, Trailblazer books, Mandie Books, and classics not included in the FAR (Far Above Rubies) curriculum, which we are currently using, and these books aren't listed as ones to be read with the curriculum. So do I count them?

A: There are a lot of "classics," so just because a book is not included in the FAR (or any other particular) curriculum doesn't mean it isn't a classic. But even if – or just because – "someone else" (even a college English prof or a knowledgeable librarian) considers a book to be a "classic," that doesn't mean I would necessarily approve it for my kids. Certain books, classics or not, I would not consider to be worth our time. They may not be "bad" or what you might consider harmful – or they *may be* – but I want to make the most of (redeem) my student's time for these years, not just read every classic just because the world considers it a classic. Kids graduating from high school will not – at least in the school setting – have read one book by Max Lucado, Elisabeth Elliot, Larry Crabb, Dr. James Dobson, Frank Peretti, Oswald Chambers, Dr. and Dorothy Moore, Gregg or Joshua Harris, Jean Lush, just to name a very few of the authors whose work has been life-changing for me and my kids! Am I making my point? ☺ Though they may seem many, there is only a certain number of hours (whether you're into counting them or not) that your student has for high school. Our highest priority has to be to ***redeem the time!*** Get out of it what *God* wants you to! And to do that, what do we have to do?! ... ***Hear God!*** ☺

~~~

Q: I am keeping track of her "extra" reading, as outlined in Barb's book, and she has read over 2,100 hours this school year, and shows no signs of slowing down. I'd like to give her school credit for this accomplishment, but was told that since it's not "educational" she shouldn't receive credit. Please offer your advice on this!

A: This *is* an accomplishment, especially where she's come from, and I would have no problem giving at least *some* credit. But I need to put it into the perspective I would personally view it from... *(Ok, ok: "...the perspective from which I would personally view it"! ☺ To be honest, I'm much more comfortable with those "dangling prepositions"!)* .... First, most of the credit I'd give her for that reading would not be for one of the "main" English credits, but rather for an elective. A *few* of the books she read could certainly go toward a credit in English Literature, specifically those books with "literary value." This is where it gets subjective. ☺ What exactly IS "approvable"? "valuable"?... For me, it has to have some sort of "take-away" value. This means that when I'm done reading the book, *something* about or within me should be affected – for the better, of course. To determine and document this very thing, I made up a form called a"Book Synopsis, which is in this section under "English." It includes the following areas of value, which the student evaluates after reading a synopsis or a book: Moral value, Historical value, Spiritual value, Human Understanding value (which means it helps you see human nature more clearly, for better or for worse), Scientific value, Life-affecting value, Tear-jerking value. No book will have all of these – at least I have yet to find one! And no book *needs* to! This list simply gives you *some* sort of criteria upon which to determine if a book is worth reading. What is determined as being "of the highest value" will obviously vary from person to person.

I personally think it's a good idea to mix 'n' match the types of books read, and the various types of values. I have learned much from a variety of books – not only ones by Christian authors. With the Holy Spirit as your teacher, speaking to you between the lines, many a book, Christian or not, can be of value. However, many books are purely for entertainment, and these I would not want to comprise the "core content" of an English Lit credit, so if all it has to offer is "Entertaining value" it will probably not make it onto my approved reading list for that class. However, they *could* be applied toward an elective. I would hope there would be books with *some* solid value besides entertainment in there though. Each parent has to determine their own criteria. ... Sharnessa accrued hundreds of hours of dance lessons, choreography, practice, and performances, but I still had other requirements I wanted her to experience or do for the "core content" of her *main* "Fine Arts" credit. She ended up with four Fine Arts credits just in dance-related areas alone! But her main Fine Arts credit was comprised of several different areas, including art appreciation; art and craft skills; dance theory; music appreciation; drama appreciation; choir; and (some) dance application, performances & choreography. Make sense? ☺

Okay, let's get on with the section!
~~~

POTPOURRI OF CURRICULUM SUPPLEMENTS

CREATIVE SUPPLEMENTS & RESOURCES TO YOUR COURSE CONTENT

If you get overwhelmed looking at this section, then you're looking at it wrong! When you go to a buffet-style restaurant, do you get overwhelmed? No, just look around at all your choices and say "Wow! I've got a lot of *choices! How FUN!"* That's just exactly how you need to view these "Learning Guides"! "Wow, I've got a lot of *choices*! How *FUN!"* In fact, I think it's even more helpful to think of them as I do my memory of the "telephone man." Do you remember, *before* the days of microchips, when a telephone man would climb up a telephone pole with this super-equipped belt full of every possible tool you could think of?!?! Whatever that guy needed when he got to the top of that pole, he needed to have *right there* at the moment he needed it. So he had everything he *might* need right there, handy, somewhere around his waist. Likewise, these have several purposes:

- to enhance, reinforce, and complement certain aspects of many things you're probably already doing
- to show you how to take a chunk of knowledge or a skill you want your child to know, and lead him into learning it for himself rather than spoon-feeding it to him
- to open up your student to a more *productive* method of learning as differentiated from a *consumptive* form of learning. Hint: workbooks are "consumptive." The student reads pre-digested material and spouts it back. There *is* a place for this type of learning, but in my book, its place is about the size of a thimble. "Productive" learning, on the other hand, involves more searching, more "real life," more thinking, and more fun.
- to make better use of the opportunities around us, in our home, our community, our state.
- to document that learning is really and truly taking place, which may or may not be useful to you later on – but I'd rather be safe than sorry. Don't get the wrong idea; not everything we do is documented! But by documenting unconventional learning experiences, we have a valid means of including and presenting them as part of our credit hours.
- to help the student learn more from an experience or book. For instance, instead of just watching a movie on the Civil War, the Movie Critique/Analysis shows how to critique and analyze it.

FLEXIBILITY IN USE OF MATERIAL

As you look through the Learning Guides in this section, you'll probably find some that you would have categorized under a different subject. We're getting into an area of flexibility here that you can use to your advantage – if you don't let it overwhelm you!

So let's take a close look at what you can do with these materials and hours. For example, it would have been easy in our Washington State History class to go far beyond the "required" ½ credit (60 or 75 hours, depending on how many hours you count for one credit) and, even then have been nowhere near ready to stop! So we could have funneled several hours out of that class into others classes, depending on the activity. Parts of *any* State History class can actually be considered U.S. and/or World History. You'll know it when you come to it.

Train yourself to tune in to specifics of what you're reading or hearing about, like: Are we talking about something that has to do with what was going on in the U.S. primarily? Or was it strictly a matter pertaining to our state? Then again, even if it is just your state, your state is still part of the United States! This is part of the "dissection" process that we'll be talking about in Section 6-B on Designing Your Classes. In any of this, just estimate hours and put them under the appropriate subject.

Another example; the "Religions of the World" study could be counted toward hours for Social Studies – specifically Philosophy of Religion, taught with God's "bias"! And/or it could be counted as Spiritual Studies. The "Open Line 'n' Mind" exercise would qualify as Spiritual Studies and/or as Life Skills under Building Family Relationships. Or let's say you make a miniature replica of a fort. That could go in either Art or History – or both. (More on that later!) Or you could write a report on the weather zones and ecology of the state and let that count for English and/or Science. (The study of weather is "Meteorology.")

The Tel-Med report (under Health) is a very versatile "funneler." It is laid out with four categorical options so that it can be counted toward a variety of these categories. None of the little lessons (pre-recorded on tape) will usually qualify under all four categories, but several will count for at least two or three. I'd compile all Tel-Med reports into one notebook and not actually count them toward anything until later on. Just keep compiling those minutes and hours, as you're doing in the other subjects. These will make good fillers for whatever subject is low on hours once you go to do your tallying. For instance, if you are having an easy time filling up hours for Science, but are short on Health hours, you'll want to funnel those Tel-Med reports that qualify for *both* of those subjects into Health rather than Science. Getting it? ☺

THE INTIMIDOMETER

There are *many* places along the road to homeschooling through high school that you are going to feel intimidation rising up in you. Thinking about college, diplomas, graduation, creating your own classes, setting up your framework, designing a transcript and/or portfolio can surely do it. Or telling your friend, parent or relative who's worked – or had their kids – in the schools for years and is as "pro-*that*-choice as one can get" that you are *(so audacious as to actually be thinking about)* homeschooling through high school is sure to get your blood pressure rising too. But I would imagine that one of the *worst* culprits is (ready?!) *Scope and Sequences!* Well, before you even *think* of looking at one of those, your confidence level needs to be on the *high* end. So, just to check that out, before we go on, let's take a look at the following "In timid-o-meter" (which is pronounced "in-tím-i-**dáhm**-uh-ter," by the way). This is a quick "self-test" to determine how high or low your confidence level is! Simply circle the number that you think accurately reflects where you are at in confidence.

Important note: this number *will* be changing (typically going *higher)* as you *learn* more, *realize* more, *recognize* and *dispel* more "lies" and "myths," get more *set free*, and above all, as you *continue to seek God* as to what you should do for high school. In fact, without you even knowing it, I bet you've risen at least two notches since you first began reading this manual – that is, ***if*** indeed this manual was pretty much your first exposure to high school homeschooling. ☺

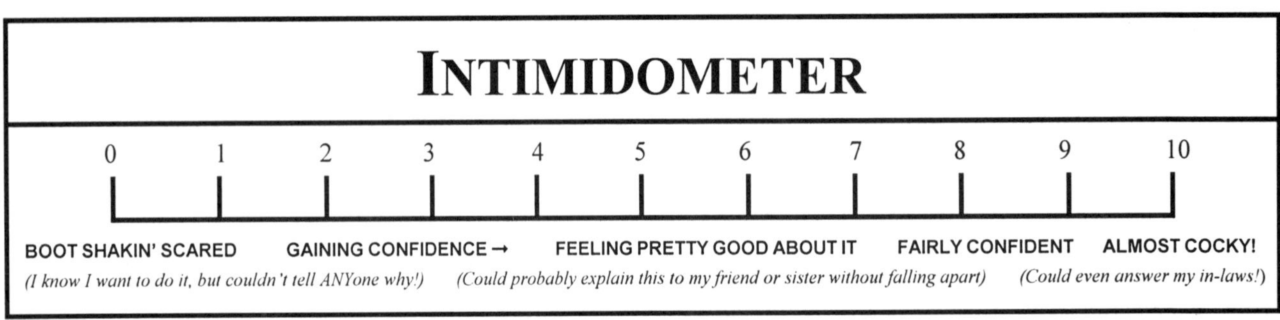

OTHER HIGH SCHOOL HELPS

There is another resource you may or may not want to obtain. It depends totally upon how your Confidence Level is doing. If it's up to at least a 7 or 8, preferably 9, then you are free to delve into this other resource option that I mentioned a moment ago, and now want to explain.

Some curriculum companies have what they call "scope and sequences"; your school district has something similar that they probably call "Student Learning Objectives," or "SLO's" for short. Or they might also be called "Instructional Objectives." These *tools* can be helpful as a source of ideas, but, one word of warning: they can also be very overwhelming and intimidating if you're not prepared! If your Confidence Level isn't at least a "7", do NOT even look at any of these! You'll have your child in regular school within 24 hours! So for you "7's" and up, ask the Holy Spirit, (the all-wise and all-knowing School Counselor) to let you know if you should even crack open the pages of one of these. If He says it's OK for you, then ask Him to be your filter as you open up these potential home(school)-wreckers, and pray for discernment. In this frame of mind and heart, and *only* in this frame of mind and heart, will you be able to glean some helpful ideas.

There are some other excellent non-intimidating resources that you can consult for direction. I mention them in the Resource (etc.) Section under "High School Related."

One thing that might help raise your Confidence Level when looking through a scope and sequence or SLO's is to keep in mind how little of the stuff in there actually sticks in the students' minds. The plan looks wonderfully impressive on paper, but because so much of it is not relevant to the student or his real life, and because the kids are overdosed with one subject, they simply don't remember even a fraction of the stuff that's poured into their heads! In fact, perhaps you have heard of . . .

"SHELTON'S THEORY OF BRAIN RECEPTIVITY"

I have developed this theory that the standard "receptivity level" of your basic, typical student's brain can be likened to an 8-ounce (1 cup) measuring cup. This is just the standard size, by the way. A person who's *really* smart (academic-wise) might be a 1½-cup measure. I generally think of myself at about ⅞, just to give you an idea of the variables. ☺ The schools, in the process of making a good, well-meaning attempt at doing a quality job, pour one gallon of "stuff" into this one-cup measure. How much does the student retain? One cup! They can pour all they want, but a one-cup brain can hold only *how much?!?*.....ONE CUP!! Or less, of course, ☺ but not *more!*

Lest you think me leaning toward the insane (not cocky; *kooky!),* I *do* have scientific proof of this! If you were to take a little scraping off any classroom floor and examine it under a microscope, you'd be able to see some actual cells of this "data overflow" phenomenon. There are *millions*, yes, even *billions* of them all over America's classrooms. Dried up of course, so no one would ever slip and fall on them. But many homeschoolers slip and fall thinking that they have to pour the same amount of data into their kids! Pour all you want, but a person will remember and utilize only so much.

(Oh – uhhh, by the way please don't mention this little scientific experiment-proof stuff to any scientist-type people you might know. Not many scientists know about it and I –well, I wouldn't want to trouble them.)

Now I do have an addendum to this theory: if a student is allowed the freedom to learn on his own while also being given relevant (because he is using it), interesting material, then that standard one-cup capacity naturally and painlessly expands! And at no added expense to you!

But no pre-packaged, pre-digested educational program, no matter how brilliantly prepared, no matter how scope-and-sequence aligned, no matter how "expert-approved," and no matter how full-color illustration-laden, will ever be able to add more than a drop or two to the "receptivity level."

I got a kick out of the Far Side cartoon showing a classroom of students, with one boy sitting in his desk with his hand raised. The caption reads: " Teacher, may I be excused now? My brain is full!"

Little did Gary Larsen know he was illustrating the "Shelton's Theory of Brain Receptivity" (which, believe it or not, I came up with a few years before he came out with that!)

A WORD OF PREPARATION FOR THESE LEARNING GUIDES

If you were to do surgery on my brain, you would be in for a little surprise. Whereas most people have nice flowing lumps, gentle folds and curvy crevices on the surfaces of their brains, mine is embedded with lines and columns. If you look closely, you can see actual chart templates furrowed into the surface membrane. Therefore, it only stands to reason that I *think* in charts. I read a Scripture verse and turn it into a chart. I hear sermons, stories, word pictures, activities – in charts! Without even trying! Sometimes it's a little aggravating, but most of the time it's fun and/or useful.

Naturally, this has affected my children's learning – since I'm a part of it – because it has affected my style of teaching, or should I say, my style of assisting their learning, and if you read this book for very long, it will inevitably affect *you* too!

The reason I have come up with worksheets as structured as these are is that there are certain areas of knowledge I want my kids to possess when they leave home. Using Learning Guides is an effective method of getting them to go searching for particular information without my hand feeding it to them. Plus, when we *write* in addition to hearing or reading, we retain more.

At first glance, these sheets appear to be typical workbook style pages. But if you look again, you'll see an entirely different focus. For the worksheets presented in this section, the student must personally go *somewhere* – and not to the previous chapter – and do the searching themselves for whatever information is requested here. No nice, neat, prepared workbook to glance back to, find key words in, and then fill in the blanks.

Now, don't get me wrong, it's not that I'm "down" on workbooks – OK, so maybe I am, and maybe more than just a little – but what I'm *mostly* down on is kids not having the fun or the education of searching out knowledge for themselves! These sheets are designed to get them thinking! *And* planning, *and* creating, *and* wondering about things, *and* looking up stuff! There's only so much time for high school. Make the most of it.

My ideas will probably even give you ideas! Is there something you'd like to get your daughter pondering or thinking about? – or your son's brain gears churning on, other than food? But simply saying "Go do a report on such-and-such" would be either too boring or too broad: "ARRRGGGHHH! Where do I start ?!" – and therefore too overwhelming. At least it sure was for me! So, at least in part, I'm creating some aspects of education that would have fit *me* better! ☺

TAKE THESE A STEP FURTHER

Many of these sheets could even be taken a step further to provide a "higher skill" opportunity, which you'll be reading about a little more in the English section of this chapter. It's one thing to *collect* information; it's a little more difficult to have to *think* about it as it's collected; and it's more difficult yet to *process* and *assemble* it into a more readable, understandable form of communication.

The information gathered on the Concert/Performance Record, for instance, could be reworded into full sentences, and these restructured into paragraphs to create an essay. Same thing with the Play Critique/Analysis, the Scripture

Verse Study, the Dog Report, and many others. But for students who are intimidated by the very mention of the word "essay," something like this is a perfect stepping stone.

NOTEBOOK METHOD

If you've done any reading on Lifestyle of Learning, you've run across the "Notebook Method." What I'm presenting here isn't the same as her method, but it's another way of approaching it in a more parent-directed way. I would never suggest replacing the other with this, I'd just use it as a supplement.

So what is this "method" I'm suggesting? Well, if you start using very many of the Learning Guides in this section, you'll probably get the idea before I even have to tell you.

Before very many weeks go by, as your student starts doing some of these Learning Guides, you'll find yourself with a growing pile of papers. Oh great, just what you need, right?! But are they really just *papers?* Look again! These are really the pages of some wonderful notebooks that your student will have researched and written himself! The question now becomes "So what do we do with them?"

ORGANIZING THEM!

I suggest starting with a few ½-inch to 1-inch thick 3-ring notebooks. If you get the Clear-vue type of cover, the student can make some interesting covers that will not only stay on, they'll stay protected. How many notebooks you start with and for which subjects depends upon your priorities and interests. The basics you'll probably want to choose from are:

- **BIBLE** - with dividers for Word Alive Studies, Scripture Verse Studies, Wisdom's 7 Pillars, Topical Studies (such as "Armor of God," "God's Actions"), Christian Denomination Studies, Name Study, and any other areas that either you or your student think up or have a passion for or personal interest in *(The sky's the limit!)*

- **SCIENCE** - with dividers for areas of study such as Chemistry, Botany, Physics, etc., and then possibly a Tel-Med section

- **HISTORY** - with dividers for each major area of study, like (Your Own) State History, U.S. History, World History, Current Events, Major Wars, and whatever other areas you'd want in there

- **FINE ARTS** - with divisions for Concert/Performance Records, Play and Movie Critique/Analyses, and Music Appreciation Worksheets

- **HOME ECONOMICS** - with dividers for areas such as Home Management, Organization, Decorating, Repairs, Remodeling, etc. Plus you might want a notebook for English with dividers for Speech, Lit, Comp, Grammar

BUT DON'T GET CARRIED AWAY!

As with anything that looks fun or interesting, it's easy to get carried away with it. The affinity of personal style that drew you to this book might be the very thing that could easily drive you to reduce all your child's learning into pages similar to these presented here. You'll not only sour his enjoyment of this type of learning; you'll also be depriving him of other types of learning experience. (See *Wisdom's Way of Learning* – "the book" that explains the *Lifestyle of Learning* message. There's so much more depth to education than what I present in this section, even though for many who have been bound by the text/workbook approach, my stuff is very freeing. This is just one small aspect; *Wisdom's Way* will help you *see* your child and their education very differently! Please don't mistakenly think that my stuff is "enough"! There's so much more freedom and understanding to be gained – which will *only* be to your and your child's tremendous advantage.)

JUST <u>ONE</u> AVENUE!

Just so you know, everything my kids learn is not delivered by this method of using Learning Guides! They also do lots of reading – on their own and with me – and have a wide variety of other experiences and projects. But this is *one* tool that works well for certain types of learning skills and experiences; ones that – well, just take a gander through this section, and you'll see!

COMING RIGHT UP!

In the previous editions of the book, all the Learning Guides were grouped together by and under their subject, with the subject heading above the first explanation for each section. And I originally had a separate page for each explanation so that it could be conveniently located directly across from the page it was explaining. Nice idea, but, for one thing, *many* of the Learning Guides could be categorized under more than one subject, so you didn't necessarily know which subject to look under!

Also, in this new revision I have *already* gone *way* over on my number of pages, even *before* starting to revise this section! This is because I've already added in all of the "Completed Class Plans" section, and that section *alone* is 80+ pages! So to keep the number of pages down to a minimum, I've grouped all the explanations together, in alphabetical order, and all the Learning Guides one after another, in alphabetical order. I hope you find this to be more convenient than the previous arrangement. The explanations begin on the following page, and all the Learning Guides will directly follow. ***So...***

Enjoy your stroll through

"Learning Guide Lane"!

LEARNING GUIDE EXPLANATIONS

AMERICA THE BEAUTIFUL VOCABULARY WORKSHEET (IV-27)

This is one song that, even though I have sung the words over and over, I have never *really* understood what they mean! And I knew that if I didn't, my kids wouldn't either. So guess what! PRESTO! Another Learning Guide sheet! ☺ Please note that this one is 2 pages. All directions are on the sheet.

ARMOR OF GOD (IV-29)

This is just one example of how you can study a section of scripture in a manner that takes you much deeper than just reading it. Using your trusty dictionary, you, parent *or* student, can dig into the Word and have yourself a tasty and satisfying "meal"! *(Actual photo of Tory using the Dictionary on page VI-D-70!)* There is probably not enough room in the boxes provided here to really do the study justice, but it's a good place for the beginner to start; the more serious student can always make up a new chart, or just take notes in his journal. Notice that the "dig word" *(underlined on the "Armor of God" sheet)* is not a word I don't know the meaning of. I know perfectly well what *armor* is, as well as *power, stand, wickedness, evil, word,* etc. But whenever I take the time and trouble to go to the dictionary and look up the meaning of the word, God always opens up new meanings of the verse to me! Part of it is that when we hear and read the Bible over and over, the sound of the words becomes stagnant, or we become "numb" to it, only because we have a preconceived idea of what that combination of words means. Take, for instance, "give us this day our daily bread." This certainly has meaning for us, but repeating the same sequence of words seems, for me at least, to cause a "numbifying membrane" to form over the words, keeping them from penetrating into new areas of my understanding, and into areas of my life that need the life they can bring. *(I hope I'm making sense!)* So, just as an example, I'll dig into what I see as being two key words of that sentence and see what happens:

give = *"to make a present of, furnish, provide free of charge, put into the possession of"*

bread = *"one of the basic foods needed to maintain life." (Wow!) Let's go a step deeper and define food which is "material taken into an organism and used for growth, repair, and vital processes, and as a source of energy; something that nourishes, sustains and supplies"*

Does that open up "give us this day our daily bread" for you a bit?! It even expands my perception of how to pray! Which leads me to the last column, "My Prayer." What you write here is obviously going to be just a condensed version of what you could pray. So just let this be a prayer starter for you that springs from what you glean from the definition exercise.

BOOK REPORT (IV-30)

This is my generic form for a book report. Nothing very in-depth, but it's good if you want just an overall view of the book or tape, and what the student basically gleaned from it. It also gives *some* sort of documentation, or if for some reason you need it, "proof" that the student read the book.

BOOK SYNOPSIS (IV-31)

This form is for use with the ***Great American Bathroom Book***, which I (we) *love!* There are at least three volumes that I know of, and each contains a multitude of 1–2 page synopses of all the great literary works. One might be concerned that a student would be satisfied with reading just the short version of a story and not take the time to read the whole book. To that, I say: 1) This simply hasn't been the case with us. Rather the synopses either help the reader know he *doesn't* want to spend time reading the whole thing, *or* they whet the appetite of the reader for *more.* And 2) *NO* one has time to read all these books anyway! At least this way the student can be familiar with the "gist" of many more of the "great books" than he'd have time to read.

CHRISTIAN DENOMINATION STUDY WORKSHEET (IV-32)

This sheet can be used as either a research form (searching in books), or an interview guide. The idea is to interview a pastor or leader of a church and have them tell us their church's position on the matters on this sheet. I suggest that, unless you know the interviewee well, it's good for a parent to be there, just for "theological safety," if you catch my drift. ☺ I have found that in doing interviews, it's easiest on the interviewer and interviewee if the whole session is audio taped. This frees the interviewer to *listen* to the speaker and stay tuned to the conversation rather than have to spend the whole time trying to slow them down and repeat to get everything down they just said. It creates more of a flowing, friendly exchange rather than being chopped up with waits while the interviewer catches up on the writing. Views on many of these issues will widely vary from denomination to denomination. You might even wonder how they could all be considered the same faith of "Christianity"! Some of these issues will be the same for every church, if it is truly Christian. I included them anyway because I felt it was important to see the similarities between the churches, not just the differences.

CHRISTIAN RELIGION TYPES WORKSHEET (IV-33)

First, you'll need at least four copies (more if you wish) of this sheet. For this study, the student first takes their main page (the one that will be the final paper) and looks up each of the four "religion types." The definitions go in the first slot in each chart. Then he calls or gets together, individually, with a few (three or more) people he knows – people who have a somewhat mature level of spiritual understanding (not a new Christian), and interviews them on *their* perception of each of these four areas. The more of a variety of people you can get, the more fun and "effective" this will be. There are no "right or wrong" answers (for the most part), but the point of this is to see how similar and different people's concepts of these terms are! So they use one page per interviewee, writing down their answers. And then for the final paper, they make a composite picture (including the differences) of each of the four types. This has a lot to do with "stereotypes" and will help the student be more objective and less subjective in his own perspective of these! ☺

COMPUTER TERMS (IV-34)

On this sheet I have listed some basic terms that one starting out in computer will need to become acquainted with. Since I have worked only with an IBM compatible computer, some (many?) of these will not be applicable to someone using an Apple or Macintosh. Not a problem! Just run to the nearest index of your Mac operating manual and write up your own (superior, right?!) list.

There are a couple of ways to approach this. One is to have the student look up the words and define them before starting actual lessons with you (or your computer-tutor friend you will be paying in their favorite pie or cookies) at the computer. *Or*, if you have a student with a learning capacity similar to mine, the definition will mean nothing (zip, nada...) until they are *shown* to me and I have seen/done/messed it up at least 28 times.

If you're looking for a way to discourage your student from ever working with a computer for the remainder of their life, frequently use such statements as "I already told you that?" "You did what? *Again?!*" and "That's so simple, even a first-grader could get it!" Such statements are 100% effective.

CONCERT/PERFORMANCE RECORD (IV-35)

This sheet can be filled in at the performance itself, or they can do it afterwards if the student's memory is reliable. I recommended affixing it to a clipboard, but my kids didn't like how obvious that was. So they at least need to have a rough draft copy of it that they can fold up and pull out to write on. Either way, they will at least need to take some notes while there, or they will not be looking for what they need to be recording – which, by the way is designed to get them *noticing* the bigger picture at such an event. Ideally, they should look over this form right before the performance begins, and then Intermission is an opportune time to write their notes. This way enjoyment of the performance itself is not sacrificed, there's enough light to see to write, and memory is fresh. Note: Do not have them do one of these for *every* concert they attend or it will dull their enjoyment of it.

COOKING TERMS (IV-36)

I don't expect my kids to be professional cooks by the time they leave home, but I do want them to be familiar with the terms on these two sheets so that Betty Crocker will make sense to them. Actually, you'll be able to find most if not all of the explanations of these cooking terms right in the Betty Crocker cookbook. (Look in the index. Many even have pictorial directions – *my* kind of cookbook!)

There are two other reasons I want them to understand these terms: I don't want my grandchildren to have to eat soggy tater tots that were *boiled* instead of *broiled*. Also, I don't want my kids' future in-laws to have reason to say to their friends: "Can you believe it?! My daughter-in-law who was homeschooled doesn't even know what *julienne* means. I'm just *sure* it's because she was *homeschooled*, don't you think?!" Actually, I don't care one iota what people think, because no matter how much your kids know, or how excellent in character and spirit they are, if someone is biased against homeschooling, all that will get noticed is what they *don't* know. And I don't have time for that nonsense. I just want my kids to be familiar with these terms so that cooking won't be intimidating to them.

This is actually one of two sheets. The terms on the second page are: Poach, Broil, Mix/Stir (any difference?), Mince, Shred, Whip, Cut in, Fry, Crush, Sliver, Blend, Toast, Sauté, Melt, Double-boil(er), and Slice.

CREATION SCIENCE RESEARCH REPORT (IV-37)

The most obvious class for this to go under is "Science," but since this Learning Guide uses the skill of summarizing, it could be used for English as well, especially if English hours are lacking and Science hours are plentiful. ... This is an example of how to get the most out of a favorite publication. You might want to do something similar for "Citizen" or "Parental Guidance" published by Focus on the Family. It's not really worth the trouble for just a one-time shot; rather this would be for a publication you want your student to read regularly. Conversely, there's no need to do this for every piece of literature your student reads! You'd use something like this for one or more of three reasons:

- to teach the skill of summarizing
- to guide the student in gleaning important information from a publication
- to provide documentation of work done (since reading alone isn't real "documentable" unless you take a lot of pictures of the student reading, and of all the books read, too.)

DINNER MAKING SCHEDULE (IV-38)

My husband has always done most of the cooking, (no kidding!), so the main purpose of this sheet wasn't to get relief for *me,* but for *him!* And to be organized in the process. With homeschooling and doing my little home businesses for the past 11 years, dinner is something he's always done at least ⅔ of the time. Since he's a better cook than I am – and actually *likes* doing it *(much more than I do)* – it's a logical-for-us and very blessed-for-me arrangement! Your "Dinner Making Schedule" can be posted on the fridge on Friday or Saturday for the following week. Everyone fills out their own slots. You can let each person have their own specific days, or do it on a "first claimed, first got the day" basis. The kids are allowed, even *encouraged*, to prepare *anything* their hearts desire. The only stipulation is that they have to get their shopping list to Dave or me at least one, but preferably two or three days ahead of when they need it.. There's also a stipulation for counting it for Home Ec credit: If they don't care if it "counts," they are free to fix something they've fixed before. However, if they want it to count, they have to fix something new. Why am I a stickler on this? Because I personally don't feel that fixing sloppy joes from a mix or B-B-Q'd hamburgers 43 times constitutes *learning*. Nor would I feel good about that being a good chunk of a Home Ec credit. I'm certainly thankful for the help and the full tummies, but I want them to be challenged and to step into new arenas. Once they master a thing – in Home Ec at least – I want them to move on! This guideline gives them a little boost in that direction.

DINNER PLANNER (IV-39)

This sheet adds a little spice and organization to the otherwise mundane task of meal preparations. The chef for the night has the entire privilege of choosing whatever ***Menu*** they want to prepare for dinner (within wisdom's boundaries!) as well as the responsibility of carrying out whatever it takes to get those choices on the table! So they need to look up and write down any ***Recipes*** they'll need to use. (Not for the simple things like cooking frozen peas; just the "real" recipes.) Plus they need to write down a ***Time Plan*** so that everything is done at the same time, and hot too! If any supplies need to be purchased, those go in the ***Shopping List*** column, and then passed on to whichever parent will be at the grocery store next. (If they forget to do this part, they'll have to "make do" – just like I *normally* have to do!) At the bottom of the page there's space for comments and evaluation by both the student and the parent – like: How did you/they/we like the dish? What would or should you do differently next time? Did the foods mix well color-wise on the plate? Was it too much work? Time spent on all of the above goes in the lower right hand corner – for safe-keeping until it's transferred to the "Class Completion Record."

DOG REPORT (IV-40)

Now this one might seem a little strange, but it proved to be wonderfully useful when our son begged and begged us for a dog. I am not a dog lover. I'm not even a cat lover – though I reluctantly gave in on that, and we do have not one, but *two*. But we felt our son needed to know what he would *really* be getting into with a dog – since this dog would be "his baby," in every sense of the word as far as *I* was concerned!

Ever since Tory first started asking for a dog *(when he was three years old)* I had been telling him he needed to find this out and that out, and then this other thing. I was secretly hoping that the inconvenience of having to go investigate these things – at the library, the vet – would either divert his attention or satisfy his craving. But he and I both kept forgetting what I had told him to find out. *(We both "are" A.D.D.)* So I devised this "Dog Report" to finally get all the required info in one place – and to get him to do some researching and writing! I didn't include it here, but there's a second page that goes with this. All it asks for is a sketch and/or a photo of the dog. (Art) He spent many good hours working on this!

To satisfy the morbid curiosity of anyone who may ask, "Well... *did* Tory ever get a dog?" here's the outcome: Yes, Tory *did* get a dog. A Siberian Husky puppy he named Dominic that was bigger than fifteen cats as a *puppy*. *(I don't think it was ever small, even as an embryo!)* In the short-but-not-so-sweet six weeks that Dominic dominated our yard and flower beds, Tory discovered (to his dismay) that love was not quite strong enough to make all the work of owning a dog – at least of *that* size – quite worth it. The dog is gone, ☹ and in the aftermath we had a flower planter to replant, the front door to repaint, and a lawn to de-fertilize. Oh well, at least we got a picture of Tory and Dave building the dog house for his Woodshop credit!* *(Uhhh... excuse me... I need to ask you dog-lovers to please stop throwing those rotten tomatoes. So I'm not a dog lover! Before you offer to adopt Tory, let me at least tell you that the decision of whether or not to sell the dog (complete with dog house and a 3-day 150-pound supply of dog food) was left completely up to Tory. He finally decided that having and training a dog – at least one this large – was more responsibility than he wanted right now. I can't say that I crie– I mean blamed him.)*

***See page VI-A–29 for a photo of Tory IN the dog house WITH Dominic!** ☺

ELECTED OFFICIALS WORKSHEET (IV-41)

I picked up a sheet at our state capital's Information Center that listed and explained the responsibilities of each of the main elected officials in our state. If you write to the Information Office at your state capital, I'm sure they'd be glad to send you any information they have on the subject. ☺

FAMILY MEETING GUIDE (IV-42)

This one is self-explanatory as far as directions go, but I wanted to add that it might be handy to make several copies of this (if you plan to use it) and keep them on a clipboard somewhere convenient where everyone knows it's located, so that *as* grievances or ideas come up that anyone wants to discuss with the *whole* family at *one* time, they can be written down immediately, and not forgotten.

When Carlianne was a five-year-old, she *loved* Family Meetings and was often the one to initiate them! And we'd even had some unpleasant ones. But she knew that things get worked out at them, and she liked that.

Anyone in the family can lead a meeting using this form. In fact, I feel it's excellent practice in leadership skills. Just make sure that the older siblings respect and heed to the younger ones when they are the leader (since it's easy to not take them seriously) or it will end up leaving a sour impression on them for a long time.

FORT CLATSOP "SEARCH-IT-OUT SHEET" (IV-43)

This worksheet on Fort Clatsop (near Astoria, Oregon*) is an example of what I call a "Search-it-out Sheet." This one is part of our "Washington State History & Government: the Search-it-out Route" course that my dad and I designed and wrote a book on! I have already made up several of this type of sheet for various museums, available in my book by the same name. I included one here so you could get an idea of what they are and how easy they are to make up! This is actually just the first of two pages, included to give you an idea. *(* Oregon and Washington State's histories are intertwined – that's why I have included Fort Clatsop even though it isn't in Washington – in case you "perfection afficionados" were wondering about that! ☺)*

If you haven't gone through the museum yourself to become acquainted with what it offers or the segment(s) of history it covers, just call or write to the museum ahead and ask for a brochure or two that describe the history in the museum – what *history* it covers – rather than the history of the museum – how the museum was built and developed. They usually have brochures on each.

Once you receive the brochure, read all the way through it to familiarize yourself with the material. Then take one sentence at a time – one that presents interesting, know-worthy facts, and turn them into fill-in-the-blanks search-it-out sentences. The brochure may even give you ideas for questions that aren't addressed or answered in the brochure. All the better! You don't *have* to know anything to make one of these up! You're going to learn something too! Just write out your question and insert a blank line in place of the part you don't know. Then let the student find out and tell you!

What's "know-worthy"? As you could probably guess from reading the rest of the book *(of course, I'm assuming you didn't just now open up to this page)*, I don't get hung up on this either! If something is interesting and seems like it would be valuable to know, I include it. However, the subject of history – even just your own state's history – is so vast, there's only so much you can include. For example, I would not consider it "know-worthy" (valuable) to memorize the dates that Lewis and Clark were born, but I would want to remember the years President Jefferson sent them out on the first expedition and about how long it took.

And as far as what's "know-worthy" regarding standardized achievement tests, that's another whole subject in itself. For one thing, I'm letting God define what's "know-worthy" for us. But even aside from that there are several versions of such tests, and you could never prepare specifically for one particular test anyway, so there's no reason to feel your knowledge is less "valid" than what they're teaching in the schools. I just want to make sure my kids: 1) get a good foundation of the overall picture of the history of our state, the U.S.. and the world, and that they: 2) cultivate an appreciation for and interest in history, and that: 3) are able to see God's hand in history, which was intentionally taken out of all history books at the turn of the century. (And, I might add, you can be sure that none of the Christian-aspect facts will be on any S.A.T.!) At the museum itself, have the student put the search-it-out sheet on a clipboard, ready to record any info they discover during the tour of the museum.

Unless you have your own personal history buff along, it's a good idea to see if the museum will provide a docent for you. If they're available, they're happy to do it. Docents, by the way, are what I would call "trained and very knowledgeable history buffs," and they are *delighted* to tell you what they know – all sorts of stories and interesting facts pertaining to the history of the museum. There is *much* more to know than meets the eye!

GOD'S ACTIONS (and Prepositional Phrases) STUDY (IV-44)

OK! Want a personal tutoring session?! *(If not, skip to the next item in this section.)* This study will benefit both the grammar segment of the brain *and* the spirit! ☺ Psalms 144 through 147 are *brimming* with active verbs of what God is busy doing. I know grammar is a painful topic for some, but my hope is that grappling with it in the Word will at least dull the pain a bit, and make it more profitable, since God's word, when it goes forth *anywhere,* never returns to Him void!

What you see here does get a bit technical, but you might be surprised at how simple some of the more intimidating grammar terms really are, if you look at them calmly, slowly, and one step at a time. Here we'll be dealing with two types of prepositional phrases. Let's start with "prepositional adjectival phrases." Take a deep breath and look at the first word. "Prepositional." *(Cover up those other two words in that term, if you have to.)* Now you already know what a preposition is, right? If not, I think of a preposition as being anything that tells you "anything about the store," like: "***in*** the store," "***at*** the store," "***on*** the store," "***under*** the store," "***around*** the store," "***by*** the store," "***with*** the store," "***through*** the store," "***over*** the store," "***from*** the store," "***across*** the store," etc., (although I don't think I left too many et's to cetera! ☺) So if this phrase *(ah! – another word to define! – hold on until the next paragraph)* starts with one of the prepositions, that's all it takes to make it "prepositional." Got a preposition at the beginning? POOF! It's a prepositional phrase!

OK, now for that "phrase" word. All a *phrase* is is a group of words that are not a complete thought all by themselves. "The group of boys." Doesn't tell us anything, does it? Like what the boys are doing (no verb there). So it's just a phrase. How about "Acting like a baby"? We aren't told *who* is acting like a baby (no subject), so, again, all we've got is a measly phrase.

Now we're ready *(I hope)* to conquer "adjectival." (That latter word is pronounced **ad-ject-tí-val**.) OK, what does an adjective do? It modifies (tells about) a noun. So if this *incomplete sentence* (phrase) that starts with a *preposition* tells us about a noun, it's a "prepositional adjectival phrase!" *(If you didn't get it, try again when you're not so tired or thinking about what to make for dinner. ☺ Please don't continue until you've taken a nap or figured out dinner.)*

OK, on the other hand, if our prepositional phrase – our *incomplete sentence that starts with a preposition* – tells us more about a ***verb*** (what's *happening*) or about an ***adverb*** (how, why, when, or where the action is happening), it's a ***"prepositional, adverbial phrase."***

All this will make more sense when we actually *apply* it to our stuff, so let's take something on the "God's Actions" sheet... "God makes peace in our borders." Let's start with the basics: ***God*** is the subject; ***makes*** is the verb; and ***peace*** is the direct object because it tells *what* is being made. Then *what* is this: *"in our borders"?* OK, what does it *start* with? "In." Recognize that as a preposition? *(Please answer, "yes." ☺)* OK, then we know we have what? A *prepositional phrase*. What does this phrase tell us *about*? ***God***? No – God isn't in our (or anyone's!) borders. ***Makes***? Or ***peace***? Well, is the peace itself in our borders, or does He *make* it in our borders? It's a bit confusing, so let's do a little test on it. Change the word *peace* and see if it still stands with *makes*: God makes soup in our borders... God makes sunsets in our borders. (If you can interchange these words and still have the phrase make sense with the verb, then that's probably what it's modifying... Now what is ***makes***? A verb, right? (Right.) So what does that make our prepositional phrase? (Modifying a verb.) Adverbial. ALL RIGHT! So what do we have here? A "prepositional adverbial phrase"! ... Now was that really so painful? I think of it as a puzzle to solve, one piece at a time rather than being generally and overwhelmingly puzzling.

Isn't this fun?! Not really? Well, to seal in what you just learned, let's do one more anyway: **"He opens the eyes of the blind."** ***He*** is the subject; ***opens*** is the verb; (the) ***eyes*** is the direct object; so what's ***"of the blind"***? Does it tell about He? (Nope.) How 'bout ***opens***? Does it tell us ***how*** He opened the eyes, ***why*** He opened 'em, ***where*** He opened 'em, or ***when*** He opened 'em? (Uh-uh.) OK, then we only have one other possibility: does it tell us ***which*** eyes? Bingo!!! *(Oh the thrill of victory!)* So ***which*** eyes does God open? The ***eyes of the blind***. Since this prepositional phrase is modifying a noun, it's acting as an adjective, soooo that means we've got a "prepositional adjectival phrase" ... Where is thy sting, oh Grammar?! *(OK, so it does produce a bit of nausea for some.)*

GROUND RULES FOR STUDYING OTHER RELIGIONS (IV-45)

If you're going to study other religions, this is a very important place to start! Directions and source are on the sheet itself. The first time through this "quiz," the kids get to have the book open as they search out and fill in the answers. When they feel they know the material, then they take an official test with the book put away. They use their "open book test" that they themselves filled out as the answer key. *(Clever, huh?! ☺)*

A homeschooling mom and missionary interjected this story as she was typing up this section for me (to help me in the midst of my "Lost Disk Trauma"). She gave me permission to print it here as long as I would pray for the girl she mentions toward the end. ☺ She wrote:

"I wanted to add a comment on your 'Ground Rules for Studying Other Religions.' When our daughter enrolled in the Christian school here, the first Parents' Night we met all of her teachers. I was all excited that she would be studying World Religions and learning what the difference was and how to counter the false teaching. During the teacher's presentation to us parents he said, "We will be studying the 'treasures' (my quotes) in other religions. The students will find these 'treasures' but they will probably come back to the 'treasure' in their own backyards. The students will probably be confused, uncomfortable and perplexed with my teaching, but that's my job." He also had on his wall, "Perplexity is the beginning of knowledge." I was so shocked that I never said a word. This was not only a Christian school, but an MK (missionaries' kids) school. One of my daughter's friends graduated from the school very confused, having more questions now and no answers. It has been two years since they graduated. This girl decided to read the Book of Tao instead of the Bible every morning because it was "profound." Just recently she is beginning to come back to Christ. All that to say how MUCH I appreciate your Ground Rules!!!!!"

(Now *you* can pray for this girl too! ☺) Isn't this amazing?!? I have heard that when they are training people to recognize counterfeit money, their main "tactic" is to have them handle ONLY real money – lots of it! NOT to handle counterfeit money! They are SO familiar with handling "the real thing" that they are immediately aware of when "the counterfeit" comes into their hands. This is an excellent analogy for Christians. The study I suggest here is not at all in-depth. The only reason that I have this class is just for a basic familiarity with, or awareness of, other religions, especially since, as homeschoolers, we do not have the same exposure that kids in the public school setting have to other religions – and I do not believe they *need* it.

HIGH SCHOOL DAYS MEMORY BOOKLET (IV-46)

Copy three, four, or five of these – depending on how many years your student plans to take to get through high school – and put them into a High School Days Memories notebook. If you like, add a few cardstock pages (which you can get at a printer shop) after each memory page for mounting photos. A few pocketed pages will keep memorabilia from falling out. Tabbed dividers labeled "Freshman," "Sophomore," "Junior," and "Senior" will make each of the years easier to access. If you use pocket pages, just affix individual tabs to the right side of each pocket-page since they are usually wider than regular sheets (to accommodate inserting regular- sized pages into them) and will cover up any tabbed divider sheets placed behind them. Finish it off with an attractive cover! You're welcome to use one of the labels here, but I bet your student could come up with something a lot more creative!

[HIGHER THINKING SKILLS]

this is not a learning guide sheet; just something I'm inserting here for your interest. The more you can incorporate "higher thinking skills" into the work your student does, the more challenged and "stretched" his mind will be. Now, before you get intimidated, this is something that you're probably *already* doing unknowingly and naturally. In fact, in Marilyn's material, she talks about this, but in a different light and with a different classification system. She defines learning as beginning with the "preparation stage," then going to the "collecting stage," then to the "thinking stage," and then to the "communication stage."

So how do you know what's "higher" level? "Bloom's Taxonomy" is a breakdown of the wide range of thinking skills, presented in graduated stages, from least difficult to most. It is as follows:

LEVEL	WHICH IS...	VERBS THAT SHOW WHAT CAN BE DONE AT THIS LEVEL
KNOWLEDGE	*Remembering*	count, know, write, draw, recall, state define, record, name, memorize, list, recite, relate, point, repeat, recognize
COMPREHENSION	*Understanding*	restate, recognize, predict, identify, review, discuss, explain, locate, estimate, tell, contrast, differentiate, associate, describe, express, compare, report
APPLICATION	*Solving*	dramatize, operate, practice, sketch, interpret, translate, employ, use, schedule, apply, demonstrate, illustrate, relate, classify, utilize, complete
ANALYSIS	*Analyzing*	contrast, distinguish, calculate, differentiate, debate, diagram, solve, analyze, experiment, criticize, question, inventory, examine, test, appraise, compare, inspect, infer
SYNTHESIS	*Creative*	plan, formulate, compose, prepare, construct, manage, set-up, organize, arrange, assemble, create, propose, design, prepare, collect, generalize, specify, integrate
EVALUATION	*Judging*	value, determine, revise, estimate, measure, select, choose, critique, recommend, appraise, assess, evaluate, judge, compare, rate, score, grade

Let's *apply* this to something! Take a look at the Book Report in this section... The info at the top is mere recording of facts; first level thinking. To "Briefly tell what the book was about" takes him into *comprehension.* Discerning whether it was "applicable to his life" takes him up into *application.* To know if the book "made an impact" on him he has to *interpret, analyze, examine,* and *evaluate* it, and his own response to it. So the report may be short, but it does utilize some "higher thinking skills." ☺ I found that I *naturally* use a lot of this in my regular conversations and dealings with my kids, so don't get hung up on it! It's just a tool – an "eye-opener" and "light-shedder." Thought you might also find this helpful for creating your own Learning Guides. *(Now, creating those Learning Guides would be at, hmmm, – the "synthesis" level, right?! Actually, I've found it's at any and <u>all</u> of the above levels!)*

HOME REPAIR, REMODELING, & MAINTENANCE JOB LOG (IV-47)

This sheet is for both you and your student to use. As project opportunities come up around the house, you can list the job here and fill in the specifics for how to get it done according to your expectations. (This is also good preparation for working under a boss and therefore could be used in Occupational Skills.) However, most of the jobs for this experience would be on more of an apprentice basis (assisting Dad) than on an assigned basis.

Specific ideas? I hesitate to give you any because it's too easy to think anything in black and white is "the standard," which is exactly what I'm trying to get away from! So only if you promise you'll use my suggestions as nothing more than ideas, I'll give them to you.

What if you just finished with a bunch of repairs or remodeling and you have absolutely NO plans for doing anything that involves a hammer or screwdriver for the next 10 years except maybe hanging a picture – and even then, it'll be only one that really needs to be hung? Well, I bet you know someone who could use some help.

Well, it's time to have some fun. Even if you have *no* plans for doing anything in this area, you might enjoy the following. Then again, if you *have* experienced any of the following, you *really* might not enjoy it!

(WARNING! If you're not into humorous digressions, please skip ahead to the next item, "Home Report"! ☺)

HOME REPAIR IDEAS:

- Unclog a drain • Put in a new sink because you wrecked it trying to fix the drain
- Unfreeze pipes • Repair (by replacing with exact replica of) blow dryer that you ruined while unfreezing pipes
- Tear out the cat-urine-scented carpet and replace the cat-urine-scented portions of floor boarding *(in all 5 places!)*
- Install a new "donut" under a toilet because you had to detach the entire toilet to extricate the toothbrush "someone" dropped into the toilet
- Remove what looks like it might be a little bit of dry rot from the base of your shower stall *(this is so that your appraisal – needed before you can re-finance your house at a lower interest rate – can hopefully be approved)*
- Once the entire (above-mentioned) shower stall *and* surrounding walls have been removed, replace the almost completely dry-rotted foundation beams of the house – which involves importing two very heavy cement blocks to hold the house up while you carve out and replace the dry rot
- Since you now have the whole bathroom floor and two adjoining bedroom floors and lower 18 inches of wall opened up anyway, hey, why not just go ahead and take out your old, existing dry-rotted shower stall and put in a bathtub instead of just a shower, which of course means taking out half of the closet in your son's room on the other side of the shower stall as well as the back corner of the adjacent hall closet
- Since the pipes under the toilet are exposed anyway, move the entire toilet over to the right 4 inches so your wife can have a longer counter which leads us to the first point in the *next* category. . .

HOME REMODELING IDEAS: *(See how easy and effective "real life learning" is?!?" Nothing contrived here!)*

- install a new bathroom cabinet • install new bathroom light fixtures • build bathroom vanity shelves
- re-texture, re-paint, and re-wallpaper the bathroom walls that were ripped out to get to that lousy spot of dry rot
- cut a rectangle out of a wall (where extra light is needed) and install a window
- design and build matching planter boxes and shutters for the exterior windows

HOME MAINTENANCE IDEAS:

- paint your home's interior or exterior *(impressed with those professional words? – I normally would have just said "inside" and "outside" but I saw those professional words on some paint cans we had around here about nine years ago)*
- add a shelf to an existing desk
- wire and install an additional electrical outlet under Mom's new vanity shelf *(since the wall is opened up anyway)*
- mount a new towel rack and toilet paper holder
- install a hand-held shower head
- install a fluorescent light over kitchen sink
- install a new dishwasher (so you won't have to use a rubberband and nail to keep the old one closed
- prepare the soil for, and plant and maintain (as in weed and prune) a garden.

HOME REPORT (IV-48)

This page was designed for the "Building Family Relationships" Unit or Course. You won't usually see this subject on a school's list of graduation requirements, but it's high on *our* list! What good is it to be able to solve an elaborate algebraic problem yet *not* be able to solve a simple family problem?!

The Purpose: As you first glance over this two-page questionnaire, you may wonder if my purpose in designing such an exercise was to try to solve all our family's problems with one clean sweep – of a worksheet! I assure you, it *wasn't*. Rather, it was to provide a tool that would accomplish two things: 1) Help members of the family take an honest, frank look at how our family functions – and how it *feels* about how it functions. 2) Provide a neutral basis for communication. Questionnaires are effective at doing this as the focus is on getting the questions *answered*, not on *arguing* over them.

Prepared for Life: In any normal family there is always some sort of balance of success and failure, harmony and contention, affirming and criticism. As each and all family members grow in Jesus (less of *them;* more of *Jesus)* the scale will tip more and more toward "success, harmony, and affirming." Ecclesiastes 3 confirms that there is "a season for everything." It is brimming with examples of opposites, or, as I prefer to call them, counterparts: "A time to weep and to laugh; to keep silent and to speak; a time to embrace and a time to refrain from embracing..." To be prepared for what will come our way in life, we need wisdom to know what is needed for the *moment*; and then the grace to do it! My desire for our family is to be heavily "imbalanced" on the fruitful positive side, and to allow God to help us deal with the negative side. I guess you could say I want for us to be "abnormal"! I want our children to leave our home with their knapsacks as heavily laden as possible with the tools they'll need to build and maintain healthy, lasting relationships with their spouses and children, as well as with those they'll be working with and neighbors to. I realize they will have to seek their own guidance from God as they enter those seasons later on, but I want to prepare them as best as I am able. I don't want them asking "Why did you teach me how to write and speak correctly, but not how to *communicate* or solve problems in relationships?!" Most of the latter will be *caught* rather than *taught*, by the way. ☺

How the "home report" works: After completing the questionnaire, the student comes to one or both (preferably both) of us to discuss the "finding." Even though they would have already talked with each family member individually when asking the questions, this is a time for us to now take a look at these newly revealed conflicts or tensions that may have been hidden until now (or have been subtly bubbling below the surface in the form of anger) and then work on resolving, healing, working through, and praying about them.

Most of the problems arising in our home require much more work of the Holy Spirit than work of our doing. Matters precipitated by our sin nature require not only training and effort by us as parents, but exposure and yielding to God's will and His Spirit. "For He is able to do far more than we could ever dare to ask, dream, hope for, or imagine!" (Eph. 3:20) I don't know about your family, but ours needs this kind of vision and hope!

[IDEAS FOR CREATIVELY STUDYING THE BIBLE]

Again, this is NOT to be found on any Learning Guides! These are just some ideas I wanted to share with you, and this is the most appropriate place...

There are many ways to go about studying the Bible. I present a few here that you may not have tried. They've been not only fun and enlightening, but healing and edifying. All of what I share here I do within the pages of my Journal. Using some sort of notebook is a must! I prefer the looseleaf type so that pages can be removed and added, allowing it to g-r-o-w with you!

✍ **Key Concepts** - I like to have a place to put references to those exciting tidbits of truth that God reveals to me that build on certain concepts, like "New Wineskins," "Hearing God," "I Possess" (a mansion in Heaven...) "What the Lord's Servant Must Be," "The Word of Truth is Able To...", "How to Grow Bigger Spiritual Eardrums," "How the Church is Built Up." Each of these titles has its own page, and as I come across verses that add to these ideas, I write them on those pages.

✍ **Word Studies** - There are several words that intrigue me and that God is helping me understand better and root deeper into me. Similar to the above idea, whenever I come across a word in the Bible in one of these categories, I write the verse down on the pages titled: "Keep," "Crown," "Anger," "Walls," etc. A "Word Alive" study could be done on any of them you want to go deeper with. There is a study on "Education" here in this section, as well as "starter-studies" on "Learn" and "Knowledge."

✍ **Cross Reference Studies** - Start with a verse that speaks to you and look up and record the cross references that are given in your Bible. (If you don't have a study Bible, it's worth getting one!)

✍ **Grammar Topic-toners** - As an exercise to sharpen my mind and feed my spirit at the same time, I have three pages in my journal that are titled:

- **Nouns** – that God's Word Says I Am, such as a daughter of the King, a priest(ess), etc.
- **Adjectives** – same idea, except these words "modify" or describe who I am (we, God's people are) such as redeemed, chosen, holy, beloved, etc.
- **Analogies** – is where I write analogies that particularly speak to me, found especially in the parables, but all through the Bible as well.

IDENTIFYING PREPOSITIONAL PHRASES (In the Bible) (IV-50)

Here is a section out of Proverbs 8:1–9, enlarged (on the worksheet) because the print in my Bible is so small. This could also be handwritten by a younger or older student, or typed up by your teen for typing practice. Plus if they type or write it up, they can give themselves more space between lines for writing in the information. The way this came about was that as I was reading these verses one day, it became obvious to me that there was one prepositional phrase after another! So I thought "Hmmm, this would make a good quick lesson on this topic!" The directions are at the top of the sheet. And the "answer key" is right below...

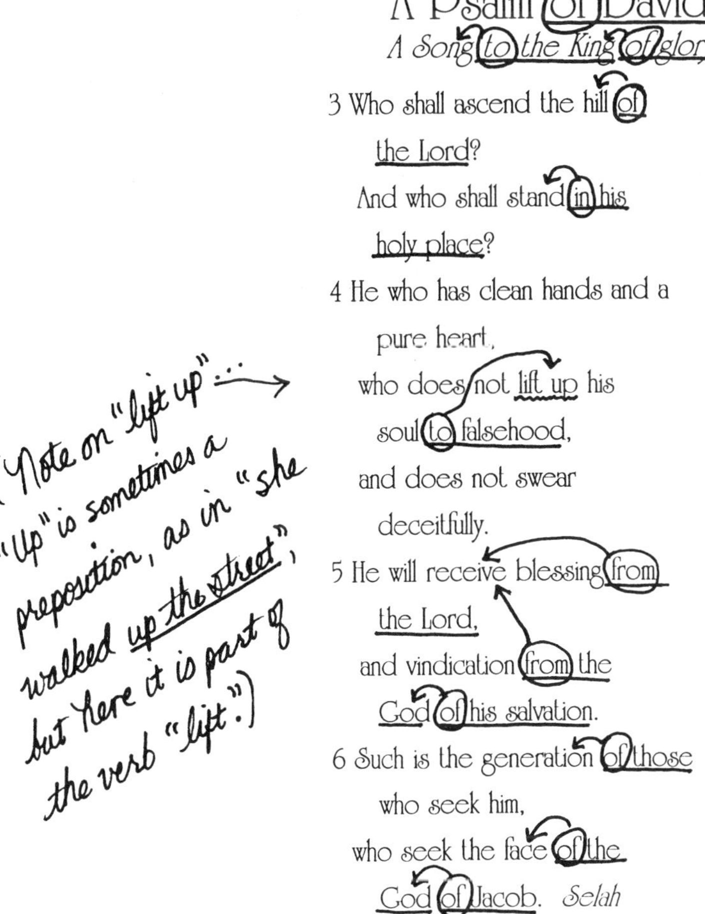

A Psalm of David
A Song to the King of glory

3 Who shall ascend the hill of
the Lord?
And who shall stand in his
holy place?
4 He who has clean hands and a
pure heart,
who does not lift up his
soul to falsehood,
and does not swear
deceitfully.
5 He will receive blessing from
the Lord,
and vindication from the
God of his salvation.
6 Such is the generation of those
who seek him,
who seek the face of the
God of Jacob. *Selah*

(Note on "lift up"... "up" is sometimes a preposition, as in "she walked up the street", but here it is part of the verb "lift.")

LOCATION DATA SHEET (IV-51)

This is from my *Washington State History & Government: The Search-It-Out Route* book, but, since most of you won't be getting that *(since most of you don't live in Washington – the real Washington, that is)*, but *could* nonetheless use this sheet, I'm including it here. This is where you record the basic information about any museums or historical sites you visit, including a picture of it at the bottom.

MOVIE CRITIQUE / ANALYSIS (IV-52)

This is similar to the Play Critique/Analysis sheet, but since there are a few differences, though minor, I just made up a separate sheet. One point I must make clear is that I do not condone watching just any crummy ol' movie for the sake of or in the name of "Fine Arts." I've never felt that just because something is an art form of any kind, license is thereby given to sin one's brains out. Nor does it sanction watching 150 hours of movies and doing a Play Critique/Analysis on each one and counting that as one credit of English! But it *can* play a significant role in several classes, actually: English, of course, but there are also some excellent videos on Science, Home Maintenance and Repair, Travel (Geography, study of the World), Computer (Occupational Skills), Child Care (Home Ec.), and History.

MUSIC APPRECIATION WORKSHEET (IV-53)

Music is one subject that's fun to put some "Heart and soul" into! Dig through some old and new tapes – yours, your parents', the library's – to find the "heart." Then for the "soul" part, use this sheet. Its purpose is to help the student become more aware of the various styles of music, the "purpose" of each, the composer's background, and to be able to differentiate between the various instruments used, the timing, and the effects of each on what the music accomplishes, how it is used, and the mood it creates. This is *not* an in-depth study of music; you'll need to look elsewhere for that, but this sheet is a tool to help unlock some doors and a place to record discoveries.

The broader the spectrum of music styles, the more comprehensive and thorough will your study be. So don't get stuck on your "favorite music" button. Music can be organized into a variety of classifications. Here are some suggestions if you are looking for a place to start:

- **Classical**
- **Country / Western**
- **Contemporary**
- **Jazz**
- **Waltzes**
- **Blue Grass**

- **Christian** – *which could be broken down further into*:
  - *Hymns*
  - *Contemporary*
  - *Gregorian Chants* (of the Catholic Monks)
  - *Gospel*

- **Nationalities / Cultural**: *Japanese, Mexican, Russian, African, Irish, Italian, Victorian, Indian, Chinese, Scottish*

- **American**: *Early, Colonial, Stephen Foster, 30's, Rock 'N' Roll, Pop, Ballads, Rock, Hawaiian, Appalachian, Lawrence Welk, Rap, Bands*

- **Operas & Operettas** *(which, as anyone with any culcha knows, are not all the same!)*

NAME STUDY (IV-54)

To do this study, you obviously need to start with a name. Your own is a pretty good choice, I'd say. You can use any baby name book to find the meaning. If you don't own one, you could possibly call the library. Ours has always been very good about looking up name meanings for me. *(I've been making Scripture Name Plaques for twenty years.)* Ask for the Information Desk, or whatever they call it. Or you might ask an OB-Gyn (baby doctor) or someone who recently had a baby if they have a spare baby name book. Baby product (like formula) companies give out free ones to pregnant patients.

There are several baby name books on the market. The one that has the best and largest listing of names, including unusual names, is called *Name Your Baby* by Lareina Rule. To be honest, the title of this books amuses me... Think about that: "Name your Baby." Like a parent who's about to have a baby hadn't thought of the idea?! I can see it now: Hank and Tricia are in a book store, just browsing through the childbirth and parenting section. They come across this book called *Name Your Baby.* With incredulous looks on their faces, their mouths drop open, Tricia and Hank's eyes lock. Almost simultaneously they exclaim, *"Oh NO! We never thought of that!! We'd better get busy thinking of a name!"* Rushing to the check-out stand just in the nick of time, they say with great relief to the clerk, "What a God-send this book is! We would *never* in this world have thought to name our baby!" I do need to give you one word of warning about this book: It contains a baby personality horoscope. I just tore out the one in mine and crossed off that word on the cover. It contains "over 6,500 names with origins, meanings, nicknames, and famous namesakes." (quoting the cover)

There are also a couple name books that you can find at Christian bookstores. (Or I can get them.) These give a Bible verse for each name, but I greatly prefer to search for my own. I've done over 1,000 name plaques and have used two, maybe three verses out of these books. I feel, and others I've talked with agree, that the verses aren't too "applicable" to the name. Back to the study...

Many names have more than one meaning. It will depend on which nationality and which source your book used. For instance, in one of my books, Mary meant "sympathetic," in another it meant "bitter," and in another it meant "fragrant." All's fair in love and names *(I just made that up)* so just choose your favorite! In *What's In a Name*, (Christian), they take the original meaning of the name and find a spiritual connotation. Many of these are not only brilliant, but they also save the poor recipient of the name a great deal of embarrassment and feelings of insignificance. ☺ Like *Melanie*. Its original meaning is "darkness." Now who would want *that* meaning? But this book comes to the rescue with "seeker of light" for an "implied meaning," using the logic that anyone in darkness would be seeking for light.

If you have a really unusual name, you can combine the meanings of other names within the name to come up with a whole new meaning. I just happen to have a couple of examples because I made up the names of both of our daughters and wasn't going to miss the fun just because there were no prior meanings. For "Carlianne" I took *Carol* which means "strong and womanly" and *Anne* which means "full of grace." So the combined meaning is "strong, womanly, and gracious." For "Sharnessa" I took *Sharon* which means "princess" and *Nessa* which means "pure and chaste" and came up with "pure, holy princess." Use a concordance (and/or someone who knows the Word) to find enhancing verses. I make sure they're encouraging, uplifting, and would be good to have for their whole life, not just as a child, for instance.

Look up the meaning of your name in any baby book. It doesn't need to be a Christian one, but if the meaning is negative, find a spiritual connotation. As you read the Word, write down verses that enhance or magnify the meaning and that encourage you.

OCCUPATIONAL SKILLS INTERVIEW (IV-55)

The purpose of this form is to provide the student with a means of finding out information about *any* occupation of interest. It can be used in a variety of ways. A student could go on a one-man "interview hunt," or go with a friend or parent.

A fun and valuable way to exchange information is in a "Careers Club" format! Several students could meet together at an agreed-upon frequency. Between them, they would decide who will find interviews with people in which occupations. This is necessary only if the group decides they prefer not to double up on occupations, therefore covering as many as possible. Duplications may even be interesting. For instance, it might be good to find out what two different lawyers have to say.

The group would need to decide how many interviews each participant will obtain in a certain period of time, for instance, one per month? Two every quarter? Also decide whether the kids will go in pairs, how often to meet, whether or not everyone wants copies of all interviews done – in which case each person would make as many copies of his interview(s) as there are students in the group, *or* the members may prefer to sign up for only those interviews he or she is interested in. My personal opinion is that it's easier to just make copies for *everyone* rather than taking *individual* orders. Plus you never know what you or a friend might be interested in down the line, or what occupation your future spouse might have!

Whom to ask for an interview? Anyone in *any* occupation that sounds interesting to you! Stretch your imagination a little! It wouldn't hurt to interview people in occupations you never thought you were interested in. It will broaden your appreciation of what other people do for a living or as a hobby, expand your horizons, plus you might discover an occupation that really would be ideal for you, one you might not have thought of otherwise!

Tips for procuring an interview: Always call ahead! Obviously, if you know the person it will be much easier to know what to say and the best way to approach them. Here's a suggested conversation if you're not sure where to start:

> "**Hello, I'm ________________**" *(Don't say "homeschooler" yet; let them get to know and be impressed with you first, then zap 'em!)* "**I'm working on a project for my (an) Occupational Skills class and I'm very interested in learning about ______________** *(their occupation).* **I was wondering if you'd allow me to act as Barbara Walters and come interview you.**"

Re-word it in a way that feels natural for you. I've found it injects some humor, plus it gives a quick picture of what you're doing. It also helps them relax a bit. For some, the idea of being interviewed is somewhat unnerving, so the "Barbara Walters" line has always been very well received. I think it kind of elevates it to a loftier standard. ☺ If they consent, find out when they are available to do this by maybe asking, "Do you have any days of the week or times of the day that are better or worse for you?" And take it from there... Be sure to keep all your interviews in a notebook!

OFFICE & BUSINESS SKILLS JOB LIST (IV-56)

Since we operate a home-based business, there have been *ample* opportunities for our teens to learn a wide variety of skills relating to the operation of an office and business. Here is a list of possibilities to give you an idea of a few of the skills you can look for either in your own home or at the (home) office of a friend to teach your teen. Keep in mind that these are what I value and use; they do not comprise a "standard" office skills list.

- **filing**
- **phone skills**
- **use FAX machine**
- **categorizing**
- **compiling information for packets**
- **have basic typing skills**
- **be familiar with basic computer skills (separate class for us)**
- **know proper addressing and mailing procedures**
- **using a copy machine with the following features, and know how to:**
  - ✔ **reduction and enlargement**
  - ✔ **duplex copying**
  - ✔ **copy 2-page original**
  - ✔ **add paper to empty cassette**
  - ✔ **use "continuous feed" for originals**
  - ✔ **collating and sorting bins**
  - ✔ **copy 2-sided original**
  - ✔ **change paper cassette origination**
  - ✔ **use "bypass" feature**

Under the "Job and Instructions" column I enter what I want done, and any instructions for how to do it, where to find it and put it when done, etc. If there is a date I need to have it completed by, that goes under the "Due Date" column. The student puts a check-mark (✔) under the " Done" column when it's completed. "Insp" means I have inspected their work, so I put a check-mark in that column once *I've* done that. And the amount of time the student spent on the task goes in the "Time" column *if* you are recording time. And if not, they can just write what kind of a *time* they had doing it: I had a *good* time, *hard* time, *fun* time, etc. ;-) *(winking face)*

OPEN LINE & MIND EVALUATION (IV-57)

The purpose of this sheet is twofold: **1)** to saturate the heart with the Word of God in order to prepare it to: **2)** go and make things *right* or *better* with a sibling. Much of this could even be done with a parent! Sub-surface feelings of resentment resulting from unresolved conflicts (can you tell I've been to Basic Youth Conflicts? Woops! I mean "Basic Life Principles"! ☺) can make life miserable around home, not to mention miss the whole point of true education! Even the younger kids can and *need* to learn how to deal with such tensions, and the younger they learn it, the less they'll have to "un-do" and the longer they'll have to practice and get good at it before entering marriage relationships. Instructions for both sections are on the sheet.

This whole area of attitudes and relationships within the family – getting along – is at the very heart of what true education is about. If we do not like what we see in our kids, then it's time...

For Mom to Ponder...

"The goal of our instruction is love from a pure heart and a good conscience and a sincere faith." (1 Timothy 1:5.) Think about this moms. How often do you not instruct in love and with a clear conscience that you are conducting yourself as God would want you to? Are you imitating God to your children, or are they getting a heavy dose of your carnal flesh? And what about your faith? Is it true, actively applied faith, or is it feigned, a mere mental agreement to doctrine, ideas and religious duty, an outward image? You interact with your children all day; they won't hold you accountable and no one else is around to see you, so it is easy to not be careful about what you do. You think that the seeds sown won't produce fruit? They will. "Do not be deceived, God is not mocked; for whatever a man sows, that he will also reap. For he who sows to his flesh will of the flesh reap corruption, but he who sows to the Spirit will of the Spirit reap everlasting life. And let us not grow weary while doing good, for in due season we shall reap if we do not lose heart." (Gal. 6:7-9) God is not mocked, whatsoever you sow, that you shall also reap.

"The fruit of rebellion, boredom, loss of soul life, loss of purpose, weak selfish character and bad relationships are all a result of how children were treated and provided for. Are their true needs being met? If mom is not dead, their true needs are being neglected, what's more mom isn't even aware of what they are! Most of mom's decisions are based upon her need to feed her own fears!"

~ *Marilyn Howshall*

ORGANIZATION SKILLS JOB LIST (IV-58)

If this sheet looks familiar, you've been paying attention! It's the same idea as the "Home Repair, Remodeling & Maintenance Job Log" except the skills are organizational in nature instead. This sheet pertains to organizing our home, car, day and office. These will include such skills as organizing work areas, desks, drawers, closets, video and audio tapes, school and office supplies, files, photos, notebooks compiled for specific interest, loaned items and books, as well as scheduling and planning, and goal and priority-setting. I don't scramble or push for ideas; as opportunities to learn, teach, or work with me on such tasks arise, I enter them on to this sheet. (More personal organization skills can be found in the "Personal" section.)

PERSONAL HEALTHFUL LIVING PLANNING SHEET (IV-59)

The idea behind this exercise is to inspire the kids to put some thought into taking better care of themselves, map out a practical plan, and then put some muscle into their method!

Youth are renowned for taking their health and vitality for granted. No amount of study will be able to *fully* turn that around, but I intended to boost their awareness of this vital gift and encourage them to do as much as they can to take care of their physical, mental, emotional and spiritual well-being as early as they can so they can remain "useful to Master, prepared for every good work" for as many days of their lives as possible. And it has worked! All three of my kids are now asking *me* why I'm eating something "fatfull" as they call it. (But I've been to Weigh Down, so my perspective has changed. ☺)

It says (in the directions on the sheet) to read and watch the assigned materials. For us, these were a variety of books and videos I found that were the direction I want to take our kids in the area of health. For a full line-up of resource ideas, see the "Health & Safety" class in the "Completed Class Plans" section.. You get to determine which resources are "nutritionally sound" and do your own assigning.

After these resources have been read or watched, they need to then pray and ask God to help them come up with a plan that will lead them down the road to life and health. Just to demonstrate to your kids the value and logic of making up such a plan, you might have them fill out one of these sheets *before* they do any reading on the subject of health, and then later, *after* they've studied the topics, have them do another one. It'll be fun to let them see how much they've grown and learned!

An example: One of the seven necessary "pillars" for physical health is "fresh air." Our physical bodies need air to be able to breathe and maintain life on earth. Next we apply it spiritually, so any of these thoughts could go under "Spiritual Application." An even greater, deeper need we have is for *spiritual* "fresh air." This we can get only from *God.* In fact, the word "inspire" means to "breathe in the breath of God"! Isn't that exciting?! Then I also found Scriptures to go with it, like "It is not mere age that makes a man wise, but it is the *Spirit* in a man – the *breath* of the Almighty – that gives him understanding." *(Job 32:8, Living Bible)* Your student may even want to continue this study and find other Bible verses or resources.

The student puts what he wants to do on a practical, physical, feet-on-planet-earth level to bring the physical aspect of this pillar to light. Like he might put *"go for a 20-minute walk 5 times a week."* The purpose of this type of goal is not to bring bondage, but balance and vision!! So if it's not met, chastising is not in order. It needs to be *internally* motivated within the student. "A people without a vision go unrestrained" and "People die for lack of vision." Vision is *good;* it just can't be "law."

(You may notice that a few letters are missing in a couple of the pillars. Not a malfunction of the computer or of my brain – I simply had to condense to fit it all on there!)

PERSONAL ORGANIZATION SKILLS CHECKLIST *(Female)* (IV-60)

There are certain personal organizational skills I don't want my kids to leave home without. So over the course of their high school years, there is an inventory of skills that I want them to either *be aware of how to do* or to have already incorporated into their lives. They can do this as the skills become appropriate for the needs in their lives. This first list is for girls. But I wouldn't leave the guys out. See the full-fledged "T.A.S." in the "Completed Class Plans" section under the "Home & Life Management Skills" class for the boys' version.

PLAY CRITIQUE & ANALYSIS (IV-61)

You might want to make twice the number of copies of this sheet that you would actually use, enough to have one for a rough draft and one for the final copy. I suggest taking the rough draft sheet on a clipboard to the play itself, and jotting down, during intermission, some quick notes about points that might not be easily remembered later. Be sure to take a program home to refer to for information on actors, scenes, etc.

PLOT WHOLESOMENESS EVALUATION (IV-62)

There are so many books and movies available to us. Many have titles that make it obvious they are off-limits for any Christian wanting to stay pure. But in the mainstream American culture which is laced with deceptions, more and more reading and viewing material is becoming unacceptable. Even movies that are rated "G" or "PG" and appear wholesome at first glance can be just as indecent as what would have been considered "X-rated" when I was a teenager! Since we don't want to just jump right into an immoral plot simply for the sake of doing a Plot Wholesomeness Evaluation, it's wise to do some checking into the book or movie first. Focus on the Family produces a publication called "Parental Guidance" that gives an overview of current movies from a Christian perspective. There are also Christian sites on the Internet that have movie reviews and evaluations for Christians. Full directions for this one are on the sheet.

POET SYNOPSIS (IV-63)

This form asks questions to get the student looking into and wondering about the person and era behind the poet, as well as taking a little closer – though not terribly in-depth – look at the work itself.

RELIGIONS OF THE WORLD STUDY SHEET (IV-64)

You can find your own sources for information on the various "religions of the world," some of which are cults or sects, at the public or a church library, or a Christian book store. A book called *So What's the Difference?* does an excellent job of explaining all the differences, and even the subtle similarities that deceive people into believing lies. We found it to be a little more detailed than we needed, but it's a good overview, and pointed out the conflicts with Christianity.

Some might have a problem with studying other religions. I think it will better prepare my kids to make wise decisions if they can find out now what these religions really believe. It gives us a chance to discuss it together, so I can hear what they are thinking and possibly help them see "light' on some deceptions or misconceptions that they may be developing – and hopefully deal with and/or cut them off *now*, rather than letting them lie quietly under the surface perhaps to be watered and nurtured by the "right" person at a vulnerable time later on, and possibly opportunity to "bloom" then!

SCIENTIFIC IMAGING TECHNIQUES (IV-65)

As we were reading *The Visual Dictionary of the Human Body* we came across pictures produced by several scientific imaging techniques, (for medical purposes) like:

- MRI Scan
- Angiogram
- X-ray
- CT Scan
- Ultrasound Scan
- Scintigram
- Thermogram
- Double Contrast X-ray (find out: how is it different from an X-ray?)
- Contrast X-ray (find out: different from an X-ray?)

You could probably just call them "imaging techniques," omitting "scientific" from the term, but I didn't want to confuse this with the New Age-y sounding "imaging." ☺

SCRIPTURE VERSE STUDY (IV-66) ✝

With this sheet, any verse of the Bible is wide open for study. Obviously this does not give the Greek or Hebrew roots, geographical facts, or historical settings, but it does enable you to get the most out of what the English language is capable of communicating. Many times God has spoken things to my heart things that have brought tears as I sat reading a simple dictionary definition of a word.

I'd even say that the more you have read the Word and the more familiar you are with it, the more beneficial this exercise will be for you! Because the longer we read the same words over and over, the less likely we are to see new meaning in them. Not that the Word of God itself is not fresh and powerful! But the Word of God is really what He's saying to us through or via the written words of the English language that are printed on the page. I'm probably sounding heretical. I hope not. My point is that when we take a few minutes to find out what the words really mean that are sitting on the page, we're in for a blessing and probably a few surprises, even with words we *thought* we totally understood!

Let me give you an example... I almost felt silly looking up the word "golden" which was describing a lampstand. First of all, I was already taking the golden lampstand to mean *me* – symbolically of course. What do you think came to my mind first? Yes, the usual "hide it under a bushel - NO!" and "set it on a hill for all to see..." But as I read this definition in just a plain ol' dictionary, God opened the Word even further: "shining, lustrous, superb, flourishing, prosperous, radiantly youthful and vigorous, favorable, advantageous, mellow, resonant." Among several other words there, I had to look up "mellow." It means: "soft and sweet because of ripeness, well aged and pleasingly mild, made gentle by age or experience; being rich and full but not garish, flashy, showy, loud or harsh." (WOW!!!) And then "resonant" which means "the increasing or prolonging of sound in one body (me) caused by sound waves from another body. (Jesus!) Woah! This about blew me away! Now those of you who haven't heard such words of love and affirmation from Jesus directly and personally to you are perhaps even a little offended that I would assume He was talking to me, *moi*, yes, Barb Shelton. My prayer for you is that you will lay down all old preconceived ideas about how God feels about you, and receive, just receive this awesome love He has for you! It's nothing I've ever earned, nothing I'm worthy of receiving; it's just totally a *gift* from Him! I don't have to do one teensy little thing to get Him to love me any more than He already does!

I hope you'll take this next statement I'm going to make in the way I'm giving it, because if you don't, it will sound terribly arrogant and egotistical. If, as you've been reading this book, you've wondered where my freedom and creativity have come from, (like 'em or not,) you've just now found out! Receiving, really and truly, completely and deeply, words of love from God himself is absolutely nothing less than transforming! If you see anything in me you like, Jesus put it there! His love-words have been His main surgical instrument to bring about the changes in me. He still has much spiritual surgery to do in me, but at least I'm on the right operating table and in the hands of the Master Surgeon! ☺

TAPE REPORT (IV-67)

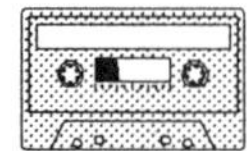

Have the student read through this *before* listening to a tape so he will know what to be looking for while listening. Use this for *any* tape your student wants or needs to report on for *any* class. (STRESS REDUCER: Keep several blank copies of these sheets on a "Tape Report Clipboard" that you keep in a special spot on a nail, maybe behind a door.)

TEL-MED PROOF-OF-COMPLETION (IV-68) ☎

Tel-Med is (or at least *was)* a tape service provided by participating hospitals and health centers. A brochure can be picked up at (or by calling) the facility that lists all the tapes available. When the volunteer answers, you simply state the number of the tape you wish to hear. He/she puts it on and you listen. It's that simple! (And no charge.) To help the student be more aware of learning while listening, and to document his/her findings, I designed this sheet. It's fairly self-explanatory, but there are a couple of points I should mention. There are four choices for "Course Category Options." This is because the subject covers all these topics. Earlier in the student's high school years, he may not yet know which category he will want to enter a particular tape into, or have it count for. So I recommend starting out with a Tel-Med (or whatever your service is called) file, which could be a "Pee-Chee" type folder (bi-fold with pockets inside) or a notebook, and storing all reports there. As more time is accumulated, it will become more evident which classes are shorter on hours and in need of these subjects, and then distribute the appropriate reports from the Tel-Med "stash" into subjects as desired and needed.

This Tel-Med worksheet is an example of one good way to utilize a community's resources. Especially if you live in a larger city, there are many others you can take advantage of. This sheet gives you a specific idea of *how* to take advantage of and record knowledge gleaned from such a resource.

However, with all that said, I need to tell you that, Donna, my "right-hand (wo)man" on the Form+U+la email list who has been helping me with the revision of this book, told me this: "Tel-Med isn't in operation any longer, at least in this area, but we used the Q & A section in front of the US West yellow pages instead. It is also tape format but covers many more subjects than just medical. e.g. Science, General Health, Home & Family, Child/Teens, Law, Current Events, Women's Health, Finances, Funeral to name a few." So I'm not sure if that particular service (Tel-Med) is available in your area, but I know there was something similar by a different name that I heard about a few years ago (and can't remember the name of!), and now Donna has mentioned her resource too. So that's why I kept this sheet here in the book. I didn't know what else to call it so I hope that the beginning of the name – "Tel" – will at least make the sheet handy for you to refer to if you do find something similar, knowing it's a TELephone service. ☺

VIRTUAL DISSECTION RECORD (IV-69)

This is a sheet that *would* have been in my book *Lab Science: the How, Why, What, Who 'n' Where Book,* had I thought of it before I wrote the book! But since it's not, and I won't be revising that for a while, I'm including it here so that you don't have to miss out. It's for use with the "virtual dissections" that you can find online! They are amazing! I'm not putting any website addresses here because I don't want to get any calls telling me *"It's not working and where else is there such a site?"* ☺ All you need to do is enter the words *dissection+ frog* or *brain* or *heart* or *eyeball,* and several sites will be listed, which you can just start trying until you find one you like. This is one of the activities I included for our "Lab Science Survey" class. It has advantages that you can't get doing the "real thing" like much more detail, getting to areas that we couldn't without a highly skilled teacher and technical instruments. *However* there are also aspects of doing the "real thing" that you cannot get by doing a "virtual dissection" – like *seeing* it firsthand, handling it, using the instruments, following directions, sketching a 3-dimensional object, etc. So I do not feel that "virtual dissections" are a substitute, but they can certainly be an excellent enhancer of the general lab experience and acquisition of knowledge!

WAR REPORT (IV-70)

If I told you that, even at my age, I still didn't know what the major wars were, what they were about, or when they (even generally) happened, would you believe me? Well, it's true, I'm ashamed to say. If I were ever to get into a historical conversation with a history buff, I would be socially embarrassed. My own pea-sized understanding of history was exactly why I came up with the "War Report." This actually works great; I make up the sheets; the kids get to hunt down the information and teach it to me! Their mission, whether they wanted to accept it or not, was to research and complete one of these reports for each of the major wars since the U.S. began. Of course, I didn't even know enough to identify what these wars were, so I asked "History Buff Extraordinaire," Keith Merritt, a family friend, if he would give me a run-down of those infamous events. Here's what he gave me, in chronological order: *(I'm sorry, I just can not and will not put little hearts beside wars!)*

ON AMERICAN SOIL:

☹ French-Indian War

☹ War of Independence

☹ War of 1812 *(One of the only two questions I ever got right on history tests was: "When did the War of 1812 take place?" You can imagine my chagrin upon finding out that it was just a joke question.)*

☹ War of Mexico

☹ Civil War

☹ Indian Wars

ON OTHER SOIL:

☹ Spanish-American War

☹ World War I

☹ World War II *(This was the other history test question I got right: "Which war came first; WWI or WWII?")*

☹ Korean War

☹ Viet Nam

☹ Desert Storm

Once completed, all these sheets can either go into a notebook or be made into their own reference booklet entitled – well, I bet you can come up with your own clever title – like "Infamous Wars" or something even wittier.

WISDOM'S 7 PILLARS TOPICAL STUDIES (IV-71)

I got this idea from Proverbs 9:1: "Wisdom has built her house, she has set up her seven pillars." I thought it would be fun to apply this idea to several areas of my life. There is no set list of pillars given anywhere; the "7 pillars" are just whatever *you* determine the basic foundation to be for any area of life that interests you, as you study the Word. A few ideas for specific areas of study are...

Wisdom's 7 Pillars For...

- A Healthy Mind
- A Self-Disciplined Life
- Harmonious Relationships
- Healthy Eating Habits
- True Education
- An Orderly Home
- Physical Health
- Intimacy with Jesus
- Training Your Children
- Building a Strong Marriage
- Spiritual Growth & Edification
- Hospitality in the Home

Get the idea?! ☺ It springs from whatever God is doing in your or your child's life; whatever you're hungry to understand in more depth. You can start by looking in a concordance for verses pertaining to the area you decide you want to study; then you get to determine, from what you read and your own observations and experiences in life, what you see as being the "Wisdom's 7 Pillars" for that area! For instance, for "Wisdom's 7 Pillars for an Orderly Home Life," based on what I (we?) have read in Marilyn's materials, you might start with:

1) **Provide a basic daily routine (not tight schedule) for yourself and your children.**
2) **Determine to work through all behavior problems NOW.**
3) **Develop in myself and my children an unselfish attitude of helping and serving.**
4) **Determine to have "a place for everything and everything in its place."**
5) ______________________________
6) ______________________________
7) ______________________________

These are just off the top of my head, so I'll let you continue with 5, 6 and 7 if you're interested in doing so. Your list doesn't necessarily have to be "all inclusive" – just what you feel are the most important aspects or principles to focus on; what God is saying to you for your family at this time. And you can get as in-depth as you want with it. If there are people that you see as being "accomplished" in an area you're wanting to grow in, consider "interviewing" them as to what they think a few of the "pillars" might be. But don't have them *do* it for you. ☺ This is just to learn from them.

Any of these studies will usually take a long time to accumulate, although if you really got serious, you could certainly do it in a short time. The point isn't speed or stretching it out, but rather getting it into your life.

There's nothing intricate about this sheet. You could certainly have written it up yourself, but I just thought that you "graph intimidados" might enjoy the luxury of having it made up for you. No fuss (with rulers). No muss (from blotchy ink). If you have a computer and you know how to do basic tables (charts), you've got it made!

Under each topic you choose, the numbers 1 through 7 are written, (you probably figured that out) with a blank space beside each. When I come across a verse in the Word that would fit under one of my topics, I write it down here. Eventually you can come up with quite a valuable study. Please note that the purpose of this type of study is not to be "correct" or exhaustive. My seven pillars for relationship might have only two or three of the ones you come up with. Dr. Dobson or Gary Smalley might have an entirely different line-up. The only criteria you should have is that the elements be life-giving to you! So, parents, I hope you won't *correct* your student on this!

As the Holy Spirit leads, ideas will spring from a variety of sources. This type of study is the most enjoyable and profitable when it's something the Lord is doing in you! – *or* in your child. An older child could do it on their own with some input and direction from you; a younger child would need to do it with you. Don't do it just to amass a pile of studies, though it's always profitable to get the Word more deeply rooted in our hearts! "His word goes forth and does not return void." Let Him lead. ☺

WORD ALIVE STUDY (IV-72) ✝

The "Word Alive Study" will be one of the most effective keys in opening the Word of God and making it come alive to you out of all that I have to offer. You could do ten of these a day and live a thousand lifetimes and not begin to scratch the surface of the depths and wonders of God's exciting truth! (I guess that's why God invented eternity!) This one activity will do more for your knowledge of God and His word and His love for you than anything else in this manual. Use it with joy! (Directions are on the sheet. Not too complicated!)

WORD (BIBLE) STUDY ON "EDUCATION" (IV-73)

If Mom is taking my "Season of Re-education and Renewing of the Mind," she will recognize this study. A student can get just as much out of this as Mom, so I included it here. ☺ It's in the Course Syllabus for that, along with similar studies on the words *learn* and *knowledge,* the definitions of which I included here for your convenience. *(You're welcome!)* I got the definitions from Webster's 1828 Dictionary – which I *highly* recommend.

I lead a women's Bible study (just this week) on "Knowledge" (see next page), and it was *very* eye-opening! If you'd like to hear the best part of it, here goes. First, read the definition of *knowledge* on the next page. Okay, now let's go to I Peter 1:5-8 that says: "**Make every effort to supplement your faith with virtue, and virtue with knowledge, and knowledge with self control, and self control with steadfastness, and steadfastness with godliness, and godliness with brotherly affection, and brotherly affection with love**." *(and now the reason!)* "**For if these things are yours and abound, they keep you from being ineffective or unfruitful in the knowledge of our Lord Jesus Christ**." If you look at the first part of that verse as a "wall" (of protection), where exactly in the line-up is *knowledge* placed in the "brickwork"? Hmmm, now where is it placed in the public school system? It's not even *in* that context, is it?!? And what does the Word say "knowledge alone" will do? Okay, now look at Romans 1:18-24. (Are you ready? This is *heavy!)* "**For the wrath of God is revealed from heaven against all ungodliness and wickedness of men who by their wickedness**" [which is "just" rejecting God and His ways] "**suppress the truth. For what can be known about God is plain to them, because God has shown it to them. Ever since the creation of the world his invisible nature, namely his eternal power and deity, has been clearly perceived in the things that have been made. So they are without excuse; for although they knew God they did not honor him as God**" [but instead completely kicked Him out of their textbooks and educational system] "**or give thanks to him, but they became futile**" [pointless, meaningless, useless] "in their thinking and their senseless minds were darkened. Claiming to be wise" [with all their degrees and "higher education"] "**they became fools**" [and now the people pay thousands of dollars to attend the schools that they teach] "**and exchanged the glory of the immortal God for images resembling mortal man...**" [thus setting *himself* up as "God" and worshiping himself, the created, over his Creator, even sinking to the ludicrous depths of actually thinking that the created could create itself!] "**Therefore God gave them up to the lusts of their hearts to impurity, to the dishonoring of their bodies among themselves, because they exchanged the truth about God for a lie and worshiped and served the creature rather than the Creator, who is blessed for ever! Amen.**"

(Please note, I'm not "adding to the Word of God; just interjecting my thoughts and responses in between inside the [brackets]. ☺) Is that passage of Scripture *amazing* or what?!?!?! Oh, Lord, have mercy on our country and upon the world! Shed the light of your truth on our darkened minds! We desperately need You, Lord! Without your mercy and grace, we are a lost and doomed people! Start with me (us), Lord, and use us however You see fit to bring the Good News of your *redeeming love!*

WORD STUDY ON "KNOWLEDGE"

DEFINITION: **"A clear and certain perception of that which exists, or of truth and fact; the perception of the connection and agreement, or disagreement and repugnancy of our ideas. We can have no knowledge of that which does not exist. God has a perfect knowledge of all His works. Human knowledge is very limited, and is mostly gained by observation and experience."**

Further definition(s): (Choose words from the above definition that stand out to you and define them.)

WORD STUDY ON "LEARN"

DEFINITION: **"To gain knowledge of; to acquire knowledge or ideas of something before unknown. We learn the use of letters, the meaning of words and the principles of science. We learn things by instruction, by study, and by experience and observation. It is much easier to learn what is right than to unlearn what is wrong."**

Further definition(s): (Choose words from the above definition that stand out to you and define them.)

WORLD MAGAZINE REPORT (IV-74)

This worksheet is based on "World," the weekly magazine publication produced by "God's World Publications." (Address is: Box 2330 / Asheville, NC 28802-2330.) Personally, I'd save it for the junior and senior years. It goes more in-depth on issues than I personally care to go. The next level down (younger) publication, "God's World," gives a good over-all picture of current events around the world – and is really more *my* level! ☺ *(I don't need excessive details in this area.* ☺) Both are from a conservative Christian Perspective.

This sheet itself is self-explanatory. However, you'll want to keep an eye on the various departments and columns. They have been known to change or be deleted after a few months.

AMERICA THE BEAUTIFUL

Katharine Lee Bates

O beautiful for spacious skies,
For amber waves of grain,
For purple mountain majesties
Above the fruited plain!
America! America!
God shed His grace on thee
And crown thy good with brotherhood
From sea to shining sea!

O beautiful for pilgrim feet,
Whose stern, impassioned stress
A thoroughfare for freedom beat
Across the wilderness!
America! America!
God mend thine every flaw,
Confirm thy soul in self-control,
Thy liberty in law!

Define each of the underlined words in their context: *(There are 28, so you've got 1 line per word.)*

thy = ___________ thine = ____________ prose = ____________________

Study the wording; discuss it with someone who can help you "interpret" the meaning, and then re-write the verses into common, understandable American-style English prose, using words from the definitions and the above meanings of "thy" and "thine":

America the Beautiful

by ______________________________________

ARMOR OF GOD:
DEFINED AND PRAYED FOR

VERSE PORTION (*Underline word to be defined*→)	Vs	DEFINE UNDERLINED WORD OF VERSE	MY PRAYER
Be strong in the Lord...	10	one having power and authority over others	Jesus, I submit to You and give You more authority over all my life.
and in the strength of His might...	10	power, resources, authority	You be the power in me to follow You.
Put on the whole armor of God...	11	protective covering that shields from harm	Jesus, be my protective covering!
that you may be able to stand against the wiles of the evil one...	11		
for we are not contending against flesh and blood...	12		
but against the principalities...	12		
against the powers...	12		
against the world rulers...	12		
of this present darkness...	12		
against the spiritual hosts...	12		
of wickedness in the heavenly places...	12		
Therefore, take the whole armor of God...	13		
that you may be able to withstand...	13		
in the evil day, and having done all, to stand...	13		
Stand, therefore, having girded...	14		
your loins with truth...	14		
and having put on the breastplate of righteousness...	14		
and having shod your feet...	15		
with the equipment...	15		
of the gospel of peace...	15		
above all, taking the shield of faith...	16		
with which you can quench...	16		
all the flaming darts...	16		
of the evil one...	16		
And take the helmet of salvation...	17		
and the sword of the Spirit, which is the word of God...	17		
Pray at all times in the Spirit...	18		
with all prayer and supplication.	18		

BOOK REPORT

Your Name____________________ Date ____________ Phone______________

Book____________________________ Author______________________

Publisher____________________________ Copyright Year__________

Where purchased or borrowed____________________________________

What type of book is this? *(fiction, non-fiction, biography...)* ______________________

For what type and age of audience is the book primarily written? ________________

Why do you think the author wrote it? __________________________________

____

Briefly tell what the book was about. *(Please fill up all these lines.)* ________________________

____

____

Which two chapters were the most applicable to your life, and/or the most enjoyable, and/or made the most implact on you? and why?

1. ____
2. ____

Quote one statement that made an impact on you, and why:

____

____

Was there a concept you didn't understand or agree with? If so, why? __________________

____

Did the book inspire you to change? If so, how? If not, was it really worth reading?

____

____

Would you recommend this book to others? Why or why not? _______________________

____

Book Synopsis

*Student*______________________________ *Date*___________ *Time/Start:*_________ *End*________ *Total* = _________

❦ ❦ ❦

Title of Book: ____

Author: ____

The ____-page synopsis is found on page ____ of the ______________________ section of Volume ___

of the Great American Bathroom Book ***or:*** ____

Is this book considered a "classic"? _______ About when (year or era) was it written? ____________

Main character(s): ____

____

Setting of the story: TIME PERIOD: ___________________________ LOCATION: _________________________

From whose perspective is the book written? *(author's; main character's?...)* ________________________________

Is there perspective ☐ **omniscient** *(author sees everything)* ☐ **limited** *(one person's point of view)* ... *How can you tell?*

____

Basic plot: ____

____

____

RATING TABLE

TYPE OF VALUE	RATING *(Circle one)*	COMMENTS
Moral Value *(integrity, high standards)*	0 1 2 3 4 5	
Historical Value *(insight into the past)*	0 1 2 3 4 5	
Spiritual Value *(Christian; see/sense God more)*	0 1 2 3 4 5	
Human Understanding Value *(see human nature more clearly, for better or worse)*	0 1 2 3 4 5	
Scientific Value *(opened my eyes to science)*	0 1 2 3 4 5	
Life-affecting Value *(changed me)*	0 1 2 3 4 5	
Tear-jerking Value *(emotional)*	0 1 2 3 4 5	

This [does / does not] look like a book I'd like to read all of. Why or why not? ______________

____

CHRISTIAN DENOMINATION STUDY WORKSHEET

Denomination____________________ Based in/at____________________

Founded by ____________________ When/Under What Circumstances?____________________

____________________ Source(s) of information____________________

What Do They Believe About... *(Use other side if necessary.)*

Jesus ____________________

The Holy Spirit ____________________

Salvation ____________________

Baptism ____________________

Prayer ____________________

The Bible ____________________

Communion ____________________

Heaven and Hell ____________________

Angels ____________________

Speaking in Tongues ____________________

The Baptism in/of the Holy Spirit ____________________

Grace ____________________

Women in Leadership and/or Ministry ____________________

The role of a pastor ____________________

The role of the people ("laity") ____________________

Drinking Alcohol ____________________

Sunday/the "Sabbath" ____________________

Homosexuality ____________________

Clothing/Attire ____________________

Dancing: In/Out of Church ____________________

Dating/Chastity ____________________

Marriage/Weddings ____________________

Having Children ____________________

Divorce ____________________

Membership & Other Rules or Guidelines ____________________

____

____

CHRISTIAN RELIGION TYPES
WORKSHEET

Name________________________________ Date ______________

EVANGELICAL: Dictionary Definition:

Generally-understood "church world" definition:

Common misunderstandings or misperceptions:

CHARISMATIC: Dictionary Definition:

Generally-understood "church world" definition:

Common misunderstandings or misperceptions:

PENTECOSTAL: Dictionary Definition:

Generally-understood "church world" definition:

Common misunderstandings or misperceptions:

FUNDAMENTAL: Dictionary Definition:

Generally-understood "church world" definition:

Common misunderstandings or misperceptions:

COMPUTER TERMS

Name ______________________________ Date ____________________

boot

delete

hard copy

software

chip

DOS

menu

file name

cursor

byte

hardware

directory

COMPUTER TERMS

Name ______________________________ Date ____________________

terminal

floppy disk

debug

F keys

database

warm boot

mouse

backup

scroll

hard drive

port

modem

CONCERT/PERFORMANCE RECORD

Student Name ______________________________ Date ______________________________

Performing Person or Group: ______________________________

Type of Performance *(i.e. Piano, Vocal Ensemble, Brass Band,...)* ______________________________

Location: ______________________________

♪ ♪ ♪ ♪ ♪ ♪ ♪ ♪ ♪ ♪ ♪ ♪ ♪ ♪ ♪

What I enjoyed about this performance: ______________________________

____

How I'd generally describe the performance to someone who hadn't been there: ______________________________

____

Describe the set: ______________________________

____

Describe the attire: ______________________________

____

Describe the props: ______________________________

____

Describe the instruments used: ______________________________

____

What I wasn't crazy about: ______________________________

♪ ♪ ♪ ♪ ♪ ♪ ♪ ♪ ♪ ♪ ♪ ♪ ♪ ♪ ♪

Where I think it would be most "appropriate": *(Underline all that describes the entire **range** of music.)*

looking at the world from on top of a mountain   trying to fall asleep   romantic

grocery store music   keeping me awake on a trip   in headphones on a walk

dance (Type:______________________ *Would you feel comfortable doing it before Jesus?*_______)

vigorous exercising   relaxing, as when doing school work, but not putting me to sleep

Other: ______________________________

I [would / would not] enjoy attending something like this again. Why: ______________________________

____

COOKING TERMS

Look up each of the following terms in a dictionary or a cookbook — preferably the latter so that you understand the meaning of the word as it pertains to ***cooking****, not just a generic meaning. Betty Crocker's Cookbook is an excellent resource for learning about cooking, not just recipes. Some are more basic than others; they're all good to know.*

TERM	DO WITH/IN	DEFINE HOW TO DO IT OR WHAT IT IS
Boil		
Toss		
Bar-be-que		
Julienne		
Grate/Shred		*(Any difference?)*
Poach		
Marinate		
Baste		
Mash		
Brown		
Simmer		
Cream		
Snip		
Thaw/Defrost		*(Any difference?)*
Fold		
Pare		
Cube/Dice		*(Any difference?)*
Grill		
Mince		
Saute		
Rotisserie		
Braise		
Scald		

CREATION SCIENCE RESEARCH REPORT

This is a report on "Acts & Facts," the monthly publication of "Institute for Creation Research." When asked for summaries, this refers to articles that educate in the area of creation science rather than relate news of their organization.

Student__________________________ Volume____ No.____ ________________(Month) _________(Year)

Highlights of news tidbits around the world and/or nation: *(State country first; then the news.)*

______________________________ - ____

______________________________ - ____

______________________________ - ____

Title, summary, and personal comments on/of this issue's "Impact" article":

(Title:) ____ - *(Summary:)* ______________________________

____

____

____

____

____

Two other main articles, summaries, and comments:

(Title:) ____ - *(Summary:)* ______________________________

____

____

____

____

(Title:) ____ - *(Summary:)* ______________________________

____

____

____

____

New vocabulary words:: ____

DINNER MAKING SCHEDULE

Monday – *Date:* ____________ *Cook:* ____________

Main Dish: ______________________________

Tuesday – *Date:* ____________ *Cook:* ____________

Main Dish: ______________________________

Wednesday – *Date:* ____________ *Cook:* ____________

Main Dish: ______________________________

Thursday – *Date:* ____________ *Cook:* ____________

Main Dish: ______________________________

Friday – *Date:* ____________ *Cook:* ____________

Main Dish: ______________________________

Saturday – *Date:* ____________ *Cook:* ____________

Main Dish: ______________________________

Sunday – *Date:* ____________ *Cook:* ____________

Main Dish: ______________________________

DINNER PLANNER

Chef: ______________________ **Date:** ______________________

Menu

Main Dish: ______________________

Vegetable: ______________________

Starch ______________________

Other (Dessert? Rolls?) ______________________

Beverage: ______________________

Recipes

For ______________________ Source: ______________________

For ______________________ Source: ______________________

For ______________________ Source: ______________________

Time Plan

____

____

____

____

____

____

Shopping List:

____

____

____

____

____

____

____

Student's Comments/Evaluation: ______________________

Teacher's: ______________________

____

DOG REPORT ON ______________________

By______________________ **Age**________ **Date**______________

Average weight and size at birth: ______________________

Average weight and size full grown: ______________________

Typical color combinations:______________________

Combination you're looking for: ______________________

How is this dog at obedience? ______________________

How is it at being a watchdog? ______________________

Typical temperament and personality: ______________________

____

Does this type of dog need haircuts? __________ If so, how often? __________

Price of haircuts: __________ At: ______________

How much of what type of food does this dog need? ______________

______________________ Price:__________

⊕ ⊕ ⊕ ⊕ ⊕ ⊕ ⊕ ⊕ ⊕ ⊕ ⊕ ⊕ ⊕ ⊕ ⊕ ⊕ ⊕

Veterinarian queried: ______________________ Phone __________

Shots needed and at what age: ______________________

______________________ Price: __________

Cost of having this dog [spayed / neutered] here: ______________

Recommendations for puppy care: ______________________

Recommendations for dog care: ______________________

* * * * *

Veterinarian queried: ______________________ Phone __________

Shots needed and at what age: ______________________

______________________ Price: __________

Cost of having this dog [spayed / neutered] here: ______________

Recommendations for puppy care: ______________________

Recommendations for dog care: ______________________

Elected Officials and Their Functions

Search-It-Out Sheet

Student______________________________ Date__________________

Write a brief explanation of the functions and duties of each of the following elected officials. *(Please fill up the lines available.)*

Governor: ______________________________
____
____

Lieutenant Governor: ______________________________
____
____

Secretary of State: ______________________________
____
____

State Treasurer: ______________________________
____

Attorney General: ______________________________
____

State Auditor: ______________________________
____

Superintendent of Public Instruction: ______________________________
____

Insurance Commissioner: ______________________________
____

Commissioner of Public Lands: ______________________________
____

Family Meeting Guide

Date of Meeting____________________ **Place**____________________

Who was present? ____

____

♥♥†♥♥

Meeting Agenda:

Open with prayer and a Scripture verse (*relating to harmony, soft answers, openness...)*

____

First, what's been going *right?* (Each person with any upcoming grievances must offer one "good" for each gripe.) *(Write any note-worthy comments here.)*

____

____

____

Grievances or concerns that we need to discuss as a family. (Make sure this isn't a matter that needs to be talked over in private with another individual. We don't want this to be a time to needlessly embarrass people.) The main idea of each of these matters should either have been written down here before the meeting, or stated as we begin our meeting. (—to make sure we cover everything that's on each person's mind, leaving noone out and making sure we don't get sidetracked.) *The grievances or concerns:*

____ *Initials*______

____ *Initials*______

____ *Initials*______

____ *Initials*______

____ *Initials*______

Any ideas for what we'd like to do next *just for fun* as a family?

____

____

____

How are we doing at having "unity of spirit, sympathy, love of the brethren (and "sisteren"), a tender heart and a humble mind, not returning evil for evil or reviling for reviling, but on the contrary blessing"? ("...for to this we have been called, that we may obtain a blessing.") *Any comments?*

____

____

Search-it-Out Sheet for

FORT CLATSOP MUSEUM

1. What U.S. Pres. sent ________________ Lewis and ____________ Clark? ______________

2. [Lewis / Clark], an officer in the ______________, was highly regarded by this President and was thus chosen for this job.

3. Clark was then selected by __________________, who was his good friend.

4. Especially for that time, they were very ____________________.

5. Why did the president send them?

   a. to find out what the U.S. had purchased in the __________________ Purchase of 18____.

   b. to find out what lay beyond the _________________ Divide (the Rocky Mountains) and out to the __________________ Ocean.

   c. to find places to occupy and farm

6. As they traveled, they took copious (tons of) notes on:

   a. the _________________ (plants)

   b. the land forms or ____________________

   c. the ___________________

   d. the various ___________________ of Indians

7. An interesting couple accompanied them on the expedition. The man's name was ___________________________ who was a ______________ *(nationality?)* trapper.

8. He was hired to ______________ game along the way for food.

9. His wife was Sacajawea, age _____ from the _____________ tribe from the Montana-Idaho tribe.

   She brought along a _____________.

10. How did he obtain her as his wife? ____

11. In what way did she have a significant effect on the success of the expedition?

    ____

12. What are the 2 (two) major reasons the expedition built and then stayed at Fort Clatsop? *(Circle the two that are true.)*

    a- protection from Indian attack d - farming

    b- to winter over—for shelter e - find a good place for a summer resort for the President

    c- to establish a permanent home f - establish a trading post

GOD'S ACTIONS STUDY

Verse	God is or does what? *(Verb)*	To/for whom? *(Direct Object, Predicate Nomnitive, or Predicate Adjective)*	Prepositional Phrases *Modifying a verb or adverb? (Adverbial) or modifying the noun? (Adjectival)*	So which is it? *Prepositional Adverbial Phrase? Prepositional Adjectival Phrase? Circle the word it modifies.*
Ps144:1	trains	my hands	for war	Prep. Adv. Phr. *(Why?)*
144:2b	subdues	the peoples	under Him	Prep. Adj. Phr. *(Whichpeoples?)*
144:7	stretches forth	His hand	from on high	Prep. Adv. Phrase *(Where?)*
144:7b	delivers	me	from the many waters	
144:11	rescues	me	from the cruel sword	
145:8b	is	abounding	in steadfast love	
145:13	is	faithful	in all His words	
145:13b	is	gracious	in all His deeds	
145:14	upholds	all	who are falling *(tricky- look up!)*	
145:15	gives (them)	their food	in due season	
145:16	satisfies	the desire	of every living thing	
145:17	is	just	in all His ways	
145:18	is	near	to all who call upon Him	
146:7	executes	justice	for the oppressed	
146:7b	gives	food	to the hungry	
146:8	opens	the eyes	of the blind	
147:4	determines	the number	of the stars	
147:4b	gives	their names	to all of them *(2 prep. phrases)*	
147:6	casts	the wicked	to the ground	
147:8	covers	the heavens	with clouds	
147:8b	prepares	the rain	for the earth	
147:8c	makes grow	the grass	upon the hills	
147:9	gives	their food	to the beasts	
147:11	takes	pleasure	in those who love Him *(tricky!)*	
147:13b	blesses	your sons	within you	
147:14	makes	peace	in our borders	
147:14b	fills	you	with the finest of the wheat *(2!)*	
147:15	sends forth	His command	to the earth	
147:19	declares	His word	to Jacob	

GROUND RULES
FOR STUDYING OTHER RELIGIONS

The following questions are based on information presented in the "Forward" of the book So What's the Difference? *They're written so that, whether you have the book or not, they are fairly understandable.*

1. Studying other religions is not to be done with an attitude of pr__________, a spirit of cr_______________, or with intentions of ill __________.

2. Since studying other faiths will naturally bring disagreements to the ________________, it is important to recognize that dis_____________________ing with someone is not the same as ____________ing him down.

3. Try to find out what you can ____________ from those faiths. While examining the _________________ it's possible to apply certain ________________ in other faiths to our own _________________. There are many things we can learn, not with the intent to change our __________ogy (= "the study of God"), but to help us examine our own ______________ and see just how deeply our faith affects how we actually _______________ and talk and __________.

4. Comparing our Christianity with other faiths will help us better understand and _________________________ our own beliefs — and since the Spirit of God is alive and active in our hearts, such a study will serve to confirm (or _______________________) our own ________________________.

HIGH SCHOOL DAYS MEMORY BOOKLET

For my ____________________ year spanning the ________ – ________ school year Age ________

The name of our home school is ____

My teacher(s) was/were: ____

My closest friends were (are) and what I like doing with each one best:

________________________ - ____

________________________ - ____

________________________ - ____

________________________ - ____

________________________ - ____

Committees or teams I was on: ____

____

Social Events I enjoyed: ____

____

Performances I was part of: ____

____

My favorite areas of study, classes, or subjects this year were (and why):

____________________ - ____

____________________ - ____

____________________ - ____

How I grew and/or drew closer to God this year ____

____

Field Trips we went on: ____

____

Special Memories: ____

____

____

____

What I did on Christmas, Spring Break, &/or Summer:

____

____

____

(Use other side if necessary.)

HOME REPAIR, REMODELING, & MAINTENANCE JOB LOG

JOB & INSTRUCTIONS	DUE?	DONE	INSP.	TIME

HOME REPORT
FOR THE MONTH OF___________, 19___

Reporter___________________________ **Age**_____ **Date**_________________________

What regular events took place in our family this month?

1. ____ - ______ times per ________________

   Involving: ____

2. ____ - ______ times per ________________

   Involving: ____

3. ____ - ______ times per ________________

   Involving: ____

4. ____ - ______ times per ________________

   Involving: ____

5. ____ - ______ times per ________________

   Involving: ____

6. ____ - ______ times per ________________

   Involving: ____

List the other major events that took place: (when/where/who)

________________________________ ________________________________

________________________________ ________________________________

________________________________ ________________________________

In what ways did our family grow in learning how to handle conflict?

____

What areas of conflict, or disharmony do we currently deal with the most? *(who, what happens?)*

____

In what ways did we grow together (closer) as a family?

____

How much of the time did you feel bored this month? _____%

Do you have any idea why?____

How frequently did we get together with friends? *(P = by Phone, W = Went on Walk, O = Overnight)* How much did each watch TV or play Nintendo? Did these people feel this was enough, just right, too much?

PERSON'S COMMENTS:

Dad P-____ W-____ O-____ TV/N-____ ____________________

Mom P-____ W-____ O-____ TV/N-____ ____________________

Sharnessa P-____ W-____ O-____ TV/N-____ ____________________

Tory P-____ W-____ O-____ TV/N-____ ____________________

Carlianne P-____ W-____ O-____ TV/N-____ ____________________

What kinds of creative activity did each person take part in?

D - ____________________

M - ____________________

S - ____________________

T - ____________________

C - ____________________

Do you see any corelation between amount of TV time and creativity?

____

____

How did you feel about the condition of the house (messy/neat?) this month? ____________________

Were you part of the problem or part of the solution? ____________________

If you were an outsider, what might you say to someone about whether they'd want to live with us for a week? ____________________

Is there anything you'd like for us to do as a family next month?

____

What family would you like us to get together with next?____________________

Why? / What would you like to do with them?____________________

If we were on a Focus On The Family radio interview, what could we say we are doing right?

____

IDENTIFYING PREPOSITIONAL PHRASES IN THE BIBLE

Circle all the prepositions in this passage; then underline all the prepositional phrases that go with each preposition; draw an arrow from the phrase to the word(s) it modifies. *(I counted 12!)*

(More advanced:) Identify the phrase as being either a "prepositional adverbial (modifies either a verb or an adverb) phrase" or a "prepositional adjectival (pronounced: *"ad-jek-ti-vl"* — modifies a noun) phrase."

A Psalm of David.

Song to the King of glory

3 Who shall ascend the hill of
the Lord?
And who shall stand in his
holy place?
4 He who has clean hands and a
pure heart,
who does not lift up his
soul to falsehood,
and does not swear
deceitfully.
5 He will receive blessing from
the Lord,
and vindication from the
God of his salvation.
6 Such is the generation of those
who seek Him,
who seek the face of the
God of Jacob. *Selah*

LOCATION DATA SHEET
for

____
Museum / Historical Site

Location ____

Spans which decade(s) ____________________________________

Main Focus *(on an Era? Person(s)? Event? Physical Aspects?)* ______________________

____

Other Minor Focuses ______________________________________

____

____

Interesting Features *(Film? Dioramas? Computer Quiz? Hands-on Activities? Life-like Scenes? Replicas? Quizzes? Exhibits?)*

____

____

____

Aspect(s) I enjoyed most ___________________________________

____

____

PHOTO(S)

MOVIE CRITIQUE & ANALYSIS

Student ______________________ Age ______ Date ______________

Movie Title ____

Director ______________________ Producer ______________________

Actors, the main characters they played, and a brief description of their roles:

______________ as ______________ - ______________

____

______________ as ______________ - ______________

____

______________ as ______________ - ______________

____

______________ as ______________ - ______________

____

Summary of the plot ____

____

____

Subplot ____

____

____

Time Period ____

____

Physical Setting(s) ____

____

Your personal critique of this movie ____

____

What was the "take away" value of this movie? *(i.e. What did you learn; how did you grow?)*

____

____

I [would / would not] recommend this movie to others because ______________________

____

MUSIC APPRECIATION WORKSHEET

Song_______________________ **Music by**__________________ **Lyrics**________________

Performed by_________________________ Conductor/Leader__________________________

Tape/Record____ Time_____________

FEEL OF THE MUSIC:

What is the "mood" of the song? What emotions surface as you listen?

____

____

Close your eyes, let the music carry you away; where does it take you? *(Stay with Jesus!)*

____

What type of setting/event/use would it be appropriate for/in?______________________________

ANALYSIS OF THE MUSIC:

Style____ Number of beats per measure______

Instrument(s) used:____

RESEARCH ON THE COMPOSER:

Born_____________________ Died_____________________ How many years ago?___________

Born in (country) __________________________ Lived/grew up in (if diff.) ________________________

What was this composer's musical background and training? ______________________________

____

What is known about his or her life? (Was it good?) (Marriage? Clergy? Children? Job? Status?) Did any events or persons affect him/her? Did any handicaps, tragedies, or hardships happen to this person?

____

Any good times, accomplishments, triumphs? ____

Would you have wanted to have him/her as a friend? (Why or why not?)

____

(IF DEAD:) How composer died; in what state was his/her mind, property, health? Did he/she know of his/her greatness?

____

NAME STUDY

Name to be studied: ________________________

Meaning(s): *(Look up several.)*

____ Nationality:____________________

____ Nationality:____________________

____ Nationality:____________________

____ Nationality:____________________

The meaning you are choosing, personalizing, or combining:

____ Nationality:____________________

Scripture verses that enhance or go with the meaning: *(Put a star next to your favorite one.)*

____

____

____Ref.:____________________

____

____

____Ref.:____________________

____

____

____Ref.:____________________

____

____

____Ref.:____________________

____

____

____Ref.:____________________

OCCUPATIONAL SKILLS INTERVIEW

Interviewer: ______________________ Date: ______________________

Business ______________________ Phone ______________________

Address ______________________ City/St. ______________ Zip __________

Name of Interviewee ______________________ Title ______________________

Skills, Talents needed for this job ______________________

____

Character Qualities needed ______________________

Type of Personality that would be best suited to this type of work ______________________

____

Education/Training required ______________________

How long does this take? ______________________

Helpful Training or Experience ______________________

The most enjoyable aspects of this job ______________________

____

The most frustrating aspects ______________________

____

The most difficult thing(s) about this job are ______________________

____

What are the hours and days? ______________________

Are you self-employed, employed by a "boss" or a corporation or business? ______________________

____

What are the opportunities for advancement or raises? ______________________

____

Usual starting pay ______________________ Maximum pay ______________________

Any benefits? ______________________

____

Vacation? ______________________

What are your responsibilities? ______________________

____

____

What is a typical day like? ______________________

____

____

Do you have people under you? over you? ______________________

____

Do you need to be "people-oriented" to do well in this job? ______________________

Why or why not? ______________________

Did you have to set up your own office? ______________ What? ______________

______________________ How expensive? ______________________

What is your advice to someone considering this job? ______________________

OFFICE & BUSINESS SKILLS JOB LIST

JOB & INSTRUCTIONS	DUE?	DONE	INSP.	TIME

OPEN LINE & MIND EVALUATION

Evaluator________________________ **Responder**______________________ **Date**______________

♥ ♥ ♥ *Preparing Your Heart* ♥ ♥ ♥

First complete these verses, and then pray them through your own heart for a while:

1) "Search me, oh God, and know my _______________. Point out anything You find in me that makes you __________, and lead me along the path of everlasting life." *(Psalms 139:23—24)*

2) "Who may go up the _______________ of the Lord? Who may _____________ his holy hill? He who has ____________ hands and a pure heart." *(Psalms 24:3—4)*

3) "Create in me a ___________ heart, oh God, filled with pure _______________ and right desires." *(Ps.51:10)*

4) "In a wealthy home there are vessels made of gold and ___________, and vessels of clay and ___________. Some for _____________ purposes, and some for ______________. If you cleanse yourself from ________, you will be a vessel for ___________ purposes,... useful to the Master, prepared for any good _____________. *(II Tim.2:20—21)*

♥ ♥ ♥ *Personal Application* ♥ ♥ ♥

Next, seek out your "responder" person and first ask if this is a good time for them to talk! Then find a quiet, secluded place where you can meet privately. Pray together, and then ask these questions— in your own words if you prefer. Don't write down the person's answers, but do have them sign at the botton when finished, just for documentation.

Is there anything I've done to offend you that I haven't taken care of? (as in apologized for or made restitution for?)

Have I been bugging or irritating you lately in areas I can do something about?

Have I been entering your room or borrowing your stuff without permission?

Do you feel I speak to you respectfully in general?

Have I been guilty of pushing you past your limit? (If so, how?)

Have I given you any un-asked-for criticism?

Have I made fun of you, mocked you, or heartlessly teased you? *(Fun teasing is different.)*

If there were an assembly line of siblings, would you have picked me? (If not, why not?)

In what ways can I be a better sister/brother to you?

*Signature of Responder*________________________________

ORGANIZATION SKILLS JOB LIST

JOB & INSTRUCTIONS	DUE?	DONE	INSP.	TIME

PERSONAL HEALTHFUL LIVING
PLANNING SHEET

Name of Planner ______________________ **Age** _____ **Date** ______________________

This planner is to be used in co*njunction with the concept of "Wisdom's 7 Pillars." After reading and watching the assigned materials, pray and earnestly seek God as to what He would have you do in each of the following "pillars." Ask Him to produce in you* ***balance****. Have a good time doing this.*

Pillars ↓	SPIRITUAL APPLICATION	HOW YOU'LL IMPLEMENT THIS INTO YOUR LIFE
WATER		
FOOD		
EXERCISE		
TOUCH		
GOOD REST		
FRESH AIR		
ELIMNATION		

PERSONAL ORGANIZATION CHECKLIST
(Female)

For ______________________________ Date ____________________

To set up, implement or do: *Notes:*

_____ Daily Planner/Calendar

_____ Desk or work area

_____ Address Book (w/ envelope)

_____ Schedule

_____ Goal and Priority Planning

_____ Purse

_____ Mini Purse File

_____ Emergency Purse Kit (w/ own priorities)

_____ Magazine Notebook

_____ Travel/Suitcase Baggies

_____ Boudoir

_____ Journal

_____ Drawers and Closet Space

_____ Keepsakes Box

_____ Personal Treasures (Heirlooms) Record

_____ Warrantees Notebook

_____ Photo Ordering System

_____ Photo Filing System

_____ Photo Storage System

_____ Photo Album(s) and/or Scrapbook(s)

_____ Stress Reducers

_____ Creative Energy

To be familiarized with: *(and implement if applicable)*

_____ Specialized Containers for Purse

_____ Sewing Order

PLAY CRITIQUE & ANALYSIS

Student ______________________________ Age ______ Date __________________

Play Title ____

Director ______________________________ Location ________________________

Actors, the main characters they played, and a brief description of their roles:

____________________ as ____________________ a ________________________

____________________ as ____________________ a ________________________

____________________ as ____________________ a ________________________

____________________ as ____________________ a ________________________

____________________ as ____________________ a ________________________

____________________ as ____________________ a ________________________

____________________ as ____________________ a ________________________

____________________ as ____________________ a ________________________

Summary of the plot ____

____

____

Subplot ____

____

____

Time Period ____

____

Physical Setting(s)____

____

Your personal critique of this play ______________________________________

____

What was the "take away" value of this play? *(i.e. What did you learn; how did you grow?)*

____

____

I [would / would not] recommend this play to others because _______________

____

PLOT WHOLESOMENESS EVALUATION

We are admonished in Scripture to flee from any appearance of evil. The problem in fleeing from it sometimes lies in how subtle it appears. It's hard to run from something you can't see! Even movies that are rated G or PG and books that appear wholesome at first glance can be unacceptable.

That's why I devised the "Plot Wholesomeness Evaluation" sheet — to help kids (and adults) evaluate for themselves whether the film or book at hand is something God wants them taking in. We have been known to walk out of theaters and turn off rented videos part way into them.

By the way, this activity differs from the Movie or Play Critique & Analyses in that the purpose of the latter two is to analyze the components or "externals" of the subject, while the aim of this sheet is to survey and measure its moral "internal" content. Like a doctor doing a physical exam on a person versus God doing a "heart exam."

Since you don't want to just jump right into an immoral plot simply for the sake of doing an evaluation of it, you should either do some checking into the book or movie first. Ask some people whose values you respect if they have seen it. "Focus on the Family" produces a monthly publication called "Parental Guidance" that gives an overview of current movies from a Christian perspective. Several Christian magazines do the same.

How does the plot line up with the following verses?

"For the grace of God has appeared for the salvation of all men, training us to renounce irreligion and worldly passions, and to live sober, upright, and godly lives in this world..." *Titus 2: 11-12*

"Drink water from your own cistern, flowing water from your own well... Rejoice in the wife of your youth. Let her affection fill you at all times with delight,... Why should you be infatuated, my son, with a loose woman and embrace the bosom of an adventuress (or watch someone else do it?!)... For her speech is smoother than oil; but in the end she is... sharper than a two-edged sword. Her feet go down to death... she does not take heed to the path of life; her ways wander, and she does not know it." *Proverbs 5: 15, 18, 19, 20, and 3-6*

____

____

How does the clothing line up with the following verse?

"Beloved, I beseech you as aliens and exiles to abstain from the passions of the flesh that wage war against your soul." *(Wisdom tells us this also means we are to abstain from igniting the passions of the flesh in others by what we wear.) I Peter 2: 11*

____

How does the language line up with the following verses?

"The mouth of the righteous brings forth wisdom, but the perverse (corrupt, depraved, disgusting, impure) tongue will be cut off." *Proverbs 10: 31*

"The tongue is a little member and boasts of great things. How great a forest is set ablaze by a small fire! And the tongue is a fire. The tongue is an unrighteous world among our members, staining the whole body and set on fire by hell. For every kind of beast and bird... has been tamed by humankind, but no human being can tame the tongue — a restless evil, full of deadly poison." *James 3: 5-8*

____

How does the moral character of the main characters line up with these verses?

"Blessed is the man who walks not in the counsel of the ungodly not stands in the way of sinners, nor sits in the (theater) seat (watching) scoffers..." (Parentheses mine.) *Psalms 1: 1*

"Bid the older women to be reverent in behavior, not to be slanderers or slaves to drink; they are to teach what is good, and so train the young women to love their husbands and children, to be sensible, chaste, domestic, kind, and sensible." *Titus 2: 3-5*

____

____

Poet Synopsis

[Author / Poet] ______________________________ Lived from ________ to ________

General Era: ________________________ Notes: ____________________________

The synopsis is found on page ______ of the ____________ section of Volume ____ of the Great American Bathroom Book or: _________________________________

Notes on his/her life: ____

What kind of an outlook did he/she have on life? ______________________________________

Did his/her poetry/work contain meter? __________________________________

____

Do you find his/her poems easy to understand? ________ Enjoyable? ______

Uplifting? ______ Thought-provoking? _______ Depressing? _______

Other: ____

Type of poetry/writing most widely known for – major theme(s) of poems:

____

Was he/she published during his/her lifetime? ____

This [is / is not] an author whose writings I would like to read more of. Why or why not?

____

Write down one line (or more) that you found [interesting / intriguing / charming / touching] and explain why ____

____

RELIGIONS OF THE WORLD
Study Sheet

FAITH BEING STUDIED: ______________________________

Is this organization a "religion" or a "cult"? ______________________________

Why? ______________________________

Who started it? __________________ When? ________ Why?/Intent ______________________

____

Who or what is the "focal point" of this faith? ______________________________

____

Does this faith have its own book of teaching? ______________________________

What beliefs does this group hold regarding:

GOD (who He is): ______________________________

____

JESUS CHRIST: ______________________________

____

SIN: ______________________________

____

SALVATION: ______________________________

____

What are some other major beliefs or practices of this group?

____

____

What attracts people to this way of believing? ______________________________

____

Could you believe in their teachings? Why/why not?

____

____

SCIENTIFIC IMAGING TECHNIQUES

Student ______________________ Date______________ Due________________

Technique: ______________________

What does the name mean or stand for? ______________________

____

____

Describe how it is done (light, heat, rays...) ______________________

____

____

____

What is it generally used for or done on? What can it detect? ______________________

____

Is it more or less effective than other techniques? For what? ______________________

____

Are there any discomforts or side effects? ______________________

____

Would you want it done to you? Why or why not? ______________________

____

Where is it usually done? (hospital? lab? doctor's office?) ______________________

How expensive is it? ______________________

When was it invented or developed, and by whom? ______________________

____

SCRIPTURE VERSE STUDY

Verse: ____

____

____ Ref.______________

Ten major words of the verse and their dictionary definitions. As God speaks to you, write!

______________ - ____

____

______________ - ____

____

______________ - ____

____

______________ - ____

____

______________ - ____

____

______________ - ____

____

______________ - ____

____

______________ - ____

____

______________ - ____

____

Go deeper with three of those definitions and journal what God is saying to you in them.

______________ - ____

____

______________ - ____

____

______________ - ____

____

Other revelations or discoveries made during this study ______________________

____

TAPE REPORT

Name________________________________ Date__________________

Tape listened to: ______________________________________

Teacher: ______________________________________

Affiliated with: ______________________________________

What you learned *NEW* from this teacher

1. ______________________________________
2. ______________________________________
3. ______________________________________

Points that weren't necessarily new, but were good reminders

1. ______________________________________
2. ______________________________________

What you most enjoyed about it

1. ______________________________________
2. ______________________________________
3. ______________________________________

Your own (brief) analysis of the speaker ______________________________________

____

Did you disagree with anything? *(If so, what?)* ______________________________________

____

Any insights you can employ? ______________________________________

TEL-MED PROOF-OF-COMPLETION

❄ ✚ ♥ ☺

Name____ **Date** ____/____/____

TITLE OF TAPE____

TEL-MED NUMBER:________

COURSE CATEGORY OPTIONS: *(Check all that apply; circle one used)*

___Science ❄ ___Health ✚ ___Home & Family ♥ ___Child Development ☺

BRIEF SUMMARY OF TAPE CONTENT: ____

____

____

THREE THINGS I LEARNED:

1 - ____

____

2 - ____

____

3 - ____

____

SOMETHING I DIDN'T UNDERSTAND OR THAT I WANT TO LEARN MORE ABOUT:

____

____

____

*(Open*_____-_____ *Days:*____________ *Phone*_______________*) TOTAL TIME*______ *RECORDED*__

VIRTUAL DISSECTION RECORD

Student Name ______________________ Dissection of: ______________________________

Website Address: ____

Company or Individual that/who created it: ____

Notes: ____

____

____

____

____

A print-out of the dissection process [is / is not] included on the next page(s).

Representative photo of this procedure from the computer screen. (Cut out and attach here:)

Photo

WAR REPORT

A Synopsis of ______________________________ *(Name of War)*

Date of War: *From:* ______________ *To:* ______________ in the ______________ Century *(fourteenth? nineteenth?)*

[Countries / Nations] involved, as well as the allies of each below:

______________________________ ______________________________

⇓ ⇓

______________________________ ______________________________

______________________________ ______________________________

______________________________ ______________________________

Who were the leaders on each side, and what was the title or role of each? *(King? Dictator? President?...)* Comment on what kinds of leaders they were. *(Were they reasonable, just, tyrannical, insane...?)*

(Name:) ______________________ *, the* ______________ *of* ______________

was ____

(Name:) ______________________ *, the* ______________ *of* ______________

was ____

Explain the primary issues and conflicts involved: ______________________________

____

____

____

____

On what (whose) terrain did the actual fighting take place? *(One? Both?...)* Why?

____

What tactics and/or forms of warfare were used? *(Guns? Spears? Bombs? Bows and arrows?...)*

____

Were P.O.W.'s taken? If so, how were they treated?

Side ______________ [Took them / Did not take] (treatment:) ______________

____

Side ______________ [Took them / Did not take] (treatment:) ______________

____

How was the conflict resolved or the war ended? ______________________________

____

What can we learn from this war? ______________________________

____

WISDOM'S 7 PILLARS
TOPICAL STUDIES

FOR: (TOPIC/AREA)	SCRIPTURE REFERENCE & KEY WORDS	INSIGHTS, THOUGHTS PONDERINGS
1-		
2-		
3-		
4-		
5-		
6-		
7-		

FOR: (TOPIC/AREA)	SCRIPTURE REFERENCE & KEY WORDS	INSIGHTS, THOUGHTS PONDERINGS
1-		
2-		
3-		
4-		
5-		
6-		
7-		

WORD ALIVE STUDY

Word being studied: ______________________________

(1828 or any) Dictionary Definition:

Seven major words out of the definition and *their* definitions

Go deeper with any words in the above definitions that stand out to you and write what you feel God is saying to you: (Hint: If you do not feel that you are hearing God, just go with whatever you feel excited about, convicted in, or edified by. Second hint: That's hearing God! ☺)

Other minor, major, or miscellaneous revelations, discoveries, insights made during your study.

WORD STUDY ON "EDUCATION"

DEFINITION: **"The bringing up, as of a child; instruction; formation of manners. Education comprehends all that series of instruction and discipline which is intended to enlighten the understanding, correct the temper, and form the manners and habits of youth, and fit (prepare) them for usefulness in their future stations. To give children a good education in manners, arts and science is important; to give them a religious education is indispensible; and an immense responsibility rests on parents and guardians who neglect these duties."**

Further definition(s): (Choose words from the above definition that stand out to you and define them.)

Bible references: (look up in concordance for any of the above words that stand out to you.)

Any additional insights, discoveries?

World Magazine Report

*By*______________________________ *Date*____________________ *ISSUE: Vol*____ *No.*_____________________*, 19*_____

What's on the Editor's mind this week? *(Editor:*______________________________*)*

In "Things They Said," what comment, by whom, made the biggest impression on you and why? *(You may summarize the comment.)*

Made by______________________________________ Why it impressed you: ______________________________

What's going on internationally?

What was this week's Commentary about? What did you agree or disagree with?

What's going on in the United States?

Anything in Arts/Media that looks like it's worth buying or seeing?

What was the most interesting blurb in "Lifewatch"? *(and why?)*

What's important is happening in the "Religion Watch" section?

Of everything you read in this issue, what most personally affected you and how?

Section Five:

GRADING AND PORTFOLIO GUIDELINES

How to determine grades for a variety of traditional *and* non-traditional classes, and ideas and help for helping your student create an attractive portfolio of his/her work.

One Homeschool Mom Discovers The Heart ♥ of Wisdom's Way of Learning

"I believe a lot of this message is about our retraining as homeschooling moms. I read WWOL because of the struggles I was having in my homeschool. I was looking for something to "fix" my kids and my school. What I found was something to fix *me!* In *Wisdom's Way* Marilyn says, "whatever emotions you experience... will be part of God's plan to lead you to begin a wonderful learning process for yourself. If you experience feelings of perplexity; if you feel shaken, overcome, stretched, or even upset, you can be assured that this is to be expected and a necessary part of your learning process." She calls it "disequilibration." I admit I have felt all of these in the past six months.

As I have submitted to God, He has revealed the motives of my heart for the things I do. Many of the things were good, even noble, yet my heart's *motives* were not good or noble. These wrong things had to go. Yet, the Lord is so good and kind and gentle. He is teaching me to say no – a new experience for me. He has provided a way out of every activity He has asked me to release.

It was only after I allowed the Lord to deal with my heart, at the cross, that I was even able to see the fruit these activities were producing in me and my family. This is an ongoing process, yet I am already seeing fruit. My son has made a plan for landscaping our backyard so it can be a family place and a place of solitude. We are in the process of doing research on soil and plants so we know what to grow. You have no idea what a miracle this project is. My nickname is "agent orange thumb." I have never met a plant I couldn't kill. My children and I are learning together. I am not teaching them out of my knowledge. They are teaching me how to learn. God is sooo good.

So I guess all this to say: *New Wine needs new wineskins.* Until we take the time to renew our minds to the Lord's mind-set in an area of our lives, we won't be very discerning. Yet when we make baby steps toward Him, he runs to meet us and show us *His* way."

~ Karen H.; Homeschooling mom of 2 for 7 years; FL

ADDENDUM: [*Here's what Karen wrote to me when I asked permission to use the above quote:*] "I would be remiss if I did not take this opportunity to thank you for all you have done for me. If it wasn't for your *Senior High: A Home Designed Form+U+la* book, I would have never looked at *Wisdom's Way of Learning.* I was totally burned out on homeschool, legalism, and home business. I couldn't handle *any* heavy message and feared *Wisdom's Way* would be too much for me in that state of heart and mind. I checked your book out of our church library – it was so refreshing, so timely, and so encouraging. From there I was able to try *Wisdom's Way.* So, you see, I am indebted to you because the Lord used you to start it all. ☺" *(Wow, Karen! Isn't it just amazing how the Lord works in such ways to bring about His will! Thank you for sharing ALL of this!)*

Grading Guidelines & Considerations

WHY BOTHER WITH GRADING?

There are several reasons that giving grades can be beneficial and profitable, even advisable in certain situations. (Just for the record, though, I personally do not recommend giving grades in the elementary years. I can only see more drawbacks than benefits.) The possible advantages of giving grades at the high school level are:

1) Grades provide documentation of level of performance should it become necessary for the student to transfer to public or private school prior to graduation.

2) The grading sheets provide a convenient way to organize and present expectations and goals, as decided upon by parent and child. Although if more detail is desired, the ACS's and TAS's do a better job of that. *(See the following topic in this section for what those are.)*

3) Grading can give a sense of "ownership." A student can have more of a sense of "possessing" his learning experience when he is able to determine his own level of accomplishment – and areas to study. However, this is more true for a more "schoolish" approach. A student who has had the opportunity and time to develop a love of learning is not going to have this need. Be aware that this book is written for a wide range of student types. ☺

4) Grades are sometimes necessary if the student is going on to a four-year college, university, or institute that requires them. However don't assume that a portfolio wouldn't suffice if suggested!

5) A student might work harder and do a better job for a grade. Sad, in one way; but, well, aren't we likely to work harder and better for a paycheck? That's not to say we *shouldn't* work just as well for unpaid work – or when no one is looking, as we are admonished in Scripture to do. But the good side of this is that once an otherwise unmotivated student experiences the joy and satisfaction of a job well done, they are more likely to realize it's actually worth the effort! So hear God on what's the best approach with *your* student! You don't want to just "indulge" the flesh, but this might just be a good way to urge the student into a higher level of production.

6) I wouldn't put all my eggs (i.e. reasons to grade) in this one basket, but insurance companies often give "good driver discounts" if a certain grade point average is maintained. If you don't have a G.P.A., it might be difficult to prove to *them* that your student really *is* a good driver.

7) Grading can provide an easy(er), more "educated-sounding" answer to skeptical friends and relatives who are worried about our kids missing out on "real" school. However, if you focus on the *character* of your child over their *academic prowess,* this will not matter one teensy bit as your and their kids eventually "turn out" – which they all *will!* ☺ And at *that* point, when your kids are excelling in character and probably a lot of other good stuff, it will be evident who was missing out on the right and the wrong stuff! *I guarantee this!!!*

DO I REALLY NEED THE GRADING GUIDELINES?

This is a really good question! I designed these *before* I had created any of the Assignment Check-off Sheets (A.C.S.'s) or the Task Analysis Sheets (the T.A.S.'s) – or even *thought* of them! While the Grading Guidelines were created to help you grade, they also had the added benefit of giving a bit more detail about the content of several classes. However, the ACS's and TAS's contain much more detail than what's here! I didn't want to omit this section, though, because the Grading Guidelines show you how to break it down for the purpose of grading, whereas the ACS's and TAS's do not. I couldn't have gotten one more thing onto most of those! So they each have their own specific purposes! You might want to combine concepts from both. Because grading wasn't an issue for me, I never felt the need to combine. If you come up with something, send me a copy! ☺

NOT AT ALL OFFICIAL!

I give suggestions for grading all the generally required subjects, plus several others that are not even normally considered classes! In fact, they might not be! More on that in a moment! Please be aware that these grading guidelines are purely a matter of (my) personal design and opinion, *not* to be considered "official" in any way! So don't tell the college admissions officer "But Barb Shelton said!" because all they'll say back is "Well, la-dee-dah – who's *Barb Shelton?!"* And I can assure you they will *not* be interested in the answer!

Don't quote me on this next one either, nor be surprised if it does happen, but it's unlikely that anyone would challenge you on it. I really don't know why anyone *would* – *unless* the person or institution you happen to deal with is antagonistic towards homeschoolers. In that case, having an orderly record of the standards for each course would probably – *possibly* – make a favorable impression. Besides, even within one school, grading procedures, guidelines and standards vary widely from teacher to teacher and class to class!

In fact, this very thing actually waters down and muddles one of the main purposes of grading for the system itself! They want to know how the kids *compare*, right? But, I have to ask – *with whom?* That's the big question because every teacher and every class is different!

Some teachers grade "according to the curve." This means the highest grade (or 3 grades) in the class get A's; the next arbitrarily-decided number or percentage of students get B's; the broadest range get C's, a smaller number get D's, and smaller yet, F's.

Translation: As one teacher explained to a high school class I was in, this means that if three students in the class get 100%, (if "3" was "the arbitrary number" he had decided on), they would get A's; if 5 of them (if 5's the number he'll award B's to) get 98 or 99%; they'd get B's; if 15 students get 92– 97, they'd get C's, and so on. This was extreeeemely unfair, and we discussed it at length with him. Because students who were actually doing "A" work were not getting that grade, maybe even getting a "C" because that teacher held fast to the idea that grading should be done on the class curve! An "A" meant the student "excelled" compared to the particular classmates they were in a class with.

While this might work to the advantage of less studious students, one repercussion of this is that while some (many?) teachers base their grades entirely on how the class does, but to them you could still get an A even if you only got an 82% *if* that was the highest grade in the class! If the "average" for the class was 56, that was a C!

Are you beginning to see why I don't put much credence in grading. Yes, these were only my limited experiences, but guess what – that's who *all* of our kids are being compared with: other kids for whom there are absolutely *no* uniform standards across the country. And within the system, they don't care about fairness across the line or across the nation! The kids are all just "cogs in the machinery"! Not mine! ☺

EXPLANATION OF GRADING GUIDELINES

The first form in this section, right after the Portfolio section, is a generic Grading Guidelines form. It is to be used as a basis for determining your own grading criterion for any classes your student takes. *Only* because I had to come up with some uniform way of referring to classes and courses and units, I came up with the following:

At the top of each of these grading sheets, you'll see "Course _________." This is the main course title, always one of *our* required subjects for graduation. "Unit/Class ________" is a more specific topic of study that's under or part of the general course. In Math it would be called a "subset." ... Here are a few examples of what I mean:

COURSES	UNITS OR CLASSES
Mathematics	Geometry, Algebra, Trigonometry, Consumer Math, or Applied Math
English	Literature, Vocabulary, Spelling, Composition, Grammar, Study/Research
Fine Arts	Community Concerts, Music Appreciation, Choir, Art, Band, Dance, Drama
Computer	Word Processing, Computer Graphics Design, Educational Programs
P.E.	Tap Dance, Aerobics, Basketball, Gymnastics, Track, Wood-cutting ☺
Home Economics	Gourmet Cooking, Sewing, Home Decorating, Home Maintenance/Repair
Occupational Ed.	Carpentry, Typing, Chef'ing, Computer, Office Skills, Teaching Dance

ONE MOM'S TESTIMONY

One mom wrote: *"I quit giving grades about my 3rd year of homeschooling. I teach for mastery. If the topic (exercise, etc.) isn't mastered, it is repeated. There are many avenues of discussion that can spin off from here - I just try to evaluate, through many different means, mastery of a subject. (competitions in various subjects also aid in evaluation) I try to not forget to allow for normal human error as opposed to mastery, and also the concept that a few things don't need to be mastered, only done. I also try to prepare the children at appropriate times (dual enrollment works well here) to fit into a world that does use a system of grading. When I prepared a transcript for my 16-year-old son to enter dual enrollment, he received all A's except in algebra (that was a B). We felt since he had mastered the other subjects, they deserved an A. In algebra he had to repeat exercises more often than we felt he should, and we felt his attitude did not reflect that of one striving for an A, hence we decided on the B."*

THEN THERE ARE THE COLFAXES

The Colfaxes are a family who lives in northern California on a ranch that they built themselves. The four boys did not do "school at home" but worked on the ranch, building, creating a solar power system, raising goats, and reading in between. All four of these very informally homeschooled boys went to Harvard.

The admissions official at Harvard where their son first son applied told Micki (the mom) that they needed to provide a grade point average for their son. She told him they hadn't kept grades, that in the homeschool setting grades were irrelevant. (They had already submitted some pretty impressive portfolio stuff and, if I remember correctly, he was all but admitted at this point.) The man still insisted that he *had* to have a grade point. Micki said "Okay, 4.0." He said "Okay," and wrote it down! ☺ Now, had their son proven to be incompetent in his studies, they might have felt bad or been in error for doing that. But he proved to do quite well.

The family's whole story is in *Homeschooling for Excellence* by David and Micki Colfax. *(We carry it.)*

ONLY PORTIONS OF CREDITS

Be aware, too, that these classes or units will not always or generally make up entire credits, though sometimes they will. For some, it may take two, three, or even several smaller units of study to make up enough hours for one credit.

Nothing to fear; nothing to be ashamed of, and not a bit of a problem! You're just being "eclectic"! ☺ We just keep shopping until we have enough. Our "Home & Life Management Skills" class is a very good example of this. (See section 6-D.) And another is Social Studies. Actually, you'll find we did this for/with quite a few classes.

If you're in need of ideas on how to fill those class hours, *Homeschooling the Highschooler* and *The High School Handbook* (see Resource Section) are packed with them! (ideas! ☺)

Since some classes will be ongoing throughout the high school years, hours will just keep accumulating in them. I'm referring to activities that are part of the student's regular daily life, like P.E., Fine Arts, Health, Occupational Skills, Typing, and some portions of Science.

Once they've earned the required number of credits for a particular class, our kids are not *required* to continue recording hours for those classes. But they are free to do so if they want to earn further credits in that or those classes.

YOU DECIDE!

So what you have to decide is this: What are *your* purposes for grading? To help motivate the student? I would rather do that from the inside out. They can do all the right stuff to get a good grade, but what's happening – or not happening – on the *inside*? There may be times when external-only motivation is appropriate, but I don't want it to be the norm. So *my* burning question is *WHO decides what's "official"?!?* ... The answer is *you!!!*

You may want to use test scores to help determine a grade. But even that is not always going to be accurate. The student's initiative, ability to search out what he needs to know, his self discipline, his ability to "knuckle down" and stick with something until it's "completed" or "well done" or "figured out" are going to have more to do with how he does in college than his test scores. So that's why those things factor in for *me* more than "mastery." But again, that depends on what we're talking about, and what your purposes are for your own student.

NOT GOING TO GRADE?

That's just fine with me! Won't disappoint me or hurt my feelings a bit! In fact, here's what one homeschooling mom, Sis Caudle (in Wash. State) shared with the moms on the Senior High Form+U+la List *(see page VIII-7 for explanation of the SHF List)*:

> When we first started homeschooling, I made out the assignment sheets and checked them off diligently. At the regular school quarter marks, I evaluated them on the work they'd done. According to the grading scale we were using, they failed most of their classes!!! :-} However, by the school's standards, they were doing pretty good, even moving up a grade level by the quarter! I made them report cards and put their fail marks on it. Now that they are graduated, they like to point out that they failed in homeschool! (When the oldest went back into high school he was on the honor roll without doing homework!!!) My "Fail" was better than their "A+"!! ... I'm just sharing *our* experiences. ☺ We no longer do "grades." The class simply isn't complete until the assignments are done (for us) whether it takes two weeks or two years, unless I've re-evaluated the class requirements. Sometimes they do something we hadn't anticipated that covers an area and we'll substitute that."

Whether or not you decide to do grades, you might want to take a quick look through these sheets anyway just for ideas, many of which are not in the ACS's or TAS's. I think you might find a few ideas that will help as you go to structure and fill your courses and classes. And if you *really* don't want to read them all, at least read the ones for "Hunting" and "Baseball"! ☺ *(Just cuz they're cool!)*

And now to answer a few Questions about Grading...

Q's and A's on Grading

***Q** What if the student does all the work for say an "A" but the quality is not excellent. Or vice versa what if they did the work of a "C" but the neatness of a "B."*

A I'd say give him (or her) a B-plus or an A-minus.

♥♥♥

***Q** Why do you have only two grades (A and B) for some of the classes you have in this section when most have three, A, B, and C?*

A For some classes, no requirements are given for a C because I felt that in those particular classes, nothing less than the prescribed amount and quality of work would be acceptable, and nothing more was needed either.

♥♥♥

***Q** Why aren't these grading sheets in with the classes? Seems like that would make more sense.*

A You'll probably end up taking these sheets and putting them in with each subject section of the student's notebook-organizer, where you'll actually *use* them. But for the purpose of initially acquainting you with the concept of grading, I wanted them in one place where you could more easily get an overall picture rather than having to jump around all over the place to find them.

♥♥♥

***Q** When you speak of "quality" I'm thinking of things like following all the instructions for presentation of the projects, typing what needs to be typed, keeping a cheerful and willing attitude (hard to "grade", I know) etc ... not just doing the stuff but doing it "with all their might." Is this "valid"? Or a good idea?*

A Good thoughts, in my opinion! Again, for *me* anyway (because I can only speak for myself), how I'd carry this out would depend totally on what they're studying or doing. The delights of their life are the main things I'd want or need to see real gusto exhibited in. As long as they *do* the other things that are not so appealing or part of their purpose in life, and do them with a good attitude and without complaining or needing to be pushed, I'd still give an A. I wouldn't expect a math-hater to do it with all her might or cheerfully – but I *would* expect her to complete everything, regardless of delight area, and to do it with a willing and *pleasant* attitude, with me not having to push her.

We'd almost have to take each child and each subject on a one-by-one basis to be able to get as specific as is needed for this discussion. My own opinion is that it's not so much a question of "what is *the* right or best way to do it?" but rather "what's right and good for *this* student for *this* class or activity at this time in his/her development?" The answers to these questions are going to differ widely from student to student; parent to parent; class to class. Bottom line? It's *really* not worth getting hung up on! Hear God!

♥♥♥

***Q** I already feel pressure to understand all the transcript, credit, and grade information. We've never kept grades before – yikes! The children's portfolios (though not required by our state) have always served as an evaluation tool for me to see how much and how well they were doing. All the official record-keeping makes me nervous.*

A Don't let it. ☺ That's part of the enemy's scheme. The pro's and con's and in's and out's are completely spelled out in this section. Not what *you* should or should not do; but things to *consider*. Most importantly, as you read *ask the Holy Spirit to speak to your heart between the lines* so that you will know what is and *isn't* for you! ☺ And then *be at peace,* like this scene: *(Actually, I just wanted to use it cuz it's pretty; had to think of some way to tie it in!)*

The Grading Guidelines in this section present a variety of ways to do it, which will give you plenty of ideas for how to proceed. Or you can just do what the Colfax's did – which was to just give him a "4.0"!!! ☺ Of course, the student does need to know something before you'd want to give them that GPA, because "the proof is in the pudding" – but that will be the least of your problems. ☺

Guidelines *for* Assembling a Portfolio

I address this section to the student, since creating a portfolio is largely a student activity.

WHAT *IS* A PORTFOLIO?

It's a collection of a student's work, intended for public viewing, that showcases what you consider to be most representative of your abilities and accomplishments. It should:

- indicate self-reflection by you on your own work and progress
- show that you are a participant in, not an object of, the assessment and selection process
- demonstrate the range of your learning and activities
- reveal progress toward meeting your educational program goals
- be able to "stand alone" – accurately representing you without you present!

The portfolio is more than just a scrapbook, and it is not merely a hodge-podge compilation of everything you have worked on all year, or through all your high school experience. If compiled progressively through the months, the task of assembling a portfolio won't seem insurmountable, nor will any vital components of the portfolio be overlooked or forgotten, or prime opportunities missed. Additionally, compiling it through the year will help you learn as you go along what is appropriate, desirable, and undesirable to be included. It's also a good way to stay focused on goals and on your progress in accomplishing them – before the end of the year draws nigh, when it can be too late.

WHAT IF I'M NOT GOING ON TO COLLEGE?

Perhaps you are not going on to college, and if my information is current, McDonald's doesn't normally require portfolios for employment. Do you then even need to consider putting together a portfolio?

My opinion, for what it's worth *(about two cents)*, is that while you may never *need* a portfolio for a specific purpose that you know about *before* the fact, there are still several benefits and advantages to assembling a portfolio (the *process* itself) and to having a portfolio (the *product*) on hand. These benefits are as follows:

- It involves you, the student, in a *positive* and *active* way in your own learning process
- It enhances and reinforces the good aspects of what you have learned
- It helps direct you toward further and better learning
- It provides a much broader, more creative picture of your learning than traditional report cards and/or standardized tests
- It tells your teacher/parent what you value
- It's there if you should ever need it.

CHOOSING A "CONTAINER" FOR YOUR PORTFOLIO

You will actually want to do this *after* you have the contents assembled and prepared because, until you do, you won't know what format or size will be best for what you have. Looking at what all you've collected, which I'll get to in just a moment, choose a cover or container that appropriately and conveniently presents your work. A few ideas are: an accordion-type file with or without pocketed divisions for the various section topics; a notebook with a Clear-vue™ cover into which you can slip a collage of photos or "representative tidbits" – mounted onto index stock for easiest insertion *(see pages 5–7 of Sec. VI-C)*; a tastefully-covered and/or decorated box that's sturdy and which can be opened conveniently. It should be close in size to the volume of your portfolio contents so there's not a lot of "dead space" giving the appearance that something is missing. You want it to look "full" – like your life is! ☺

WHAT ABOUT MY "LIFE-SIZE" PROJECTS?

You've probably already realized that some of the projects, activities, and work you've done won't fit into anything you can carry, except perhaps with a U-haul truck, which is actually quite true with a horse! No problem! Before that science projects gets disassembled, that gorgeous cake eaten, that 4-H cow you raised from a darling baby calf made into steaks *(oooh, sorry!)*, or that authentic Medieval costume you labored so hard on donated to the National Museum of Medieviology *(I think I just made up that word)*, have someone take pictures or a video of you with it, in it, beside it, or doing it, so it will be evident that *you* were actually personally involved in it!

A word to the clueless about mounting photos: Index stock, also called "cardstock" is much better and more sturdy than mere paper. I have always used rubber cement because of how easy it is to remove, but some say that it's not good for pictures in the long-run.

You might want to encase your photo sheets in plastic sheet protectors to safeguard against sneezes and spills. Plus they offer a snazzier, more attractive appearance.

Perhaps you wrote a lovely piano piece; audio tape it and include the cassette in your portfolio. Video tape your ballet, gymnastic, singing, or drama performance – just portions of them – your best speech at Toastmasters, or highlights of the track meet events you took part in.

And a word to the wise: *Keep these short!* Professionals won't be nearly as interested in a lengthy (though totally awesome) Home Movies session as Grandma and Grandpa would be! ☺

SO WHAT SHOULD BE INCLUDED IN MY PORTFOLIO?

☐ A Table of Contents

☐ A letter addressed to your potential audience about yourself, your work, and the portfolio

☐ Your *best* work that represents or shows the following:

- Communication skills: Reading, Writing, Speaking, Foreign Language
- Science/ Computation/ Math
- History/Philosophy/Social Studies
- Arts, as in Literature, Music, Theater, Dance, etc.
- Marks of a successful learner

☐ A "Notation Sheet" attached to each selection (signed and dated) explaining why you selected it. What were your criteria for including it? What do you see as being of merit or value?

UH-OH – WHAT DO YOU MEAN BY "SUCCESSFUL LEARNER"?!

So you were doing fine with all these content items, perhaps even getting a little excited about the thought of doing this, *until* you came to this?!? Relax. I don't think it'll be as challenging as you might think. Here's a list of what to look for in this area:

- Self-disciplined and diligent, even when understanding of the material isn't coming easily
- Knows how to seek out answers
- Creative and innovative; a problem solver, especially when "the going gets tough"
- Open to suggestions and editing (humility)
- Organized in study habits, processes, and style of working
- Organized in his physical learning environment (not necessarily always *neat* though)
- Not afraid to make mistakes
- Takes pride in work
- Willing to collaborate and work with others when called for
- Doesn't get immobilized, overly discouraged, or quit if something isn't working
- Willing to do work over to make it better; Isn't satisfied with mediocre work
- Self-motivated; Takes initiative; Doesn't always wait for direction(s)
- Curious; wants to know more (is a "life-long learner")
- Works well even when others aren't watching
- Views learning as something to pursue for oneself, not something to do for a teacher.

No sweat! You probably have most of these down-pat, right?!? If not, these are some excellent qualities to possess for *life!* These also happen to be *exactly* what employers are looking for in employees; colleges in prospective students; men in prospective wives and women in prospective husbands!

By the way, you do not list these in your portfolio; it would just behoove you to have tell-tale *signs* of these sprinkled throughout your work. Maybe even via a letter of recommendation of some sort. (not the "job type"; just a personal letter of praise if someone was so kind as to write such a thing! Or use even an *excerpt* from a letter.)

*(**Note to Parents:** These are excellent things to casually talk with your student about during the year; qualities to reinforce, encourage, and work together toward. They're really just about "character development"!)*

These, dear Student, are character qualities that you learn *only* "en route" – as you're *doing* the other stuff! They aren't independently learnable. And if you feel (know) you're a little short on a few of these, first, don't feel bad; no one I know is adept at all of them, least of all *me!* And second, ask God if He wouldn't mind building them into you. In fact, I'm *sure* He'd be glad to oblige! I'm not just making this up; read James 1:5 here:

> **"If any of you lacks wisdom** (*or* self discipline, *or* diligence, *or* initiative, *or* a love of learning, etc.), **let him ask God, who gives generously** (as in His grocery store is *huge) **to all men** (students too, I bet!)* **and it will be given him** (or her.)"

You mean we don't have to make all these self-improvements *ourselves?* ... Nope! That's not part of the deal we signed up for when we came to Jesus! *(Yay!)*

YOU CAN'T JUDGE A BOOK BY ITS COVER – OR *CAN* YOU?

At least they *say* you can't. But it happens all the time. Someone *will* be "judging" you – your character, and your work – by your portfolio in a very tangible way. Your portfolio will be giving someone *some* kind of picture and impression of who you are.

As Christians, we know that "God is our Judge" – the only one who will count in the end, when all is said and done. But in the *process* of all being said and done, as long as we're on earth, Man *will* be "God's Little Helper" in whatever arena he sees fit, whether appointed by God or not, and whether God-hearted or not.

This portfolio business, then, falls under a different kind of "judging." One that we may or may not prefer to, or actually submit ourselves to, and that we ourselves might even give very little credence to. But it's none-the-less an accepted and bona fide tool for those who must make decisions about us to use. So take the time and make the effort to make yours a *good* and honest impression!

CREDIT WHERE CREDIT IS DUE!

The information in this section was compiled, organized, and enhanced quite a bit by me, but the basic content was graciously thought up and released by Elaine Brouwer who, at the time I took her workshop on this, was the Administrator of Covenant Christian School and Co-director of Alta Vista College which produces K through 9th grade curriculum for homeschoolers. Through the college, Elaine offers *(or at least she did back then)* workshops to homeschoolers on learning styles, multi-intelligences, how to teach with integrated units, and student evaluation, including portfolios.

Without these tips from Elaine, these portfolio-presenting pages of this book would have resembled the sky in a blizzard – great doodling paper!

But *with* these tips from Elaine, who is an expert in the field of "portfoliology" ("port-fō-lee-ól-ō-gee" – *I just made up that word too!* ☺), you have the tools to produce a portfolio that will accurately represent the high quality of the one submitting it. One that will raise someone's eyebrows, (someone whose eyebrows you'd like to raise) and elicit the same comment from their mouth that God said on several occasions during His "Creation Days":

"Wow! This is good!"

(or something similar; actual wording may vary.)

The Grading Criteria Sheets

are coming up right after an excerpt from an "Uncle Eric" book!

Grab a cup of your favorite beverage and enjoy!

Well, as much as you can "enjoy" stuff like this!

Another Lifestyle of Learning Insight...

From Marilyn's Book:

Develop a Lifestyle Routine

"I found that the only schedule that works in our home is the one God gives me for each day. Therefore, the best way to accommodate this level of flexibility and still have order is to define blocks of time in which certain activities occur instead of the traditional and rigid hourly schedule that only works in institutions. I divide my day into five or six blocks of time and these can be arranged or rearranged in my day any way I need them to be according to our current schedule. They do not necessarily maintain the same beginning and ending times each day. There are too many unknown interrupters and legitimate brief detours that will take time away from completing tasks within the prescribed time-blocks." ❤

The following is a quote from ***Uncle Eric Talks About Personal, Career and Financial Security,*** by Richard J. Maybury. This is one of a series of books by this author, including *Whatever Happened to Penny Candy?; Whatever Happened to Justice?; Are You Liberal? Conservative? Or Confused?; Evaluating Books: What Would Thomas Jefferson Think About This?;* and *Ancient Rome: How It Affects You Today.* All are thought-provoking, understandable explanations about history, economics, and the law that junior high through adults will find fascinating and eye-opening. Used with permission from Jane Williams, owner of Bluestocking Press which publishes these books. Call them at 530-621-1123 to request a free product catalog of the 1,000+ items they carry that focus on resources of a similar nature. *(We are currently checking into arrangements for carrying the "Uncle Eric" series of books.)* Their address is P.O. Box 2030 - Dept. SR / Shingle Springs, CA 95682-2030.

Excerpt from

Uncle Eric Talks About... Personal, Career and Financial Security

A few months ago I was watching a pro football game between the San Francisco 49ers and the New York Yankees. In one of the plays, the quarterback stepped up to the 60-yard line, took the snap (hike), swung the bat, hit it over the right field fence, and it swished right through the basket for two points.

Now what do you know about what I just said? You know *it did not happen.* But you weren't there, you didn't see it. How do you know it didn't happen? Because you know something about how football, baseball and basketball work. You know the rules. You know what goes together and what doesn't. You know the likely sequence of events. So you *know* it could *not* have happened the way I said it did.

Even though you weren't there, you know I'm not telling the truth. ... The data did not fit the model. That gives you a choice. Either the data are wrong or your model is wrong. Well, which do you think it is? Did you think your understanding of football must be wrong? You doubted *me,* right? You didn't question your model, you questioned the data.

Humans are very reluctant to throw out their models. They'll throw out data very quickly, but not models. This is because models are so important. ... Models are how humans understand their world. A **model** is a mental picture of how the world works. Sometimes models are called **paradigms**.

Most of us don't realize it but we devote a lot of time and effort to testing and improving our models, or paradigms, so we have a lot of faith in them. Models are rugged. They aren't infallible, but they're tough. I believe one of the most important things you can teach your children is that *models are how we think, they are how we understand how the world works.* As we go through life we build these very complex pictures in our minds of how the world works, and we're constantly referring back to them – matching incoming data against our models. That's how we make sense of things. ... One of the most important uses for models is in *sorting* incoming information to decide if it's important or not. ... In most schools, models are never mentioned because the teachers are unaware of them. Teachers teach what they were taught – and no one ever mentioned models to them, so they don't teach them to your children. For the most part, ... children [in school] are just loaded down with collections of facts that they are made to memorize. Without good models or paradigms, the children have no way to know which facts are important and which are not. The teacher doesn't have a way to know either.

> **What's so important about George Washington?** *He was the father of our country.* **What does that mean?** *He was the first president.* **Big deal, so he was first.** *He won the Revolutionary War.*

No, he didn't. He was a military disaster. If his strategy had been followed consistently, the war would have been lost. Washington is important mostly because of something very few of us have ever heard of – the Newburg Crisis. In 1783, many people wanted a new government, and the army officers wanted to set up a dictatorship and make Washington the dictator. Washington not only turned the offer down, he said any soldier who followed the orders of this new government was wrong. The soldiers' job was to protect liberty, not necessarily to obey orders. *Under Washington's encouragement, the troops mutinied and deserted* – and that kept America from becoming a dictatorship. But no one teaches about this in the schools. Why? I don't know. I suppose they don't want the children to learn that Washington encouraged the troops to disobey orders, he wanted the troops to be dedicated to a *higher law than any government's law.*

The principle of higher law is not taught in the schools – or at least in very many schools – which is why a great deal of what *is* taught doesn't make much sense. There's no model. At age 18 the students graduate and all they remember about George Washington is that he chopped down a cherry tree.

*By **Richard J. Maybury***

GRADING CRITERIA
AND COURSE ASSESSMENT

s: ____________________ **Course:** ________________

__________ **Grade(s)**______ **Teacher/Advisor**________________

ADING: *(Students may choose which grade they wish to work toward.)*

" — STUDENT MUST:

____

____

____

____

❑ *TO EARN A "B" — STUDENT MUST: Do the above, plus or except:*

____

____

____

____

❑ *TO EARN AN "A" — STUDENT MUST: Do the above, plus or except:*

____

____

____

____

ASSESSMENT: ____

____

____

____

Signature of Instructor:______________________________ *GRADE:*____

GRADING CRITERIA
AND COURSE ASSESSMENT

Unit/Class: "Algebra" **Course:** "Mathematics"

Student____________________ **Grade(s)**______ **Teacher/Advisor**________________

CRITERIA FOR GRADING: *(Students may choose which grade they wish to work toward.)*

❑ ***TO EARN A "C" — STUDENT MUST:***

Complete 90% of Saxon Math "Algebra 1/2" book.

Earn a minimum of 75% on class work.

Quality/neatness standard of work turned in must be "Average."

❑ ***TO EARN A "B" — STUDENT MUST:*** *Do the above, plus or except:*

Complete 95% of Saxon Math book.

Earn a minimum of 80% on class work.

Quality/neatness standard of work turned in must be "Good."

❑ ***TO EARN AN "A" — STUDENT MUST:*** *Do the above, plus or except:*

Complete 100% of Saxon Math book.

Earn a minimum of 90% on class work.

Show teacher "computing" page after completing work.

Quality/neatness standard of work turned in must be "Excellent."

ASSESSMENT: ____

____

____

Signature of Instructor:____________________________ *GRADE:*____

GRADING CRITERIA AND COURSE ASSESSMENT

Unit/Class: "Baseball" **Course:** "Physical Education"

Student__________ **Grade(s)**_____ **Teacher/Advisor**__________

CRITERIA FOR GRADING: *(Students may choose which grade they wish to work toward.)*

❑ *TO EARN A "C" — STUDENT MUST:*

Attend all games and practices (so far as possible with scheduling).

Show good sportsmanship at games and practices.

Be willing to accept direction from the coach.

Be *almost* as happy after a "loss" as after a "win."

❑ *TO EARN A "B" — STUDENT MUST: Do the above, plus or except:*

Have a helpful servant's attitude toward team members (like taking a coat out to a guy on base).

Be a good example of a guy who loves God – not succumbing to coarse jokes or mocking or belittling others (on either your team *or* the opposing team).

Demonstrate *improvement* (not mastery) of weaker areas as the season progresses.

❑ *TO EARN AN "A" — STUDENT MUST: Do the above, plus or except:*

Pray for your team members and coaches.

Be an active encourager, especially of the team members who are least skilled and most. discouraged. "When others are in trouble (or discouraged, or ashamed and embarrassed in front of team members) *you* be the one to help them out."

Take the initiative in getting your uniform into the wash.

ASSESSMENT:__________

____

____

*Signature of Instructor:*__________ *GRADE:*____

GRADING CRITERIA AND COURSE ASSESSMENT

Class: "Building Family Relationships" Course: "Life Skills"

Student____________________ **Grade(s)**______ **Teacher/Advisor**__________________

CRITERIA FOR GRADING: *(Students may choose which grade they wish to work toward.)*

❑ ***TO EARN A "C" — STUDENT MUST:***

Read When You Can't Get Along by Gloria Chisholm and fill in answers (in the book) and reflections.

Read Chapters 1, 2, 3, 6, 10, and 11 of Family Fun and Togetherness.

Complete at least 2 "Home Reports." *(more if time during unit).*

Complete an "Open Line & Mind" evaluation with each person in the family one time during the unit.

❑ ***TO EARN A "B" — STUDENT MUST:*** *Do the above, plus or except:*

Show diligence and thoughtfulness in formulating answers to all assignments.

Complete at least 3 "Home Reports." *(more if time during unit)*

Complete "Open Line & Mind" evaluations **2** times during unit.

Initiate, organize, and then lead one Family Night.

❑ ***TO EARN AN "A" — STUDENT MUST:*** *Do the above, plus or except:*

Complete at least 4 "Home Reports."

Complete the "Open Line & Mind" evaluations **3** times during unit.

Organize and lead a Family Meeting. (– using the Family Meeting Guide)

ASSESSMENT: ____

____

Signature of Instructor:______________________________ *GRADE:*____

GRADING CRITERIA AND COURSE ASSESSMENT

Unit/Class: "Composition" Course: "English"

Student______________________ Grade(s)______ Teacher/Advisor__________________

CRITERIA FOR GRADING: *(Students may choose which grade they wish to work toward.)*

❑ ***TO EARN A "C" — STUDENT MUST:***

Complete all course material.

Earn a minimum of 70% on tests.

Quality/Neatness Standard = "Average"

Produce 2 drafts of assigned work. *(= a "Rough Draft" + a "Final Draft")*

❑ ***TO EARN A "B" — STUDENT MUST:*** *Do the above, plus/except:*

Earn a minimum of 80% on tests.

Quality/Neatness Standard = "Good"

Participate in School Biz "Creative Writing Class."

Produce **3** drafts of assigned work. *(+ a "Second Draft")*

Have a good attitude toward evaluation and editing of work by teacher.

❑ ***TO EARN AN "A" — STUDENT MUST:*** *Do the above, plus or except:*

Earn a minimum of 90% on tests.

Quality/Neatness Standard = "Excellent"

Participate enthusiastically in School Biz "Creative Writing Class."

Produce **4** drafts of assigned work. *(+ a "Third Draft")*

Work open-mindedly with teacher in the editing process; receive suggestions and corrections favorably.

ASSESSMENT: ____

____

____

Signature of Instructor:______________________________ *GRADE:*____

GRADING CRITERIA
AND COURSE ASSESSMENT

Unit/Class: "Computer: Word Processing" **Course:** "Occupational Skills"

Student____________________ **Grade(s)**______ **Teacher/Advisor**____________________

CRITERIA FOR GRADING: *(Students may choose which grade they wish to work toward.)*

❑ *TO EARN A "C" — STUDENT MUST:*

Be able to: access DOS and then the desired program, "boot up," use disks with appropriate drives, and exit the DOS environment.

Enter a document into the WordPerfect word processing program using the commands and moves for cursor control, be able to "Center," "Delete" and "Undelete," "Indent" (L and R & L), "Print," use "Bold," "Underline" and "Page Format," (including setting margins and page length), and be able to "Retrieve," "Save," and "Exit" a document.

Earn a minimum of 70% on written tests.

❑ *TO EARN A "B" — STUDENT MUST: Do the above, plus or except:*

Additionally be able to use the "Block" and "Copy" (cut and paste), "Switch Document," and "Search" features.

Earn a minimum of 80% on written tests.

❑ *TO EARN AN "A" — STUDENT MUST: Do the above, plus or except:*

Complete all of the above requirements with greater proficiency.

Have a basic understanding of the "Reveal Codes" operation and be able to use it to correct a simple dilemma, be able to use "Spell-check" (to correct spelling and count words in a document) and "Thesaurus" (to find more creative synonyms), and to "Format" a disk.

Earn a minimum of 90% on written tests.

ASSESSMENT: ____

____

Signature of Instructor______________________________ *GRADE*:____

GRADING CRITERIA AND COURSE ASSESSMENT

Unit/Class: "Gourmet Cooking" **Course:** "Home Ec"

Student____________________ **Grade(s)**______ **Teacher/Advisor**__________________

CRITERIA FOR GRADING: *(Students may choose which grade they wish to work toward.)*

❑ *TO EARN A "C" — STUDENT MUST:*

Complete all course material, following cooking directions.

Earn a minimum of 70% on class work and tests.

Quality/Neatness Standard of Finished Products = "Average"

❑ *TO EARN A "B" — STUDENT MUST: Do the above, plus or except:*

Complete all course material, following cooking directions.

Earn a minimum of 80% on class work and tests.

Quality/Neatness Standard of Finished Products = "Good"

Kitchen must be left neat and clean.

❑ *TO EARN AN "A" — STUDENT MUST: Do the above, plus or except:*

Complete all course material, following cooking directions.

Earn a minimum of 90% on class work.

Quality/Neatness Standard of Finished Products = "Excellent"

Kitchen must be left neat, clean, and sparkling.

Student must prepare at least 2 of the dishes in host/hostess manner.

ASSESSMENT: ____

____

____

*Signature of Instructor:*____________________________ *GRADE:*____

GRADING CRITERIA AND COURSE ASSESSMENT

Unit/Class: "Hunting" **Course:** "Life Skills"

Student_____________________ **Grade(s)**_______ **Teacher/Advisor**_________________

CRITERIA FOR GRADING: *(Students may choose which grade they wish to work toward.)*

❑ *TO EARN A "B" — STUDENT MUST:*

Go on all required hunting expeditions with Dad.

Work with Dad at:

- gutting out the animal
- quartering and skinning it
- packing animal to vehicle
- hang animal in garage to age
- butchering the meat
- wrap and label

❑ *TO EARN AN "A" — STUDENT MUST: Do the above, plus or except:*

Have a good attitude on hunting expeditions.

Set out hat, gloves, gun, boots, socks, coat, long-johns, and food the night before.

Hang wet clothes to dry in front of heater without being asked.

ASSESSMENT: ____

____

____

____

Signature of Instructor:________________________________ *GRADE:*____

GRADING CRITERIA AND COURSE ASSESSMENT

Course: "Lab Science"

Student____________________ **Grade(s)**______ **Teacher/Advisor**_________________

CRITERIA FOR GRADING: *(Students may choose which grade they wish to work toward.)*

❑ ***TO EARN A "C" — STUDENT MUST:***

Complete all required course work *(science experiments/lab projects).*

Earn an average of 70% or above on tests.

Complete one Science Project of own choosing *(with teacher approval).*

Course work and notes must be compiled into a notebook.

❑ ***TO EARN A "B" — STUDENT MUST:*** *Do the above, plus or except:*

Complete all required course work fairly well.

Earn an average of 80% or above on tests.

Complete two Science Projects: one of own choosing; one of teacher's choosing.

Notebook must be nicely done.

❑ ***TO EARN AN "A" — STUDENT MUST:*** *Do the above, plus or except:*

Complete all required course work very well. *(be thorough and neat).*

Earn an average of 90% or above on tests.

Complete three Science Projects: 2 of own choosing; 1 of teacher's choosing.

Notebook must have an attractive, well-made cover, tabbed dividers and title page.

ASSESSMENT: ____

____

____

____

*Signature of Instructor:*______________________________ *GRADE:*____

GRADING CRITERIA AND COURSE ASSESSMENT

Unit/Class: **"Music Appreciation"** **Course:** **"Fine Arts"**

Student__________________ **Grade(s)**______ **Teacher/Advisor**________________

CRITERIA FOR GRADING: *(Students may choose which grade they wish to work toward.)*

☐ ***TO EARN A "B" — STUDENT MUST:***

Attend the 6 assigned "Community Concerts":

* "Dallas Brass" (trumpet, horn, piano, trombone, tuba, drums)
* "Vienna Strauss" Vocal Ensemble
* "An Old-Fashioned Holiday Sleighride" - The Maldonado Singing Players
* "DeWayne Fulton" - Harp Virtuoso, Electronic Sampling Keyboard
* "The Texas Boys' Choir"
* "Helicon" - Instrumental Ensemble (wooden flute, hammer dulcimer, penny whistle, guitar cittern, fiddle)

Do 20 "Music Appreciation" sheets covering a wide diversity of music styles and composers *(See "Explanation of Music Worksheet" in "Potpourri" section.)*

☐ ***TO EARN AN "A" — STUDENT MUST:*** *Do the above, plus or except:*

Do 30 "Music Appreciation" sheets *well.*

Attend the concerts with enthusiasm (even if you have to fake it).

ASSESSMENT:____

____

____

Signature of Instructor:__________________________ *GRADE:*____

GRADING CRITERIA
AND COURSE ASSESSMENT

Unit/Class: "Nutrition" **Course:** "Health or Home Ec"

Student_____________________ **Grade(s)**_______ **Teacher/Advisor**__________________

CRITERIA FOR GRADING: *(Students may choose which grade they wish to work toward.)*

❑ ***TO EARN A "C" — STUDENT MUST:***

Watch "Fit For Life" video and report on the 3 concepts that were most interesting and applicable to you. *(200 words minimum on each concept.)* Ask Aunt Kathy (a scientist) (or find out from a book on digestion) why the food combining aspect is not scientifically accurate.)

Read assigned chapters of The McDougall Plan. *(– which is simply basic God-designed nutrition; nothing new! They just explain the processes and make it clear to us **why** God knew what He was doing.)*

Read (only) assigned chapters of Fit For Life *(not the food-combining part.)*

❑ ***TO EARN A "B" — STUDENT MUST:*** *Do the above, plus or except:*

Read assigned portions of the Back to Eden Cookbook.

Make 6 Health-supporting recipes for the family including one of each of these categories:

* Main Dish
* Lunch Meal
* Un-sweetened snack
* Milk Substitute
* Dessert or Cookie
* Muffin

❑ ***TO EARN AN "A" — STUDENT MUST:*** *Do the above, plus or except:*

Make 4 additional *health-supporting* recipes in the categories of your choice.

Write a short report on each of the recipes including the title, how you liked it, how family members liked it, whether or not you'll make it again, changes you might make next time to improve it (other than adding fat and sugar!)

ASSESSMENT:____

____

____

Signature of Instructor:______________________________ *GRADE:*____

GRADING CRITERIA AND COURSE ASSESSMENT

Unit/Class: "Performing Arts" **Course:** "Fine Arts"

Student____________________ **Grade(s)**______ **Teacher**____________________

CRITERIA FOR GRADING: *(Students may choose which grade they wish to work toward.)*

❑ *TO EARN A "B" — STUDENT MUST:*

Attend all required performances (concerts, plays, movies).

Have a willingness to learn, not just be entertained.

Complete one Concert/Performance Record, Play Critique/Analysis, or Movie Critique/Analysis for each performance watched or attended.

Compile all reports into a notebook.

❑ *TO EARN AN "A" — STUDENT MUST: Do the above, plus or except:*

Show thoroughness and neatness on all reports.

Be willing to go out for pie and coffee (or salad and herbal tea) after performances and be seen in public with your parent(s) and actually act like you're having a good time.

The notebook must have a tastefully done (hand-made or computer-'n'-you-generated) cover.

ASSESSMENT: ____

____

____

____

____

Signature of Instructor:____________________________ *GRADE:*____

GRADING CRITERIA
AND COURSE ASSESSMENT

Course: __"Physical Education"__

Student____________________ **Grade(s)**______ **Teacher/Advisor**________________

CRITERIA FOR GRADING: *(Students may choose which grade they wish to work toward.)*

❑ *TO EARN A "C" — STUDENT MUST:*

Take part in at least 4 different types of P.E. activities such as walking, aerobic dance, weight training exercise (not the "bulking" type), muscle-toning exercise for the entire body, swimming (continual), bicycling, bench aerobics, walking and "arm work-out." walking.

Watch one video on fitness walking tips and techniques (Sally Struthers did one that is excellent!)

Read the chapters in Fit or Fat on health and fitness-promoting, *correct* exercise

Your exercise routine, in whatever form you choose, should include:

- ☐ **Warm-up of the muscles that will be used in work-out** *(5 minutes)*
- ☐ **Your chosen exercise work-out** *(20 minutes minimum)*
- ☐ **Cool down** *(5 minutes)*
- ☐ **Put away equipment used**

Learn how to monitor heartrate, and DO it during work-outs

❑ *TO EARN A "B" — STUDENT MUST: Do the above, plus or except:*

Take part in (try) at least 6 different exercise forms, trying each 2 times.

Be aware of the following information:

* To maintain level of fitness and weight: work out 3 to 4 times/week
* To increase fitness level and lose weight: work out 5 to 6 times/week

❑ *TO EARN AN "A" — STUDENT MUST: Do the above, plus or except:*

Take part in at least **8** different exercise forms, trying each at least 2 times.

ASSESSMENT:____

____

*Signature of Instructor:*________________________ *GRADE:*____

GRADING CRITERIA
AND COURSE ASSESSMENT

Unit/Class: "Religions of the World" **Course:** "Social Studies"

Student____________________ **Grade(s)**______ **Teacher/Advisor**__________________

CRITERIA FOR GRADING: *(Students may choose which grade they wish to work toward.)*

❑ ***TO EARN A "C" — STUDENT MUST:***

Read the book (with teacher in class): What's the Difference?

Write up a "Religions of the World Study Sheet" report on each of the 10 major religions.

Complete a "Religious Terms" sheet.

Complete a "Ground Rules for Studying Other Religions" sheet.

❑ ***TO EARN A "B" — STUDENT MUST:*** *Do the above, plus or except:*

Give more detail in writing up the "Religions of the World Study Sheets."

Study the "Doctrinal Statement" of three different groups and write a short (200-word) essay about *why* groups have them.

❑ ***TO EARN AN "A" — STUDENT MUST:*** *Do the above, plus or except:*

Read the booklet "Mormonism" *(before doing the 1-page report on it)*

Find out what the terms *evangelical, charismatic, pentecostal,* and *fundamental* mean in the "church world" and complete that sheet.

Find out about these Christian denominations *(via books and/or interviews)* and write at least 50 words on each of their basic beliefs:

- Seventh Day Adventists
- Presbyterian
- Lutheran
- Church of Christ
- Baptist
- Episcopal
- Non-Denominational (Like Shekinah)
- Disciples of Christ
- Methodist
- Assembly of God
- Church of God
- Foursquare Gospel
- Full Gospel
- Roman Catholic

ASSESSMENT: ____

____

____

Signature of Instructor:______________________________ *GRADE:*____

GRADING CRITERIA AND COURSE ASSESSMENT

Unit/Class: "Self Esteem & Understanding" **Course:** "Life Skills"

Student____________________ **Grade(s)**______ **Teacher/Advisor**__________________

CRITERIA FOR GRADING: *(Students may choose which grade they wish to work toward.)*

❑ *TO EARN A "C" — STUDENT MUST:*

GIRLS — Read Mothers and Daughters Together with Mom.
Read So You're a Teenage Girl with Dad.

BOYS — Read Mothers and Sons with Mom.
Read Becoming a Man with Dad.

❑ *TO EARN A "B" — STUDENT MUST: Do the above, plus or except:*

Do the above with enthusiastic willingness, taking the initiative to pursue the reading-together times.

Write a book report (form) on each of the two books read.

❑ *TO EARN AN "A" — STUDENT MUST: Do the above, plus or except...*

Write a list of the "Top 10" character qualities or strengths you feel you possess and how you feel God wants to develop them further and possibly use them.

Write your "Bottom 5" weaknesses and what needs to "break" or happen in you for God to change them; how can you cooperate with Him to mature in these areas?

ASSESSMENT: ____

____

____

Signature of Instructor:____________________________ *GRADE:*____

GRADING CRITERIA AND COURSE ASSESSMENT

Unit/Class: "Spelling/Vocabulary" **Course:** "English"

Student____________________ **Grade(s)**______ **Teacher/Advisor**__________________

CRITERIA FOR GRADING: *(Students may choose which grade they wish to work toward.)*

❑ *TO EARN A "C" — STUDENT MUST:*

Complete all course material.

Earn a minimum of 70% on class work and tests.

Quality/Neatness Standard = "Average"

❑ *TO EARN A "B" — STUDENT MUST: Do the above, plus or except:*

Complete all course material.

Earn a minimum of 80% on class work and tests.

Quality/Neatness Standard = "Good"

Do 6 Vocabulary Word Analyses/Month.

❑ *TO EARN AN "A" — STUDENT MUST: Do the above, plus or except:*

Complete all course material.

Earn a minimum of 90% on class work.

Quality/Neatness Standard = "Excellent"

Do 12 Vocabulary Word Analyses/Month.

Keep ongoing list of "Words Not Understood" in personal reading, look them up, record meanings neatly, keep in notebook.

Participate in 4 Spelling Bees.

> Note: The "Vocabulary Word Analyses" are not included in this manual. All we do is look up a *challenging* word, write its pronunciation, how it is divided into syllables, and the part of speech it is, and then define it. All this is obtained from the dictionary.

ASSESSMENT: ____________________________________

____

Signature of Instructor:________________________ *GRADE:*____

GRADING CRITERIA AND COURSE ASSESSMENT

Class: ______________________ State History **Course:** Social Studies

Student ______________________ **Grade(s)** _______ **Teacher/Advisor** ______________________

CRITERIA FOR GRADING: *(Students may choose which grade they wish to work toward.)*

☐ *TO EARN A "C" — STUDENT MUST:*

Make a basic cover for the notebook *(just a label).*

Visit and do a "Search It Out" sheet for each of the Museums on our itinerary.

Do a Museum/Site Data Report + 1 photo for each place.

Find **4** developments on each of the 20 Decade-at-a-Glance sheets *(under the topics of your choice).*

Do one "Focus Report" *(from the list of Report Ideas).*

☐ *TO EARN A "B" — STUDENT MUST: Do the above, plus or except:*

Make a "classier" cover for the notebook.

Take **2 to 3** photos at the Museum or Site.

Find **6** developments on *each* of the 20 Decade-at-a-Glance sheets *(under the topics of your choice).*

Do **2** Focus Reports. /// Work and Notebook look *good.*

☐ *TO EARN AN "A" — STUDENT MUST: Do the above, plus or except:*

Make a *photo collage cover* for the notebook.

Take **4** photos at the Museum or Site

Find **8** developments on each of the 20 Decade-at-a-Glance sheets.

Do **3** Focus Reports. /// Be actively interested & cooperative.

ASSESSMENT: ____

____

Signature of Instructor: ______________________________ *GRADE:* ____

GRADING CRITERIA
AND COURSE ASSESSMENT

Unit/Class: "Tact, Grace, & Social Skills" **Course:** "Life Skills"

Student____________________ **Grade(s)**______ **Teacher/Advisor**__________________

CRITERIA FOR GRADING: *(Students may choose which grade they wish to work toward.)*

❑ *TO EARN A "C" — STUDENT MUST:*

Read How to Start a Conversation and Make Friends by Don Gabor.

Write a paper entitled "The Ten Tips I Found Most Useful" in which you write the tips themselves and then explain how you actually used them in your own experience.

Find, clip and read 2 of Elizabeth Post's columns on Etiquette in Good Housekeeping Magazine. How does her advice line up with the Word of God?

❑ *TO EARN A "B" — STUDENT MUST: Do the above, plus or except:*

Interview 2 people *(agreed upon)* who are exceptionally gifted in the area of tactfulness and grace. *(Use sheet; add own observations.)*

Do a word study *(10 references)* on the word "grace" in the Bible. *(See "Ideas for Creatively Studying the Bible.")*

Find, clip, read, and compare 5 of Elizabeth Post's columns.

❑ *TO EARN AN "A" — STUDENT MUST: Do the above, plus or except:*

Interview **3** people *(agreed upon).*

Do a 20-instead-of-10-reference word study on the word "grace."

Find, clip, read, and compare **8** of Elizabeth Post's columns.

ASSESSMENT: ____

____

____

Signature of Instructor:____________________________ *GRADE:*____

GRADING CRITERIA AND COURSE ASSESSMENT

Unit/Class: "Tap Dance" **Course:** "Fine Arts" or "P.E."

Student____________________ **Grade(s)**______ **Teacher**__________________

CRITERIA FOR GRADING: *(Students may choose which grade they wish to work toward.)*

❑ *TO EARN A "C" — STUDENT MUST:*

Attend all classes *(unless absence is excused).*

Participate with willingness to learn and cooperate.

❑ *TO EARN A "B" — STUDENT MUST: Do the above, plus or except:*

Practice outside of class 1 time/week. *(Extra performances or helping class friend learn routines count as practice time.)*

Earn an average of at least 80% *(or "good")* on in-class dance evaluations by instructor.

❑ *TO EARN AN "A" — STUDENT MUST: Do the above, plus or except:*

Practice outside of class 2 times/week.

Earn an average of at least 90% *(or "excellent")* on in-class dance evaluations by instructor.

Choreograph one dance per year.

ASSESSMENT: ____

____

____

Signature of Instructor:____________________________ *GRADE:*____

GRADING CRITERIA
AND COURSE ASSESSMENT

Unit/Class Typing **Course:** Occupational Skills (or Elective)

Student____________________ **Grade(s)**______ **Teacher/Advisor**__________________

CRITERIA FOR GRADING: *(Students may choose which grade they wish to work toward.)*

☐ *TO EARN A "C" — STUDENT MUST:*

Complete all course material.

Type at 20 words per minute *(adjusted for errors).*

☐ *TO EARN A "B" — STUDENT MUST: Do the above, plus or except:*

Complete all course material.

Type at 30 words per minute *(adjusted for errors).*

Work must be neat.

☐ *TO EARN AN "A" — STUDENT MUST: Do the above, plus or except:*

Complete all course material.

Type at 40 words per minute *(adjusted for errors).*

Work must be neat and compiled attractively in notebook or report cover.

ASSESSMENT: ____

____

Signature of Instructor:__________________________ *GRADE*:____

Section Six:

PERSONALIZING ALL THIS STUFF!

A peek at how everything
you're reading in this book *looks* on a
real live family via a couple of candid glimpses
into the Shelton household, and an overview
of the four sections to follow (6-A – B)
that will help you as you begin to
personalize all this stuff!

A homeschool mom wrote this to me: "Your book was recommended to me when I sent out a plea for help with my ninth grade boy using "XYZ Curriculum" this year. (Not enrolled now, thank God!) *He's ready to throw in the towel and forget about school. (Me too!) Your book is a literal "Godsend" to me. The weight has been lifted from my shoulders to God's. I really have hope that we can do this. As long as my son will give us a little more time. He is very discouraged and was crying yesterday wanting to go to the public school because he is 'stupid' and 'lacks social skills.' I will not let him go, but he's very agitated."*

Well, I was very concerned about this boy and his mom, and felt so bad for them, but I knew that, in this case, Sharnessa could be of more encouragement than I! So I asked her if she'd be willing to write to this boy. To my joy, she said yes! And she did! And she gave me permission to share it with others... (Sorry this is so small and cram-packed in! It was the only way I could fit it in!)

Encouraging Letter to Reluctant Homeschool Boy

Hi! I'm Sharnessa Shelton, Barb Shelton's daughter!

My mom sent me an e-mail letting me know that you (the son, ☺) don't want to home school - you want to go to public school, but your parents won't let you. Listen, man, ☺ your parents know what they're doing. Trust me! I homeschooled from the beginning, (age 5) all the way through to graduating from high school, and I am *soooo* thankful my parents did this! I *loved* homeschooling! And I'm not your "sit-at-home-and-do-nothing" kind of person! It was seriously such a blast! Instead of being stuck in a classroom for 6 hours a day, and then an added 1-3 hours of homework, I was able to focus on the areas I loved and was passionate about! I mean, I still had to fill all the credits required by my mom, but we got to fill them with such awesome, hands-on experiences! We went to seminars, workshops, historical sights and museums all around Washington (and OR and Calif. too!), conferences, on field trips, watched videos, and more – it was incredible! Plus sometimes we did classes with other homeschooling friends!

But the main thing that I was so thankful for, like I said earlier, was that I got to focus on what I LOVED, which is the Arts, dancing and singing. (I also love history and government.) Here's one example: When I was 16, I directed, choreographed, and produced a Christmas production, and I mean *production!* It wasn't a nativity scene thing (which is cool, but this was a 30-minute show!) at our church, and I was in charge of 25 kids my age! It was awesome! And you know what? I got to count that as part of my school requirements! I was able to learn in such a creative, free, hands-on way, and that's something that public school kids are never (or seldom) able to do. They're stuck in their classrooms, reading dry, dull textbooks – their desire to learn is just sucked right out of them! All through my homeschooling experience my public school friends SO wanted to home school! I felt so bad for them!

Actually, I did go to a public school, one day every year with a friend, ☺ and I have to tell you, it was a joke! My mom was really worried that when I visited, I wouldn't want to home school any more, but man, I came back even more sure that I NEVER wanted to go! I told her, "It's a nice place to visit, but I wouldn't want to try to get my education there!" Kids were talking, eating, sitting on desks – it was pathetic! I mean, I'm a VERY social person, but that's just the thing, school should be for learning, and socializing should be for later! Now, if you're worried that you'll become a "social reject" or something – as people are typically afraid will happen to homeschoolers – forget it! I had SO many friends! (Still do! In fact, now at 21 she's been in the weddings of four of them! One flew her from NY, and the other flew her from London! JUST to be in their weddings!) Most of them were public schoolers! So don't think that just because you home school, it means you won't have any contact with people.

That reminds me, a little side-note here. Have you heard of the pop group "Hanson"? You probably hate them. (Most guys do! ☺) But did you know that they homeschool?! They have since the beginning! My point is – look at what they're doing! They're doing what they love! Because of their parents taking the time to teach them at home, they were able to focus on their passions! And not only that, but I've seen them in interviews, and all three of them are so mature, eloquent, and sweet! And they're Christians, too!

Just keep up your friendships, have some sort of an activity (what do you love to do?), and maybe get involved in youth group, if their values are good. School's not supposed to be a social place anyway! Also, just enjoy hanging out with your family! My main times to be with friends were at church, youth group, and on weekends – also, I took dance lessons. So PLEASE don't let any social issues get in the way. This is honestly going to be a very precious time in your life. Your parents know what they're doing. Personally, from my experience in homeschooling, I believe that I came out of high school with so much more than any of my friends. I came out of it with the desire to learn even more, with an awesome relationship with God, with strong relationships with my family (you'll realize how important that really is), with more maturity (than my peers), with solid character, strong morals, values, and standards, and with true friendships. (Just ask my mom and dad, they'll "amen" every word of this!) I came out of it totally satisfied, not feeling at all like I had wasted 4 years of my life - instead, it was a time of preparation. I wasn't in high school living from weekend to weekend, and anticipating getting out like other kids do to the point where they aren't even able to enjoy the moment. But I did! I have NO regrets. See, I think that by homeschooling, I was able to escape the superficiality of the public school scene and build on the areas and friendships that would (and have) last(ed). I am so blessed to have the parents I have. Seriously. (And my mom is not paying me to say this! ☺) Not many are willing to give up their time so their kids can have the ultimate experience in education, and I have many friends who wished theirs would've. Don't take it for granted. It's the greatest gift you'll ever receive!

Sincerely, Sharnessa Shelton ^_^

PERSONALIZING ALL THIS STUFF!

FOR WHOM THIS SECTION WAS WRITTEN

By now (if you've read the rest of this book up to this place) you're probably asking: "Okay – what am I supposed to *do* with *all this stuff?!"*

Since you asked, I just happen to have some ideas. I divided them up and put them in the upcoming order of Parts A through D for two good reasons: They'll make more sense, especially if that's what you're trying to make out of all this right now. What follows will also give you an anchor in the storm, if you're feeling a bit tossed and torn by such questions as "HELP! Where do I start?!?"

But before I launch you into a bunch of technical stuff, I want you to see what it looks like played out in the form of a real family...

WHAT'S A TYPICAL DAY LIKE FOR YOUR FAMILY?

Looking at all the forms in this manual intimidates people. I don't know why; it just does. For some reason they think that when all this stuff is translated into their daily lives, the regimentation of the army will pale in comparison. This couldn't be farther from the truth!

Their reaction is probably similar to the one I originally had when I looked at the Bible, not that I would ever begin to compare the Word of God with this book – although much of what you see in here was His answer to *my* plea for "HELP! Where do I start?!?!" When I first thumbed through the Bible, it seemed overwhelming. "I've got to read and know and *do* all this stuff! I'm not gonna get to have *any* fun! What a drag!" But as those of us know who are entering deeper into the "abundant life" that Jesus died so that we might take part in, we now realize that God's word to us is not binding! Far from it – it's *freeing!* Likewise, the framework I present in this book is intended to be *freeing!*

So before we head into how to put it all together, I want to give you a picture of what it looks like *in our home.* Of course, it's going to look much different in yours. But seeing a glimpse of how we operate might make you relax a bit about how all this is actually lived out by real people. To demonstrate it, I'll share with you a "typical" day in the life of the Shelton family. However, I do have *two* rules for reading the following Shelton Family Back-to-Basics Schedule. If you *start* reading it, you *must* hereby solemnly agree to:

1) ***read what immediately follows it, but not until after you...***

2) ***finish reading it*** *(or before going on in this section)* Otherwise, NO FAIR! And I guarantee you'll be *verrrry* sorry if you don't!!! All right, let us begin! ☺

The Shelton Family Back-to-Basics Schedule

"Rise 'n' Shine" time is 5 a.m. sharp, and I don't have to tell the kids to get going on their Quiet Times because that's already on their schedules posted on each of their bedroom walls. (Each schedule is printed on their favorite color – to match their room – and then laminated.) Breakfast is 6 a.m. sharp, with everyone showered, fully dressed, hair combed, room cleaned and dusted (you know how that accumulates in the country), pajamas put away, and, of course, everyone bright-eyed and bushy-tailed. All of the to-do items just mentioned are, of course, checked off in the corresponding box of their age-appropriate Daily Report. We spend the bulk of breakfast time reciting Bible memory verses learned during the previous seven months – the "heavenly number" you know. "Morning grog" is just not allowed; life's too short.

At 6:30, Household Chores begin with instructions obtained from the Family Job Chart, of course. Because the forms are so God-ordained and so enjoyable to look at, everyone happily pitches in to do their part.

At 8 a.m. we break from jobs and gather at the couch for devotions. Each one excitedly shares what the Lord showed them during their chore time which, of course, they've already journaled before coming to the couch. Devotion Time is always pleasant because it's so spiritually applicable on a practical level. No one would dream of arguing or insisting on having their own way or the best seat. (Besides there's a Seat Chart to take care of that.) The 1½-hour Devotion

Time just whizzes by with all involved, sharing their own personal insights.

At 9:30 we have our Fresh Juice Break. While the juice-maker person (we just refer to the Juice Making Schedule to find out whose turn it is) makes the juice (from the apples grown out back in our orchard) in the juice machine (purchased with money earned from a family project of collecting aluminum cans off the side of the road), the others seize the moment to brush up on prepositional adverbial phrases or explain the Pythagorean Theorem to one of the younger ones who hasn't quite grasped it yet. Then we have five minutes to drink the juice and clean up.

This revives us giving us the energy needed to launch into individual delight-directed learning times. These are totally spontaneous, of course, though interests must be initiated at 9:45 promptly and ended at noon, with two 10-minute breaks at 10:25 and 11:10.

Lunch, made the night before by the Lunch Maker, and packed away into lunch boxes, is eaten listening to some sort of educational cassette tape. The most popular ones in our family are French (4th year) or our multiplication drills set to boot-camp marching songs. (a delightful tape we produced for a delight-based Occupational Skills project last year. All proceeds go to the local homeless shelter.)

Afternoons, beginning formally at 1:00, are more leisurely with time for P.E., working on various aspects of the family business, finishing the solar roof heater, adding chapters to the children's books we're writing (for a unit in English, which is also counting for French, Spanish, and German because the older children are translating them as they go into those three languages); or working on the new addition to the house.

If it isn't raining (which is seldom in Washington State) we have plowing, planting, weeding, or harvesting out in the 40-acre garden. Who does what depends on whose turn it is according to – you guessed it – the Garden Job Chart! (You're getting the hang of it, aren't you?!)

Between projects, we squeeze in an hour for Family Orchestra Time. To prevent arguments and tiring of doing just one instrument all the time, we draw names for who will play each of the instruments: harp, accordion (yes, even the toddler!) oboe, trombone, violin, piano, and tympani, since each of us is adept at all of them.

I'd finish telling you about the rest of our day, but I think that if you'll just take a look at the Dinner-Making Schedule, the Dinner Jobs to be checked off on the age-appropriate Daily Reports, and the Pictorial Bedtime Chart, that'll tell you more than you ever wanted to know.

HOW LONG DID IT TAKE YOU?

...to figure out that the above is *total* non-reality for the Shelton household? No fair guessing if you know us or just dropped by yesterday!

There are actually a few *elements* in the above scenario that *do* exist. Heavy on the "few"! ☺ After reading the rest of the manual you'll know which ones they are. Okay, I'll give you a hint: they *don't* include the timpani, a garden, the juice, the toddler (our youngest is now 12), drills, or sharp promptness. In fact, whatever aspects intimidated you – up until you figured out I was only kidding – probably don't exist. (However, in all fairness to our son who just turned 20, he is exceedingly faithful at his Quiet Time!)

I *do* do charts, which you've probably figured out by now, and I've actually color-coded several of them just because I had the colors on hand and it's fun. ☺

OK, SO WHAT'S A *REAL* DAY LOOK LIKE?

This is going to be a bit difficult because I'm going to have to get vulnerable now. ☺ I'm going to *intentionally* be vague in my references to time; because as soon as I tell you what time I get up or how long we spend at this or that, it'll be easy for you to compare or feel what we do is a standard, for better or for worse! (most likely *worse!)* And I don't want to put you (or me!) into bondage by doing so.

I'm also going to have to protect the identities of certain parties for reasons that will unfold along with our "typical day." (**Quick 1999 update:** I need to let you know that the following was written back in 1993, when all three of our children were home. This is just one "snapshot" of our lifestyle over the many years of Dave being on shift work. Not only did our day look different three months before and three months after what I present here, it looks very different now, with Dave on straight days with weekends off! Yay! A long-awaited, most-welcome luxury for our family!)

One Actual Basic Schedule of the Real Shelton Family

(Remember, this was in 1993)

We're currently on Swing Shift, which means Dave gets home at about midnight. Since he's not ready to go to bed then, he likes to work for an hour or two on the new addition (we're converting our garage into a stor-

age room, a work area, and a bedroom for Tory so I can have his old room for an office. (If you saw our dining room, living room and family room right now(in 93), you'd know ***why****.) Because the house is so quiet I use this goldenly silent time to work on my writing at the computer. Dave and I get to bed at 2ish, which explains why the crack of dawn does its thing without us on this shift.*

Tory's up by 8 or 8:30. He groggily heads for the heating vent in the livingroom over which he arranges himself under his blanket so that the hot air stays in his billowing blanket instead of reaching the thermostat which he has turned up so he'd have plenty of hot air in his blanket. Here he has his Quiet Time.

When I get up (and turn the thermostat down), I get into my exercise duds, have my "tall glass of warm water" as Dr. Raymond Moore recommends (the best "system regulator" there is! Gently wakes up your system and prepares it for food. Try it!) and then find a quiet corner and spend some time getting myself tuned in to Jesus, or, as they say, "getting my day hemmed in prayer so it's less likely to unravel." (For me, there's almost always ***something*** *that unravels anyway, so prayer keeps ME from unraveling!!)*

After that I go in and wake up "sleeping beauty," who, just a few short hours ago, was "waking night owl" (Name withheld, of course, to protect the innocent). And to be honest, she ***is*** *innocent! Let me digress... Who made up the commandment that dictates "Thou shalt rise early and be doing school by 8 a.m.? I haven't seen it! Shar– (woah!– just about gave it away!) this "night owl" is one of the most disciplined, self-motivated healthy-inside-and-out people I know; and the time she gets up (late) has nothing to do with it! One of the benefits of homeschooling is flexibility and freedom to let our adolescents get the extra sleep they need! (End of digression.)*

We have no set schedule for ***any*** *of our day; just set "stuff to do" in certain time-blocks. For the kids, morning is when they have their Time with God, do their jobs, eat breakfast, do P.E., and get themselves showered, dressed, hair combed, etc. Each individual chooses what order works best for them. Tory frequently works with Dave out in the garage, building the addition, saving jobs and P.E. for after Dave leaves for work at 2:10 p.m. If they get all this stuff done early (I don't hold my morning breath), they might get going on reading or math, but that usually comes later.*

I generally use mornings to get office work done. (I get an average of 100 homeschool-related calls a month; about 25 of those are requests for information packets for new homeschoolers. Plus I do my Home-Based Ed Course stuff. This adds up to a lot of time with manilla envelopes, address labels, tape, and scissors, not to mention the copy machine. [1999 Update: We now get an average of 500 homeschool-related contacts a month, including phone, email and snail-mail.] Besides office work, I, too, get my exercise in the morning, sometimes with Carlianne, and then usually a walk with her right after exercising, when I'm all warmed up. (I'm always cold!) Dave, after a couple of hours working in the garage on the new addition in the morning, comes in and fixes dinner – no, not lunch, a full-fledged dinner. (He's a much better cook than I am – and I like it ***just fine*** *that way!) So dinner's anywhere between 12:30 and 1:45; earlier if we have something to do in the afternoon. Which we* ***do*** *have every Tuesday – either Homeschool Skating or our School Biz Chapter meeting. Dave usually gets in a game with Carlianne, some snuggling and/or joking with Tory and Sharnessa, and a cup of coffee with me before he leaves for work at 2:10p.m.. (Yes, he's for real. No, he doesn't do seminars. Yes, he's been asked and I'm trying to talk him into it!)*

Let me stop right here and make an important point: I am perfectly content with the fact that, up until now, we haven't done anything that you might call "school"! "Schoolish-type lessons do enter the picture later, but I believe that this IS "school" in the most significant meaning of the word! Part of the Bigger Picture, which includes (self) discipline and on-the-job character training.

Gramma's Day is every Wednesday, when she takes all three kids in the afternoon for a treat of their choice, plays or just "be's" with them at their home, fixes them dinner, and then we pick them up later.

If it isn't Tuesday or Wednesday, afternoons usually include a reading time where I read to Sharnessa and Tory. We get Carlianne going on something so we can have an unbroken, or at least ***less****-broken time. Later they do their individual "school work" meaning their own reading (not usually textbooks), Math, and/or projects. At some point Carlianne spends time with Sharni and Tory individually – fun projects, a game, or reading that either I've assigned them to do with her or that they choose. She and I also have our own time for reading and doing little projects. I continue with office work and/or writing, depending on the need.*

By the way, our reading times have a certain structure to them that you might find interesting. We have an average of three 20-minute sessions each day; four sessions some days; two sessions others. These are usually done all in one stretch of time. During these we read for 20 minutes in each of our "chosen books." Like, for instance, right now we're reading Beyond and Back and the Bible for "Spiritual Studies," The Story of Liberty for "History," Home Made Health – for "Health," etc. As interesting articles come across our path, we'll substitute those for one of the books.

Also, something I consider to be a vital and meaningful part of our homeschool experience and their "education" is discussing the various hassles, trials, difficult or awkward situations with people, frustrations, and hurts that they and or we experience on a day-to-day basis. Time-wise, many of these discussions get funneled into "Life Skills." You can read all you want in self-help, relational books, but what better ***real-life*** *training can you get than walking through the "hard stuff of life" with a parent who's (trying to stay) tuned in to the Lord and His ways?! They're learning more about "Tact, Grace & Social Skills" in a* ***practical*** *way than they could ever learn* ***in theory*** *from a textbook! IMHO*

Once a week Sharnessa has her tap dance class. During that time Tory works out at a fitness center – which is P.E. for both of them. (Or Fine Arts for Sharnessa.) But since P.E. needs to be done more than once a week, they both have other activities they do – like working out with a "safe" aerobics video or for Sharni, walking. And Tory likes to go on a bike ride with another homeschooled boy. They both enjoy trampolining at our neighbor's.

At dinner time the kids and I have "lunch" since Dave's at work and we already had dinner at lunchtime. (Remember, we're on Swing Shift.) In the evenings the kids catch up on things they didn't finish during the day, which is pretty usual, since "interruptions" seems to come up every day.

I ***do*** *actually have a pictorial bedtime ritual chart for Carlianne which we have pretty much memorized by now. But every so often, when she's finding more and more excuses to get up (water, bathroom, teeth, a hug...) or when one of the kids helps her get her ready for bed, it's nice to have it for reference. We can just send her to it and help her keep (and* ***us****!) on track.*

Sharni and Tory are ***supposed*** *to go to bed at about 10:00, but since they're both night owls like their mother, it's hard to actually pull this one off. I like to sit on Sharni's bed with her and give her a backrub before she goes to sleep. (****She*** *likes it just as much!) I like it because a backrub will elicit all sorts of conversations out of her; the later it is the more she'll talk. So I take full advantage of this. Tory's not as much of a backrub fanatic, but he likes to snuggle and talk. Being the same height as me now, he doesn't fit on my lap too well – though I sometimes force him anyway. ☺*

Which brings us full circle to Dave getting home at about 12 midnight. By this time I'm usually back at the computer and the kids are in bed, sometimes even asleep. And thus ends – and begins – another day in the basic, regular, <u>*real*</u> *life of the Shelton family.*

❤❤❤❤❤

[BACK TO THE PRESENT...] At the time I wrote that, as I said, we were on each work shift – Swing, Days, and Graveyard – for three months at a time, which meant that just as we were getting used to the lifestyle of one shift, it was time to change to another! In fact, since originally writing the above "Typical Day" six years ago, we've now changed to permanent Day Shift (with many other shifts in between) – which is a *totally* different "Typical Day" from the one I just described! Especially since everyone is now six years older! Don't worry, I won't lay any more "typical days" on you! Two was plenty, I'm sure! Besides, my purpose isn't to give you a year-by-year schedule of our lives, but only an example of how it looked on and in our family – and the degree of flexibility we have. (And enjoy!)

Do you see why I couldn't get into any set routine? Getting into a "rut" actually sounded very attractive to me at times! *(And probably would have been GOOD for me!)*

The fact is that many of you are in a position of having an irregular schedule as well, though it may be due to a new baby in the house, a family business, multiple children, a child in need of special care, or just the fact that you don't operate well within time slots. Since my husband had been on this type of shift work for all of our married life, I'd never even had a *chance* to be "regular"! I didn't have a clue until just last fall (of 1998) what it would feel like! And that was 22 years into our marriage and having *graduated* two of our three children before I found out!

The key to being content in such a lifestyle situation is to be content, flexible and innovative in the particular state-of-schedule you're in, rather than look at how other homeschoolers (or schools) do it and comparing yourself unfavorably – and *unfairly* – with them. "School at 8 a.m." has *never* been right for us! If it's right for you, more power to you! But if it *isn't,* I say even *more* power to you! ***Jesus*'** power, of course! Because you need it! – *and* His wisdom, too, to bring you to the best place for *your* family!

And now for the specifics of Personalizing All This Stuff

WHAT YOU'LL FIND IN 6-A, B, C, AND D

Just for the record, each of the four sections that follow (right after this one) should actually have their very own number. However, when I first wrote the book, each of these was much smaller and part of the bigger picture of "Personalizing All This Stuff." Since then they have greatly grown to the point that they really should have been re-named sections 7, 8, 9, and 10. However, sections 7 and 8 had already been established and mentioned in many times throughout the book and in several reviews, and the thought of re-doing all that *(and correcting all the references to, and messing up the reviews of them)* felt totally dismaying to a one-(wo)man writing staff, namely *myself*. Thus – *sigh* – we still have the original 6-A through D page set-up. I'd like to introduce you to each of these sections one by one...

6-A: FORMULATING YOUR FRAMEWORK

In these first three sections, 6-A, B, and C, we start with the "big" and go down to the "small." So that means that in the first section, 6-A, you will set up your framework. Or at least you will *read* about setting up your framework. ☺ You will probably want to read the *whole book* before actually *doing* so because your framework is going to reflect and help you implement your vision – or your "heart 'n' soul" – regarding the education of your child. And reading this *whole* book (and the others I recommend) will help you formulate and develop that vision.

I hope you aren't offended at the step-by-step, simple-illustration mode of presentation I use in the latter part of this section, where I show you how to set up the time aspect of your framework, *if* you are utilizing that. Please don't think it's because I'm assuming you're simpletons! This stuff is just a lot easier to put into practice – *once you figure it out* – than it is to explain. I'm presenting it this way more for my own sanity.

Plus I had overwhelmingly positive responses from my eighteen original Homeschool Parent Testers *(back in 1993)* who tested the original version of this book and made such comments as: *"These sections tie all the rest of the book together and make everything else you present totally useable!"* Mary Schofield (author of *The High School Handbook)* said: *"Whew! Great! Great! Great! Well explained – I got it! I will be referring to this for others to use in my own workshops."*

If at the end of this section you say "Hmmm – this wasn't all that complicated! I wonder why she made such a big deal out of it" I will have succeeded. Because if you knew how much brain taxation and confusion and praying went into laying it out to you in a logical, understandable manner, and the condition I was in during the several-week period of doing so, you'd be surprised I even finished writing it! And if you knew the condition of my brain on *any* typical day, you'd be even *more* amazed that someone like me even thought of attempting such a feat. Someone should have warned me:

**"Do NOT attempt this stunt!
It is intended for
EXPERIENCED Stuntmen *only!*"**

I was stretched far beyond what I thought was my "maximum stretchability"! I've stood *(sat* at my computer) in awe of how God has clarified and illuminated my thoughts! Any credit and glory goes to my Master Education Consultant, Jesus! It's so true that "His power shows up best in weak people"!

So once you are feeling ready – and not a moment sooner(!) – come back to this section and get the practical help and examples for setting up your own framework, which I call the "shelves" or the "skeleton" upon which everything you do will sit or function.

6-B: DESIGNING YOUR CLASSES

Next, in section 6-B, we look at how to create classes, the "academia niches" into which *ALL* the "stuff" of your child's education will fit. You are free to take this to whatever extent you desire, whatever extent seems right and needful to you.

One woman who reviewed this book and evidently did not really *read* it said that this approach of utilizing all of life seemed taken to an extreme. In fact, I'd like to take a moment to address a couple of her comments here... *(Please humor me; this has been a source of pain and irritation, mostly because when misleading statements are made public in a widely-read publication, the author has no recourse to address, as in respond to or correct them.)* She *(unnamed reviewer from unnamed publication)* said:

> "Mrs. Shelton offers a hefty dose of her "real world" (non-textbook) teaching philosophy and includes sample fill-in sheets to help you get started: book reports, museum reports, play critiques, Bible studies, and menu planners to name a few. Hunting trips, baseball games, and listening to pre-recorded health tapes from a local hospital are all fair game for 'school.' This approach, however, seems taken to an extreme. When organizing your purse counts for high school credit, why bother keeping track of credits at all? I don't think every Jello-making session should count for Home Ec."

Had she truly *read* my book for its *heart* and not merely skimmed it for the external form, *(which I have found is very typical for reviewers who are being paid to turn out reviews of products that they themselves are not in need of, or do not "connect" with, in essence the same as "reading someone else's mail")* she would have realized that I present a much bigger picture than merely organizing your purse or making Jell-o. She must have missed the following *(quoted directly from the first edition of Form+U+la that she reviewed)*...

> "...The kids are allowed, even encouraged, to prepare anything their hearts desire. The only stipulation is that they have to get their shopping list to Dave or me at least one, but preferably two or three days ahead of when they need it... There's also a stipulation for counting anything

for Home Ec credit: If they don't care if it "counts," they are free to fix something they've fixed before. However, if they want it to count, they have to fix something *new*. Why am I a stickler on this? Because I personally don't feel that fixing sloppy joes from a mix or B-B-Q'd hamburgers 43 times constitutes learning. Nor would I feel good about that being a good chunk of a Home Ec credit. I'm certainly thankful for the help and the full tummies, but I want them to be challenged and to step into new arenas. Once they master a thing – in Home Ec at least – I want them to move on! This guideline gives them a little boost in that direction."

Hmmm, perhaps she should have read the book before reviewing it. Perhaps she would have understood that my intent was to give many varied examples for a wide readership.

And as for "organizing your purse," had she looked further she would have seen that *that* was just one small part of a "Personal Organization" unit of Home Ec. – which, by the way, I have taught as part of a community college class on organizing. I was requested to do that part by a mom who saw how I organize my purse, and that aspect was *very* well received by a class of *adult* women. *(Sorry, it's not included in this book other than the name of the activity itself; it's part of a seminar I used to do on-site only, but have since ceased because it involved tearing apart my entire house. Another book another year.* ☺ *)*

Well, thanks for letting me vent. ☺ Lest you think me hyper-sensitive, *("Wow, that wasn't so bad!" you might be saying),* I should tell you that there was much more than this – which I won't go into here. I'm thankful that "hers" was the only negative review my book has ever received. But the particularly difficult part of it was thinking that people who actually needed and could benefit from it would be turned off to it before ever even *seeing* it *(because of this one review)* was unbearable. God, the *redeeming* God that He is, used it to do a refining work in me though, *(felt quite similar to getting a wart burned off)*, and to show me that *He* is the One who would open and close the doors for my book, not any homeschooling business or ministry. (*Now* my words "To the Reviewer" at the beginning of this book, page "ix" *(that's 9 in Roman numerals for those who have rusted up since school* ☺) will make a little more sense! ☺

Anyway, as for planning classes – *other* than but including Home Ec ☺ – this section will give you lots of ideas. You'll get three different approaches for doing so, and lots of examples.

6-C: ORGANIZING YOUR STUFF

In this section you'll get tools for organizing *everything* – including your student's "High School Organizer Notebook"; a file for your records and "Learning Guides" *(from sec. 4)*; your "High School Resource Shelf"; your "Input Box"; your student's "Work Pending" file; your student's individual topical notebooks. You'll even get instructions for turning this manual into a handier 3-ring notebook form!

6-D: COMPLETED CLASS PLANS!

This section is new to the manual, though a few "pre-cursors" appeared in the previous edition. Having taken both Sharnessa and Tory all the way through to graduation since having written the original edition of this book, I now have *completed* class plans for every class we did. Most are actually for Tory's classes as it wasn't until the last couple of years that I came up with the final version of the class plan format that I did – that of having the hour check-off chart at the bottom of each class plan. So his are more "state of the art." ☺

As always, though, I have to tell you that *his plans were created mostly as we went along.* If, at any point along the way, you think I planned all this ahead, you will be misled and overwhelmed, and I don't want *either* of those to happen to you!

NOT AN "ALL OR NOTHING" DEAL

If after reading the hourly portion "Formulating Your Own Framework," you feel that figuring credits by the hour is not for you, don't assume that this book won't work for you. You can still use just about everything else contained herein *without* using my approach to computing credits.

In fact, in Section 7, "Lifestyle of Learning Applications," you'll see a much freer, more creative approach to learning and education. Even if you like the hourly approach, I still highly recommend taking a look at that section just for the additional insights into true education.

Once you've formulated your framework, be open to the changing needs of your student, as well as your own deepening understanding, broadening vision, and growing capabilities. Notice that the "Master Framework Plan" is printed *(and filled in by you)* on paper, not etched in *stone!* ☺ When you copy that sheet, just make up several extra's so you'll be ready to do some adjusting if, down the path a ways, you and/or your student would like to make some changes in credits and hours. Flexibility, even *en route,* is one of the boons of this system! ❧

Well, are you ready to start Personalizing All This Stuff?!

Section Six - Part A:

FORMULATING YOUR FRAMEWORK

Here's where you will first get introduced to a variety of frameworks, and then – when you're ready – you'll get to roll up your sleeves, get up your nerve, and start setting up your own framework to accommodate your student's bents and "style"!

Excerpts from a letter from a concerned sister regarding...

Misunderstanding the Lifestyle of Learning Message

Many homeschooling moms "out there" have totally misunderstood the Lifestyle of Learning message. I'm not sure how this has happened because I think Marilyn did a great job on her book. The problem seems to be that many moms think the Lifestyle of Learning approach is about throwing out all curricula, sitting back, and letting the kids do whatever "delights" them.

Now, you know what that will be if these kids: 1) have been in public school before, and/or 2) have not been trained to be "spiritually-minded." These moms, not understanding what the real message is, essentially do nothing to direct the children. Not understanding the difference between "delight-directed" and "flesh- directed," many moms flounder around not knowing what to do because the kids are not interested in any learning things. Husbands, not seeing any fruit, become a bit impatient with the lack of results. Most unfortunate of all, however, the children have no direction.

If our life, our homeschool, is not bearing fruit, we are doing something wrong. But many moms, realizing their lives *aren't* bearing good fruit, still don't know where to take it from there. Many of us are the product of public schools; it takes time to break out of this mold, and many people need specifics to know how to do so. "How do we implement this?" "What does this look like on a day- to-day basis?" The concept of finding a vision is so abstract they don't know where to begin. I think your work provides them with "eyes-on" examples. I understand that Marilyn does not want people merely copying what she does. And I agree with that. But, I also think many moms need to be led by the hand for a little bit, before they can venture off on their own.

Many homeschoolers begin with the best of intentions, but get lost with the worries of this world. Curricula, test scores, and requirements become the focus and topic of every conversation. Marilyn's message – and yours – brings homeschoolers' focus back to God. Without a vision the people perish. Right!? But where does that vision come from? God's word! People need to hear what God has to say about education, knowledge, wisdom, and the worries of this world. That's why your work is a ministry. It's not just about how to do high school our way and get away with it. It's about getting back to God's vision for His people.

You, my dear friend, have a wonderful way of making that Lifestyle of Learning vision practical and do-able for those struggling to grasp it, who might otherwise just abandon it. Your articles on Accreditation, Testing, Honors Diplomas, and Dual Enrollment are wonderful examples of how you bring us back down to earth but out of "the world" at the same time and remind us of our real reason for "education".

– Luz Marina Lovins

[**NOTE FROM BARB**: The "Honors Diploma" article that Luz mentioned is on the back of Section 2's title page. "Accreditation" is interwoven into the content of Sec. 2 itself. "Delight-Directed Learning," which will help clarify what Luz talks about here (above) is on the back of Section 8's title page. The article on "Testing" is in the *Homeschool Guide-a-log*.]

FORMULATING YOUR FRAMEWORK

VARIOUS FRAMEWORKS FOR HIGH SCHOOL

Before we jump into the latter portion of this section that deals with setting up a time-based framework, I want to first discuss several basic ways to design a framework for the high school years. The only one of these that I designed is the Time-Content Based framework, and even that I did not "invent," but rather took what I wanted from "the system" what would work for me, and developed it into a reproducible system that homeschoolers could set up and use in their own homeschools, independent of any institution. So this is why you will not be receiving a diploma from outside your home; it is *your* "baby"! ☺ And I would also add: do not take your completed plan to a public or private school and expect them to endorse and give a diploma for! Just as you would not want to issue a diploma ("approval") of a public school plan you did not approve of or have close supervision of, neither do they want to do so; and should not be expected to do so. When we leave the system, we need to leave our need for the system's approval. We get to trust in God, instead, to open the doors *He* wants opened for us post high school.

None of these framework approaches is completely exclusive of characteristics of the others. In fact, all of these approaches contain elements of each other and overlap in nature. So using one type of framework does not exclude incorporating features of the others. In fact you will find yourself doing so without even trying; you just didn't know it had a name! ☺ Nor is one framework any more "formal," "academic," or "natural"; *that* depends entirely on <u>how</u> you fill it.

If you need to have a clearly defined means of establishing a "credit," or if you want a means of determining goals and how and at what point they will be considered "done," then you will need to create *some* type of framework. This will serve as a "skeleton" on which to hang the "meat" of your high school experience; criteria (determined by *you*) to complete so that, at some point (designated by *you*) your student will be "done." You'll consider your student's style and how he thinks and operates in deciding which one to use – and which format will best suit his or her needs!

DIFFERENT FROM MY WORKSHOP

Those of you who have heard my workshop on "Choosing Your Own Framework and Filling It With Good Stuff" will remember that I have presented three different frameworks: the "Project-Based Approach," the "Time-Based Approach," and the "Point-Based Approach." Since then I have seen that all these were in need of some major adjusting and re-structuring including my own! In fact, I also ended up seeing more than just the three frameworks! In one sense it's a little more complicated, but I believe it will also give you more options, and a broader sense of how to *look* at all this and then, finally, how to apply it.

AN OVERVIEW OF OPTIONS

What I'm doing here is presenting my own approach *(in greater detail than the others, for obvious reasons)* as well as my perspective on how I view other approaches as functioning. By their inclusion here I am not condoning any as being the best; in fact, I don't think there *is* a "best" framework; only a "best fit" for *you.*

I'm just telling you here what I would tell you if you came to my home for the afternoon and said "I don't know what to do for a high school framework! All the possibilities are confusing and overwhelming! Do you have any insights that can help me decide?"

The first "insight" I want to share with you is to *be thankful* you have some options! We had none, being "pioneers" – which we didn't even know we were until about twelve years into it; we just thought we were "loners" trying to figure out what in the world to do!

Next, before I go on, I have to make the disclaimer that I am not an official spokesman for *any* of these approaches – other than my own. Having delved into this high school homeschooling arena for several years now, I even still am *not* an expert and don't even pretend to be one. But I *have* come across several approaches that others are using to homeschool through the high school years, and just as I do in my seminars, I present the various options that I have found "out there" and pass along introductions of them to you here so that you can have a smorgasbord of possibilities to choose from as you make your decisions. I liken this to the *Compact Classics* (a.k.a. *Great American Bathroom Books)* which have two-page synopses of many of the "classics." Steve Anderson did not write the classics; just two-page synopses of many of them, for the purpose of giving the reader a good idea of what the book is about, whetting appetites where the book and the reader would make a good match; closing doors (and saving time) where they wouldn't. That's really what I'm doing here. My explanations of each approach are just that – *my* explanations. My intent is to help you find a "right fit" for you and your family, and therefore I present you with a taste of each of the approaches I have discovered here. But if you intend to pursue or implement any of them, you will need to gain an understanding of the author's own writings. Even if you use mostly my (this) book, you will benefit greatly from the perspectives of these other homeschoolers! You'll glean different gems, insights, and freedoms from each one's unique experiences, styles, and backgrounds! Don't just limit yourself to a straight diet of *me!!!* ☺

So with this in mind, allow me to introduce to you several basic frameworks for high school. I'm going to present my very basic explanations of each one first, (starting below) in alphabetical order for your convenience. Then I'll present samples of some *(not all)* of the frameworks in the pages that follow. The "Master Framework" for the approach will be on the left, and one class sample for that framework on the right.

CONTENT-BASED
FRAMEWORK

This framework focuses on completing the ***content*** of the "classes," which can consist of a broad range of learning activities, chosen by student and parent, in whatever depth and quantity of any aspect that is desired. When the prescribed content for a class is completed, the credit is earned.

Content of any and all types can be included, such as reading (real books), hands-on and real-life experiences, lessons, traditional and non-traditional learning, field trips, service to others, life skills, etc.

"Challenging" a course, where a student takes a test to see if he knows the material of the class, would "fit" in with this approach, but is not its primary focus or function, as we are wanting to set up a *framework* for a whole lifestyle for high school, not merely completion of the content. The idea is not just to amass credits quickly; I want the "hours of our days to be profitable." What's happening *during* the whole process of "high school" relationally, character-wise, etc. is a major *part* and consideration of the *content* of the education – and therefore the value of the whole experience. (This is actually true for *all* the frameworks.)

The sample for this is under the "Time/Content-Based Framework" because I have since combined the concepts of the "Time-based" and "Content-based" approaches into the one "Time/Content-based Framework." If you prefer one or the other, simply use the top portion of that form (which is the *Content* aspect*)* or the bottom portion (which is the *Hours – or Time* aspect).

GOAL-ORIENTED PROJECT-BASED
FRAMEWORK

☆SEE SAMPLES (6–7)☞

In this framework, the focus is on completion of *projects*. Specific goals are outlined for each class, and presented on that class's form. What a "project" consists of can be interpreted within a wide range of possibilities, including everything appropriate to the student's life, from books to hands-on projects.

Mary Schofield, author of *The High School Handbook,* is a master at providing endless ideas for structuring classes with specifically outlined, attainable goals. These can simultaneously be used to determine a grade for the class.

As far as the quantifiable aspect of determining credits is concerned, this framework is similar to the time-based approach, except that ***points*** are used instead of hours. The number of points roughly parallels the equivalent amount of time: i.e. 150 points equals 150 hours. (Mary gives specific suggestions that she feels are acceptable; you can decide on whatever amount – for points *or* time – that you deem appropriate.)

The planner (Mom or Dad) needs to have *some* prior concept of how much time various activities take, but this does not take long to acquire, especially with all the help available in her book (and mine as well) – which you can use as general guidelines. As Mary explains in her book, she does not advocate "clock watching"; she feels "it all comes out in the wash." This book is full of goals for a wide variety of classes, including a wide variety of projects.

TIME-BASED
FRAMEWORK

This is the framework that was originally presented in this book. Actually, it *is* still in here (later in this section), but I have now modified and added to it, and now call it the "Time/Content-Based" framework. (So see that one for the sample.) It is based, very simply, on ***time***. For us, one credit equals 120 hours, but a I present ample discussion on how you might go about determining the best choice for you. The basic premise of this framework is to fill a certain number of hours and earn a credit, not just with *anything,* but to fill the hours of our days with good, pleasant, and profitable things. Projects, experiments, reading, field trips, etc. all go into filling up this time.

The purpose is not to just "fill up time," as it could easily appear, but to grow in being discipled to Jesus, personally and within the context of "family" that He placed us in. "Education" is the process of this taking place, not a mere "filling of our minds" with knowledge. We need to be equipped for the "work(s) He prepared beforehand, that we should walk in them." It takes "tools" to acquire this "education" as well as "character." And all this takes "time."

TIME/COMPETENCY-BASED
FRAMEWORK

☆SEE SAMPLES (8–9)☞

In this approach, the "competencies" for each class are linked to content and/or performance objectives. Co-author Diana McAlister calls it "a more traditional, liberal arts approach." (Candice Oneschak is the other co-author.) The framework is traditional, but the objectives-based "contracted learning" can be individualized to incorporate projects, traditional textbooks, unit studies, independent study. You create a traditional-looking "box" inside of which may be quantifiable, non-traditional *or* traditional stuff. Their book, *Homeschooling the High Schooler,* presents many baseline or "minimum core" expectations as well as non-core and college-prep electives for the classes, listed by subject.

In some ways this approach is very similar to mine as well as, in different ways, to Mary Schofield's in *The High School Handbook*, but this book has its own unique focus, which is why I wanted to include it here. (The sample forms in this section are not exactly how they appear in their book. Same exact content, but I re-formatted it for this book.)

TIME/CONTENT-BASED
FRAMEWORK

☆SEE SAMPLES (10–11)☞

This is really just a "combo" of, you guessed it, the Time-based *and* Content-based approaches! It's the one we ended up with because, as we got further into it with Sharnessa, I realized that just "doing the ***time***," filling the ***hours***, was not *really* "all" I wanted – as far as having her fulfill specific requirements. *(As a pioneer, I got to figure it out by trial-and-error!)* There were certain things I wanted to be learned, which *unfolded as we went along*. So this is why ***content*** is now included in our requirements as well as ***time***. And this is reflected in a list of required stuff to do on our "Completed Class Plans," (also known as "Assignment Check-off Sheets") *as well as* the "Hour Check-off Chart" at the bottom of that same form. *Now* the "stuff" *and* the "time" now have to be completed in order for the class to be "done." Mom decides on "core content requirements" which are based on the "basics" she wants her kids to know or be able to do, as well as the student's own interests and bents. And then the student fills in the remaining time with personal pursuits and interests, approved by Mom. Most of our courses *(as I have probably said elsewhere, but need to say whenever I mention this approach)* were roughly and sketchily planned at the beginning – by *design,* not laziness. And this was added to, changed, and deletions made as we went along. *Then,* very close to the end, it was all made into the nice plans you'll see at the end of Section 6-D *(also called "Assignment Check-off Sheets")*, *after* they were completely done with them! ☺ You would not necessarily need to make yours look this spiffy – unless you have a *reason*. *(Or just want to!)*

UNIT-OF-LIFE PROJECT-BASED
FRAMEWORK

☆SEE SAMPLES (12–13)☞

For those wanting to pursue a non-traditional Lifestyle of Learning approach to high school (to just plain "education," really), this is one way you could structure it. In the chart for this approach is an outline of the various projects that, at that time, comprised Marilyn Howshall's daughter, Kathryn's, "Master Plan" based on several projects Kathryn was doing over a period of several years, during the high school "season." Take some time to study it to see how Marilyn broke it down. This not necessarily what she ended up with as she further developed her thoughts on this, but the purpose of this "Master Plan" is merely to give you a "starting point" to show you how you could possibly organize and structure a high school level education done in a more "lifestyle of learning" manner, a thought that might be daunting to even the most creative chart afficionado. ☺ You can take it from here!

Then on the page after that Marilyn has outlined the "requirements" for just *one* "Unit-of-Life Book Project" that was listed in the upper left column of the "Master Plan" – and this is Kathryn's "Bird Book." The Howshalls have now produced a booklet, by the same title, that describes this project in detail – not just the project itself, but what's *behind* it!

In this kind of system, when the designated ***projects*** are completed, you would consider the "credit" earned – *if* you even feel the need to assign credits, which you may very well not have any reason to do. If you don't like or need to keep track of time or points, this type of framework is a good option.

A self-motivated, delight-directed student will do best with this, whereas one who is "just trying to get by" won't have the self-motivation or creativity to develop the projects or take them to completion. So if your student does not enjoy learning, this is not the best choice. There is *so* much more to it than just this, though! *Wisdom's Way of Learning* gives you the *whole* picture, and *how* to deepen the learning value of your high school experience, even if you prefer a more traditionally structured framework! Section VII offers an intro to it. *(Please do not skip it!* ☺)

Several Sample Frameworks coming up on the next few pages!

The remainder of this section presents, step-by-step, how to set up the time aspect of your framework.

Section 6-D has a plentitude of ideas for how to *structure* and *fill* the classes, as well as how to create the format for doing so.

Schofield Family Graduation Requirements

As it might look with credits recorded at the end of each year.

SUBJECT	9th	10th	11th	12th	REQUIREMENTS
Bible	Fund. of Faith 5	Christian Apologetics 5	Christian Living 5	Christian World View 5	20 credits (5 per year)
English	Fund. of Lit & Comp 10	Elements of Lit & Comp 10	American Literature 10	English Literature 10	40 credits (10 per year)
Mathematics	Algebra 1-2 10	Algebra 3-4 10	Geometry & Trig. 10	Calculus 10	20 credits
Science	Creation Science 5	Intro to Chemistry 5	Field Biology 5	Physics 5	20 credits
History & Geography	World History 8	World Hist 2 U.S. Hist 5	U.S. Hist 5	State History 2	22 credits (10: world, 10: U.S., 2: state)
Government & Economics		Government 1	Government 2	Government 2 Economics 5	10 credits (5: govt., 5: economics)
Physical Education	Soccer 5	Swimming & Soccer 5	Beg. Karate 5	Int. Karate 5	20 credits (5 per year)
Vocational Education	2		1	2	5 credits
Home Economics & Family Living	Cooking 2 Fam Living 1	Auto Care 2	Fam Living 1 Sewing 2	Home Maintenance 2	10 credits
Typing & Computers	Typing 5	Computer Science 2	Computer Science 5	Computer Scienee 3	10 credits
Health & Driver Education		Driver Ed 2	Health 3		5 credits (3: health, 2: driver ed.)
Foreign Language	Latin 1 5	Latin 2 5			10 credits
Fine Arts	Survey of Fine Arts 1	Survey FA 4 Int. Piano 7	Survey FA 2 Int. Piano 3	Survey FA 3 Adv. Piano 10	10 credits
Electives	Logic 5	See Fine Arts, Math & Computers			23 credits
Total Credits:	64	65	59	64	**Total Credits Required:** 225
Cumulative Total:	64	129	188	252	

Project-based
WORK

Course Standard

(Course description on page 99; course design on page 137.)

Course: **Sewing**	Credits: 5
Student: Jane Brown	Grade: B - 83.5 %
Semester: Spring 1987	

Grade based on: Three projects required to pass, four for an A.
Basic skills listed below must each be used in at least one project.

Scoring as follows: 28 points per project:
- 7 = finished appearance
- 7 = pattern & fabric selection; layout & cutting
- 7 = work attitude, habits, sewing methods, care of equipment
- 7 = basic skills from list

16 points for fourth project:
- 7 = finished appearance
- 3 = pattern & fabric selection; layout & cutting
- 3 = work attitude, habits, methods, care of equipment
- 3 = improvement on basic skills

Basic Skills Required: (list project which included it.)

1	Hand hemming	2	Machine hemming	3	Darts
1	Basting	3	Buttons & button holes	2	Zipper
1	Casing & elastic	2	Gathering	2	Waistband
3	Sleeves	2	Matching plaids or print	3	Collar
2	Pockets	3	Top stitching	3	Facings

Projects and grade:

Project	Finished Appearance	Pattern, fabric, layout, cutting	attitude, habits	basic skills used	Project Grade
1. Skirt	6	5	4	5	20 — 71.5%
2. Skirt	6.5	6	5	6	23.5 — 83.9%
3. Blouse	6	7	5.5	6	24.5 — 87.5%
4. Top	7	3	2.75	2.75	15.5 — 96.9%
Total Points	25.5 91.1%	21 87.5%	17.25 71.9%	19.75 82.3%	83.5 points 83.5% = B

High School Framework

as presented in

Homeschooling the Highschooler

by Diana McAlister and Candice Oneschak

SUBJECT	MINIMUM	COMMUNITY COLLEGE	COLLEGE PREP
ENGLISH	3 (years)	4	4
SOCIAL SCIENCE	2.5 – 3	3	3
SCIENCE	2	3	3 – 4
MATH	2	3	3 +
P.E./HEALTH	2	2	2
OCC. EDUCATION	1	2	2
FINE ARTS	1	1	1
FOREIGN LANGUAGE	0	0 – 2	2 +
ELECTIVES	5.5	1 +	[as desired]

Typical Course Progression for Graduation Requirements:

Grade Nine
English – Composition / Grammar
State History / Elective
Science / Elective
Math
PE / Health
Fine Arts

Grade Ten
English – World Literature / Composition
World History / Geography
Science
Math
P.E. / Health
Elective

Grade Eleven
English – American Lit. / Comp.
U.S. History
Science
Elective
Elective

Grade Twelve
English Lit. / Composition / Elective
Contemporary Issues / Civics
Occupational Education
Elective
Elective

SOCIAL STUDIES: CONTEMPORARY WORLD PROBLEMS

The Student Will Be Able To:

- Read and understand news from contemporary sources at various levels of sophistication *(i.e. U.S. News and World Report, local daily paper, N.Y. Times, Wall Street Journal, Atlantic Monthly, U.S.A. Today, local and/or national TV news, etc.)*
- Write a position paper in support of his/her view regarding a particular issue
- clarify the finer points of an issue by writing a position paper from the opposite point of view to one they hold
- write a position paper on a topic of the teacher's choice, from the perspective of the teacher's choice, in the manner of the teacher's choice. *(This may duplicate on of the above writing assignments.)*

~~~~~

Course Plan/Objectives:

❑ Write 50 article summaries over a 3–5 month period. Each summary must:
- have attached to it a copy of the article being summarized
- state the essential facts of the article or report *(who, what, when, where, why, how)*
- be accompanied by the student's response to the article or report.
- come from a variety of sources including Atlantic Monthly, local daily newspaper, Wall Street Journal, Newsweek/U.S. News, and U.S.A. Today.
- Note: It is desirable, but not required, that the student "track" a story over several days or weeks.

❑ Discuss each article or report with an adult who initials the article summary.

❑ Write a minimum of three position papers on issues or stories previously encountered: One paper must be written from the student's perspective on the issue or report, one paper must be written from the opposite point of view the student holds, and one paper must be written as directed by the teacher consultant or mentor.

❑ Write a comparison of news sources, evaluating each source for depth, accuracy, newsworthiness, ethics, and/or bias.

TIME: 75 HOURS = .5 (½) CREDIT
~~~~~

Master Framework Plan

Student: Tory Shelton

Year of Graduation: 1997

Date compiled: 8-20-92

Listed to the right are all the courses and classes the student will be taking, as well as the required number of credits to graduate from this home school and requirements (if any) for individual subjects for promotion from grade to grade.

We define "one credit" as being 120 hours

To be promoted from one grade to the next, the student will earn a total of 630 hours

The student is taking 5 years to complete high school.

To stay on track from year to year, the student needs to complete:

About 504 hours per year.

About 14½ hours per week.

About 3 hours per day.

MY COURSE LINE-UP

COURSES & CLASSES	Req'd to Graduate	Req'd per Year?
English	4	1
Math	2	½
Soc. Studies =	(2½)	–
• WA St. Hist	½	–
• U.S. Hist & Gov	1	–
• World His. & Gov	1	–
Science	1	–
Lab Science	1	–
Occup. Educ.	1	–
Life Skills:	1	–
• Tact, Grace & Social Skills	–	–
• Self-Esteem	–	–
• Fam & Marr. Rel	–	–
Bible / Spiritual	2	½
P.E.	2	½
Health	1	–
Typing	½	–
Fine Arts	1	–
Home Econ.	1	–
TOTAL CREDITS =	21	

✓ ASSIGNMENT CHECK-OFF SHEET ✓✓✓✓✓✓✓✓✓✓✓✓✓✓✓✓✓✓✓✓✓✓✓✓✓✓✓✓✓✓

Class: **Typing / Keyboarding**	
Student: **Tory W. Shelton**	*Class Value:* **½ Credit = 60 Hours**

Resources / Instructions: Use "Mavis Beacon Teaches Typing" interactive computer tutorial as the main "text" of the class. Complete all other work prescribed below.

BASIC REQUIREMENTS CHECK-OFF LIST	Check Here When Done
Learn and practice correct typing form according to instructions for proper finger placement in Mavis Beacon program.	
Use the various typing games and exercises in program interchangeably, as you see fit, to improve weaker areas.	
Read the "Owner's Manual" instruction booklet for Mavis Beacon Teaches Typing.	
Achieve the WPM *(Word Per Minute)* proficiency of your choice: ______WPM	
Practice typing/keyboarding skills by typing up own compositions and letters.	
Type a business letter of your own (or for Mom) using one of the four correct business letter formats *(as shown on pg. 38, 42–45 of the Owner's Manual)*	
Do the 12 Practice Exercises on pages 61 – 72 *(Use chart here:)* #1: ○ #2: ○ #3: ○ #4: ○ #5: ○ #6: ○ #7: ○ #8: ○ #9: ○ #10: ○ #11: ○ #12: ○	ALL 12 DONE

HOUR CHECK-OFF CHART
(Check off hours as they are completed)

60	59	58	57	56	55	54	53	52	51	50	49	48	47	46	45	44	43	42	41
40	39	38	37	36	35	34	33	32	31	30	29	28	27	26	25	24	23	22	21
20	19	18	17	16	15	14	13	12	11	10	9	8	7	6	5	4	3	2	1

Master Plan for Kathryn's Education

Unit-of-Life Book Projects

Time Span	Date Completed	
		"Bird Book"
		"Book of Horses"
		"Spider Web Book"
		"Texas State Book"
		"Washington State Book"
		"Book of Butterflies"
		"Book of Flowers"

Life Skills Preparation

Time Span	Date Completed	
		Applied math course
		Driver's license/travel skills
		Home Economics
		Planning meals/shopping/cooking
		Sewing/mending/stitchery
		Home beautification & maintenance
		Outdoor Projects
		Vegetable gardening
		Yard beautification and maintenance
		Animal Husbandry
		Raising peacocks
		Raising chickens
		Care, healing, and training of Happy

Life-Message Preparation

Time Span	Date Completed	***Reading Program***
		Read books about America's Christian History
		Read biographies of Christian ministers
		Read other books for spiritual growth
		Study Program
		Study in "Big Red Books"
		Indepth word Bible study

Career Preparation *(Video Production)*

Career-Based Skills & Services

Time Span	Date Completed	
		"Howshall Home Family Video" Project
		"Nature Projects Instruction Video"
		"JMJ Productions Video"
		"Washington State Video"
		Videotape/Edit/Mix LOL Meeting

Career Preparation *(Writing and Illustrating)*

Career-Based Skills

Time Span	Date Completed	
		"Book of Compositions"
		"Book of Sketches"
		"Book of Paintings"
		Typing course
		Computer basics
		Desktop publishing basics

Career-Based Products or Services

Time Span	Date Completed	
		"Nature Projects" book
		"Insect Collecting Kit"
		"Six Categories of Record Books"
		"True Adventures with Nature"
		2-page story books

Unit-of-Life
FRAME-

Career Preparation *(Horse Breeding)*

Career-Based Skills

Time Span	Date Completed	
		Read books/magazines about horses in general
		Read books about veterinarian care
		Read books about horse breeding
		Do volunteer work at a horse stables
		Do volunteer work at a veterinarian
		Design, illustrate, and compose six instruction sheets about six separate aspects of horse care; add these to "Book of Horses"
		Utilize the veterinarian supply catalog with own resources

Career Preparation *(Wildlife Rehabilitation)*

Time Span	Date Completed	***Career-Based Skills***
		Do volunteer work at a veterinarian under the direction of an expert in wildlife rehabilitation.

From Master Plan:

- [x] Unit-of-Life Book Projects
- [ ] Life Skills Preparation
- [ ] Career Prep—Video Production
- [ ] Career Prep—Writing & Illustrating
  - [ ] Skills [ ] Products or Services
- [ ] Career Prep—Horse Breeding
- [ ] Career Prep—Wildlife Rehabilitation
- [ ] Life-Message Preparation

Description of Requirements

Project Planning Sheet for:

"Bird Book"

Project-based WORK

TOTAL # OF BIRDS ______	Contents Page	Feathers	Drawings	Photos	Data	Additional Info.	True Stories	Time Making Pgs.
SONG BIRDS ______ Species								
GAME BIRDS ______ Species								
DOMESTIC BIRDS ______ Species								
BIRDS OF PREY ______ Species								
WATER BIRDS ______ Species								
EXOTIC BIRDS ______ Species								
OTHER BIRDS ______ Species								

Check the items included for each bird.

MISC. TASKS	# Pgs.	Hours Spent
Set up Framework of Book		
Sort Feathers		
Make Title Pages		
Make Contents Pages		
Book Cover/Spine		
True Stories		
Poems & Songs		
Metaphor		
Bird Nests & Eggs Section		
Bird Pellets Section		
Instruction Sheets Section		

FORMULATING THE TIME ASPECT OF YOUR FRAMEWORK

WHAT TO EXPECT IN THE REMAINDER OF THIS SECTION

Have you ever filled out an income tax form? Wait, wait, wait!! Before you bail out, let me assure you that this isn't going to be *that* bad! It's just *similar.* You'll be filling in lines and boxes with your own figures, and then doing a little math. (So keep your calculator handy.) When all is said and done – and filled in – what you'll have in hand will be *(drumroll, please...)* ***YOUR VERY OWN HIGH SCHOOL PLAN!***

To accomplish this monumental feat (tho' it's really *not* all that monumental; it just might *seem* that way right now), I'm going to take you through several steps, one by one. It'll be kinda like navigating a plane onto a runway through dense fog, except not quite as hazardous. And I assure you it will be less terrifying than either the snowstorm or the tax form.

SPEAKING OF FOG

If your brain is foggy for the first two hours of the day, or late in the evening, do not choose that blurrier time of the day to work on this section. I also suggest waiting for a time when you have some peace and quiet and can think "in a straight line," even if you have to go somewhere else to do it (like out in the car). Make sure, also, that you have a stretch of uninterrupted time, or you'll get extremely frustrated. This isn't just an interesting little project you're working on; it's the next several years of your child's life! You *need* to be able to devote your whole mind to this, or you'll feel like chucking the whole thing into the woodpile.

WORSE BEFORE BETTER

It's going to get more confusing before the fog clears. Just plan on it. That's because we're dealing with *flexibility within structure.* An excellent combination, but not an easy one to demonstrate or figure out.

However, I encourage you to stick with it! The end result will be well-worth your heroic efforts! As Marilyn Howshall urges, "When you take the time to study this manual ***you will be set free to homeschool through the high school years with confidence to implement the vision God has given you for your children's education."*** I couldn't have said it better myself, and that's why I didn't try. Things of value take time, but are worth the effort. This is one of them.

So get your calculator, a pencil, your pilot's wings, a cup of ____________ *(insert name of favorite beverage),* and let's get going!

Step 1

Purpose: To provide a foundation

The very first step (and, of all the steps, please don't do this one out of order) is to read through the Table of Contents and the first five chapters of this book. Not just you, but your high schooler as well! It would be *ideal* to read it out loud together so you can discuss the concepts presented as you go. (This is how we do several of our academic subjects anyway, sitting on the couch together.) You *all* need to be familiar with the content of the manual.

And, Dad, I sure hope you're in this thing too! If your wife and teen start getting set free from traditional educational shackles and start catching the vision God has for your student, but *you're* still back at the starting line, there's going to be a lot of confusion and conflict on the race track. In fact, if you just sit back and cheer them on: "Yeah, sure Honey, go for it!" you will end up with two very different concepts of just exactly what the finish line *is! They* will be dashing for that one over there, while *you're* saying "Hey! Where do you think you're going?! **This** is the goal line over *here!"*

This is not good. "A house divided against itself cannot stand." Yeah, but this isn't that big of a deal, I hear someone saying. Oh yes it is! We aren't talking about mere academics here! We're talking about the last phase, the most crucial and decisive in my opinion, of training up our child in the way they should go. As a parent, I am thrilled to have the privilege of participating with God in such a fantastic project! I consider it a high and lofty calling.

As you read, earmark the consumable forms you intend to use. I suggest using a small post-a-note hung off the right edge of the page so you can see it when the book is closed. (Paper clips work too, but they damage originals.) Speaking of which, DO NOT WRITE ON ANY OF THE ORIGINALS in this manual, *even* if you plan to use them as is. Once they're "consumed" you won't be able to use them for another student (or ever again for the *same* student.)

Really read through the wh-o-o-o-o-le book. My hope is that you won't simply gravitate to the structure and project ideas offered in these pages, but that you'll take time to study and consider the *whole picture.* The forms are helpful; they provide structure just as bones give our bodies form. But the good, liberating (in Jesus) stuff God has put in my heart, which is much more life-giving than the forms, is like the *muscle* that gives *strength* to the body. And I've got a feeling you'll both want and need all the strength you can get!

With God as our Witness, we do hereby solemnly state that we really and truly did read every word of this manual. (– even the "prelude pages" before Section One.)

☐ **Manual read by Dad:** **Signature:**______________________________

☐ **Manual read by Mom:** **Signature:**______________________________

☐ **Manual read by Student:** **Signature:**______________________________

(If any Dad/Mom/Student trio actually completes this part, and I mean *all three of you* read the *whoooole* book, *and* photocopy this page with *notarized* signatures, I will *(and I'm not kidding)* send you personalized Name/Scripture bookmarkers, hand-and-computer-crafted by "yours truly." *(and her 486DX)* Be sure to tell me all 3 of your first names! I'll even leave the rosebuds off Dad's bookmarker, but it'll be hard!)

Step 2

Purpose: To decide how many of which subjects

The fun part! With your student, decide which classes, and how many credits of each, will be required to graduate from ***your*** homeschool. Don't clog your brain trying to figure out how to *fill* them for now! That's coming up in Part B.

As you consider and plan, keep in mind that there are only so many hours in a day. Requiring 29 credits may allow for everything you *want* your child to learn, but it would probably take till he's 57! My feeling is that we are laying a *foundation* here; not trying to make them learn everything there is to know! *That* will take more than a lifetime – it will take *eternity!*

The blank rows you see below are not there to *all* be filled in! They're just available if you need them. Likewise, the subjects I have filled in are to be kept there *only* if you want them there! This is *your* show and *you* get to call the shots!

COURSE/CLASS TITLES	RECOMM. NO. OF CREDITS FOR COLLEGE	BASIC RECOM. #	YOUR STATE'S GUIDELINES (if you care)	YOUR HOME SCHOOL'S REQUIRMT
ENGLISH	4	3		
MATH	3 - 4	2		
SOCIAL STUDIES: U.S. HIST	1 - 2	1		
____________ ST. HISTORY	.5	.5		
WORLD HIST & PROBLEMS	1 - 2	1		
SCIENCE	1 - 2	1		
LAB SCIENCE	1 - 2	1		
OCCUPATIONAL SKILLS	1 - 2	.5 - 1		
FINE ARTS	1 - 2	.5 - 1		
BIBLE/SPIRITUAL	—	(Mine. 2 - 4)		
FOREIGN LANGUAGE	2 - 3	—		

SOME FREEING FREE ADVICE

For greatest harmony and best results with planning your subjects, parents should keep in mind that this is the *student's* life — and the *student's* future, both of which the *student* will hopefully and wisely place in God's hands.

And dear Student, even though you are maturing, you'll be wise to remember that you are not yet an adult and that it would behoove you to heed your parents' suggestions. They know you better than you might think or hope, or even feel comfortable with! If you don't see eye to eye on some matters that concern you and that possibly even make you angry or extremely irritated, you have a perfect opportunity to watch and see how mighty God is! Not *necessarily* because He'll change your parents' minds over to your way of thinking, though that is one *possibility*, but because, *as you submit your will and your life to Him by submitting to your parents* He will work through your circumstances to bring about His will in your life. (I know I'm preaching. Sorry. I couldn't help myself. It's only because I've seen God do *sooooo* much with messes and lives and (self) wills, and hopes and dreams that have been surrendered to Him!)

I want to mention that once our kids turn 18, major decisions are theirs, although as long as they choose to live under our mossy roof, each of their decisions will have two stipulations to line up with: *(– and, NO, I'm NOT saying "...stipulations with which to line up"!)*

1) It must be in alignment with our morals and standards, and
2) It must be something we can live with comfortably – not disruptive to our lifestyle, or demanding of our time, money or resources *(though we will surely be willing to bless and work with our kids where there is a need.)*

But back to now! *Parents and students:* You'll *both* be wise to remember that what we're dealing with here is *homeschooling,* where parents are not outsiders or mere observers. They have much of their hearts, souls, and finances wrapped up in their children's education, not to mention their very lives. So if neither parent nor child insists on being the "taskmaster," harmony will be freer to reign. But the matter of *who will have the last word should a "stale mate" occur* needs to be addressed and decided upon long before it has the opportunity to become a problem. Hopefully you will be able to determine *together* and *peaceably* what subjects will be taken. It will help to keep in mind that this is much more than mere curriculum planning; it's a great opportunity for parents and students to be intimately involved with God in working out the child's goals and aspirations for *life!*

THE ULTIMATE "TEAM EFFORT"

My prayer is that God will give abundant grace to each of you on your "Board of Directors" (under the "Big Boss," God) as you consult Him and work jointly on this "team effort" project! If there are some non-like-minded parents and teens in your house, it could be the *ultimate* in team effort, as well as the beginning of a new experience of trusting and hearing God! I'm so glad He's not a respecter of persons and that He doesn't play favorites! (— because if He did, I know I'd *never* have been on His M.V.P. First-string Line-up!) He doesn't choose some to get to hear His voice and do His stuff, leaving others excluded. We all have the same exact opportunity to "Seek and you will find; knock and the door will be opened; ask and it (the answer for you) shall be given..."

TO BE (A CLASS) OR NOT TO BE?

There will probably be all sorts of topics you want to cover, so how do you determine whether or not to call "it" a "class"? The rule of thumb I came up with for *us* is that if I want them to have at *least* a full ½ credit of "it" then "it" becomes a class and will be listed on a transcript. For instance, I want them to have a full ½ credit of "Tact, Grace & Social Skills," so that's a *class.* But "Gourmet Cooking" will be just a *unit of study.* How many hours exactly, we won't know until we get further into it, but I know it will be fewer hours than ½ because there's lot's more stuff they'll need for their "Home Ec" credit and I don't want Gourmet Cooking filling half of the time – unless one of them were to decide to go into that for a career. But in that case, the time could count toward "Occupational Skills" instead.

MORE FOOD FOR THOUGHT, SINCE WE'RE IN THE KITCHEN

How do you decide whether to have "Typing" and "Computer" be part of "Occupational Ed (Skills)" since they could be counted either as separate classes or *under* Occ. Skills? For us there were three determining factors: 1) whether or not I had enough other stuff to make up a credit of Occ. Skills. *(We did...)* 2) whether or not I wanted my kids to have the words "Typing" and/or "Computer" on their transcript *(looks good on a resume, too!–) (I did...)* and 3) whether or not we had room for the Typing and Computer in our line-up of classes. *(We simply* ***made*** *room! These are two skills that will take them far in both the job world and in their own personal lives.)*

And speaking of "food for thought,"
here's an excerpt from the book ***Dumbing Us Down***
by John Taylor Gatto that I know you'll appreciate – well,
at least it'll make you think!

"Don't be fooled into thinking that good curriculum or good equipment of good teachers are the critical determinants of your son's or daughter's education. All the pathologies [sicknesses, deficiencies – in this case, in the educational system] we've considered come about in large measure because the lessons of school prevent from keeping important appointments with themselves and with their families to learn lessons in self-motivation, perseverance, self-reliance, courage, dignity, and love – and lessons in service to others, too, which are among the key lessons of home and community life." ***– John Taylor Gatto***

Step 3

Purpose: To determine what you mean by "1 credit"

Next we'll determine how many clock hours *you* want to make up one credit, for those classes that are "time based." I present plenty of food for thought in Section Two on "Requirements for Graduation & Promotion," if you want my input on this. 150 hours? 120?

____ Clock hours = 1 credit

Step 4

Purpose: To determine at what point you're promoted from grade to grade

OK, are you ready to do some math? ... First enter the number of clock hours you came up with above onto Line A. (See, I told you this is like doing an income tax form!) Now you're ready to go back and get that total number of credits needed to graduate from *your* home-school. Enter *that* number on line (B). What you have now, on Line C, is the total number of hours your student will spend on "studies." Sounds like a lot, I know, but between now and Graduation Day (if you even have one) your teen will have *also* spent 13,140 hours sleeping (at 9 hours/night), about 1,825 hours eating; your daughter 2,190 hours on personal (shower, hair, teeth, make-up(?) – figured at 1½ hours a day total) while your son will have spent a grand total of 18 minutes on such mundanities. No, actually, at 30 minutes a day, he will have spent 730 hours. So in that context, your Line C figure takes on new meaning. If it doesn't, and smelling salts are beginning to sound pretty good, *run* – don't just *mosey* on to the next step.

_____ (A) Your number of clock hours per credit

× _____ (B) Your total number of credits required for graduation

(1's digit of "B" × "A")

(You know, the usual multiplication procedure)

\+ ____________ *(10's digit of "B" × "A")*

____________ (C) Your outrageously intimidating total number of hours for high school studies

Step 5

***Purpose:* To determine the number of completed hours needed to be promoted from grade to grade**

This is one of the simpler operations. All we're doing here is taking the total number of hours required to complete the high school years and dividing it by 4, for the 4 grade titles of high school. Once the first block of this number of hours (and other specific requirements you come up with) is completed, the student moves from Freshman to Sophomore; the next completed block merits him the title of Junior; the next moves him into Seniority; and the fourth and final block, Graduation.

_______________ (E) This is the number of hours required to be promoted to the next grade

4) _______________ (C) Total number of hours to complete all required credits

Step 6

Purpose: To determine how to pace yourself on a daily basis

Before we can figure how many hours a day to spend on studies, we need a general idea of how you want to spend your days at the various times of the year.

Do you want to go year-round, spending pretty much the same amount of time each (week)day on studying? Or do you want to go year-round, but do less in the summer and at holiday times and more the other months? Maybe you prefer to skip the summer entirely, although, unless you quit living altogether, you will probably be doing *something* that would count toward credits in one or more classes. For instance, swimming, rafting, or bicycling would be great for summer P.E. options. Then there's Bible and Spiritual that you won't abandon just because it's summer. And creative-type projects, depending on what they are, could go under Occupational Skills, Home Ec, Fine Arts, etc. I think you get the idea.

This can get a little messy here because there are too many possible combinations to be able to design just one single formula that applies to all. So let's go back to the previous page and grab that Line C figure, the total number of credits you'll be earning. Since it already has a name, "Line C," let's just go ahead and call it that here too. Since every number that gets put inside a division bracket (the dividend) *must* be divided by *something* (the divisor), we will have to come up with one. I'm going to let you pick it! (the dividend, that is.) For best results, you might want to choose the number of years your student will take to get through high school. But wait! Before you pick that number, take a quick look at my idea chart designed especially for those who suffer from "divisaphobia" (fear of choosing a divisor) just because it will profoundly affect the next several years of their student's life.

YEARS	APPROPRIATE STUDENT TYPE
3	fast-paced, extremely self-motivated; possibly wants to graduate with certain friends or simply has places to go that don't include high school
4	the "norm" – which isn't necessarily *normal,* just more common or usual. (Four years is just fine)
5	the perfectly normal student who doesn't want to have to cram 4 years of study into 4 years of time. (I highly recommend this route for guys – like our son!)

(Step 6 continued ☞)

Yes, this division problem *does* look amazingly similar to the problem we (OK, *you)* just did on the previous page. So why are we doing this again? Because if you are doing high school for any number of years other than four, all the previous calculation told you was how many hours you need to be promoted from grade to grade. Since I could think of only four grade titles, 4 seemed the appropriate number to divide by. But now that we might be dealing with a 3 or a 5, we'll be coming up with some different numbers that will tell you something different. Specifically, they'll end up telling you how many hours you need to work a day on accruing credits IF you want to graduate when you intend to graduate.

So, if you don't *care* if you graduate on target, skip ahead to the next section.

_______________ (E) This is the number of hours you need to complete per year

______ (D) _____________ (C) Total number of hours to complete credits

(Space to do actual division calculating)

Step 7

Purpose: To determine how to pace yourself in the various subjects

You're now ready to wrap up the final details of your plan! And I have just the place for you to do this (as you might have expected). You've already done most of the work for this by now, so don't let your blood pressure rise as you look at this final work page. In fact, if you have a headache, P.M.S., or someone just ate your last Girl Scout mint that you were hiding and looking forward to, don't even start. Another day, another time, another (better) frame of mind.

This exercise has two high callings: One is to make sure you don't get to the beginning of your senior year of high school and discover you (your student) still have 2 credits of English, 2 credits of Math, and 2½ years of Social Studies left, with all of your more creative, enjoyable stuff behind you. There's certainly no problem with the fun stuff being done, but part of the beauty of this system is that it gives you the flexibility to experience the more enjoyable classes *all the way through* your high school years. I'm not saying that English, Math, and Social Studies aren't enjoyable, because there are *many* creative ways you can fill these hours that *are* enjoyable – although Math is probably more difficult to get innovative with. But if you leave certain subjects till the end, there's probably a reason – and it usually *isn't* that you wanted to save the best till last!

The other purpose is to give you a clearer picture of how close you're getting to being promoted to the next grade. What you come up with here will help you figure out where you're at when you do your tallies at different points throughout the year. Once you get your figures, I'll finish explaining it and give you some examples.

OK, let's get down to business. The first thing you'll need to do is fill in your major subjects in the upper section. By "major" I don't mean your favorite; I mean the ones that you'll be doing 2 or more credits of. I couldn't even begin to guess this part because there are so many variables, so you'll have to do the writing yourself. Sorry.

In the lower section, fill in all the other classes; those that are less than 2 credits. Now just fill in the "Number of Credits You Require for Graduation" *("Oh so **that's** what that means!")* with the figures you came up with in Step 2.

Multiply that by the number of hours you have decided to equal one credit. (That was Step 3.) This gives you the total number of hours that course will take for the entire time of high school. To whittle that down to a per-year figure, divide it by the number of years the student will be taking to complete high school.

Now we get to the part that's the most highly variable. To reach your credit goal by the end of the year, you'll need a *general idea* of how much to do per day and/or week, not necessarily both. A weekly goal may be sufficient, especially if the number of available hours varies from day to day. So feel free to leave one of those two columns blank if you don't need it. How do you figure those hours? This will depend *completely* on how you plan to distrubute your studies through the year.

So now it's time to decide if you'll do studies year-round, or take off for the summer or holidays or vacations. One way to do this is to decide how many weeks a year you want "off."

Let's say you want 12 weeks for the summer, 2 for Christmas (or Hanukkah) vacation, 1 for spring break and another 2 weeks for holidays and "snow days." The total number of weeks off would be 17. You'd subtract 17 from 52 (weeks in a year) and end up with 35.

Next step: In 35 weeks of school you'd have 5 days of school a week times 35 = 175 days of "school." Now divide your "NUMBER OF HOURS PER YEAR TO STAY ON TRACK" figure by 35, and you have the number of hours you need to apply yourself to studies a week. And/or divide it by 175 and you have your total number of hours per day!

If you want to do studies year-round, fairly evenly, that's even simpler: you just divide the number of hours needed per year by the number of days you want to do studies.

Maybe you'd like to go year-round but have a lighter load in the summer to make the rest of your year a little lighter. Start with the number of hours you'd be willing to put in per *day* in the summer. Let's say it's 2 hours per day which adds up to 10 hours per week. Multiply that by 12, the number of summer weeks, and you have 120. Deduct that from your total hours needed for the year, and then divide *that* total by the number of days you want to study the remainder of the year – which might very well be the same number as above, 175.

Since I have no way of making up a single formula for all these variables, just use the space on the opposite page to do your figuring. Be sure you stay in the lines! You'll be counted down for sloppiness! (Actually, I gave you a box just so you wouldn't feel guilty about writing in the book!)

OK, one last thing you might want to know is the number of hours you'll need to spend on *all* the subjects put together. Obviously, you just add all the totals up. This is the figure that would go at the top of your Daily Report. Remember, this isn't "law"! It's just a way to help you plot your course so you get where you want *when* you want to get there. It's like taking a trip. If you know you have 1,000 miles to cross in a one week period, you probably ought to figure out many miles you need to go in one day and at what speed. (Whether you have frequent rest-stoppers in your vehicle will make a difference too!) All this pesky math is just to help you *pace* yourself!

Here's an excerpt from a Mother's Day card I received from Sharnessa shortly before this book went to press. I share it to give a little glimpse of the personal side of our relationship, amidst all the forms and frameworks! ... And yes, I do feel tremendously blessed!

Dearest Momma,
I want to thank you for being a totally amazing Momma. You truly are. And I realize it more and more as the days go by. I am so blessed that you are not only a brilliant mom, butan awesome friend as well. ... I know you've worried at times that you've made mistakes, but firstly, all parents make mistakes. And secondly, you have both been so good at apologizing when you've been wrong, and taking time to explain and talk things out when either of us had been wrong. That's one of the many things I have learned from you guys. And a very vital thing at that! So thank you, Mom! Thank you for doing the best anyone could ever do. Thank you for taking so much time with me. Thank you for giving me room to grow. Thank you for leading me to Jesus' feet. Thank you for being my friend. Thank you for being such a beautiful example of what a mother should be. Thank you for admitting your faults, and in turn teaching me to do the same. Thank you for not compromising and settling for the easier route. Thank you for teaching me to be a pioneer. Thank you for the freedom to be who God called me to be and to go where He's called me to go. Thank you for not letting ME settle for anything less than God's best. Thank you for your precious hands – the hands that rubbed my back, ☺ stroked by face, wiped my tears, disciplined me, held me. I love you, Momma!

Sincerely, Sharnessa

Ultra-Personalized
PACE DETERMINER
Worksheet

(This isn't a permanent record, so you don't even have to write neatly – just so you can read it!)

COURSE/CLASS TITLES	NO. OF CRED'S YOU REQ FOR GRAD	NO. OF HOURS YOU'VE SET PER CREDIT	TOTAL NO. OF HOURS CLASS WILL TAKE	NO. OF YEARS CHILD WILL TAKE TO DO HI SCH	NO. OF HOURS PER YR. TO STAY ON TRACK	NO. OF HOURS (about) PER DAY or WEEK TO STAY ON TRACK Hrs/Week	Hrs/Day
		×	=	÷	=		
		×	=	÷	=		
		×	=	÷	=		
		×	=	÷	=		
		×	=	÷	=		
		×	=	÷	=		
		×	=	÷	=		
		×	=	÷	=		
		×	=	÷	=		
ALL OTHER SUBJECTS COMBINED:		×	=	÷	=		
					Total time needed per day:		
					Total time needed per week:		

Step 8

Purpose: To get a picture of your whole framework – at a glance

Now that you've done all this figuring and computing, you need one central location where you can look at your entire framework at a glance without having to shuffle through all the pages of this section. This, then is your "Master Framework Plan." Just go back through the section page by page and fill in all that applies.

I can't stand blank space! So here's a note I received that might encourage you if you are in the same boat...

Dear Barb,

I saw your name on the AOL Homeschool/High School message board and am assuming you are the author of the book Senior High, A Home Designed Form+U+la. I just want to say thank you for writing this book! My oldest is in 6th and I am looking ahead to high school. I have only just discovered that there is a better way to educate than the "traditional" way. As I've been looking into how to homeschool him through high school, I had begun to feel that I had no choice but to return to the "old" (dry, boring ☺) way of educating him in order to earn the credits he will need for a diploma. I just picked up your book yesterday and stayed up very late last night reading. I couldn't put it down because it addresses my needs right now. Your book has given me hope that I will be able to find alternatives in how he learns through high school. I had nearly reached the point of just giving up and putting him in a Christian school. I don't know what we will finally decide, but your book has refreshed me and given me hope. Thank you so much...

God bless!
C.

Master Framework Plan

Student: ______________________________

Year of Graduation: ______________________________

Date compiled: ______________________________

Listed to the right are all the courses and classes the student will be taking, as well as the required number of credits to graduate from this home school and requirements (if any) for individual subjects for promotion from grade to grade.

We define "one credit" as being

____ hours

To be promoted from one grade to the next, the student will earn a total of

_____ hours

The student is taking _____ years to complete high school.

To stay on track from year to year, the student needs to complete:

About _____ hours per year.

About _____ hours per week.

About _____ hours per day.

MY COURSE LINE-UP

COURSES & CLASSES	Req'd to Graduate	Req'd per Year?
TOTAL CREDITS =		

Master Framework Plan

Student: Tory Shelton

Year of Graduation: 1997

Date compiled: 8-20-92

Listed to the right are all the courses and classes the student will be taking, as well as the required number of credits to graduate from this home school and requirements (if any) for individual subjects for promotion from grade to grade.

We define "one credit" as being

120 hours

To be promoted from one grade to the next, the student will earn a total of

630 hours

The student is taking 5 years to complete high school.

To stay on track from year to year, the student needs to complete:

About 504 hours per year.

About 14½ hours per week.

About 3 hours per day.

MY COURSE LINE-UP		
COURSES & CLASSES	Req'd to Graduate	Req'd per Year?
English	4	1
Math	2	½
Soc. Studies =	(2½)	–
• WA St. Hist	½	–
• U.S. Hist & Gov	1	–
• World His. & Gov.	1	–
Science	1	–
Lab Science	1	–
Occup. Educ.	1	–
Life Skills:	1	–
• Tact, Grace & Social Skills	–	–
• Self-Esteem	–	–
• Fam & Marr. Rel	–	–
Bible/Spiritual	2	½
P.E.	2	½
Health	1	–
Typing	½	–
Fine Arts	1	–
Home Econ.	1	–
TOTAL CREDITS =	21	

CONGRATULATIONS!

Having made it to the end of this laborious section, you deserve a blue ribbon! But those were too expensive, so I hope you'll settle for a certificate of completion! You'll also have to fill it in yourself! Of course you know your *foremost* reward is the fact that you now have a framework within which you are free to reach your goals – and have an awful lot of fun on the path!

Certificate of Completion

is hereby awarded to

____

for persevering efforts to complete all of the stupendously laborious, enigmatic, and exacting computations as necessary for completion of this heroic feat of skill and intelligence.

Signature of Witness ______________________________

Here Tory, at about age 14, I think, is camping out with his (heavy on the "his") new puppy, Dominic, in the dog house that he and (his) Dad built. ☺

The full story on this very educational episode of our lives is on page IV-8 under "Dog Report," and the "Dog Report" form itself is on page IV-40.

BUT
DOES IT WORK?

This is a really good – not to mention *common* – question. One that I'm sure is lurking around in your mind and heart. It's similar to the "Does your Diploma mean anything?" question, yet gets into another whole issue. To answer that question, I first have to ask *you* a question: *What do you mean by "work"?*

If you mean *"Will our diploma or program be accepted by any employer or college official we submit it to?"* I have to honestly tell you that I really don't know. I'm not going to try to pull the wool over your eyes by saying "Hey! Do my senior high program! It *really works!"* Because, in the realm just mentioned, there are no guarantees! As I've said elsewhere, when you decide to homeschool, you are bucking the traditional system that our society is firmly entrenched in.

However, if you want to know whether it "works" in relation to building character and a deeper relationship with the Lord in your student... Or whether it will give him a love of learning or equip him to know *how* to learn for the rest of his life... Or whether he'll be prepared for the "real world"?...

Again, I can't answer. Because *that's up to you.* All I'm doing in this manual is presenting you with a flexible framework and some nutritious groceries – just tools, really. What you build with them is *totally up to you!*

Does it work? You get to answer that question! And I'm excited for the results to get back to me! I'd *love* to hear from you in a few years – with your glowing reports and success stories. I have a feeling that's *all* I'm going to get. Not because my book is so great, but because when you link God with a loving parent, *all things are possible!*

– *Barb*

Section Six - Part B:

DESIGNING YOUR CLASSES

Ideas and approaches
to help you in creating your student's
classes and utilizing all of their life's
learning in the process! Be ready
to "see" them and their life and
interests differently!

What is Real-life Learning?

KEY

P = in Potpourri section ❖ G = in Grading Guidelines section ❖ L = in Lifestyle of Learning books

Service in community ✂☛ Making a quilt ✂ Planning a vacation on computer ✈ Gardening: Food, Floral, Landscaping ❁ Hunting: taking an animal from field to freezer (or jerky or sausage, etc)(G) ➬ Attending a city council meeting ✯ Serving on a committee you have an interest in ❤ Planning and completing a project(L) ❧ Planning an appropriate exercise program for yourself (P) ☻ Making a High School Days Memory Book (P) ✎ ✚ Taking a health-related workshop at a hospital ✉ Working at the family business ☺ Having an adequate season of being free to focus on and develop an interest (L) ☼ Camping ❂ Being involved in a play in any capacity ❀ Listening to and studying music (P) ♬ Touring museums or visiting historical sites ❦ Watching a movie or play with historical, human-understanding, or aesthetic value (P) ❄ Cooking or baking an exotic gourmet dish (P) ✄ Taking a craft or sewing class at a fabric or craft shop ✲ Tutoring or reading to young children ✡ Studying an overview of the major religions of the world (P) ☻♬☺ Performing in a recital, play, concert ❤ Making up a 7-aspect health plan for yourself (P) ♪ Lessons: instrument, gymnastics, dance, voice, choir, skating, band, symphony, golf, sport ⌂ Watching the process of a house being constructed over a period of several weeks ✣ 4-H or Scouts meetings ✌ Visiting a nursing home, a shut-in, hospital ❤ Studying and cooking healthy dishes ✿ Leading a family meeting (P) ☞ Doing home repairs (P) ⌂ Feed nursing home patients ✞ Helping remodel or build an addition to the house ➡ Learning how to sew ✄ Collecting food for a food barrel Gleaning ❦ Helping with "Meals on Wheels" ☹ Going through difficult situations with parental guidance and the Holy Spirit ✞ Reading "real books" ✉ Writing a letter to the editor about something you are concerned about ✏ Attending workshops and seminars of special interest ⌚ Organizing a closet, room, shelves, warrantees (P) ✳ Attending concerts and performances ♬ Building a dog or cat house ✰ Finding out the conflicts and outcomes of the major wars of the U.S. (P) ➳ Attending a Creation Science seminar ✾ Participating in a running/jogging./bike marathon ✍ Evaluating the wholesomeness of a song or movie (P) ◘ Taking a computer class or learning computer terms (P) ✼ Listening to a special speaker at church ✩ Going on a mission outreach trip ☎ Participating in a Toastmasters (Speech) Meeting/Class ☺ Attending a Basic Life Principles seminar ✞ Preparing a meal, including planning the menu, finding recipes, making the shopping list, and cooking it (P) ❄ Browsing an art show at a library, museum, park ❀ Attending a Master Gardeners meeting ✆ Interviewing people in various occupations (P) ✍ Learning the various cooking terms (P) ☺☺☺☺ Helping a neighbor or relative take care of younger children ♥ Finding out the meaning of your name and a verse to go with it (P) ✞ Doing a creative Bible Study (P) $ Calculating the costs and care of getting a pet(P) ✑ Interviewing pastors, elders, deacons on their denomination's beliefs (P) ✓ (Re)searching the Internet for a topic of (valuable) interest ✠ Visiting (and really experiencing) a science center ☎ Helping man the phone lines at a Christian radio telethon

DESIGNING YOUR CLASSES

Deciding how to design your high school classes can be overwhelming if not downright scary, but I want to encourage you regarding this process. Take it one step at a time, and most importantly, let those steps be ordered by the Lord. He has a *good* plan for each family, *and* for each *child* in each family! I believe you can even have a good time through the process of seeking! I have considered it a treasure hunt!

Perhaps you don't quite yet share my enthusiasm, but you'd *like* to. This section contains some concepts and tips that I hope will raise your level of enthusiasm and confidence as you begin *(continue?)* to see how easy it really is. Most of what your student does for high school will not be something they do *on top of* the rest of their life; it *is* their life! In the last section we focused on getting your general framework set up and organized, the "shelves." In this section we'll focus on taking the "real life learning" – *see ideas to left(!)* – and the academics, and creating the "classes" out of it ***all.***

But remember, as you go through all these ideas, we don't want just an "external form," but the *heart.* And we want to expand our *freedom* and *vision*.

GETTING "UNPROFESSIONAL-IZED"

Professional teachers know what "works" within their own system, but they do not have a corner on knowing what subjects are essential and relevant for one individual's life! – namely *your child's!* And what we mean by "works" depends on what you're wanting to accomplish. Here's a quote from *Dumbing Us Down* by John Taylor Gatto who was voted "New York State Teacher of the Year" for 1991:

> "...If we're going to change what's rapidly becoming a disaster of ignorance, we need to realize that the school institution "schools" very well, though it does not "educate"; that's inherent in the design of the thing. It's not the fault of bad teachers or too little money spent. It's just impossible for education and schooling ever to be the same thing. It's absurd and anti-life to be part of a system that compels you to sit in confinement with people of exactly the same age and social class. That system effectively cuts you off from the immense diversity of life and the synergy of variety; indeed it cuts you off from your own past and future, sealing you in a continuous present much the same way television does. It is absurd and anti-life to move from cell to cell at the sound of a gong for every day of your natural youth in an institution that allow you no privacy and even follows you into the sanctuary of your own home, demanding that you do its 'homework.'"

(Cross reference on the subject of "homework" on page I-7 of this book, starting at the bottom of the left column.) Speaking of being "cut off from the immense diversity of life and the synergy of variety" and from "your own past and future," here's just one example that I'm pulling from the dusty files of my own past history...

STATE HISTORY

I'm sure Washington state is not the only state that has an abundance of museums and historical sites. These can be wonderful means of getting a sense of our past in our own region; how life was lived by both "Native Americans" and the early pioneers and settlers, how our present was shaped, what brought people across the ocean and "out west," and how they got here.

To be considered a "valid" course in the public school arena, state history must include some elements that I agree are valid and appropriate: like a study of the constitution, government, and the process of lawmaking. These are undoubtedly essential elements of knowledge for any student who's going to be a contributing citizen of a state to have under their belt – not mastered and memorized, but to have a basic awareness of how it all works.

But in our state, when I was in high school, there was one requirement I considered a complete waste of time! – that of memorizing all the thirty-plus counties in our state *and* their county seats! While there is some value in memorizing one's *neighboring* counties, who needs this mass of useless information cluttering their brains and taking all the time to get it lodged in their brain when there are so many other *valuable* things to spend our *valuable* time learning, especially when there are *maps?!* There are only three possible reasons I can imagine for such information being considered worthy of consuming the kids' precious time:

1) Being confined to the limits of a classroom, maybe the curriculum planners couldn't think of anything else to finish filling that *"x number of hours"* with. Our little homeschool class, on the other hand, visited several museums around the state and wondered how we'd ever have time to visit as many of them as we would have liked!

2) Perhaps someone thought it would impress parents for the kids to memorize all those counties. But I'd rather have my kids impress *God* and study to show themselves approved by *Him,* and wash their brains at the same time by memorizing thirty words of the Bible. That would certainly impress *me* more! – and *edify* them!

3) Or maybe someone pointed out that the skill of memorizing is a good discipline for the brain. This is true. But with the need for continual renewing of the mind, especially in our culture, wouldn't knowing thirty words of Scripture bring more fruit than knowing thirty counties of a state? This isn't to say that *nothing* secular is worthy of being memorized. I'm just making a point.

SO WHAT *IS* MY POINT?

To turn you into a rebellious rabble-rouser? No! Although I admit a heated streak *does* rise up in me when I see parents feeling confined to a less-than-excellent or non-relevant curriculum because they believe the "experts" have all the knowledge and authority – or at least more than *you* have – to determine what's "valuable"! Let's hear a bit more from John Taylor Gatto on this very thing:

> "I've come to believe that genius is an exceedingly common human quality, probably natural to most of us. ... I began to wonder, reluctantly, whether it was possible that being in school itself was what was dumbing them down. Was it possible I had been hired not to enlarge children's power, but to diminish it? That seemed crazy on the face of it, but slowly I began to realize that the bells and the confinement, the crazy sequences, the age-segregation, the lack of privacy, the constant surveillance, and all the rest of the national curriculum of schooling were designed exactly as if someone had set out to *prevent* children from learning how to think and act, to coax them into addiction and dependent behavior. ... I dropped the idea that I was an expert, whose job it was to fill the little heads with my expertise, and began to explore how I could remove those obstacles that prevented the inherent genius of children from gathering itself. I no longer felt comfortable defining my work as bestowing wisdom on a struggling classroom audience."

I very respectfully submit to you: those we have regarded as "the experts" are experts only at "schooling the masses" and perpetrating the public school system! You have the same access to knowledge as they do! And you and your children can learn together! What you feel weak in you can find help for as there is a *wealth* of knowledge and resources, but *you* can decide what *you* believe are priorities for your child to know! The "experts" have *no* concept of God's bigger picture of and purpose for education! With a teachable heart, a willingness to be taught and led by the Holy Spirit, who is your "Master Education Planning Consultant," and some study of what *true* education is, you *are* capable of determining your own high school experience.

I'm not just "anti-anything-having-to-do-with-school"! Many textbooks are good. Many of the classes your child will be missing in high school would have been "good." But you can't know how they would be taught, in what "light" or with what slant. Nor is there *any* guarantee that they would have been any *better* than what you would have chosen and/or designed. As Cathy Duffy, author of the *Curriculum Manual for Junior/Senior High* said:

> "Although there are some excellent teachers and classes in schools, there is no guarantee that any child is going to absorb all that wonderful class content listed in the course outline. Often students drift through classes just marking time and completing minimal requirements."

There are many things I would not want my children to be exposed to at this age, before their thought processes are developed and refined enough, their minds renewed enough to discern between truth and deception. And even though there are many wonderful people and even God-loving Christians in the public school system, there is *a great deal of* untruth there also, primarily because God has been kicked out of the system completely.

How can science be founded on truth when its very Designer is disregarded, pre-empted by "the Big Bang," "chance," and "natural selection"?!? How can the truth about history be known when every detail concerning the Author and Finisher of it was ousted from its books at the turn of the (last) century?!?

SO SHOULD YOU OR SHOULD YOU NOT...

...be concerned with meeting state guidelines in determining the content of your State History class, for example, or of *any* other class there might be regulations for? I would say: *Only* if you plan on re-entering the public school system before graduation. But since our plan was to homeschool all the way through to graduation – and that's what this manual is geared to help *you* do too – I wasn't at *all* concerned about meeting state requirements for filling my classes, I included the general recommended topics previously mentioned because *I myself felt they were valuable aspects to include.* And that is *IT!* ☺

"BUT HOW DO YOU *KNOW* YOU'LL MAKE IT TO GRADUATION?"

Or, more specifically, if your mind works like mine, "What if I die or get some terminal or debilitating illness? *What will I do with my half-educated kids then?!"* I figure that if I thought this way on everything, I wouldn't do *anything!* It would keep me from entering all of life's major experiences and roles:

- I wouldn't have gotten married. *"What if* we end up in divorce? It would be too painful! Too humiliating!" I just couldn't live through that!"...and...
- I wouldn't have had children. *"What if one of them died? – I couldn't bear it!"* Or *"What if we got divorced; I couldn't bear putting them through all that heartbreaking pain?!"* ...and...
- I wouldn't have started this book. *"What if people don't approve of it? Which, of course is very likely, but why did I write it? To obey Jesus and help you get set free? Or to get approved by man?*

MORE DEBILITATING THAN ANY DISEASE!

The "What-If's" debilitate us! They'll keep us from going after our dreams every time! I want to see God's plans for our family *fulfilled* in our home, not quenched! I love Isaiah 8: 11–13

> ***"For the Lord spoke thus to me...and warned me not to walk in the way of this people,"*** (and the public school system is one of the "Top 10" ways of this people!), ***"saying: 'Do not...fear what they fear, nor be in dread. But the Lord of hosts, Him you shall regard as holy; let Him be your fear, and let Him be your dread. And He will become [to you] a sanctuary..."***

I looked up "sanctuary" and found that it is a "haven; a place of refuge, covering, security, and protection." Oh, and guess what the antonym for this is? *Prison!* God's message for His precious children is to ***be released from fear! All*** fear! As long as that work is not complete in us, we are in a "prison" of sorts.

It's not God's heart for His precious children to live in fear of *any*thing! Not state requirements. Not Instructional Objectives. Not the local school district! If our state's law ever changed, I *could* be in an extremely dangerous position. Because I do seminars and write, I'm among the most visible homeschoolers in this area. Just as pastors and the educated have been the prime targets of religious oppression in other parts of the world, I would probably be among the first to be hauled off to prison. I heard of one situation – I can't remember where or when, but it was earlier this century, *(I told you I was historically illiterate)* where anyone who wore glasses was automatically arrested and executed! That's not my idea of fun, but I firmly believe God called me (us) to this, so the consequences don't matter. And I mean that with all the fortitude and grace I have in Jesus. In past years I have gone through a good deal of turmoil over this. But God has brought me to a place of total peace, because my life is in His hands. And I've found He is totally trustworthy. That doesn't mean I won't go through any adversity *(which I can say with confidence because I already have)* but it *does* mean I can trust God through and in anything I go through.

So why do I say all of this to you now as you begin your class planning? Because even though you may be out of the system, if the system is not rooted out of you, your class-planning will be greatly affected. Getting more specific, they'll be hampered, restricted, and limited. If you are not free, then, like a powerful magnet, the textbooks and workbooks will draw you to themselves as being "the best and only choice" for (too) many of your class components. (If you skipped ahead to this section without reading especially sections 1 and 2, please go back and read them before continuing with this section. ☺)

DO YOU *REALLY* KNOW HIM?

If you're already a Christian, skip to the next paragraph. If you don't yet know the Lord, this is a prime opportunity to do so. Invite Him in, deep into your innermost heart, the place you don't let even your spouse into – which is as it should be, because there's a place within each of us that was designed to be the dwelling place of the most-high God *himself. No one* else can fill this place, because it was created for Him alone. God wants to meet you personally, right where you're at, like you've never imagined possible!

Even for us Christians, there's *more!* No matter how mature we are in Him, because He is so vast, there's always *more!* Do you realize we get to go from "glory to glory" right here on earth?!? And I know from personal experience that there's always *more* for Him to heal in me, to fix in me, to change in me, to add to me, to release me from. And one of those things He wants to release you and me from is *fear. Fear* of the state, particularly in planning our children's classes. *Fear* of what our relatives will think or say if we do something different. Let Him take your fears. He knows exactly what to do with them. Once you do *(and it will be a process, so just expect it)*, *then* and *only* then will you be ready for this section of this book. In fact, believe it or not, you'll actually have a wonderful time with it!

I don't want to be disqualified from the "land flowing with milk and honey" – the place with the *best* groceries! – because of the same malady that kept Moses out: the "What-If's"! *"What if* water doesn't come out with only one tap?" ...*"What if* I can't speak before Pharoah?" No thanks! *I'm* going for God's *highest* quality groceries, and therefore have chosen to let Him cover my *(many)* shortcomings, change what needs to be changed, and remove my fears. Fear and the "What-If's" will only prevent us from entering the Promised Land and experiencing God's very best for us – for our children and for their education!

We documented everything we did so that as our "vision" gained clarity (in more ways than one), and as we moved down the road, all would not be lost if we had changed direction, but would still have been "translatable," if necessary, into "traditional" form.

TAKING IT ONE YEAR AT A TIME

I have to tell you that He finally showed me that I had to give up my position of "we're taking this one year at a time." He had allowed that for the first several years, and it was probably even the right thing for me back then. But there came a time when I realized He was taking me further in my walk of faith with Him, and *for me* taking it one year at a time was now actually no longer based in "wisdom" or "caution," but was a lack of hearing Him and having enough faith and courage in His power to do that which He was now leading me to do! He was no longer saying "take it one year at a time," but "take it to completion." Had He told me at the very beginning that I was going to be doing this all the way through to Graduation, I don't think I could have handled that! And He knew it! This is not necessarily true for *you!* I'm just sharing how He worked in *me.* He deals with each of His children as individuals.

There's a saying: "Where God leads, He provides." So I finally came to the place where I was able to say "Yes, this is what we are doing all the way through to completion!" We jumped in head-first – with heart close behind – onto the path I believe God prepared especially for *us* to walk on. It was actually very freeing and joy-and-peace-producing! My faith was now in God – not in how homeschooling went!

If something didn't "work," then we just realized we just needed to change it – and seek God for *how* and *what* needed changing! And this wasn't necessarily the materials I was using, but also my *values*, my *perspective*, and my *approach.* This is where *Wisdom's Way of Learning* will help you in rethinking what you *value* about education, your *perspective* on learning, and therefore your *approach* to planning your classes!

TRUST YOUR "HOLY INSTINCTS"!

If you know God, then as you begin to create your classes, keep in mind that you have access to *all* wisdom! The most valuable, real, life-producing kind! So by "holy instincts" I mean that voice God planted within you to help you guide your children into adulthood. It's not your voice; it's *His* voice, and you need to know that you *can* hear it. Jesus said,

> ***"My sheep hear my voice, and I know them, and they follow me..."*** (John 10:27)

Yes, there are elements of classes you most likely won't want to miss – at least ***I*** don't/didn't.[1] I want(ed) to make sure our kids had/have the skills necessary to survive in the "jungle" and to be able to train for and enter the career they want. But there are other classes that aren't *necessarily* necessary! Some may be a total waste of time for *your* child. For instance, that extra third (*or* second, *or* fourth...) credit of Math may be unnecessary for the plans God has for your child. Taking it anyway probably wouldn't *hurt* anything, *(although the student may think so,)* but those hours may be better spent at other pursuits.

The idea is to do "enough" for the proficiency you feel led to attain. Getting a good foundation in English is something I would not consider to be a waste of time, unless it were filled with "junk." I believe there isn't any profession for which good grammar isn't needed. How many "No check's cashed" signs have you seen taped on restaurant cash registers? *(And if you didn't notice that error you might want to brush up on usage of apostrophes!* ☺ *)* Be open to the fact that God may still have something better in Mind. Being the Potter, it's *His* privilege *and* job to form the clay. So I guess we're not fully able to trust our "holy instincts" until we are able to trust *Him*, and our ability to hear Him. If the ears of our "instincts" – our ability to hear and sense the voice of God – aren't yet cleaned out and "de-crusted," they aren't too "trustable," are they?

So if you aren't "there" yet, do you just *not* start planning your classes? Do you have to wait until you *are* able to hear God's voice better? *(Much better?!)* NO! *This* is your *practice ground!* This is the "classroom" in which we get to learn!

As one mom said: "I have been making little changes over time, as the Lord leads. With each step I have been learning to hear His voice better, but I was never sure that each thing was from the Lord. So I just took the risk and did it. This caused me to realize that I needed to know Him better." If you're at a place where

[1] *It's kind of awkward referring to all this at a point where I have now graduated two, and yet still have another who has not entered high school. So my values are actually past, present, and future!* ☺

where you know you need and want God to do some de-crusting and purifying of your spiritual eardrums (which is a place I *continually* find myself!) it's not a difficult thing! You don't have to take "Hearing God 101," This actually IS "Hearing God 101"! On-the-job training! You *do* have to take *action* though. The Word of God tells us specifically what we need to do at this point. "Do not be a hearer only, but also a *doer*."

Why don't you get out a pen, open up to Isaiah 30:15, and do another little exercise I've prepared? I love this kind of stuff and *hope* you do too! It makes the Word of God come alive to me even more fully!

> **"For thus said the Lord God, the Holy One of Israel, 'In ________________**
>
> **and ________________ you shall be saved,'"** *(out of what specifically, for you? Fear? Lack of courage? or confidence? ______________________________)* *(and it continues:)*
>
> **" 'In ________________ and ________________ shall be your strength.'"**

FILLING YOUR ROOMS WITH PLEASANT RICHES

You may want to use some of the many ideas that I have given you in this book for designing your own classes, especially the ones in Section 6-D; there are *plenty* in there! But my hope is that my "stuff" will not only give you ideas, but more importantly, "license" for coming up with your *own* "stuff," and the freedom to feel totally confident in using it and following the leading of the Lord! *(He's probably been trying all along to do this, but you have possibly not been hearing His voice because your spiritual eardrums have been too crusted over with wrong ideas about education, and paralyzed with fear!* ☺)

Of course, college-bound students have some additional things to consider, especially if going into specific fields. But ask for God to help you make it all enjoyable! Listen to Proverbs 24:3:

> ***"By wisdom a house*** (home school?) ***is built, and by understanding it is established; by knowledge the rooms*** (courses?) ***are filled with all precious and pleasant riches."***

WHAT ABOUT DOUBLING UP ON HOURS?

I want to discuss this matter of applying hours to more than one class.[2] For instance, let's say you do a speech, perhaps for a homeschool group presentation, on some aspect of History. Do you count it as English? As History? Or as ***both****?*

This is a good question, one I'm frequently asked, and it's not easy to answer. In fact, I *can't* answer it for you; but I can pass on to you some insights I've gleaned in just thinking it through, in considering dual enrollment programs, and in talking with two administrators of two reputable home study "schools."

One such administrator I spoke with about this believes that hours should not count for more than one class, stating "We are striving for *excellence.*" An attitude among parents who agree with this view might be: "I want to have as many hours of "opportunity" with our children in these last years before they leave home as I can! I don't want to just fill requirements as fast as we can so we can get on with the good stuff of life; this *is* the "good stuff of life' – for this season. "

Yet *another* such administrator I asked about this said that they *do* allow doubling up on hours, and with absolutely no qualms. In some cases, the calling upon a student's life requires certain training and only minimal study in some areas, so getting the credits for those minor areas in the quickest possible way is completely in line with how God is leading them. So they would feel that if a student produces a well-written composition on, for instance, the Civil War, they have spent three hours improving English skills *and* they

---

[2] I deal with this topic of doubling up on hours elsewhere in the book, so forgive me for any repetition here. But here's the deal: I have *no* idea where it's mentioned; I just did a search of the many files this book is composed of, and I can't find it anywhere. I, my lone self, am basically *the* entire "staff" for producing this book, author, production manager, editor (though I've had some help with the editing!), lay-out design, etc., So I have no other brain to consult for the whereabouts of the discussion of this subject. This section (above) isn't a reprint of it, so I'm sure I make some different points, but I really don't know! ☹ And with our last book having been sold this morning, and a completed book being 3 weeks away from when they receive the originals, I don't have time to look further. If you happen to come across it, though – and I'm sure you will – don't bother to tell me where it is cuz this is the last revision of this book that I'll be doing for a *loooong* time! ☺ ... And one other thing; I don't like how this footer looks, split up and uneven, but I tried *everything* I could think of , and just could *not* change it!

have spent three hours learning history. (Or you can calculate it in what ever way seems right to you.)

Also, part of why I felt comfortable with doubling up on a portion of the hours, as far as how "the world" would regard it, is how dual enrollment programs operate. (This is called "Running Start" in Washington state.) In these programs, credits earned in the junior and senior years of high school count not only toward graduation, but also toward an A.A. degree at a community college! – assuming the appropriate classes were taken for both, of course. If they're doubling up in counting two whole *years* for both a high school diploma *and* an A.A. degree, certainly we can double up on a few hours without feeling we are cheating!

And then there's also the concept of "challenging a class" or taking a "CLEP test." *(Hope I have that worded correctly).* When I was in college it was called "challenging a class" – when you took a test to see if you (already) knew enough of the content of a class to pass it and thereby get credit for it. *(You still had to pay for the credit as if you'd taken the class.)* It's one option if there's a need.

Another related thought in favor of doubling up on (some of the) hours is that there are *plenty* of hours to be filled as it is; the student should be as free as possible from burdensome academics to be able to go after other pursuits that are just as worthy of their time. Although I must add that there was hardly anything our kids did that I couldn't figure out a place for in the line-up of subjects. ☺ So it depends on your goals.

The questions you have to ask *yourself* are: *"What are our goals for education? What purpose do we see it as having? How do we want to fill these years? What do we value? When do we want to be done? How will we define "done"? Where are we headed? How does God want to use me/us? What are my giftings? My weaknesses? My strengths?"* Only you and your student can answer these questions. And your answers will help you determine what's right for you.

"All things are lawful, but not all things are profitable." (I Cor. 6:12) While it may be "legal" or "acceptable" to count hours twice, is it really the best "way" for your child in light of what God has for them? It may very well be that it is! And it may very well be that it *isn't*. ☺ Here again, guess what – *<u>You</u>* get to hear God on what's best for *your* situation and *your* student! *(If you weren't before, you should be pretty good at seeking and hearing God by the time your student graduates!)*

And to those who might be feeling a bit shaky about making such a decision: Don't *(as in **do NOT**)* assume that you haven't got enough of what it takes to make a wise decision! Look at some guidelines; find out what others are doing, and why; get some ideas from curriculum books. But don't *(as in **do <u>NOT</u>**)* regard or revere these as "law"! Mostly, *pray!* Know your own goals. And ask God to expand and sharpen your ability to hear Him – specifically on what *His* agenda is for your classes for these formative years. (A "season of re-education and renewing of the mind" is unto this very purpose, by the way!) Then trust God to speak. You know He *wants* to! All He needs is a willing heart – to hear! And if hearing Him on matters as detailed as this has been a weak point for you, what better place than your high schooler's life and education to begin!?

TRANSLATING "REAL LIFE" EXPERIENCE INTO "CLASSES"

This is one major aspect of senior high homeschooling that many would like to be able to figure out, and actually have the potential to do so, but they find the translating-it-into-classes part a bit mind-boggling at first. Just like beginning a new job, having a baby, or moving into a new house. In fact, it's actually similar to all three of those! *(Sans (without) the pay and diapers!)* Too often parents are overwhelmed and frustrated into grabbing for the easier, "quicker," yet less *real,* less *meaningful,* less *fun* route of "textbooks" and prepackaged curriculum, and call it "good."

I'm not here to argue about whether or not prepared curriculum is "good," because a lot of it *is* good. And a certain amount of it will probably be desirable and/or necessary, at least for the skills of Math and some English. But why settle for something less than what I call "peaceful, attainable excellence," especially when your kids are already involved in *doing* a lot of the "stuff" and, with just a little input, would be able to translate it into a system that would be valuable and useful to them?!? *(Yes they'll still need more academics, but don't <u>start</u> there.)*

FILLING HEADS OR PREPARING LIVES?

There's more to *learning* than just filling our kids' heads with "scope-and-sequence-approved" curriculum via classes that a "professional" would approve of. When you get set free from the illusion that "Instructional Objectives" or "Student Learning Objectives" possess more wisdom concerning your child and the direction he's headed for in life than you and his God possess, you're in for a richer high school experience! One that will produce a student who is better prepared for life, who will know how to *think,* and who will know how to *learn for himself* once he's out of the "school" setting! John Taylor Gatto, having been in the teaching business for 25 years, gives us further insight *(starting at the top of the next page)* that is *vital* for us to be aware of as we approach our homeschool class planning. To me it underlines our need to *not* reproduce in our home what is already *not* working in the schools:

"The first lesson I [as a teacher in the system] teach is confusion. *Everything* I teach is out of context. I teach the un-relating of everything. I teach dis-connections. I teach too much: the orbiting of planets, the law of large numbers, slavery, adjectives, architectural drawing, dance, gymnasium, choral singing, assemblies, surprise guests, fire drills, computer languages, parents' nights, staff-development days, pull-out programs, guidance with strangers my students may never see again, standardized tests, age-segregation unlike anything seen in the outside world....What do any of these have to do with each other? Even in the best of schools a close examination of curriculum and its sequences turns up a lack of coherence, full of internal contradictions."

This approach – offered here in this book in conjunction with gaining a grasp of the "learning tools" as described in the *Lifestyle of Learning* message – has the built-in feature of equipping the student to fill, *himself,* whatever holes may have been left in his high school classes. The side benefit of this is "educational gap insurance" for Mr. or Mrs. "I-Don't-Feel-Competent-to-teach-a-high-schooler."

So *this* system frees us up to experience anything and everything we see fit that falls in line with our goals. At the same time, it allows our students' experiences to not be merely "enhancements" or supplements to an already-overloaded schedule, but to be the actual *content* of many of our classes. It takes a little *thinking* and a little creativity, but we were made in the image of an all-knowing, *creative* God. In layman's terms, that means "Like Father, like son!" Not just *me!* But *you* too! So if we have not His creativity – and even a speck of it will take us many miles – it's only because we ask not.

HOW TO FIGURE HOW MANY HOURS?

When first starting out, you *won't* know exactly how long your various activities will take, but you have many examples in all my "Completed Class Plans" to at least give you some general ideas of what to expect, time-wise, of what we did. Eventually you'll start getting some pretty good ideas of how long certain things will usually take. But keep two things in mind: 1) You don't need to be *that* specific; a general idea is plenty. And 2) *Many* activities, whether they're ones I present in Sec. 4 or ones you find yourself, have no limit; you can do them for whatever amount of time you desire or however long it takes. *(Just don't be bound to the clock!)*

Books, on the other hand, will take a fairly predictable amount of time. All you have to do is figure the time it takes to read one page (or read 5 "average" pages and divide it by 5); then multiply that by the number of pages in the book, subtracting the number of blank pages.

Several of the activities I present in the Potpourri Section *(Sec. 4)* are almost limitless in the number of hours that can be spent doing them. Or, using *any* resource as a starting point, the student could write a report, design a model, take notes, write an outline, etc. Finding resources will be the *least* of your problems! Once you get your framework set up and your class plans started, you'll be free to fill it all in *as you go along* – i.e., be "led by the Spirit" – as the Lord brings "good stuff" your way!

WHAT IS A "BIBLICAL" EDUCATION?

Many think that having every piece of their curriculum be "Christian" is the extent of a "good Christian education." They think that their grammar textbook, for example, should have the practice sentences be Christian statements. While this can be a way to have some good content, it is *not* the essence of Christian education!

The *nature* and *design* of the education must be God's idea to begin with in order to be truly "Christian." The traditional school system was never designed by God, so merely "Christianizing" some of its content will not make it "Christian."

This is where Marilyn Howshall's writings will help you to "renew your mind" in this area of finding out what *true* education is. *("Wisdom's Way of Learning" is the place to begin. See Section 7 for an intro.)*

But for starters, begin to take advantage of the opportunities that arise many times a day to talk about God, His values and His will in this situation. Seek His leading in situations that you or your children are going through, repent to Him and/or each other when you're wrong (leading your kids to do the same), receive His healing for hurts, and pursue His heart and will in trying situations. Then *with your children* trust that God is working all this together for good because you love Him and are called according to his purposes.

We are training our kids in the "nurture and admonition of the Lord," imparting not only His Word and His precepts, but, more importantly, His *heart,* showing them by our own lives how mere humans can get real with God on a personal, practical basis. Without this aspect, we'd be giving them nothing more than "hollow religiosity." It's messier, more vulnerable and humbling, harder, and riskier, ***but it's the only way that will get you the fruit you desire!***

And in case the question is burning in your mind, I *do* teach *(or learn along with them, or make sure they're somehow taught; I don't teach or know it all myself)* the academic skills of Readin', Ritin' 'n' 'Rithmetic – though when they were younger I didn't teach these *nearly* as heavily as they are taught in school. And in high school they are still not our *entire* focus, as they *are* in the school setting.

I used our classes not as "knowledge that puffeth up," but rather as raw materials for God to eventually use as He pleases, when and where He pleases. If I have prepared them academically *only,* without imparting, as well as I can, the desire to know God intimately and passionately, then they will not possess a willingness to allow Him to use their skills for His glory and His Kingdom. The *heart* is so critical!

Beyond that, I have laid down any hopes, goals, or aspirations I might have had for *them* and am trusting them and their futures to God. My job as a parent is to do what I can in training them to be "useful to the Master, prepared for any good work." I don't know how God plans to use them. I can only take seriously my call to point them to Jesus and, by my life and example, make that something they see *working* in me, that's *real,* and that, therefore, they would *(hopefully!)* want, too. The bottom line? I believe the true, Biblical, Christian nature of education lies much, much more in the *reality of God in their teacher's and their own life* than in the text or content of the curriculum employed.

THREE "CLASS CREATING" APPROACHES

These three names for how to approach putting your classes together are really nothing more than astute-sounding titles for some very common, ordinary everyday processes. I just wanted terms that expressed to *you* what I meant in a clear, concise manner, which is a challenge when you're talking about something so broad with so many applications. *I just don't want you to be intimidated by them!* In short, the three approaches to devising your own classes are as follows:

- The **"Collect-and-Construct Approach"**
- The **"Priority Appropriation Approach"**
- The **"Dissect-and-Distribute Approach"**

In the pages that follow I'll be taking each one of these approaches and explaining them in more depth, giving charts, examples, and probably lots more detail than you had ever hoped for, imagined, dreamed of, or needed! ☺ But for those of you who just happen to want or need it, *you got it!*

What Mr. Gatto says here was actually intended more for the younger ages than for high school, but for one thing, many of you homeschoolers with high schoolers also have younger children, (not to mention grandchildren coming up in a few years!), but also the basic principle behind what he shares is just as true for high school – or ANY age – learning!

> **"...'How will they learn to read?' you ask, and my answer is 'Remember the lessons of Massachusetts.' When children are given whole lives instead of age-graded ones in cellblocks, they learn to read, write, and do arithmetic with ease, if those things make sense in the kind of life that unfolds around them."**
>
> ***– John Taylor Gatto***

(I need to get the page lay-out (left and right) adjusted here before we go on, therefore you get...)

Another Lifestyle of Learning Insight...

From Marilyn Howshall's Booklet:

Get to Know Your Child

"The child must have the privilege of developing many interests; to be taken seriously by parents who will provide the needed resources so the children can develop expertise in their fields of interest, and time needed to develop collections and abilities. There are fundamental requirements for developing self-motivation and vision in children. If the academic provision is largely a subject focus where parents are concerned about subject "gaps," then I believe the result could be the child's ignorance of his purpose and an empty lifestyle that will follow him into his future; thus the *true* gap! The development of many interests from a young age also develops in children purpose at home. They become productive, contributing and caring members of the family and family purpose. They enjoy their life at home because that is where they have developed strong purpose. They do not have to begin wandering or leaving home prematurely to look for a life! When they do leave, it will be because they know where they are going and with what purpose they go there, and not because they are just eager to go." ❤

THE COLLECT-AND-CONSTRUCT APPROACH

The "Collect-and-Construct Approach" is where you begin with a class title you want to fill with content that *at this point* is unknown. This one is the easiest to understand and implement. I encourage you to branch out from the usual, "mainstream" resources, though. Instead of reverting to the "age-appropriate" text or workbook, try some other more creative, novel resources. See our Completed Class Plans in Section 6-D, and the Grading Guidelines in Section 5, and the Learning Guides in the "Potpourri" in Section 4 for starter ideas. Every one of those Learning Guides will fit into one – usually more – of your classes!

PUTTING TOGETHER A HEALTH CLASS

In order for you to see the "Collect-and-Construct Approach" in operation, I'm going to show what kinds of resources (not all) could be put together for a course of one subject: "Health."

- ***Home-Made Health*** by Dr. Raymond and Dorothy Moore is full of practical wisdom on living a healthy, *balanced* lifestyle.
- ***Fit For Life*** by Harvey and Marilyn Diamond contains much knowledge I want my children to have in the area of food, what it does in the body, and what we need to eat for maximum health – without over-doing it or being legalistic. *(I don't necessarily agree with everything, especially the idea of food combination; but as with anything, you take what you like and leave what you don't!)*
- ***Fit or Fat*** by Covert Bailey explains thoroughly, yet in a way even ***I*** can understand, how the body works in relation to fat, muscle, exercise and metabolism.
- To help our kids personalize what they read in these books, I designed a "Personal Lifestyle of Healthy Living" planning sheet. *(In "Potpourri")*
- Another resource I designed is "Wisdom's 7 Pillars for Personal Health." It's a study originating from my personal Bible Study time that relates the seven physical elements necessary for balance and health with their spiritual counterparts. The concept of "Seven pillars" originated in Proverbs 9:1 where it says: *"Wisdom has built her house, she has set up her seven pillars."* Isn't that appropriate?!? I love it!
- In a set of Student Learning Objectives *(SLO's, which I obtained from our school district for $10)* I found some good AIDS material that I will use to teach our children, but with my own moral slant.

And so we have a good start on a Health class without using one textbook! Do you think this child will have as good a grasp on "Health" as a public school student? I hope you're starting to get "freed up"!

A FEW EXAMPLES

On the opposite page is a chart that contains ideas for various kinds of resources you can *collect* to *construct* 3 other classes, in addition to Health. It's not comprehensive, but a place to start. Whether or not you'd need to add more depends on how much time it takes to complete what's here and what else you come across. I ended up not using or deleting things here on this list and adding others that I preferred!

FRIENDLY SOURCES FOR RESOURCES

As people find out you're homeschooling, some of them will probably want to bless you with materials they think are educational. Look at their gesture purely as interest in what you're doing and "benevolence," and not as a put-down, insinuating that you are deficient in teaching ability. Just smile and be appreciative. ☺ You may acquire some pretty fun stuff! One very old "reader" *(book, not a person!)* that someone gave me ended up being the same one I had used in school *many* years ago, and what a treat to read those stories to my kids! *(This was when they were younger.)*

My dad became one of our best "defenders" *(after we'd been doing it a few years and he saw good fruit in our kids)* and has had many discussions about homeschooling with his friends and cohorts at their daily coffee meetings. On several occasions, some of these friends have given Dad a section out of the Sunday paper or an educational magazine to pass on to us! My mom has gotten lots of neat stuff for us at thrift shops over the years; several people have given us shelves, and someone else gave us a box of notebooks! ☺ I've just been thrilled with and thankful for it all!

THE "COLLECT-AND-CONSTRUCT" APPROACH

CLASS	RESOURCES (COMPONENTS)
LIFE SKILLS	Read ***How to Start a Conversation*** by Don Gabor Read ***When You Can't Get Along*** by Gloria Chisholm Learn skills out of the "Brigance Assessment" Read ***Old Enough to Know*** by Michael W. Smith Read ***Where Do I Draw the Line?*** by Michael P. Farris Read ***Just Do It!*** ("Learn how to recognize God's laws – what you hear, what you see, is what you do") by Buddy Harrison Read ***Responsible Living in an Age of Excuses*** by Kurt D. Bruner
FINE ARTS	Attend Community Concerts and do "Concert/Performance Record" Craft classes at School Biz Attend and do a Play Critique for the play "Fiddler on the Roof" "Nutcracker" Ballet performance View Crafter's Display at the Mall Take Choir Classes with School Biz Christmas "Angel Dance" program Do "Movie Critique/Analysis" of movies watched Attend Grampa Edtl's L.C.C. Choir Christmas Performance View Art Display at Library
WASHINGTON STATE HISTORY	Visit Fort Nisqually and do a "Search-it-Out" sheet Visit Bonneville Dam and do a "Site Data Sheet" Do "Decade-at-a-Glance" sheets for all 21 decades of Wash's history Visit Maryhill Museum and do a "Search-it-Out" sheet Do a "Purpose of Studying History Think 'n' Worksheet" Visit Fort Simcoe and do a "Search-it-Out" sheet Do a "How a Bill Becomes a Law" Search-it-Out sheet Visit Yakima Nation Museum and do a "Search-it-Out" sheet Do a "special project" of your choice (see list of ideas) Visit Fort Clatsop and do a "Search-it-Out" sheet Fill in Museum/Site Data Sheets for each museum visited Visit Fort Vancouver and do a "Search-it-Out" sheet Do an "Elected Official" Search-it-Out" sheet
HEALTH	Read ***Home Made Health*** by Dr. Raymond and Dorothy Moore Do "Tel-Med" reports Read ***Fit For Life*** by Harvey and Marilyn Diamond Read ***Fit or Fat*** by Covert Bailey Do a "Personal Healthful Living Planning Sheet" (Wisdom's 7 Pillars) Read ***Home Built Discipline*** by Dr. Raymond and Dorothy Moore

A MIND MAP FOR THE VISUAL LEARNER

If you're a visual learner, you're in luck because I am too; if I can see a picture of it, I catch on much more quickly. Therefore I couldn't *possibly* overlook an opportunity to satisfy my co-visual-learners similar need!

If you're a kinesthetic learner, you can just *feel* the page, like braille! And you auditory learners can ask someone to read it aloud to you. *(Sorry I wasn't able to include a tape!)*

MINOR DIGRESSION ON "BEST FRIENDS" (SKIP AHEAD, IF NOT INTERESTED)

First of all, this digression has absolutely nothing to do with anything I've said so far. I include it because I had a little space here and felt it may be of interest to teens, especially girls. (Guys don't care about this stuff!) Possibly moms, too; for those who want to take an extra minute to listen... *(Names have been changed to protect the guilty and keep me out of court.)*

(WARNING: Get out the Kleenex box...) I was the "third wheel" all through my growing-up years. This isn't self-pity or distorted memory; just *fact*. Specifically, I was invited to do things with Cindy *only* if she had already asked Suzie, and Suzie couldn't make it. This didn't do a lot for my self-esteem. In fact, over the years I was eventually deeply wounded at always being someone's "second best friend." Even if it wasn't said outright, *(which it usually was)*, it was still *reality*.

Of course, I have friends who I'm *closer* to at different times in my life, but even through my adult years I have been careful to not use the term "best friend" with any of my "close friends" because it's so comparative! Each friend I have is totally unique and precious – "best" *in her own way!* I know how I feel if someone I'm extremely close to says something to me about their "best friend" and it doesn't happen to be me they're talking about. So to keep others' feelings from being hurt, I just have lots of "terribly cherished" friends! ☺

Droplets shown on opposite page ☞ were hand-cut-out ✂ by an *actual* home-schooled high schooler!

This is a good place for a quote from page 9 of Cathy Duffy's book: Christian Home Educator's Curriculum Manual for Junior/Senior High:

"Blessed Are the Flexible For They Shall Not be Broken"

"Flexibility is a mandatory characteristic of successful home schools. I can guarantee that you will encounter many situations where you will have to alter course, put something on hold, or deal with unanticipated problems. If you are set in your mind that NOTHING is going to interrupt your plan for home education, you are setting yourself up for a nervous breakdown.

"...While most of us tend to strive too hard and judge ourselves too harshly, there are those among us who need less flexibility and more accountability and discipline. Some of us drop all planned academic work every time a field trip comes up – whether or not the field trip is worthwhile for our child."

(From Barb) So what's the bottom line? Balance?
No, as usual: Hear God! He knows what you are in need of!

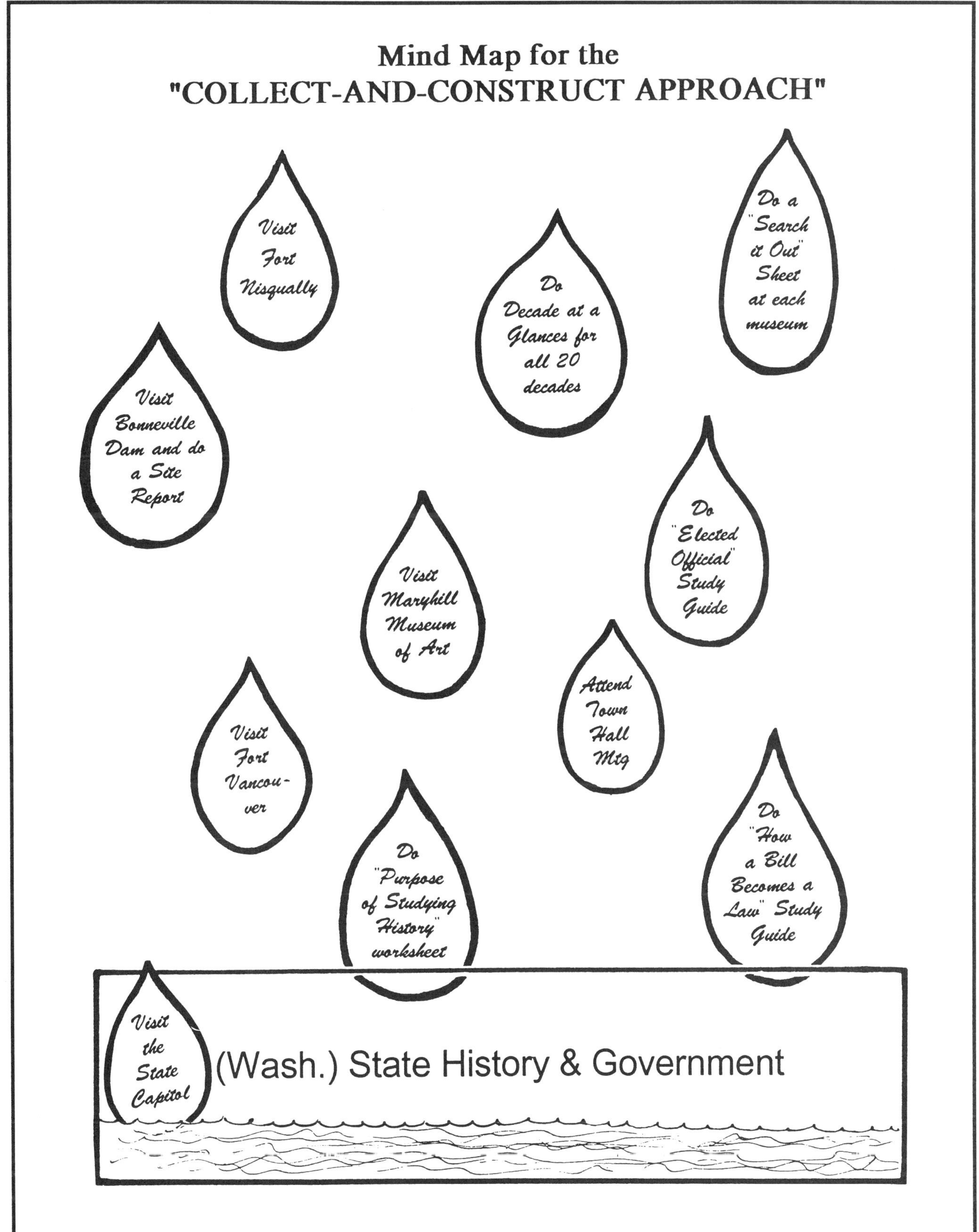
Mind Map for the
"COLLECT-AND-CONSTRUCT APPROACH"
Visit Fort Nisqually
Do Decade at a Glances for all 20 decades
Do a "Search it Out" Sheet at each museum
Visit Bonneville Dam and do a Site Report
Do "Elected Official" Study Guide
Visit Maryhill Museum of Art
Attend Town Hall Mtg
Visit Fort Vancou-ver
Do "Purpose of Studying History" worksheet
Do "How a Bill Becomes a Law" Study Guide
Visit the State Capitol
(Wash.) State History & Government

THE "PRIORITY APPROPRIATION" APPROACH

In the Priority Appropriation Approach you simply determine what priorities you and your student deem important for him or her to be *truly* prepared for life, funnel those skills under a class title, and then gather resources and make a plan to teach these skills. A "priority" can be something they do, a character quality, a skill to be learned – *anything* at all that you feel will better equip them for life. It can be very simple, such as being able to balance a checkbook; or it can be more involved, like being able to manage a home. It can be spiritual, practical, home or work oriented. This is where the parent gets to do some "value imparting"!

EXAMPLE: "RELIGIONS OF THE WORLD"

One priority or value I have for my kids is that they have a basic awareness of the differences between the various Christian religions and between the Christian faith in general and the "cults." The book, *So What's the Difference?*, is a Christian-based book that explains the main religions of the world and how they compare to Christianity. It also tells how to talk with people of different non-Christian faiths about God. It was both fun and challenging to structure the content of this book into a class *unit* I called "Religions of the World." It was originally going to be a whole "class," but we didn't want to spend that much time – one or even one-half credit – on this area. I feel that if our kids are familiar with what these cults *really* teach *before* they leave home, they are less likely to get lured into one of them *after* leaving home! That doesn't mean we want to over-expose them; just make them generally aware. This class could be categorized under "Social Studies," "Spiritual Studies" or "World History & Government."

FREE TO FOLLOW THE HOLY SPIRIT

I wanted to be free to use whatever God brings our way and places on my heart to teach our kids, to be able to take advantage of all the excellent resources available in and outside of our home. Resources like the books mentioned in Section 7 on *Lifestyle of Learning*. Resources like what "Focus on the Family" and "Institute in Basic Life Principles" *(formerly known as "Basic Youth")* have to offer. "Resources" like "community service." *(see below)* "Resources" like "Youth Alive" and "Jesus Northwest." *Unplanned* "resources" *(a.k.a. "opportunities")* like those awkward or painful situations that we have to seek God in if we want His will in our lives and if we want to become more like Him. Such resources are the *heart* of our "curriculum," not just "extras." They are the "meat." The *real* stuff of life! The choicest part of our curriculum!

As high schoolers with lots to do, our kids are likely to think they don't have time for such "extras." They may not be motivated to do or take part in them, no matter how rich *you* see them as being. But you still need to lead them in this direction because this is where the Holy Spirit will have the most access to give you good groceries for your kids.

Yeah, they'll get some great stuff out of the books and experiences too (assuming you are making progress in following God's direction on which stuff to get,) but I think you'll find that the most exciting and relevant stuff will be less "schoolish" in nature. So it might help with "student motivation" if, as you get ideas, you make a dash for the appropriate Class Planning Sheet and get it written down in black *(or blue?)* and white.

Diana McAlister, co-author of "Homeschooling the High Schooler" on...

Community Service

While much of the natural development of adolescents involves young people in personal searches for experience and wisdom, that same development often allows an adolescent to create a focus on himself, his desires, dreams and needs. During this time of gradual separation from parents, students may benefit from regular and consistent volunteer work within their community to assist in balancing their focus on issues and ideas outside themselves.

Community service is a dynamic means to train students in balancing their responsibilities to themselves with their responsibilities to others. Raymond and Dorothy Moore, strong proponents of community service for homeschooling students, suggest a combination of academics, physical labor and ministry as a means to nurturing more balanced adults. If a student is looking at a five-day week, scheduling 2 days of academics, 2 days of work experience, and 1 day of service or ministry during that week is reasonable.

The "Priority Appropriation" Approach

Priority, Value, Skill	Class(es) or Course(s)	Resources/Ideas
Be able to balance a checkbook	Math, Life Skills, Business Skills	Work with parent at balancing checkbook; then do it with assistance; when ready, do alone
Be able to manage a home and "keep house"	Home Ec, Life Skills Occupational Skills	Do all skills involved at least 3 times each through course of living at home
Be able to plan and prepare meals	Home Economics, Occ. Skills	Plan and cook one balanced meal menu per week
Be able to deal with difficult people	Tact, Grace & Social Skills, Life Skills	Discuss real life hassles, listen and watch as parents do it, read
Be able to work under a boss	Occupational Skills	Learn how to work hard, submit, make an appeal, follow through, not complain
Be able to plan, plant, prune and keep a flower garden	Horticulture, Home Ec, Occupational Skills	Work with parent(s) and/or grandparent and/or family friend
Know the basics for leading a healthy life	Health, P.E.	Read Home Made Health and Fit For Life; Do a "Wisdom's 7 Pillars for Healthful Living" Plan
Be able to choose a mate wisely; and stay married	Marriage Preparation, Life Skills	Read material from Basic Life Principles
Be able to pay bills and manage finances	Life Skills, Occ. Skills	Do it with parents; Keep own finance and budget record; Read You the Buyer
(Your ideas...)		

WHERE TO FIND "PRIORITIES"

This will be about the *least* of your challenges. You've probably already got all sorts of ideas forming, but if you'd like to expand your concept in this area a bit more, stay with me! ☺

Mary Schofield has a page in her book called "Schofield Family Basic Skills List Excerpts." As I read through this, I realized that this is really their family's "priorities," and Mary has already "appropriated" them into the classes she considers them to fall under! And she graciously allowed me to include this page of her book right here! So you'll find that on the very last page of this section! ☺ (And you'll find lots more good stuff like this in her book! See why it's good to have more than just one author's perspective?!)

It really boils down to allowing what's already in you to be "unearthed" and seen as being of greater value, and also just keeping your eyes open for opportunities all around you. They're probably coming into your home and lives regularly and you may not be seeing them as learning opportunities for your kids.

For instance, do you get the state-of-the-ministry letter from Dr. James Dobson each month? He almost always writes on topics that are timely and interesting, material you may very likely want your kids to hear about. If so, it becomes a – *yup, you guessed it!* A *"priority"!* ☺ Which academic niche it will get funneled into depends on the *content* of the letter. It may go into "Spiritual Studies," "Current Events," or "U.S. Government," for example.

My husband, Dave, read an article in Reader's Digest on teenage drivers that he wanted Sharnessa to read since she's learning to drive. That went under "Health."

I found a Reader's Digest article called "How Average People Excel" which could go under either "Life Skills" or "Self-Esteem and Understanding." *Teaching Home Magazine* had an article in their "Focus on Teens" issue that delved into the whole dating issue. That could have gone under "Building Life Relationships" or "Spiritual / Biblical Studies."

I found a little booklet called "Schools: Government or Public?" that went under "U.S. History and Government." *(Unfortunately it's out of print!* ☹)

We got a booklet in the mail on financial stewardship – *that* went under the "Economics" section of our Social Studies class.

I got an audio tape at a homeschool convention on watching TV – and the dangers and concerns therein – by Pastor Reb Bradley *(who has a wonderful multitude of excellent resources on rubber-meets-the-road topics of vital interest to Christians)* – *that* went under "Spiritual Studies."

Refer to the Completed Class Plans section (6-D) for a host of other ideas. You'll have to look *closely* as they are woven into the fiber of the whole class. In fact, you might not even be able to recognize them. ☺

"PRIORITY APPROPRIATION" MIND MAP ☞

Here I have taken different *priorities* and *appropriated* them into various classes. Pretty simple!

(By the way, yes, I *am* fully aware of how elementary this exercise is. But, hey, that's what you get with a visual learner writing this book. ☺)

> *"Now if anyone builds on this foundation with gold, silver, precious stones, wood, hay, straw, each one's work will become clear, for the Day will declare it, because it will be revealed by fire; and the fire will test each one's work, of what sort it is..."* (1 Cor. 3:12–13)

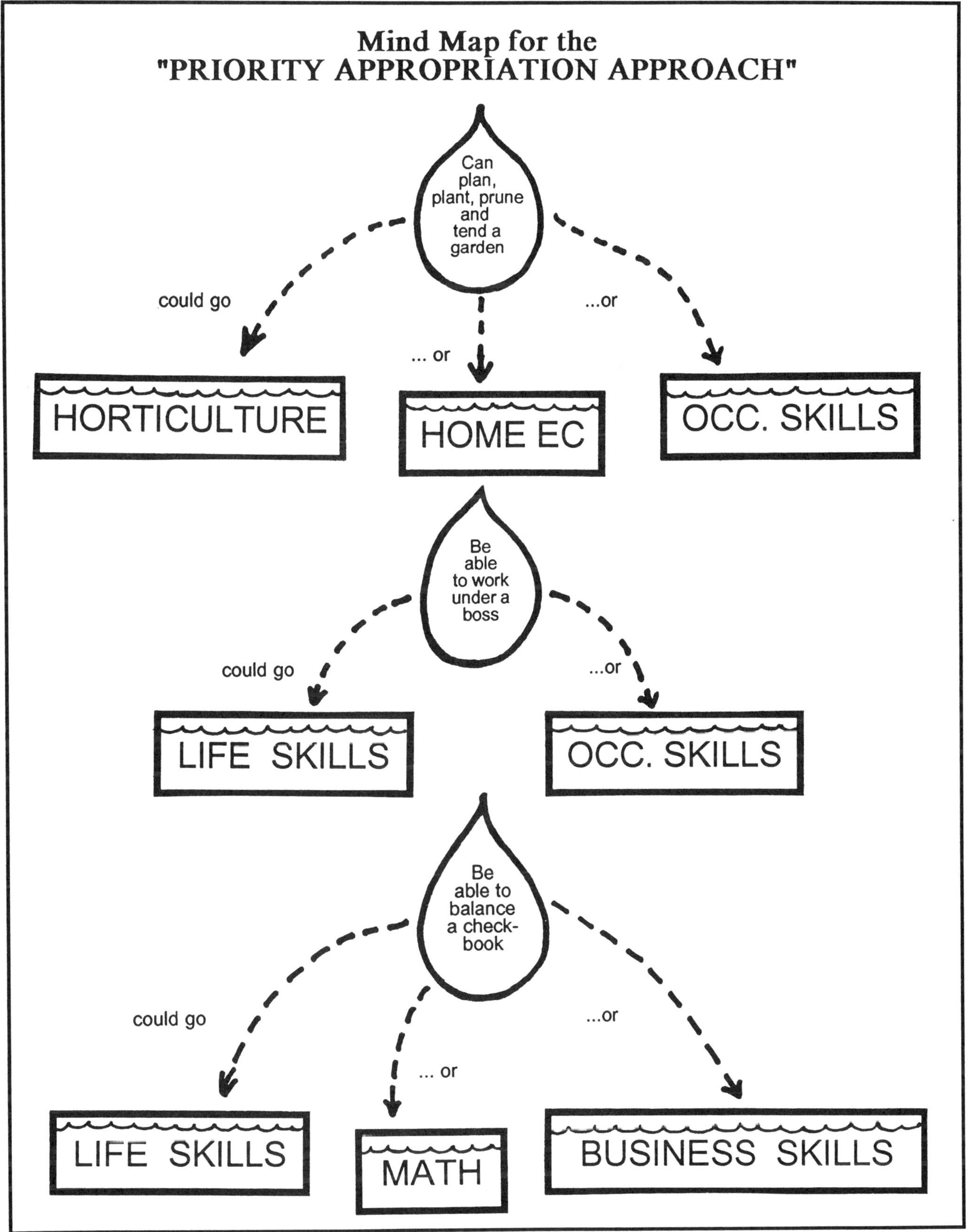
Mind Map for the
"PRIORITY APPROPRIATION APPROACH"
Can plan, plant, prune and tend a garden
could go
... or
...or
HORTICULTURE
HOME EC
OCC. SKILLS
Be able to work under a boss
could go
...or
LIFE SKILLS
OCC. SKILLS
Be able to balance a check-book
could go
... or
...or
LIFE SKILLS
MATH
BUSINESS SKILLS

THE "DISSECT-AND-DISTRIBUTE" APPROACH

Next we'll look at how to implement the "Dissect-and-Distribute" approach to devising classes. Keep in mind that this is just one example, given for inspiration, not to make you feel you have to come up with a blueprint for adding on a bedroom to your own home to make your child's high school experience complete and valid! *(Remember, that's what we're getting away from – the idea of someone else defining what is "valid" and best for us!)*

In the "Dissect-and-Distribute Approach" you start with an experience of the student and then figure out how to dissect, or break it down, and funnel it into various classes. No *single* experience will ever avail itself to *all* the subjects, but it's not *supposed* to, so don't waste time trying. Your child will be involved in a variety of experiences – unless heavily overloaded with traditional curriculum – so the Lord will provide ample opportunity for many experiences to surface in their own sweet time and in their own unique way, and as He sees the need arise. *("My God shall supply all my needs...")*

ONE DOWN-HOME EXAMPLE

We were in the process of converting our two-car garage into three smaller rooms: a work area for my husband *(which he later gave up and turned into a badly-needed second office; mine was too small (and messy) for both of us to work in!);* a bedroom for Tory; and a storage room. I know Dave would get a chuckle out of the first words of my first statement, "*We* were in the process..." "*Who* did the converting?" he'd ask. OK, he and Tory, our then-13-year-old son, did the work; I confess that my only part in the "work" was that I was the official Floor Plan Consultant, Coffee Maker, and Picture Taker (of every step of the process) – which I think was a pretty fair distribution of responsibilities – fair for *me,* at least, don't you think?!

Since Tory was very involved in assisting Dave in the work, I wanted (him) to glean as much as I could from the experience. Here's how we took this "constructive" experience *(get it? Construction?!)* and broke it down into "classes," the hours of which eventually were woven into his "Home and Life Management Skills" class *(like Home Economics, but broader).*

I took a close look at what all he was doing in the process, "dissected" it into components, and did a little embellishing and enhancing. Then I distributed it all into classes. Thus the "Dissect-and-Distribute" title. Starting off the list is the most obvious class of:

OCCUPATIONAL SKILLS – Or **Carpentry,** to be more specific. Don't forget the time spent in planning, and choosing wood and supplies. And then looking closer, there are also...

MATH skills that are used in calculating the area of the room for floor-boarding, insulation, and sheet rock; figuring the perimeter of the room for foundation boards; and in computing prices, individual and total. Making a to-scale drawing of the lay-out on graph paper is also a practical Math skill involving measuring, area, and percentage. Calculating *just* the area of the closet would be an algebraic application.

And just so you know... We used Saxon Math for his main Math credits but these above-mentioned skills *could* have fallen under a separate class or unit called "Applied (or Practical) Math" which is possibly even more important for *life.* Memorizing algebraic equations and knowing how to work a calculus problem are good skills, but if they don't know how to apply math concepts to their daily life, *big deal!*

ENGLISH – If you wanted to get some "English" value out of this, you could give the assignment of compiling all the facts, details, and steps of the process, into a *usable* (translated: *legible and detailed)* booklet. Have them do it well enough, including enough "tricks of the trade" and photos, that they would actually be able to *use* the booklet as a manual someday down the road when they would want to do a construction project in their own home or help a friend with theirs.

In fact, this was why I took all the pictures. Dave would frequently call me out to the garage where he and Tory were working and direct me on what aspect to get a shot of next. I had intended that all these photos would be glued onto cardstock pages and made into a booklet. That never panned out, it was a project ***I*** would have liked him to do, but it just wasn't something he was interested in doing. (He did plenty of other stuff.) But at least we *have* the pictures, which he – or anyone, for that matter – could refer to, need be.

The potential for more-advanced English skills lies in the creative writing. To get "high school" credit for their work, they had to be willing to let me edit it. If not, they weren't learning much – or if they were, I wouldn't know it – and therefore it wouldn't "count." (Good motivation to let me edit.) Specifically, I looked for ways to challenge them and take them further in their current level of skill in sentence construction, grammar, spelling, and vocabulary. *(I have reference books such as "Elements of Style" and "On Writing Well" to help in this area.)*

The "Dissect-and-Distribute" Approach

EXPERIENCES	BREAK DOWN INTO	IDEAS, SKILLS, and RESOURCES
Helping Dad add on the new bedroom	Occupational Skills	For the occupation of carpentry — Dad: be sure to not just do everything yourself, but to include teen in actual process, from purchasing the wood and screws to why boards are a certain distance apart, to the final finishing touches of trim and paint.
	Math	Work on calculations for boards, square feet, purchases, area of floor and carpet, wallpaper calculations.
	Art	Draw a to-scale top view of the room complete; draw finished model sketches of each wall of the room.
	Photography / English	Take pictures of the entire process, from start to finish, and assemble into a neat, concise, step-by-step instruction booklet that can actually be used by someone else, or by the student him/herself someday when they want to do this again, but can't remember the details.
	Computer	Format the above booklet using computer (word processing, graphics, cover).
	Home Remodeling (or Home Ec)	Any of the above may be categorized under this class title instead.
Getting a new dog	English	Using encyclopedias and/or other reference books, complete a "Dog Report."
	Math	Calculate the entire cost for keeping a dog for the first year (which will uniquely include price of dog, spaying or neutering, shots, de-worming, etc.), estimated cost of damage to property, and then for the second year, which will include mostly food, health care, and hair cuts.
	Woodshop	Build a dog house (which may include similar classes as for above experience.)
	Office Skills	Keep neat and accurate records of all of the above.
	Art	Draw pictures of the dog at "puppy," "adolescent," and "adult" stages (purchase art book on how to draw animals; there's a definite skill and technique involved.)
Christmas "Angel Dance" Performance	P.E. Fine Arts Computer Occupational Skills:	The 3-hour practices are physically demanding. This is an expressive dance, similar to ballet. Make up flyers on the computer to distribute as invitations. *(For a high school sibling to do:)* Videotape the event.
Visiting a Museum	Photography (Fine Arts) English History (Social Studies)	Study professional photography tips to get appealing, well-planned photos inside the museum and out. Do further research on the site to do a "bonafide" report. Obviously the content of the material is historical.
Helping with Family Business	Office/Business Skills Typing Math Occupational Skills	Phone contacts, filing, record-keeping, copying, collating Business letters, typing practice, record sheets, invoices Pricing, data entry, profits, receipts, balancing checkbook Any of the above could count as Occ. Skills.

SPEECH - (under one of our English classes) A presentation could be made at a Homeschool Kids' Day or Home School Presentation Night. For an exercise such as this to be considered "high school level," I would simply get a book at the library (or homeschool resource store) on "Speech" and incorporate any "higher level" skills and elements that challenge the student.

I was given a Speech textbook (big and thick!) by my brother-in-law and was a bit overwhelmed at first glance. Such books can seem intimidating, but remember that *any* book is nothing more than a tool in your hands! You never want to give such a book even an inch, or it will take a mile! And there you'll be, feeling like a dummy, wondering where in the world you ever got the ridiculous notion that you could actually take a *high school* student through to graduation! No, you have to be clever enough to *fool* such books! It's not hard. You just thumb through it, acting like you're just casually browsing, not letting on that you imagine it knows more than you do – which, of course it does on this subject, but you don't have to let *it* know that! With an air of calm aloofness, glance at a page here, a page there, and when you come across something that looks like your child might be at a good point to learn, or you want them to be someday, *casually* glide a flat bookmarker into that page. Notice I said *flat.* If it's thick, chilly drafts of intimidation will be free to leak out. Then come back to it later, when you are good and ready to actually present it to your teen.

ART - Drawings of each of the walls of Tory's room needed to be made so that we could figure out where to put a baseboard heater, where to position the door, and where to cut the window out of the wall. For these drawings to be of any actual use, they needed to include his furniture and be drawn to scale.

COMPUTER - The text of the booklet your student makes could be entered into the computer, making the editing process much quicker and more palatable because this way the whole thing won't need to be re-done if there are mistakes (which there *will* be); just the edits entered. Title pages could be made for the various sections, such as: "Planning," "Purchasing," "Preparing" (or "First Things First"), "Building," "Decorating," and "Finishing Touches." Using graphics adds another dimension of computer skill *and* enhances the appearance. A basic cover could also be started on the computer, and then photos, a sketch, or any other creative motifs your student might come up with added around the computer-generated base.

AND ONE FINAL MIND MAP ☞

This is the one I just *know* I'm going to get chided for by those looking for a way to denigrate my work. (It wouldn't be too difficult!) I can just hear it now on the 6:00 News:

*"Top news story of the day: Homeschool Author Barbara Edtl Shelton has the audacity to call her "mind map" on the concept of getting a new dog 'high school level'! [A shot of this very "mind map" to your right flashes onto the upper right corner of the television screen.] "Yes! Homeschoolers purchasing this manual and em**bark**ing on high school at home needn't worry about having a **ruff** time with studies or being **dog**ged by the demands of college-prep courses! With this book they are free to concentrate on the more **dog**-gone fun aspects of life and education!"*

Come on! Gimme a break!!! ... The reason I chose this particular experience was to show you just how non-schoolish and every-day an experience can be, yet still be capable of yielding some academic results, or giving opportunity for them to arise – if you just put a little ingenuity into it.

MORE FREE ADVICE *(OK, I GUESS YOU **DID** PAY FOR IT.)*

If all this still overwhelms you, cry out to Him! It may be difficult "debriefing" and "un-systemizing." Just expect it to feel uncomfortable for a season. Marilyn Howshall calls it "disequilibration."

The traditional system may be so imbedded in you that any attempt to pull out of it – or to pull *it* out of *you* – may even feel to you like *adversity!* (I liken it to pulling out a sliver!) Adversity isn't *just* losing a job, losing a loved-one to sickness or death, or losing your home to fire. Not that those don't qualify for the Top 10! But sometimes "adversity" can be part of the process of losing your *self* to God – along with whatever in the arena of "school" has given you a sense of security! *He* wants to become your entire security!

I want to reassure you that *this process, how ever painful it might be, is OK!* In fact, it is *good.* Distress, trauma, being shaken to our very core – these are not things we are encouraged to embrace in our American, microwave, instant-everything, if-it-feels-good-do-it culture. It grates against our American (or just *worldly)* grain to be in the least bit uncomfortable! That's why, when I read the following verse in Isaiah, my flesh cringed a bit. *(OK, a **lot!**)* See what *you* think about this:

> ***"He will surely be gracious to you at the sound of your cry; when He hears it, He will answer you."*** (Great so far, but get this:) ***"And though the Lord give you the bread of adversity and the water of affliction, yet your Teacher will not hide himself any more..."*** (and now it gets even better!) ***"...but your eyes shall see your Teacher. And your ears shall hear a word behind you, saying, 'This is the way, walk in it,' when you turn to the right or when you turn to the left.'"*** *(Isaiah 30: 20-21)*

Mind Map for the "DISECT-AND-DISTRIBUTE APPROACH"

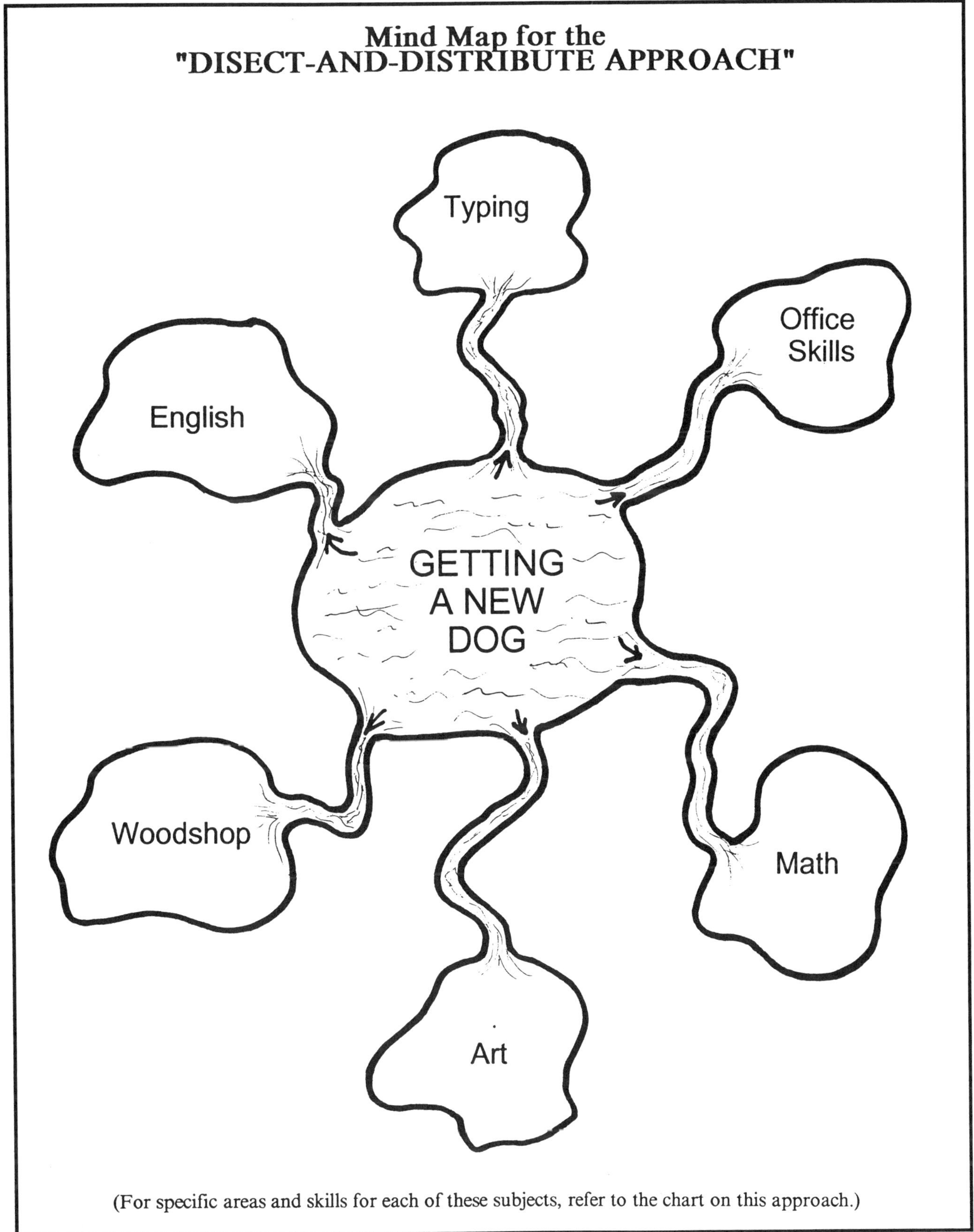

(For specific areas and skills for each of these subjects, refer to the chart on this approach.)

Is that how closely you'd like to hear the voice of the Lord?!? Me too! But I would sure have preferred for the part about adversity and affliction to have been left out! But it's part of the "total picture." The more I walk with the Lord, the more I'm experiencing the deep, peaceful, joyful fruit of having let Him take me through difficult times. I've found that "trials" aren't as horrible as they're cracked up to be – but only because I've found that, in clinging to Jesus through them, God is a lot more awesomely and wonderfully *real* than *He's* cracked up to be – by sinful, fallen man, who'd rather trust in *anything,* including *himself* and the world's unfruitful ways, than trust in *God!*

Everyone on earth is going to walk through storms, but when we let God guide us through them, and comfort us in them, and when we just let ourselves *be* with Him in the midst of them, it's an entirely different experience than what the world, on its own, goes through! Also, the more open we are to the Lord, the softer and more receptive our heart becomes to His word and His direction, the quicker we'll get to the other side!

WHAT'S "THE OTHER SIDE"?

First let me say that I don't see "the other side," *in this case,* as being Heaven, though that is certainly the *ultimate* "other side" – far surpassing any joy or glory we'll ever experience here on earth! But it's not the gold streets, the huge mansions, the breathtaking scenery, or even the yet-unheard-by-human-ears music that make Heaven the glorious place it is. It's *God's presence!* And that is something I *know* *(firsthand)* we can start experiencing *right here on earth!*

It's true, we see and feel only *dimly* here on earth, as in a mirror *(a really lousy, old, chipped mirror that might have been found at a thrift store)*, but ***"when a man turns to the Lord, the veil over his mind is lifted."*** *(See II Cor. 3:18.)* That means my precious, forbearing family and friends are *not* stuck with the "old model Barb"! We all get to be *new* creations! And one of the standard basic features of a new creation is *new eyes* to see with! ("with which to see" ☺)

And for those of you going through a particularly difficult time, please remember this verse: ***"What the enemy intended for evil (destruction), God will turn for good."*** It has been a tremendous comfort and encouragement to me!

WHAT'S WAITING FOR US THERE?

This verse in Isaiah 32:15–18 describes the blessed life awaiting us on the "other side" of seeking and waiting on God for His direction: When ***"the Spirit is poured upon us from on high"*** the first thing mentioned is that ***"...the wilderness becomes a fruitful field..."!*** So if you feel like you're in a barren desert when it comes to planning your classes, you're in for some delicious fruit if you are willing to allow the Holy Spirit's creativity and grace to rain on you – to reign *in* you – and pour though you! You're not in this thing alone! There's more! It also says that:

> ***"...Righteousness will dwell in the fruitful field. And the effect of righteousness will be peace, and the result of righteousness, quietness and trust forever. My people will abide in a peaceful habitation, in secure dwellings, and in quiet resting places."***

Wow! *I'd* say that's well-worth some laying down of my *self* and some old, crumbling edifices in my life!

NAME THAT CLASS!

One last point I want to cover is that of how to name and file your classes and non-traditional learning activities. Math and English present no challenge, obviously, but what about the day-to-day stuff that doesn't have a name?

Your activities will be different from *ours,* but I hope that by giving you a few examples from our life, you'll get the idea...

Sharnessa asked me "What would I call re-doing our tap recital outfits? And what class would I put it under?" None of the three homeschooled girls in her tap class liked them – they *were* pretty weird – so they totally re-designed them! And then she added, We decided to call the activity itself "Costume Reconstruction." Now this isn't a class, so the title is no big deal, but she wanted an intelligent-sounding label for the activity itself. As for the class, it went under "Sewing" which is under the course "Home Ec."

Another example: Tory read me something that came on the liner in his Fruit-by-the-Foot container, called "Top 10 Reasons for a Long Face." It was absolutely hilarious! *(Of course, I laugh easily; you may not have thought it so humorous.)* I all of a sudden got the idea for him to send it in to "Reader's Digest." If he were the first to submit it, *and* they accepted it, he'd get a little cash out of the deal, so we had some good motivation! I said, "You know, Tory, if you were to type this up as a list on a sheet of paper – it was just on this long strip of gum liner – you might have a better chance of it being accepted." He agreed. Part of what he had to do was to look up in the magazine where to send it, which department, their required format, etc. I figured all this ought to count for *something,* so we called it "Publication Submission Procedure." It could go under "English," "Life Skills" or even the typing part of it under "Typing."

And one more example: In putting together the final copy of the first edition of this manual, I needed help in re-doing the Table of Contents. So I asked Sharnessa to go through the originals *(all 330 of them)* one by one and check to make sure the correct title headings were on every single page that I had them listed under in my *original* Table of Contents. It took an hour, so she had one hour of "Book Assembly" to go under "Office Skills"! ... *Getting the hang of it? Just be creative!*

Mary Schofield, author of
***The High School Handbook*, shares her...**

Insights On Credits

Sometimes homeschoolers balk at the idea of tracking credits. It can seem to undermine the whole concept of home education to have to keep records in such a public-school-looking way. It should help to remember that tracking credits *does not* mean that you have to stop your wonderful home school lifestyle and begin "school at home." As stated previously, credits are simply used to translate all the things you have done in your home school into the commonly understood education language.

In addition, knowing the language of credits is very useful as a planning tool. If you have an idea of how much study of a particular area will usually translate into a semester's worth of credit, then you will be able to plan your own studies effectively. While most parents who are learning the process are concerned that considering credits will force them into doing *more* work, the opposite is usually true. Knowing how to figure out credits typically helps the home school parent to tone down a bit on requirements.*

I do not advocate clock-watching while homeschooling. It is far more important to me that my children have learned something than that they have spent an hour sitting at a desk. *(Note from Barb: I agree! I just use time as a "framework.")* ... I simply use the time computations for credits as a *planning tool* when I'm setting up courses." *(And she covers this thoroughly in chapters 4, 5, and 6. I introduce her ideas in sec. 6-A.)*

Please remember that if your student is enrolled in a *public* school program, he is a public school student, subject to the requirements set by the public school in your state. This book deals with home schooling privately, that is, *apart from* the government or 'public' school system.

**More on Mary's approach to creating a framework in Sec. 6-A!*

Coming up next! Check these out:

(They're an important part of this section!)

- ✔ **Article on Dual Enrollment by B❤** *(with another perspective from a homeschool mom)*
- ✔ **The Schofield's Basic Skills List** – ***for "Priority Appropriation"***
- ✔ **Q's and A's on Designing Classes**
- ✔ **"Delight-Directed Study" by Gregg Harris** *(including insights*
- ✔ **"Delight Directed" (learning) by Marilyn Howshall**
- ✔ **"Brief Intro to Delight-Directed Learning" by B❤**
- ✔ **Conversations (online) on Figuring Classes, Hours and Subjects**

"DUAL ENROLLMENT": A HOMESCHOOL OPTION?

For those who may not be familiar with it, this program *(called Running Start in WA)* is where a student who is a Junior in high school may take classes at their community college, and at the end of the 2 years, assuming they've taken the right classes, can earn both a diploma from high school *and* a two-year AA (Associates in Arts) Degree from the college. Tuition is sometimes paid for by the state; books are paid for by the student.

In some states, the student must be enrolled at a local public high school because funds from the state are channeled through the high school to the college. So homeschooled students may theoretically enroll at the local high school, not take even a single class there, but take all their classes at the college. I say *theoretically* because when I interviewed the Dean of Admissions at our community college, I quizzed her specifically about this. and she stated this was all correct. But problems may occur for homeschoolers when those running this program at the college *or* high school either don't fully understand the technicalities, or have biases against homeschooled students and simply opt to make up their own regulations for them that may make it more difficult or impossible for a homeschooler to participate.

Even if, in some states, the tuition is free within the context of this program, there are several "hoops" to jump through before you are at the "free" stage. It's not an automatic thing. You need to check into the regulations and procedures for enrolling in this program. There are certain requirements and deadlines to follow. I just don't want anyone thinking they can simply walk into the registration office of their local community college and say "Hi, I have a high school student here who'd like to take a few classes (for free)."

Next, let me say that while this is a good option for some, it is vital that you consider much more than just the apparent advantages as you make a decision. Let me give you a "bigger picture" to consider as you decide whether or not this option is for you.

If you participate in this type of a program in the way most of them are set up to be used, your student would enroll for a "full load" for each term of the last two years of high school, which are concurrently *also* the first two years of college. It is possible to take just a class or two. What it really amounts to, though, is "early college enrollment."

Yes, this is one option *for* homeschoolers, and a very attractive one at first glance. However I don't know that I'd really call it a *viable* option, because *viable* means "living." While, on one hand, the tuition for two years of college is paid for by the state, there are hidden "prices" to be paid by those who participate.

The first one that, surprisingly enough doesn't seem to dawn on people, is that if a student enters this program, he is no longer homeschooling! He is going *away from home* for his "education," to attend *college,* even if he's still living at home! This alone nixed it for me! My years with my kids are already too short! I'm not selling my birthright for a "bowl of alphabet soup"!

Consider that their main influence would now be the community college! Is that *really* what you want? There is much to the "education" of our children than merely academics, but classes at and homework from the community college, not to mention the influence of the environment, can't help but overshadow and dominate every other influence and activity in their lives, even just from the perspective of *time* alone! I have recently put together what I call "Wisdom's 7 Pillars of True Education" – only *one* of which is the academic aspect! *(See end of Sec.1)*

Now, if you prepared your child for early college entrance, carefully planned ahead, discipled your child, and prepared him/her thoroughly for the college influences and worldly philosophy, and *know* (that you know) that this is God's leading and timing for your student, it's one thing. But to just hear about this option and say, "Wow! Early (and possibly free) college!" without considering your *vision* for the *whole* child, not just the academic and financial aspects, is short- sighted at best. Nothing is really free; you *are* paying a price for this avenue. The price is that *you will lose two of the best years of influence in your child's life.* Are you ready to pay that price?

Personally, I don't think that even most eighteen- year-old high school graduates are truly ready for the worldly, anti-Christian thinking that abounds at colleges, *even* community colleges, and *even* at many Christian colleges, let alone a 16- year-old entering college via this program. This doesn't mean they don't exist; I just haven't met one – not even my own offspring, and they are pretty strong in their faith. Spiritual maturity is my chief concern here; almost anyone can be *academically* ready for college at an early age, but *spiritual* maturity is entirely different. I think it takes many years for a person to gain enough maturity to enter the college scene and not have it affect their faith. I'm not even sure *I'm* ready! *(And I'm 46!)*

So "Dual Enrollment" is not a "homeschooling option." Once you enter this program – at least on a full- load basis, you are no longer homeschooling. So in reality this is merely "an option that is open to homeschoolers," as opposed to being a "homeschool option." It is really just "early college entrance."

If that is truly the direction you feel God is taking your student, great. But I urge each parent and child to earnestly seek God on this and to not allow the enticement of early college or free tuition to rank high – if at all – on your list of motivating factors. If community college is part of God's plan for your student and money is a challenge, as it is for most, I believe God will provide another way. And even if that "way" is the student working for a year before entering college, so be it. God can, if given the opportunity and the *whole* heart, make the most of *any* situation. Sometimes there's a price to pay to follow God and not "sell out" to the world's "deals."

Our son, after graduating last summer, is currently doing a year of "in-depth studies and discipleship" at home, and is also working to earn money to attend college this coming year. I believe that he would not have been ready, even if he (or we) *did* have the money this last year. So even if he had been *given* full tuition, I would not have wanted him in that setting at that point. His convictions were not strong enough or "set" or rooted deeply enough in him. And a faith in the One Living God that is not deeply rooted can and *will* be uprooted in the secular college scene. Tory's studies this year are preparing him for the anti-God views he is going to come up against at the community college he'll attend next year, not to mention throughout his *life!* He's studying *Understanding the Times* by David Noebel, several books with a spiritual focus, the *Experiencing God* workbook (which we are doing with another family), books on the "evolution vs. creation" controversy, "Secrets of a World Changer" by Jeff Myers, several excellent resources by Marshall Foster, and Y2K studies, just to name a few. He studies four hours a day, and sets his own pace through the week to help prepare him for that aspect of college work.

Our job is pretty simple: Obey God. "My sheep know Me, and hear My voice." And where God leads, He also provides. The risk of the world snatching away the faith, heart and values of our children is too great to choose this option merely and primarily for get-college-done-early or financial motivations.

"For what does it profit a man to gain the whole world, and yet lose his soul?"

Perspective from another homeschool mom...

As we investigated the 'community college' alternative, we ran into some of the same objections to the community college as we had for the local public school. The community college was just an "older sibling" (a little more sophisticated) of the public school situation. There is however, less pressure to be involved in a particular peer-influence group, as there are many students of different ages. Another factor to consider – I wasn't excited about my daughter being in classes with men in their 20's and 30's! Might be easier if I had a son. Having attended the same community college, I remembered how "godless" a place it was. I didn't believe it would be a positive influence on my daughter's character. Is it worth our while to sacrifice the influences on our daughter's character in order to gain academic knowledge? I found the answer to that question was "no", for now. Maybe when she is older; I don't know. I take this homeschooling thing one year at a time. I do know that if your child has 30 hours of credit at the community college level, most 4-year college admissions directors aren't interested in seeing a high school transcript, whether they did it at home or anywhere else. (No GED is required for her to take classes; they have their own tests: one in English – one in Math). And also, if your child goes part-time to the community college (usually less than 12 hours a semester) they may still retain eligibility for freshman scholarships at colleges. Plus, in most instances, they get credit for some of the classes they took at the community college. However, if they have gone full-time to the community college (usually 12 hours or more) they may then become a "transfer" student after obtaining 30 credit hours and may become ineligible for all those college scholarships. There are some scholarships (merit and need) for transfer students, but not as many because most of them have been given out to incoming freshman. So, just keep these things in mind if you decide to enroll in a community college. The wise thing would be to contact the admissions office of the college you intend to graduate your child from and find out if they have any particular requirements that might affect your child's status should you enroll your child in the community college courses. Most college admissions counselors are only concerned about where your child is getting their lab work for the science classes. Everything else seems to be done pretty well by homeschoolers, except this one area. My husband and I both used the community college before transferring to universities and are grateful it was there for us because of it's affordability. In fact, that's where we first met. But, we are not convinced that it is the best place to put our high schooler while she is still under eighteen.

I encourage all of you to investigate every other alternative (even starting your own tutoring classes) before enrolling your child in a community college. We ended up enrolling our daughter in a homeschool academy that a group of us homeschooling parents started that year (that met just once a week) for subjects that required expertise that we did not have; such as Latin, Spanish, Biology and Advanced Math. These were mostly taught by homeschooling parents who had majors in the subjects and possessed a passion for the material they covered. The Biology class consisted of mainly labs and all the equipment was professional science lab quality. It was truly a blessing. We discovered that God did not lead us down this path of homeschooling for all of these years to then abandon us in the high school years. He is truly with us always and in everything. You just never know how God works in your desires for your children until you check it all out personally. God will lead you!"

– Betsy Lewis, homeschooling mom for 12 years; Annapolis, Maryland

Schofield Family
Basic Skills List Excerpts

Bible

- Have a knowledge of salvation and how salvation is obtained
- Know how to find information in the Bible related to any topic of interest
- Be able to use a concordance, commentary, and other Bible study aids

Character

- Know the importance of keeping his word
- Consistently consider how his actions will affect others before he acts
- Know how to answer softly

Economics

- Understand Biblical principles of the free enterprise system
- Be able to balance a checkbook and pay bills
- Know how to make and follow a budget

Fine Arts

- Be able to read music on at least a basic level

Geography

- Be able to locate the major countries and cities of the world on a map
- Know how to read a street map
- Know how to give good directions

Government

- Be able to "think governmentally" – discerning who or what is in charge
- Be able to understand the different spheres of government established by God
- Be able and willing to submit to proper authority
- Be self-governed
- Be able to read and understand ballot issues

Health

- Know CPR and basic first aid
- Know what a healthy diet is

Home Economics

- Know how to follow a recipe
- Have basic sewing skills
- Know how to paint a room
- Know how to clean really well
- Be able to set up appointments, gather information, etc. by telephone
- Know how to organize a file cabinet, warranties, family records, etc.
- Be able to handle basic home repairs
- Be able to change a diaper, burp a baby, or entertain toddlers for an afternoon

Literacy & Communications

- Be able to read anything at any level
- Be able to locate any information in a library
- Know how to lead a meeting
- Know how to give a speech
- Be able to write a business letter
- Know how to write a paragraph on any topic

Mathematics

- Be able to figure out a waiter's tip

Science

- Know not to mix bleach and ammonia
- Know not to put gasoline in a washer
- Have a basic ability to garden

Vocational Skills

- Have basic typing & computer skills
- Know how to give a good job interview
- Be able to make a resumé
- Know how to use basic carpentry tools

Other

- Be able to change a tire
- Know how to say "no" nicely
- Know how to delegate
- Know what to look for when making major purchases like cars, furniture, houses

23 ❖

Q's and A's on Designing Your Classes

(I discovered this 3-page collection of Q's and A's just hours before the book was supposed to go to press(!) and the rest of the book was already completely formatted! So I couldn't weave it into the rest of the book. If anything here is redundant (i.e. you read something similar elsewhere in this book) please forgive me. I (Barb) am the entire writing staff and at this point I don't have time to go and check this out any closer than I have. Thanks for your grace.)

***Q** I have been using the Weekly Assignment sheets and recording what we do each day, but not with points or hours. I feel like we are going along well, but I'm not sure to know what I want us to master and how I will know we have finished. How do I know, for instance, when we have completed enough to call it a completed course in English I?*

A This is where I personally feel that having *some* kind of point or hourly system comes in handy. Even just a vague idea will help. Someone suggested keeping track only to the half-hours – which I think is a great idea! Less cumbersome, and you've still got a ballpark idea, which is all you really need! But, if you *really* detest or can't handle the idea of "time," the ideas for classes in section 6-D of this book and in Mary Schofield's *High School Handbook* will give you an idea of how much work constitutes what *we* consider to be "one credit's worth of work" (or whatever the credit quantity happens to be), both of which are loosely based on the hourly system. The reason for the need for some sort of "time-ly" system, how ever loose you decide to make it, is that there's only a certain number of available hours in the "high school season." ☺ That can range widely, of course, depending on how slowly or quickly you want to take to go through high school. So you're already actually working within a certain time- frame, (i.e. 3, 4, or 5 years.) Also, for ideas for content (but not as much for the "whole class" plans), see *Homeschooling the High Schooler* by Diana McAlister and Candice Oneschak. Cathy Duffy's *Curriculum Guide for Jr./Sr. High* is very helpful in choosing resources as it gives reviews on hundreds along with her insights.

As for how to know when you've "completed enough" of *what*, this is where you've got a lot of freedom and latitude – which is probably what feels uncomfortable. ☺ My requirements will be different from yours, which will be different from someone else's. It depends on:

☞ what you consider to be important which, of course will based on your own values and value system.

☞whether or not college looks like a real possibility. This will affect (alter or temper) some of the requirements you decide on, negate others, and necessitate others. Whatever you do, though, don't let the prospect of college negate, squelch, or override the fun stuff – the "stuff" of your child's "real life" and calling!

☞ your student's strengths and weaknesses. For instance, if your student has already done a great deal of reading of literature, then the Literature portion of one of his English credits will not "need" to include as heavy an emphasis as for someone who has had very little in that area. Yet, because of the following reason, you may still want to include more.

☞ your student's interests and bents. (Continuing the above example...) Your child may not "need" more Lit to satisfy what you or the state *(about whom I am not concerned; God did not give the state the responsibility to educate my children; the state does not know my child; the state does not have the values that I have or that God has; the state is not producing truly educated people – but don't get me started on that!...* ;-)...) might deem "adequate," but if that's an area of interest for your child, you can feel free to accommodate that! My advice is that you might not want a greater emphasis on Literature to be to the exclusion of other important elements. In other words, still provide a good foundation in grammar, vocabulary, composition and whatever else is important to you (lots of ideas in the books I've mentioned!), and then let the *extra* Lit. be, possibly, their third or fourth credit in English.

❤❤❤

***Q** I don't have the assignments planned out in any way; I just record what we do. Any ideas or advice on whether this is a good idea or not?*

A That's pretty much how I started out, except that on the "Class Content Planning Sheet" I had just jotted down a few basic ideas at the beginning, to get them going. Then we both (student and I) came up with more ideas as we went along, many right out of real life, like editing a letter or book I'd written (for "English Comp"), an awkward situation with a friend that we had to work out (for "Tact, Grace & Social Skills"), a magazine article that we ran across on how a pair of Siamese twins functioned ("Science") and then later a program on TV (maybe on 20/20) that showed how they were surgically separated. ("Science," but it could also have been "Social Studies"), Sharnessa interviewing Josh Harris for my Sr. High Video (interviewing skills for "English") and then testifying in court on socialization for a homeschooling family in which the ex-husband did *not* want the mom to homeschool their daughter *(which also ended up in the video, and went under "Social Studies." See full story on page VI-C– 1!)* Back to your question...

There isn't a bit of a problem with doing it this way – affectionately called "winging it" – *unless* you have some specific goals that, at some point, you want accomplished, or are feeling a need or reason to have. If not, then just *go for it* as you have been doing!

❤❤❤

Q ***Part of my problem is there is so much we want to do that I always have a nagging feeling of not doing all we could do. Can I avoid that?***

A I *really* felt this, too, as we neared the end of high school for our two oldest! ☹ It's sad, in some ways, but it's a fact of homeschooling life! ☺ That's why it's good to have done more thinking and praying and reading on all this *before* you ever get to high school! (But we pioneers didn't have that luxury. ☺) God is covering it all anyway! And probably the best thing He's done is to show that the point (*His* point, for education) is not to do it all, read it all, or teach it all, but to teach our children how to learn, as well as to make sure we retain – or *unearth and revive* – their love of learning. Then they will be *lifelong* learners.

❤❤❤

Q ***Barb, when you set up course requirements, how do you go through the process of deciding what to master and how much to include? I have your book which I've read once quickly and keep it at hand to look up things.***

A This is actually a multi-part answer. ☺ And I think I've already partly answered that above (and elsewhere in this book), but there are a few *other* things I want to say here... Having done some reading (in the high school and Lifestyle of Learning books), I started to get a feel for what all was *possible,* not just "content-wise" but "true-learning-wise." I believe that God helped me "sift" everything, (including all the resources I came across) in my mind. He helped me to allow what we *didn't* need to just "fall away" in my mind/heart from feeling like it was a priority, and made other things feel important to include. I also looked at my student's interests, strengths and weaknesses, jotted down a few basic ideas onto the Class Planning Sheet and started on those. Then, as we went along, I allowed the Holy Spirit to lead me, though it never felt "heavenly" if you know what I mean. It just "happened" and I felt drawn toward this or that, or else I felt like "Nah – we don't need to do (as much of, or any of) that." God will speak to your heart, ideas will come to your mind and "stand out." None of it should feel panicky or heavy. Just "next." Does that make sense? ☺

Also, each class is done completely differently, and I don't try to do them all at one time. We go through seasons of "getting the vision" for each class wrapped up, but it was all developing slowly and one piece at a time over the whole four (or three or five) years. Note: ☺ Following the leading of the Holy Spirit can be "messy" and not real comfortable until we start realizing that we really *can* hear God, that He really *does* have specific things in mind for *your* child's high school experience that encompass much more than academics, and that it just isn't as hard as we make it or think it is. ☺ I highly suggest you read this book again, not-so-quickly this time ☺, stopping frequently, especially where you feel overwhelmed, inadequate, or confused – to *pray through whatever you are reading*, or whatever is being dialed up as you read. Also read the other three books I recommend; they were *tremendously* helpful to me!! And this is assuming you have *already* read Marilyn Howshall's books.

You'll avoid "that nagging feeling" as you let the idea of "doing it all" fall away. You *can't* do it all anyway! Focus your heart more on discipling your child, (see *Shepherding a Child's Heart)* and on developing a love of learning and the learning tools rather than as much of a focus on content (*Wisdom's Way of Learning*). As you pray, I believe that it will all come together for you more easily than you think. ☺ The problem lies in our "wineskin." We try to hold the new wine God wants to pour into us in our old wineskin. It won't work. We leak, crack, stretch, and, if we don't turn in our old wineskin for a *new* one from Him before too long, we'll eventually burst. *Verrrry* messy! ☺ The getting transformed (by the renewing of our minds) to God's idea of education for our individual children is the very redeeming we need for our old wineskin! If we do that first, the rest will come much more naturally.

As for how do I know how much to master and how much to include? I really didn't "know" that when I first started out, and, frankly, I still don't *know* now that I've graduated two! ☺ I used the above-mentioned books, but then my own ideas for my own kids started to surface as I went (go) along. Certain things seem more obvious to me in certain areas, which are my strong areas. For subject areas I'm weaker in, I lean more heavily on the suggestions in books in those topics. Please note!!! Most of my ideas for our classes did (do) not come at the beginning, but *as we went (go).* But, if you keep in mind that you don't have to teach everything, but rather teach them how to learn for themselves for *life*, it will help you relax more.

But, even so, I know what you mean, we wonder how much they should know of each of the different subjects, especially if going on to college. Because every school district is different, you have a *lot* of leeway here! There is nothing standardized. Even in the two Learning Objectives books that I read through here in our area *(there are two school districts)*, they had different write-ups for the very same classes! Obviously there are overlaps, but there's still a lot of room for you to do what feels or seems right to you.

On one hand, so much leeway can make you nervous and wondering *what* on earth (literally) to choose! Which is *right?!* What will they need in *life?* In *college?* I have good news for you: *It doesn't matter nearly as much as you might think.* Any student in the public school system is going to go through the classes of just that *one* school district he/she's in. And just because they took the class and aced the test doesn't mean they *really* learned anything! We watched a video on "How to Be a Superstar

Student" (which had some good points, but it's mostly how to succeed in "the system"), and the teacher noted one student who got a 4.0 all through high school. She took Chemistry and memorized all the facts and data, and aced the exam. When she got to college she *failed* Chemistry because she had not been concerned about understanding the *processes* of Chemistry – she had only filled her head with data and spit it back out for the test! She aced the system, but failed where it counted.

I think the following three articles/excerpts by Marilyn Howshall, Gregg Harris, and myself will give you some valuable food for thought as you consider how to design your classes. These are not ideas *for* classes, but will give you much insight for the "heart 'n' soul" of your classes

If ideas for classes do not come naturally to you, get one or hopefully *all* three of the high school books I have mentioned. I did not perfectly – or even closely– duplicate any of the class ideas in them, but used these books for ideas and reference. They were tremendous "idea starters," and "stress relievers." But mostly – what do you think I'm gonna say(?)...

Hear God!

Another Lifestyle of Learning Insight...

From Marilyn's Booklet:

The Lifestyle of Learning Approach

"Many of the principles outlined in the four "Wisdom's Way of Learning" books and supplemental material [*all the rest of her books that continue to develop the concepts in WWOL*] have been applied for thousands of years without ever having labels attached to them. Godly child training, families minding their own business, and parents tending to the needs of their own children, discipling them to the Lord, apprenticing them, etc. – these are all ancient and biblical educational principles. However, when one hears the term "educational approach"or "method" the tendency in this day and age is to ask, "Where's the curriculum?" or "What will I be using this year?" A focus on irrelevant-to-the- student subject content that is delivered through a lifeless process is a lower order approach to education and one you will not find in the Lifestyle of Learning message. [Its] emphasis is on using the content of real life as God provides it – real life such as the family's unique circumstances, lifestyle provisions, abilities, giftings, talents, skills, and interests of family members." ❤

"Delight-Directed Study"

Gregg Harris

A delight-directed study is like a wonderful fire in the mind of a student. It starts small, but as it grows, it begins to consume vast amounts of information until it bursts into a roaring blaze of insight, understanding and creativity. It takes on a life of its own.

In a delight-directed study, a child's interests are fanned to flame and supported in ways that increase his interest in his studies. The child's delight is the spark that ignites everything. Once established, like a fire, it is self-sustaining. The student begins to study for his own personal satisfaction, and the fruits of his study begin to flow outward to others.

This approach is especially helpful for the child who has been "burned out" on school. It helps restore his love for learning. But delight-directed study is more than just a method of remedial instruction. As we shall see, it is the foundation for all true scholarship. Once the basic concepts of delight are understood, the approach itself is easy to implement...

Is this Anything Like Unschooling?

At first glance, delight-directed study, with its emphasis on enjoying study, may sound a bit like the unschooling approach mentioned earlier. *(See Gregg's book, The Christian Home School.)* Unschooling, developed by the late education reformer and author, John Holt, emphasizes the child's freedom from adult control. It takes a more or less non-directive approach toward instruction. Like Rousseau, Holt viewed adults, and especially parents, as the major defilers of children. The result of his liberal bias abandons children to their own limited resources and further disarms parents in the face of child rebellion. There is little place for discipline in unschooling. *(Remember, Gregg is talking about "unschooling" in its original context; not necessarily what some Christians have now "re-defined" it to be, within a godly context.)*

Delight-directed study is child-responsive, but still parent-supervised. Dad and Mom remain fully in charge, and discipline is a constant part of the mix. Delight-directed study strategies are more responsive to the interest of the student, without being indulgent. Rather than allow the student to study whatever he sees fit, however he sees fit, delight-directed study urges parents to guide their child's studies and establish clear accountability for his work. Also, whereas unschooling is deeply humanistic and therefore disdainful of the Scriptures, delight-directed study is based on a distinctively Christian worldview and has substantial support in the Bible.

The foundation of delight-directed study is the *Goodness of God.* While most of us agree that God is good, we may not realize just what that means to our day-to-day experience. But the Bible is clear that God is good. "How can I repay the Lord for all his goodness to me?" (Ps.116:12) The word *good* means "beneficial." A good God does everything He does for good reason. That means there are real benefits in every aspect of God's activity. When God observed, in the first chapter of Genesis, that everything He had made was good, it meant that everything served a good purpose. ...

But when it comes to schooling our children, we Christians doubt God's goodness. In a fit of uninformed conservatism, we bring out the "hickory stick" and demand that our students submit to their instruction without regard for their enjoyment or pleasure. School attendance is compulsory. Teachers must be stern and mean. We suspect that something must be wrong if our students are actually enjoying their studies, because we didn't. "It's not supposed to be fun," we say. "Studying is a discipline."

Is it really? Granted, children need discipline for their disobedience, and they are as likely to disobey in school as anywhere else. And there is, of course, a general need for self-discipline in their studies, just as there is in their development of skills in music, art or sports. But nowhere else do we find it necessary to make everything so joyless and compulsory. In every other area of human need, the wholesome pleasure of satisfying what is needed draws the person into the activity.

Why is there so much emphasis on teaching young people to endure hour after hour of boring, disjointed and generally uninteresting activity? Why is schooling such a deadening experience for so many, even in high school and college? Is it possible that the main objective of our school system, with its passion for responding to school bells, blindly following instructions, and fitting in to the social pecking order, is not academic at all, but rather preparation for the labor force? Could this be the education of pawns? John Taylor Gatto, former New York City and New York State Teacher of the Year, ... believes it to be so..."
(And for more on this you'll need to read the rest of this excellent chapter on "Delight-Directed Study" which is planted in his excellent book, *The Christian Home School.* ☺)

"Delight-Directed"

One of the "Seven Natural Vital Signs of the Learning Process"

Marilyn Howshall

"Delight yourself also in the Lord, and He shall give you the desires of your heart." Psalm 37:4

"Train up a child in the way he should go [and in keeping with his individual gift or bent], and when he is old he will not depart from it." (Proverbs 22:6 The Amplified Bible)

To delight in something is to *take a high degree of pleasure or satisfaction of mind in something.* Delight is of a permanent nature as opposed to fun, or a sudden burst of joy which lasts for the moment. There is no moral nature to delight. It is neither good nor bad. Rather it is the *object* of delight that will determine its moral quality or educational value. Children will delight in something whether it is fruitless, diversive entertainment or something of educational value. With this in mind, delight as a *vital sign* must be viewed with the goal that the thing delighted in has dimension, substance, quality and purpose.

The child's personal interests in life should not be looked upon as having mere fringe educational benefits, but should be recognized as central to his development. These interests can give direction to his future studies, and provide current opportunities for rich spiritual lessons. Likewise, the parent can draw from the child's interests for the content needed to develop his language skills. As one interest is allowed to develop, it will begin a natural overflow into other areas of interest and into a more mature expression as the child develops. *You will not reap a learning delight in your children if you attempt to prematurely expect all their activities to yield a polished product.* Quality products that are a true outgrowth of the child's learning pursuits require time to surface.

It is for this reason that we need to enable and release our children to go the distance with an interest of value and of their choosing – one that has the potential of long-term delight. The dabbling approach to learning will never lead anywhere. Each child must be encouraged to pursue an interest with his whole heart, beginning a natural process of growth. Developing a quality skill or knowledge in an area of interest at a young age will give the child something of value to share and with which to serve others. This then develops into vision for individual life-purpose.

Delight's counterfeit comes in the form of *anything* that is fun for the moment. Enjoying or simply being agreeable to an activity that has been parent-generated is not the same as taking delight in it. Often the parent is merely making stabs in the dark in her effort to provide educational experiences for her child. In fact, too many activities of this nature, in which the child never goes beyond surface curiosity or interest, may actually keep the void in his life from being recognized and effectively addressed.

Destructive forces of boredom may be present within the home. If this is true, then it may be necessary to begin tempering less desirable interests and indulgences, while at the same time directing activity into more fruitful areas. This can be a challenge with adolescents whose value systems are already set. Approaching change within the context of their personal relationship with God is the best place to begin. A pure, teachable heart is essential to developing a learning delight and learning process – first in ourselves, them in our children."

Note from Barb: "Delight-Directed" is just *one* of seven "natural vital signs of the learning process." The other six are described in Book 1 of the Wisdom's Way of Learning 4-book series, *The Science, Art, and Tools of Learning.* In this particular book, Marilyn also discusses: "The Biblical S.A.T.'s of Learning"; "The Process and Product of Learning"; "The Three-Stage Learning Process"; "The Heart of the Learning Process"; "The Five Learning Tools"; "Exchanging Old Wineskins for New" (this chapter is by Kyle Miller). The general concept of "delight-directed" learning is thoroughly and graciously addressed throughout all of Marilyn's writings, from many angles that address not only the "approach" but also the heart of it and how to implement it.

A Brief Intro to

Delight Directed Learning

Barb Shelton

Many homeschooling moms attempting to implement *Lifestyle of Learning* principles into their homes and lives, or even just trying to operate with more freedom *(some have not yet heard of L. of L.)* often find themselves confused, frustrated, and full of questions. One such mom said "I am trying to implement more delight-directed learning with my son, but he seems to get nothing done if I don't give him a schedule to follow."

Early on, I had many "mis-impressions" of not only "delight directed learning" but also of "lifestyle of learning." As moms who have been in the traditional school system all our lives, we are very much in need of a fuller understanding of what "delight-directed" means, not to mention what "education" means! This is part of the "renewing of the mind" that needs to take place in each of us if we are to enjoy God's freedom and fullness in truly educating our children.

First, just to clear up the *usual* first *mis*-impression, "delight-directed" does *not* mean "flesh-directed." It's easy to confuse the two, but they are really at opposite ends of the educational spectrum. "Delight directed" primarily means that as much of the child's education as possible is built on his/her interests and delights rather than on generic text/workbooks and/or a curriculum company's scope and sequence.

It does *not* mean you just let your kids indulge in doing anything and everything that delights them! Godly child training is involved, which includes a disciplined – *not* haphazard – lifestyle. Surprisingly to many, Marilyn's family usually operates within a flexible routine. It is not rigid or packed with artificial methods of learning. Rather, their home and lives are orderly, and they accomplish *much*. There is beautiful fruit in their lives, and their home is peaceful. (Read *Develop a Lifestyle Routine* for much more insight and practical help on this.)

Getting to know your children is also of prime importance in approaching "delight-directed learning," which entails *seeing* them with new eyes. These "delights" we are talking about are not just toward *things* – like Legos or dolls or bikes or birds or computers or... *whatever*. We need to look at our children more closely, beneath the surface, and begin to understand what makes them tick. To do this, we may need new "eyes" and new understanding. What is it *about* whatever they naturally gravitate toward doing that is *really* motivating them? What is truly the *inner delight* of their life? The answers will give you clues as to what God created them for; not just to be a parent and procreate, which can certainly be *part* of their God-ordained purpose, but there's *more*!...

What unique "thing" did God breathe into your child that He wants to bring to life, perhaps unearth from beneath that pile of curriculum, and develop for *His* purpose – part of His reason for creating them and placing him on earth!? God has "works that He prepared beforehand that we should walk in them." What are those unique "works" He prepared for *your child?* For *you?* As *you* begin to understand this more fully – and walk in it *yourself* as you enter a deeper relationship with God – you will begin seeing *everything* differently! And it will naturally overflow to your children! New delights in *all* of you will begin to blossom! New purpose! New interests! Let God take you by the hand and open a whole new world to you and your family. But you have to begin with letting go of the "old."

[Be sure to read the section on "Delight-directed Learning" in *Wisdom's Way of Learning.* Actually, it is vital that the *whole* book is read (several times, even) as you embark on this journey, as taking off with only a superficial understanding of "delight-directed," *(such as what I've just presented here in these three article/excerpts)* will *not* get you where you want to go. There is a much bigger picture which, as you begin to understand it, will bring a new and wonderful freedom to both you and your student! ☺ It only gets better!]

Conversations on Figuring Classes, Hours and Subjects

The following is a collection of posts I saved from the SHF List *(Senior High Form+U+la List; see* *page VIII-7 for an explanation)* **that I think you'll find interesting and helpful, especially once you really get into *doing* the class stuff yourself. Some of these are just a little hard to follow because sometimes there is more than one response to a comment or question, but keep reading; you'll figure it out. ☺ It's very similar to a group of women sitting in a room chatting together; it's just all on computer! ☺**

[***Debbie** shared this interesting fact:*] "According to the NEA's own research, of the 900 hours that high school students are credited for, only 200 are spent "on task." "On task" means time spent teaching the required subject. The rest of the time is spent changing classes, having lunch, getting settled for the start of class, preparing to leave a class, etc. Of those 200 hours "on task," how many hours is an individual student actually paying close attention, ready to learn what is being taught?" This is from an article in "The Learning Edge" published by Clonlara in Michigan."

---

*(In response to the above came the next question from **Kathy** in CA, and then the two posts that follow ...)*

"Yes, I get the "Learning Edge" also. Are you saying that I should include having lunch and driving to and from co-op classes in my Clonlara time sheet to fulfill the 180 on task hrs that they require?"

[*Response to the above from **Debbie**:*] "What is most important to me is what my children are learning, not that they are spending a specified amount of time engaged in study. Reading Barb's Form+U+la book and Mary Schofield's book *(the High School Handbook)* really liberated me from the timer and the clock, and set me free to allow my children to really *learn*. All I am saying is that we shouldn't be in bondage to the clock. Public school kids are getting a lot of credit hours for what?? We need to make sure that our children are getting credit for what they are doing too... we just have to make sure it is presented in a way that colleges can understand."

[***Viola Kay M.** answered:*] "From my experience of teaching in a good Christian school your statement (about on-task hours) is true. I taught 7–12 grades... it took 15 minutes to get them to get their books and paper and pens out and settle down. Then I taught 20 minutes really with them all listening and then they spent the last 15 minutes watching their watch and thinking about getting out of class to go to the next one or so they could speak to their friends once they got out the door. 99% of the children were there to socialize – NOT to learn; and they'll tell you so!"

[***Kathy** further explained:*] "Clonlara* is like an umbrella school (for homeschoolers), but they allow you to do your own thing and just keep your records. They are very clear in their requirement of 180 clock hours. We must turn in monthly time sheets that tell the clock hours spent on each learning experience. The learning experiences can then be jumbled together to form courses when related experiences reach 180 hours. We are not talking real-time hours here or on-task hours, but actual clock hours learning, and they must add up to 180 per credit. If you take a community college course that is just one semester but would add up to a full year high school course, they will accept that as one credit. They also accept just about anything like work experience and internships and apprenticeships. ... A lot of good stuff here but I still feel stressed."

[*The website address for Clonlara as of May 1999 is: http://www.clonlara.org/index.html
If that no longer works, just do a search for "Clonlara" on the Internet. I do not have any other info on them.]

[***Conna** had said in a different post:*] "...While studying world history this year, my son has done many reports and mini-biographies. We count those as English even though the subject matter is History; he needs the credits for History and English so we use what we need where we need it."

---

confused me. For instance, my daughter is involved with a Shakespeare summer camp each summer. She would certainly have enough hours from all those summers for an English course and/or Drama. Or would it be better to just giver her extracurricular hours?"

[*To which* **Linda** *responded:*] "Doesn't Barb talk about just keeping up with all these credits and when you get ready to do a transcript, decide where you need to put the points? It's easier to see where it all goes when you get to the end."

[**Conna** *answered Joanne:*] "We don't count the same hours twice although I know some people that do. [*Barb: I discuss this in a couple places here in this book.*] Since we usually have something else to read or do pertaining to History, we count that as the History hour and the other as English. I just decide what hours will go where."

[*And to this,* **Kelli** *responded:*] "Barb calls this "Dissect and Distribute." She explains how to do it [in section VI-B]. Now I'll explain how *we* handle this. [*From Barb: OK, at this point Kelli explains the system she designed to accommodate her own family's style. Without examples (which are impossible to give by email on the SHF list), this is going to be a little hard to grasp, but it may be "just the ticket" for you, so if it sounds at all like something that might work, it'll be well-worth the effort to try to get it.* ☺] I have made up a system on color-coded index cards. On one side of the index cards are 40 boxes to color in. On the other side of the card is the subject (like Language Arts) with all the sub-topics listed that qualify for Language Arts (spelling, grammar, literature, reading, etc...) When the kids finish a project, we fill in squares under whatever subject we need the credit *most.* BUT we also write *(inside the colored-in square)* things like "H" *(which stands for history)*. That tells us that that particular hour *could have* qualified as history. At the end of the year, we can trade squares between the subjects, and balance them out how ever we want. For example, if we have too many Language Arts, but not enough Science, we check the Language Arts card to see if we have some Science squares we can switch over the science. Then we fill out a new LA card to reflect the change. Each card, when completely filled in, represents ¼ credit. That makes it easy to count credits. If there's a form that was filled in to represent a particular "hour" of credit *(for example, there was a book report filled out or a report done)* then we write *(also within the colored-in square)* what form *(each form has a code)* was filled out to verify that hour. It isn't as difficult as it sounds. This system will be available for sale *(labeled and color-coded index cards and file box included)*, along with all forms and instruction manual. It will be helpful in putting together your portfolio. However, you can do it yourself without too much effort." *(Contact Kelli by email at: ndonald@infinet.com or at: Kreative Minds Technology / 1833 Reddington Rd. Newark, OH 43055 740-522-6832.)*

[*And to Joanne's question,* **Patti** *in FL answered:*] "Hey Joanne, I would count the summer(Shakespeare) play in several areas... I would count some of the hours as English. After all, she is immersed in Shakespeare. I would also count a lot of hours to Fine Arts as she is learning and participating in play production – much better than just reading about it. Then I would have the time spent learning about the historical aspects as history. Perhaps add a historical novel about the time period of Shakespeare or the time period of the play. Don't be a purist about counting "exact" minutes. I do a lot of estimating (with the benefit of the doubt going to fewer hours), and still have way more hours than needed for credits – and more knowledge is retained."

[*And this is what* **Kathy** *in CA asked, along the same lines:*] "I would like some suggestions for dividing up a four-day Mission Trip that my son is taking to Mexico. Here's what I'm thinking: **English Composition**: He is giving a sermon on "The Woman at the Well" and he has to research and write it out. **Speech**: Delivering the Sermon. **Spanish**: speaking with the people. **Geography**: Traveling to Mexico??? or maybe **World Cultures**??? **Religious Studies**: Listening to and learning from the leaders and other speakers. What have I forgotten? Would you rearrange anything? I always have trouble with these trips because so much goes on. Do I include PE when he plays basketball and soccer with the children there? The entire trip counts for Community Service also because we double hours for any School subject that would also qualify for Community Service. I would be interested in how someone else might divide up the times on such a trip."

[*From Barb: I didn't save any responses to the above post, but she had some great ideas even without further input!*]

[*And Joanne Turner added later on:*] "When we did our transcript this year, what we actually did was give Jenny one credit for Shakespearean literature and ½ credit of drama and speech (from a toastmaster's course along with all those summer workshops." *(To which Barb says: Excellent idea, Joanne!)*

The previous conversations are a good example of what I think is one of the best benefits of having other high school homeschooling moms to bounce ideas off of! *(This was via the Form+U+la email list.)* ***As you can see, some of the most valuable input comes not from me,*** *(I think I put in three whole words at the very end there!)* ***but from other homeschooling moms who are DOing it!*** ☺

Section Six - Part C:

Organizing Your Materials

Step-by-step directions
for how to organize all your STUFF,
restructure this book into a 3-ring binder,
and help your student create a handy
and attractive High School
Organizer Notebook.

The Wisdom of Noah Webster

Here are few quotes of Noah Webster, taken mostly from his book called *Advice to the Young: A Moral Catechism.* The forward of this book is by David Barton; the bulk of the book by Noah Webster, who wrote the original "Webster's Dictionary," now known as the "Webster's 1828 Dictionary to differentiate it from today's versions which are a far cry from what Mr. Webster originally wrote. As David Barton said in the forward: "Unfortunately, since its original publication in 1828, Webster's dictionary has undergone extensive censorship to remove its Christian nature; and although the most popular dictionary in America continues to bear his name, it no longer reflects the spirit of the original." I *(Barb)* would say this is an understatement!

Make a quick run over to section 6-D, pages 25 and 26 and read the "1828 Dictionary" definitions of "Education," "Learn," and "Knowledge." *(I'll just wait patiently while you go look there, because the rest won't make sense unless you read those; at <u>least</u> the one on "Education"...)* Well, did what you read sound *anything* like what you would read in a *contemporary* dictionary?! ... Here are a few quotes of Noah Webster, the first one from *Letters* and the others from *Advice to the Young:*

"In my view, the Christian religion is the most important and one of the first things in which all children, under a free government, ought to be instructed... No truth is more evident to my mind than that the Christian religion must be the basis of any government intended to secure the rights and privileges of a free people."

"My young friends, the first years of your life are to be employed in learning those things which are to make you good citizens, useful members of society, and candidates for a happy state in another world. Among the first things you are to learn are your duties to your parents. These duties are commanded by God, and are necessary to your happiness in this life. The commands of God are, "Honor thy father and thy mother."— "Children, obey your parents in all things." These commands are binding on all children; they cannot be neglected without sin. ... It is the duty of parents to feed, clothe, protect and educate [their children]; and for this purpose it is proper and necessary that parents should have authority to direct their actions. Parents therefore are bound by duty and by right to govern their children; but the exercise of this right is to be regulated by affection. ... Hence it is a primary duty of children, and as much their interest as it is their duty to "Honor their father and their mother." This honor not only forbids the child to disobey his parents but it forbids all rudeness and ill manners towards them."

I included the above for the benefit of any "wayward" or rebellious children who may need to read (and understand) it, and for any parents who might heavy-handedly lack affection in the administration of their parental duties. I would suggest to any who are thinking of homeschooling, or who are homeschooling but are frustrated with the fruit they (don't) see in their children that they begin <u>here</u>; by establishing <u>godly</u> control in their home. Without this, educating your children will be not only impossible but pointless, as this <u>is</u> the foundation of true education.

"Among the first and most important truths which you are to learn, are those which relate to God and religion. *(Notice that he did not say Math, Geography, History or Science?!?)* As soon as your minds become capable of reasoning, or excited by curiosity to know the causes of things, you will naturally inquire who made the world, who made you, and why were you made? You will understand, by a moment's thought, that the things around you cannot have made themselves. You will be convinced that a stone or a mass of earth cannot have made itself, as it has no power in itself to act or move; it must then have had a creator, some being that had power to act or move, and to bring the stone into existence. ... Now when you think that of all the substances about you, not one can have been its own creator, and when you see the vast multitude of things, their variety, their size, their curious forms and structures, you will at once conclude that the Being who could make such things must possess immense power, altogether superior to the power of any being that you see on earth. You will then be led to inquire who is this Being, and where is he. ... The first and most important point to be decided in your minds is that God is your Supreme or Sovereign Ruler. On this point, there can be no room for doubt; for nothing can be more evident than that the Being who creates another, has a perfect, indisputable right to govern him. God has then a complete right to direct all the actions of the beings he has made..." *(Note from Barb: Think of all the societal problems that would be solved and prevented if we all believed and followed this!)*

ORGANIZING YOUR MATERIALS

It's pretty obvious that, with all the stuff you'll be working with, things could get messy and out of control in a hurry. The seven areas you'll want to get organized before you go much further are: this manual (the one you're reading); a file for your "Learning Guides" (from sec. 4) and records; your "High School Resource Shelf"; your student's "High School Organizer Notebook"; your "Input Box"; your student's "Work Pending" file; your student's individual topical notebooks – though you'll add to that as you go along. And I just happen to have instructions for "all the above" in the remainder of this section!

The following story has absolutely nothing to do with this section; I'm just (as you've probably noticed by now) weaving in stories and excerpts that I think you'd enjoy and/or be edified by! "White space" is nice, but since our time (and space) is limited, I'd rather fill it! ☺ Enjoy the story of...

Sharnessa's Day in Court!

I had been asked by a homeschooling mom to be an "expert witness" for her court case, and Sharnessa went with me to observe a court procedure. *(This was for her "Wash. State Government" class).* Asking Sharnessa to testify was a spur-of-the-moment idea I had gotten on the way to the courthouse! But the frustrating thing was that once Sharnessa conceded to being a witness, she then had to stay out of the courtroom, as the opposing (prosecuting) attorney would not allow witnesses *(including me)* to be in the courtroom during all of the preceding testimonies! And the case went on *all day!* We got there at 8:30 a.m. and left at 6:30 p.m.! I went in first at about 5:15 and she went in right after me at about 5:30! So she (we) didn't see much of the case, though we got to watch it later on video, but she *did* get the experience of testifying – and wow did she do a good job!...

I was on the edge of my seat ready to shoot the opposing attorney (uh, shoot *rubberbands,* that is) at the opposing attorney who was, in my motherly eyes, a real jerk. *(Several others there shared my opinion.)* Hope the word "jerk" doesn't offend you, but I can't think of a more appropriate word to use! He was conniving, rude, manipulative, and "stinko" – and yet through it all, Sharnessa stayed cool and calm and gave *excellent* answers, even though he'd had the homeschooling daughter in tears several times that day! (Of course, he was much more rudely probing and insulting to the other girl since she and her mom were the ones on trial. He'd try to get the person on the stand backed up against a corner *(his job, of course)* – which Sharnessa just wouldn't let him do! She came right back at him!!! Respectfully and calmly. For instance, one time he asked her how much school she did in a day *(I felt like we were on trial!)*, and she thoughtfully answered "*I'd say about three hours.*" To which he accusingly queried "So do you think you're getting as much of an education as kids in school who are there for 6 hours?!?" *"Yes, I do."* "Well how can that be, if you're only spending half the time they are?!?!" *"Because I've visited classes with my friends for a day and seen what it's like there."* "So you think that by visiting for *one* day you know what school's like!?" *"Yes, I got a pretty good idea."* "So how do you think you're getting a better education?!" Sharnessa, kind of chuckling, *"Well, do you actually want me to tell you what the typical classroom is like?"* "No... I'm through questioning the witness."

My own personal bias *(which others agreed with)* is that Sharnessa was one of the key witnesses who helped win the case for this mom. ☺ And yes, she won! Much to the disgust of the ex-husband and the attorney!

Step 1

Purpose: To make the manual as convenient as possible: Transform Your Softbound Form+U+la into a 3-Ring Notebook

Why go to the trouble? It doesn't make a bit of difference to *us* if you do this or not, but there are many sheets to copy in this book (exactly *how many* depends on which ones you want to use "as-is"), and there are over 150 forms available, and having them in a 3-ring notebook makes it easier to just pop them in and out of the notebook for copying.

To make the softbound version handier: Once you get past how destructive this sounds, you can actually cut the whole book apart using an exact-o knife and ruler, 3-hole punch the pages, and put them all in your own binder. (I've also heard printers do this.) Tabbed dividers make finding everything much easier. You can also cut the cover carefully along the edges of the whole cover and insert the front, spine, and back covers into a clear-vue notebook for a snazzy effect. If you want to have access to what's on the INside of the front and back covers, *(which I'm sure you WILL – both are unbelievably fascinating),* then, you guessed it, just make copies! *(Before inserting the covers into the clear pockets would be easiest!* ☺)

Tips for your Masters: One suggestion: You don't necessarily *have* to 3-hole punch the sheets that you'll be using as masters. Instead you could just keep them separate in a manilla folder. If you really want all the pages "in context" in the book, you have a couple of options:

OPTION #1 - You can copy your originals (even enlarge them just a bit if you want since, for trimming purposes at the printer, the softbound pages are a bit smaller than a regular sheet of paper, probably 103% would be about right), keep those separate, and then punch the actual book pages and put them back into the notebook where they belong. (My typical style in my books is to have the master page on the right and the explanation of it on the left. It'll make more sense once you see the book. However, there are a number of exceptions to watch for.) Or...

OPTION #2 - You could, *instead* get those plastic sheet protectors, the kind where you can slip the sheet into the pocket in the opening at the top that has an extension along the side with 3 holes in it, so the sheet itself isn't punched. The only drawback is that these protector sheets will stick out about ½" farther than the rest of the pages. But the advantages are that everything will be in the right order (in context for reading), AND your masters will NOT have hole punches in them.

Does all this sound like sheer drudgery? Well, it is. ☺ But let me share with you a note I received from a gal online (making it an *e-mail,* actually) that might be encouraging:

> Hi Barb, I wanted to let you know of the delightful morning I had today taking your book apart and putting it into a binder. Sounds odd, you're maybe thinking. My teens thought so too. Let me explain... The humor you've added throughout is WONDERFUL! I want to put aside the rest of my life and just read and laugh. Little phrases (in the book) popped out as I was transferring pages and I'd stop and say "Girls, listen to this!", laughing as I was trying to read it to them. Maybe it was one of those things that you had to be there, but now I have new enthusiasm for homeschooling. There's so much I want to learn and experience (my daughters, too, of course), and so little time left with my teens! Thank you!" – *Lorraine in Plainwell, MI*

Hope you have as much fun putting this together as Lorraine did! ☺

Step 2

***Purpose:* To set up a resource center in your home – so you and your kids can find your stuff quickly and easily!**

Your student needs to be able to find books, resources and the learning guides without having to rummage through a bunch of boxes or piles. Also, both you and your student need a place to file and store new resources as you find them and bring them home. Our resource center, easily accessible to both our kids and myself, consists of two areas that were very simple and inexpensive to set up.

1. **THE FILE** - This contains individual files for *each* of our learning guides, transcripts, daily reports, extra pages for the student's organizer notebook (like Planning Sheets and Class Completion Sheets), and a file for each student individually labeled "Sharnessa's Work -Pending" for learning guides that are ongoing or that they simply haven't finished yet. What kind of files are best to use? This is individual, but here are my comments on each of the two main types:

   manilla – cheapest; easy to re-label; can be stored in a file drawer or box

   pendaflex – more expensive; you need a special type of file drawer (or just the right-sized box) to store them, but they're more adjustable for growth. Some people wouldn't use anything else, but I personally don't feel they're necessary. The manilla tri-cut file folders have pre-creased lines along the bottom that allow for expansion. (With all the files I use, I'd be broke if I used pendaflex.)

2. **THE SHELVES** - Find some shelf space where you can store the books you'll be gathering. Even a box on its side will do. All books that the kids will be using, even way down the road, are in this section. (I'm afraid that if I stored them *away,* I'd forget we even had them until their 10-year Reunion!) Plus you never know when they'll be ready for them.

3. **THE "INPUT BOX"** - This is where we put brochures or notices for new resource ideas and opportunities we find. It merits its own step, but I wanted to mention it here since it's part of the whole picture. In fact, it sits right next to the file.

Step 3

Purpose: To provide a place for your student to work out of

At some point, you'll need to set up another notebook *just* for your student. *(If you have more than one high schooler, make one notebook for each.)* The material in this manual gives you the general plan and guidelines; what goes in your student's notebook is his *own* plan. Here are the headings to include, *(– you'd write these titles, or abbreviated forms of them, on the divider tabs)* and then what to put in each of the sections.

- **Requirements for Graduation & Promotion:**
  - Our Requirements for Promotion
  - Our Requirements for Graduation
- **1 Divider for each Class and/or Course Title:** (See below)
  - 1 Class Content Planning Sheet (filled out, obviously)
  - Grading Criterion Sheet for that class (unless you want all these in one section called"Grading." *I prefer to have all information for each class in one place.)*
  - Class Completion Records *(start out with one; add to it as needed)*
  - Any brochures that will help you in planning for that class. *(such as for community concerts, an upcoming Creation Science seminar, etc.)*
- **Class Content Planning Sheets** *(just a place to store blank ones so they're there when needed)*
- **Class Completion Sheets** *(blank)*
- **Yearly Records** *(divided into Freshman, Sophomore, etc.):*
  - Student's copy of Transcript *(kept at beginning of section)*
  - Record of Hours & Credits completed for each year *(1/year)*
  - Tally Sheets *(1 or more per year)*

DIVIDERS FOR CLASS TITLES

At the end of this section you'll find several pages of Course and Class titles in boxes. These are for you to use in making up your own divider pages. Further directions are given "back there."

Step 4

Purpose: To put a face on the student's notebook

JUST AN OPTION!

As with everything in this entire book, what follows is *not* any kind of a requirement or standard. Let me re-phrase that: *Do not feel you have to produce a cover like ours (presented on these next two pages)!* The sole purpose of presenting this idea is to give you an idea if you are fresh out!

MAKING A COLLAGE COVER

A collage makes a very attractive and interesting cover for your High School Planner/Record notebook. Plus it's easy and fun to put one together. You'll have to put a little time and money into some photos, but it's well-worth the effort. If a picture speaks a thousand words, ours is pretty verbose! — spewing 17,000 words! (We did the front, side, and back!)

Find, or have someone get some pictures of you busy at all the things that make a composite picture of your life, including both academic and fun activities. You'll probably use anywhere from 5 to 10 photos for a one-page layout, depending on how small the usable parts of the photos are.

We use the white notebooks with the clear plastic pocket over the whole cover. The opening is at top where the cover sheet slides in. Get a piece of stiff paper — either cardstock, posterboard, or construction paper. (I prefer black because it sets off the photos best.) Make sure your paper is the same size as the notebook cover *before* you start! Most are not the same size as a standard sheet of paper, so you may need to cut and/or add to your paper to make it the correct size. (If you add a strip, overlap the two just a bit.)

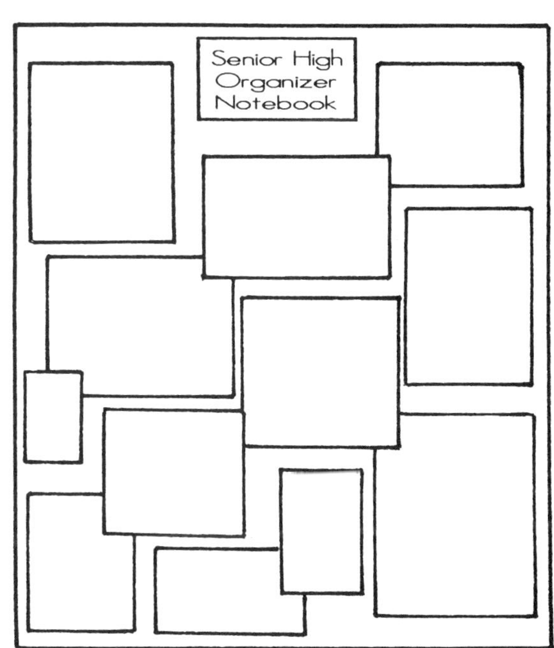

In large, neat letters, write your title on a separate piece of (white) paper (or a label). Or you might use press-type letters, OR do them on a computer print program. Try to keep the size down to maybe 2½" X 5" maximum, to leave plenty of room for your collage photos.

Next, gather all your photos (and any decorative touches) and lay them out so you can see what you're working with. Cut them down as small as you can (so you can use as many as possible), yet leaving enough background for the subject of the picture to still be in context.

Lay all of these on your prepared paper, trying different arrangements. There are actually many ways of doing this: Overlap corners, if you wish, letting the "dead space" of a photo be the "under" part. (But don't cover up an important part of a photo.) Or you could have *no* background showing, overlapping all pictures on each other on all sides. (This is my least favorite way because it looks more like a mass of color and shapes rather than featuring your individual photos.) I prefer (and I'm sure you want to know my preference) to overlap many edges and corners, leaving plenty of (black) background between the pictures. You might prefer to have NO overlapping. This looks nice too, but is more difficult and doesn't allow for as many photos.

Once your arrangement is finished, (and don't forget to stick in that title!) use rubber cement to glue them down. Since your collage will be under clear plastic, you need only a small dab in each corner - no need to coat the whole back. I recommend rubber cement *only,* (never glue sticks or Elmer's glue,) because it allows total flexibility. If you make a mistake, or want to replace a photo later — even months or years later — it will lift right off without damaging the picture OR the paper! (except construction paper.) Once the photo is pulled off, the rubber cement residue (hard) just rubs right off with your finger! (Other glues will pull the backing off the picture every time.)

Now you're ready to *carefully* insert the whole collage into the top of the pocket, making sure, as you slowly push it down inside, to watch that each picture makes it in along the top edge. It could easily catch on the top and will bend before you realize it. If you wish, do the same thing for the side binding and the back! -- especially if you have more photos you'd like to use! (We just couldn't decide, they were all so good! So we just used 'em all!)

And there you have it, a beautiful cover for your notebook. And don't forget to count it toward your Fine Arts (Art) hours!

Senior High Organizer Notebook

Senior High Organizer Notebook

Because I just *knew* how much you would be hoping for a sample to help you fully understand the concept of putting together a collage cover, ☺ I went ahead and made one up for you! ⇨ Some of these photos wouldn't necessarily be appropriate for a cover for a High School Organizer Notebook, so I gave this page a different title.

(Okay, you're right; this was really just a good excuse to show you a few of my favorite family photos!)

❤

SHELTON FAMILY MEMORY PAGE

❤

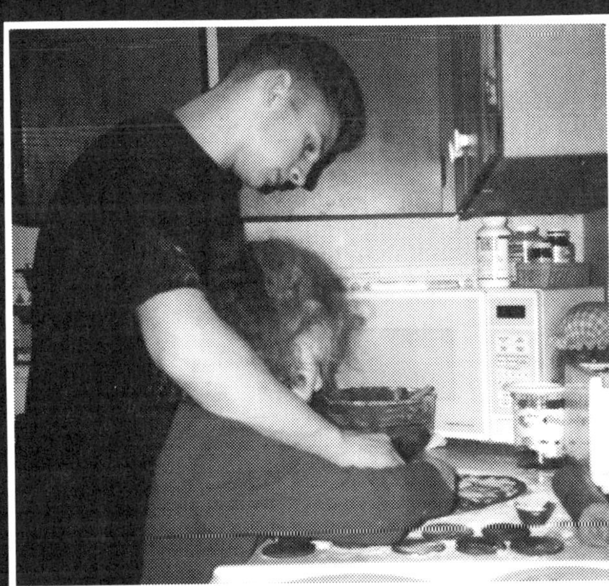

Step 5

Purpose: To set yourself up with forms

Go to a high-quality copy machine and run off *just a few* copies of the forms you know you'll be using "as is." Do not succumb to the temptation to copy a huge pile "since I'm here." I guarantee you'll regret it! (It took me only four or five run-ins with myself to finally get this through my head!)

If you are going to alter any of these forms: Before going to the trouble of making up a whole new one, just use some Wite-out on the parts that don't apply to you, and fill in your own "specs." Then use the form this way for a while, kind of like a "probationary period"! You do NOT want to make massive copies of ANY form until you've had time to make it "tried and true" with your kids. I have yet, in my entire ten years of homeschooling, to do a form "right" the first time! It takes using it a while to find out the quirks and refine your form to perfection. If you're like me, you're just anxious to get it done, but you will save yourself tremendous trouble and regret if you start slo-o-o-o-w and give yourself time to come up with something that will *really* work for you. Someone once said: "There's no such thing as a good writer; only a good RE-writer." That applies here as much as anywhere!

After the "probation period," go ahead and make up as many copies of your finished forms as your heart desires *(and your shelves or files can accommodate),* leaving *some* room for books. You'll probably need:

From the Record-Keeping Section:

- ☐ 1 "Class Content Planning Sheet" per class
- ☐ A few "Class Content Planning, continued" sheets to have on hand for "overflow"
- ☐ 1 Daily Report Sheet per week your homeschool will be "in session"
- ☐ 1 "Class Completion Record" or "Check-off Sheet" per class (Make a few extra of each type, until you find what you like for each class.)
- ☐ 1 copy per year of the "Tally Worksheet"
- ☐ 1 "Staying On Track Record"
- ☐ 1 Transcript for the entire high school season

> I don't know whether the paper this book is (*will be, at the time I'm writing this)* thick enough to keep the print on the other side from showing through. If it *does* show through when you copy, here's a nifty little tip: Place a sheet of black paper *behind* the page you're copying! It works *great!* ☺

Plus out of other sections:

- ☐ 1 "Grading Criterion Sheet" per class
- ☐ Copies of any of the Learning Guides in "Potpourri" section you want to use
- ☐ 1 "Master Framework Plan" for the entire high school season (Make a few extra's, just in case.)
- ☐ Any course, class or notebook cover titles you want to use

Step 6

Purpose: To set up a central location for new input

As time goes on, and more and more "stuff" comes into our house, I'm finding some wonderful opportunities arising. And I mean for more than just learning how to file or discard. There are "happenings" in our community, read-worthy material to pass on to my kids, excellent treasures to weave into the fiber of our "curriculum." But being organized enough to take advantage of them can be a bit tricky.

I finally came up with a plan that was so *simple,* I myself could not believe I had not thought of it sooner! I call it our "INPUT BOX" for lack of a better term. But I suppose you could call it a "Resource Box," "Idea Box," whatever would make you think to actually *put* the stuff *in* it!

When I go to do my weekly planning of our kids' daily reports (see the Record-Keeping Section), I just pull this box out and browse through it, using some of the stuff, or saving it for another week.

Since not everything has a brochure or is in booklet, book or letter form, I needed a place to put *concepts.* Thus the "Input Box" chart on the next page. Here's what ours has in it right now, just to give you an idea for what I mean:

- a cassette tape, "The Writing Christian," that we are listening to *(for "English")*
- a notice for a Town Hall Meeting that Bob Williams (a Christian legislator) will be leading *(for "State Government")*
- a flyer for "Brown Bag Concerts" telling where the Irish Storytelling will be *("Fine Arts")*
- the Oct./Nov. '92 issue of Teaching Home magazine – it has an article on a biblical approach on dating and courtship I want to read to my kids *(for "Spiritual Studies")*
- a monthly letter from Dr. Dobson. *(for "Current Events" or "Spiritual")*
- our "Input Box" sheet, explained on the next page. Here's a label for your box:

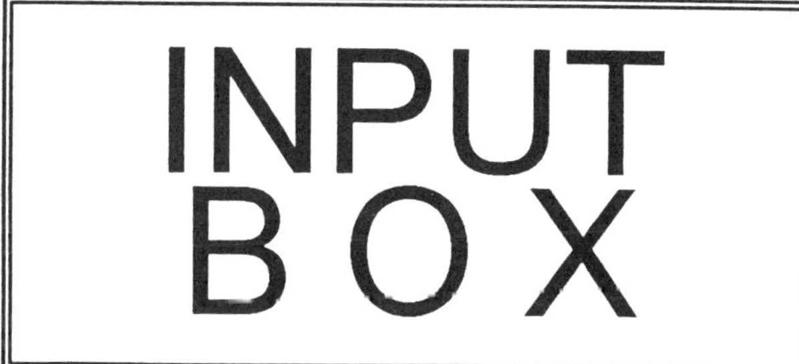

INPUT BOX SHEET

You will be coming up with ideas you want to implement at some point down the road, but there won't be anything tangible about them to put in the box. Out of sight, out of mind for me. So that's why I came up with this sheet. A few examples of what I mean:

Basic Idea	Source	Phone?	Date or Deadline?	Other Notes
Irish Stories Pres.	Daily News	577-3356	March 16	@ McClelland Arts Ctr.
Community Concert: Harp	Season tickets	—	May 28	@ Theatre of Perf Arts
AIDS Seminar	Church	423-1300	Apr. 14	@ Shekinah Chr. Ctr.
Geneology Center	Sharni heard	—	(anytime)	S. wants to go
Mario Murillo	Posters	—	Apr. 9 & 10	@ Zion Christian Ctr.
S's Tap Recital	Tap class	—	June 4	@ Theatre of Perf. Arts
Home School Pres. Nt.	School Biz	—	May 21	@ Lv. Ch. of the Nazarine

Sharnessa, Dave, Carlianne, Barb & Tory Shelton

INPUT BOX

(for the intangibles)

Basic Idea	Source	Phone?	Date or Deadline?	Other Notes

Step 7

Purpose: To provide places for all the "stuff" your student will be producing

"WORK PENDING" FILE

Unless your kids are not like mine, you can expect to accumulate a mass of papers. To decrease blood pressure and extend your life by at least three years *(unless Jesus returns sooner),* start a "Work Pending File." Since we have two high schoolers, one is labeled "Tory's Work Pending" and the other "Sharnessa's Work Pending."

As soon as, let's say, Tory *doesn't* finish a paper, it goes in this file, where it stays safe and protected from the "Do-we-really-need-this?-Let's-just-throw-it-away" scrutiny of Dave No-Clutter Shelton. The other advantage is that it can be found and *finished.* Which may or may not actually be an advantage, depending on who's talking.

INDIVIDUAL TOPIC NOTEBOOKS

OK, let's say our student actually gets a few of those Learning Guides – or *any* written, non-workbook material completed! What do we do with them now? I've already mentioned this in the Potpourri Section (4) under "**A NOTEBOOK METHOD**," but since this book is so vast and there's so much to take in, I want to mention *individual topic notebooks* again here (in case you skipped the conversational portion of that section and headed straight for the fascinating Learning Guides.) As I said there, what you really have is the pages of some unique, written-by-your-student *books!*

See pages IV-4 and 5 (in the Potpourri Section) for details and ideas on how to assemble these.

These two pictures of my Honey 'n' me were actually "try-outs" for the inside back cover. For obvious reasons, the one on the left didn't make it. (In case you couldn't guess, it's because Dave looks like, well, an *egghead* here – which he doesn't at *all* look like in real life!) And in the one on the right he appears to have a double chin, which he *doesn't.* However, I actually look *good* in both pictures! — which is highly unusual for me! I also like being in my Handsome Honey's arms (in the picture *and* in real life!) So I obviously just *had* to use these pictures *somewhere* in this book!

Step 8

Purpose: To make your class plans and records easier to find

If you want some ideas for those course and class titles, you've come to the right place – or step! For my purposes, a *course* is the overall title of a credit, as listed in the requirements for graduation. *(This is explained more in the Glossary at the very end of the manual.)* Then there are also the *class* titles which may or may not be entire courses. They tend to be a more specialized, narrow topic of study. For example, the *class* "Wood Shop" is found under the *course* "Occupational Education." As such, Wood Shop would satisfy the graduation requirement for "Occ. Ed." Likewise, a *class* in "Choir," or a *class* in "Drama" or a *class* in "Music Theory" would satisfy the requirement for a *course* in "Fine Arts." ... Courses are in all-caps and bold; classes are not.

BIBLE/SPIRITUAL	Cooking	**P.E.**
Biblical Studies	Gourmet Cooking	Aerobics
Spritual Learning	Home Improvement	Exercise
ENGLISH	Decorating/Remodeling	Mountain Biking
English Composition	Home Maintenance & Repair	Sports
English Literature	Home Management & Organization	Walking
Grammar/Punctuation	**LIFE SKILLS**	**SCIENCE**
Speech	Building Family Relationships	Biology
Spelling/Vocabulary	Marriage Preparation	Chemistry
FINE ARTS	Matters of Life & Death	Geology
Art / Crafts	Personal Organization	Lab Science
Dance:	Self Esteem & Understanding	Natural Science
Band/Orchestra	Tact, Grace & Social Skills	Physics
Choir	**MATHEMATICS**	**SOCIAL STUDIES**
Drama	Algebra	Ancient Civilizations
Music Appreciation	Applied Math	Civics
Music Theory	Calculus	Current Events
Performing Arts	Consumer Math	Economics
FOREIGN LANGUAGE	Geometry	Educational Foundations
French	Trigonometry	Geography
German	**OCCUPATIONAL EDUCATION**	International Relations
Spanish	Auto Shop	Philosophy of Religion
Latin / Greek	Business Skills	Political Science
HEALTH	Career Interviews	Psychology
AIDS Education	Computer	Sociology
Healthful Living	Metal Shop	U.S. History & Government
Nutrition	Office Skills	World History & Government
HOME ECONOMICS	Typing	*(Your Own)* State History & Gov't.
Carpentry	Wood Shop	Art of Living (under Home Ec.)

Ellyn Davis on...

Relationship Skills, Life Skills, Academic Skills

"In I Corinthians 13 in the Bible, the apostle Paul writes about love and explains how a person may know and do wonderful spiritual things, but if he has no love, it all counts for nothing. We feel a similar analogy can be made about life skills. We may use the best curriculum, know all of our history names and dates, be able to write grammatically correct papers, and even work calculus problems, but what really counts over the long haul of life are relationships and life skills...

Even secular educators are beginning to talk about the necessity for what they call "emotional intelligence" and their studies are finding that people with the highest standardized test scores are often the least likely to be happy and successful in life. In fact, there is mounting evidence that people who have high academic intelligence (who excel in school) often fare poorly at living life. They know how to "crack the system" when it comes to getting good grades in an artificially contrived environment such as school, but they lack the internal resilience to tackle the tough relationships and issues of life. Our goal, as we have home educated our children, has been to find a balance between relational skills, life skills, and academic skills..."

– Ellyn Davis of the Elijah Co.

Section Six - Part D:

COMPLETED CLASS PLANS

And now our own
Completed Class Plans – which
will look quite different on *your* family,
but which I'm presenting ONLY as
examples and bouncing-off points!
I HIGHLY recommend that you
NOT start in this section!

(It will only overwhelm and confuse until and unless
you've read the rest!)

A Panicky Mom's Q's
& D❤nna's A's

(another homeschool mom who's "been there done that")

[PANICKY HOMESCHOOL MOM:] ***"Well, now that I have read Form+U+la, I am totally in a panic; not that I wasn't already shaky before reading it. I feel completely inadequate – feelings I have been fighting for the last couple of years concerning homeschooling – and am ready to throw in the towel, go to the nearest curriculum catalog and purchase a complete high school curriculum with textbooks, tests and record keeping service. (Not that we could afford that either!!)"***

[DONNA RESPONDING:] "Don't panic and don't do any of the above! ;-) I'm here to tell you that you *can* do this, even with only 3 ½ years left and with a limited budget. You have to remember that Barb has been doing this for a while and what you are seeing in her book is the culmination of 6 years of high school experience (of her own kids). (Not to mention 10 years of what she calls "creative, flexible homeschooling before that!) You should not expect to "begin" where she is now. And she doesn't expect you to, either! She just gives us the tools and insight; we take it as we can and as the Lord leads us. So just take it a little at a time. Here's one thing that I did. I picked a subject that Barb had a ACS/TAS sheet on. I used hers to get ideas for mine. Design yours "in pencil" – meaning "not written in stone and can be changed." ☺ Begin with that and add to it or change it as circumstances develop. Use Form+U+la as your "guide" and feel free to change it to fit your family! Also, remember that what looks like bits and pieces of nothing actually will come together to form a complete picture when you get to the part of recording it all. I didn't record my sons activities until he graduated. I was very concerned that we wouldn't have enough to give credit but SURPRISE! When I actually began to gather it all together – we had more than enough!

[PANICKY:] ***I feel so unprepared for this. There are so many more things that I wish I could teach my daughter, and continue guiding her carefully through the path of life, but she's accumulated more than enough credits, and she wants to get on with her life plans. I wish I could go back and start over, knowing what I now know, because I think we could have fit in so much more, and we could have done a much better job of preparing her for it.***

[DONNA:] I felt the same way when it came time to graduate my son! I had to keep reminding myself that I didn't learn all that I know now by the time I graduated from high school!! Our goal should be to give our children such a love of learning and the tools to continue to learn that they will continue to educate themselves. Rest easy in the knowledge that you have done these things and let them take over for a while :-) Have you two discussed a continued season of education after high school similar to what Barb and Tory are doing?

[PANICKY:] ***"Not to mention I am nowhere near capable of putting together a course of study like this for her! It would take me years!"***

[DONNA:] "First of all, why do you feel you need to put together a complete course right now? All you need for now is some sort of framework or a general idea of what you feel is important for your child to know. Build that course over the next 3 to 4 years. One of the beauties of this type of homeschooling is that you don't have to do all of a particular course in a semester. You can do a course over a period of the rest of her high school years, then put it all together on a record.

[PANICKY:] ***"I am totally deflated. I have prayed for wisdom and guidance."***

[DONNA:] "Now is the time to sit back and KNOW that HE is in control. If this comes from the Lord, He will provide all that you need to complete His goals. This is the most difficult part of the whole thing! I know! :-) You need to listen now - be calm in your spirit - forget (or at least put aside for now) the world's ideas of educational requirements. Let Him show you what and how He wants you to complete your child's education. This is the basis of YOUR re-education – and it costs you NOTHING except control! Give that over to Him."

– Praying for you, *Donna Heck*

(Barb's assistant on the Sr. High Form+U+la email list)

Completed Class Plans!

WHAT YOU'LL FIND IN THIS SECTION

For the most part, on the right side of the page you'll see either an "ACS" (Assignment Check-off Sheet) or a "T.A.S." (Task Analysis Sheet), both of which I'll explain momentarily...

True to my typical formatting style, you'll (usually) find the forms on the right and the explanations of these on the left. In these I'll give other ideas and possibilities for the form, and comment on things that you wouldn't know just by looking at them.

However, because of formatting challenges and extra pages of explanations or extra ACS's or TAS's, this will be switched around in a few cases. ☺ So will my formatting style from page to page. There are a variety of reasons for this, which I can't really explain; I just hope you'll be able to overlook the imperfections and inconsistencies and know that I am doing my best with the time and skills I have available to me at this point in my life.

WHAT ARE THE ACS's?

"ACS" is the abbreviation for the title I've come up with for my class plans: "Assignment Check-off Sheet." (The TAS's will be explained in just a minute.) These ACS's are my actual class plans (one sheet per class) except that much of what we did ended up being recorded "after the fact," not before, as the word "plan" might have hinted at. ☺

However, there *will* be more planned ahead for Carlianne (only 12 as of this writing) now that I have done high school with two others – who were very different from each other. So I now have a much broader range of experience and ideas from which to draw! There will still be much room for her unique interests to take root and develop, but I now know more what I want for the "basics" – all of which are included in the ACS's – including the unique-to-each-child particulars as well as the general, "It-doesn't-matter-if-it's-delightful-to-you-or-not, you-still-need-it-for-life" basics!

WHAT ARE THE TAS'S?

"T.A.S." stands for "Task Analysis Sheet." This is what I made up for very specific tasks that I wanted completed or learned, like for "Vehicle Maintenance," "America Online & Email Skills," "Word Processing Skills," and "Carpentry Skills," just to name a few.

Specifically, these sheets gave me a very simple, check-off-able way to say "This is *exactly* what I want learned" instead of just something general like I want you to "Learn how to use a word processor," or "Know how to maintain a car," or "Learn how to use the Internet," or "Learn to cook," or "Get your life organized," or "Learn how to do basic carpentry work," or "Learn how to maintain a home" – *all* of which are included in this section! These give both parents and students the specific parameters and/or lists of things to be learned so that both can know what's expected, what the goals are, and when a general task or area of learning is considered "done."

You won't need these for every class! In fact, many classes won't call for them at all! They are just a tool – a very *optional* tool – for those wanting or needing it in very specific areas. As you examine these forms, you'll know more what I mean and whether or not you need them.

USING THE ACS & TAS SHEETS

I would imagine that most of you are not going to be using these sheets "as is." But if you do find that a class is pretty much what you would do, you can just Wite-out™ any portions of our info that don't apply to you, and nicely print in yours. Then when you go to finalize your own classes, you can re-do it up nicely. Those of you who are adept at chart-making on your computer will want to start from scratch, I'm sure! *(Note about the term "Wite-out": Yes, that's really how they spell it! Shame on them! But I actually use "Liquid Paper" which is the same thing. However whenever I say "just Liquid Paper it," people give me a blank look and say "Whaaaat?!?!")*

MY OTHER REASON FOR DOING THESE...

...forms was because I'm a "historian" and I love to have detailed descriptions of what I've done! If nothing else, just to have it down on paper! (I was inspired by all the detail that Lewis and Clark put down on paper on their trek across America! My trek through the wilderness of homeschooling through high school may not be of quite as much significance as was theirs through the mid-to-western pre-United States, ☺ but it *is* a trek, and I *am* going "where no man has *(or I should say "few men have")* gone before"!

So I'm writin' it all down as I go! It may be for myself; it may be for my kids who want to know more as they start having their own children; it may be for *you*! ☺ But at any rate, if I've said it once *(and I know I have)* I've said it a thousand times, *you* get to hear God on how much *you* need – or don't need – to document, OK? ☺ I show you what I do, and how to do it, but *you* let God show you what's for you, and what's *not*. Your needs, style, and preferences may be entirely different from mine!

NOT A BLUEPRINT!

I just have to say that it really feels very scary – in a vulnerable way – to put my stuff out for the world to see (and scrutinize) because I know people will undoubtedly have judgments, opinions, and misgivings about it, will feel intimidated by these forms because even after all my warnings, they *won't* read the rest of the book first, or else they'll feel that what my children have done is either "too much" or "too little." All I can say is that *my stuff is what it is,* just my stuff. No more, no less. I hope you'll just be able to take it for what it is – just my own ideas – and use it for ideas and inspiration as you see fit. Nothing that you see in this section – or in this *book* for that matter – is intended to be a blueprint for you any more than other people's stuff is a blueprint for me. ☺ ❧

And now for some...

Q'S AND *A*'S ABOUT THESE FORMS

Q: Did your kids actually read all those books you listed on the forms? Or did you read them aloud together?

A: We did both! And, looking closely at many of them, you'll see it says "read selected portions" which means that not all of the book was read. It may have been that I thought only certain "selected portions" were valuable to them, or that we started reading it and then realized it just wasn't what they needed, or could understand, or whatever.

~~~

Q: I can't imagine my kids reading all those books especially on homeschooling.

A: Then don't have them read them! ☺ What works for one family – or child – doesn't for another.

~~~

Q: My daughter who just turned 14 has just discovered reading. She used to dread reading, so last year I backed off and didn't require her to read *anything!* Is this a good idea or not?

A: It's a very *good* idea! One thing I discovered is that who a child is at age 14 is very different from who they are at 18. There is *so* much more maturity and interest! And I had the blessing of having Tory for an extra year! What a blessing!!! He did much more reading in his last two years of high school than the first 3 (or 4)! (Note: We started high school in his 8th grade year, but because of stuff we were going through, he didn't really do a lot. It was a "survival year" in many ways. ☺ And no, I *don't* feel he would have been better off in school. Not at *all!* He still had his daily disciplines, but we/he did much less than I would have liked. It's a long story; listen to my audio tapes if you want more of that. ☺)

Q: (This is actually more of a testimony, continued from the previous "Q".) We continued our family read-aloud time and my son and I continued to model our love of reading. We would discuss books and she would see our excitement and I guess she figured she was missing out on something so she started checking out books from the library. As soon as she finished her first novel by herself, it gave her such confidence and she started tackling more books. She's at the point where she can read a young adult novel in a day.

A: This is such a *perfect* example of how it's supposed to work! Plus the fact that you modeled it for her! (Her daughter has *way* surpassed me! Of course, that's not saying much! I was always a slow reader until recent years. I've now worked my way up to "almost average"!)

~~~

Q: How do I deal with my child resisting doing what I think they should, especially if it's not that much?

A: It's one thing for a child to resist too much work being laid on them, but it's another to resist something just because it isn't what they'd prefer to be doing at the time. Only the parent can discern the "heart of the matter." We *really* have to hear God on all this because there's a very fine line here. We are to train our children, but that can be taken way too far, into a very heavy burden being placed on them – all for the sake of "discipline." I just received a letter yesterday from a mom who said that her husband would have liked their homeschool to operate military style!!! This is *not* God's heart, so my suggestion is that this mom do some loving, non-pushy, gracious talking with her husband, reading pertinent excerpts to him that she finds in her reading, and then *pray!* Part of my job as a parent is to disciple my children, helping them go to the cross of Jesus, which means they need to die to their flesh. We want learning to be delight-directed, but not "flesh directed."
~~~

Q: When I first saw the ACS's, I felt very overwhelmed and inadequate. I know that is not your intention and you don't want us to clone your education for your kids.

A: I want to amen this! It is absolutely *not* my intention, and I do NOT want anyone to try to clone our education for their kids! This is the very reason I greatly hesitated to even put what we had done out for the public to see. While I knew it would help many, I also knew it would overwhelm others and make them feel inadequate – which grieves me as the heart of my whole ministry is *enabling* and *equipping* parents, freeing them up – not *binding* them up! If this is what you are feeling, please read this next part carefully: In this section is an abundance of sheets that will either be a blessing or a curse to you. If you have not read the rest of this book up to this point, or grappled with the issues presented therein, and/or *(preferably "and")* listened to my "Out of Fear & Burnout" and/or read much of the Homeschool Guide-a-log, please do not even *think* of looking at this section! You don't need it yet. It is a *tool* for those considering what to do for high school. You need at least a forming vision for your children before you need these form, as they are intended only to give you ideas for down the road. Possibilities. Food for thought. Brainstorming material. What I present here is *not a blueprint, nor even a recommendation!* And it is *certainly* not "the best way" or "the only way" to create these classes!!! ☺ I hesitate to even let these out because of the way people will inevitably receive them, so that's why I ask – no, I *beg* – you to not even look at them until you sense "it is time." And you will *know* when that is."

~~~

Q: How much of the material do you pick out, and how much do your kids pick out?

A: Tory *wanted* me to pick out almost everything for him, although as he got older I tried to wean him away from needing my input so much. Sharnessa chose many of the books (and studies) herself.

~~~

Q: Do you put what you *both* choose on the same sheet?

A: Yes! Everything, whether chosen by me or them went down on the form the same. And I didn't feel it was necessary to indicate whether she/he or I chose it.

~~~

Q: This looks like so much stuff to do! Did they ever have time to themselves?!

A: Believe it or not, the "amount of material" they did never amounted to more than 3–5 hours a day. There was *plenty* of time to pursue personal interests.

Q: I started to think how does this fit in with the Lifestyle of Learning concept and can't help but wonder: is it too close to falling back into "requirements"? (Someone else asked a similar question:) I know I am supposed to rely on God to lead me in what to do everyday, but I am struggling with what I have read of Marilyn's material and some of the forms and classes that you put together. I know your ministries complement each other, but when you had Tory do the things on those forms, were you leading a Lifestyle of Learning approach? I personally love to learn and want (to have) my children to explore many avenues, but I see the value in delight-directed, long-term commitment to a pursuit. I want to design classes to 'round' out their education. You know, the things they won't pursue on their own. There are just some things they won't do on their own and I need encouragement on how much to "make" them do! Marilyn's message is soooo spiritual. My family has been changed because of it, but I desire to document and see progress. I think my children do too. Since you wrote Form+U+la have you changed your mind about how much input you had in their classes?

A: What we are doing doesn't "fit in" with LOL; it *is* our (and no one else's) Lifestyle of Learning. ☺ Ours is different from Marilyn's. (*And* from what *yours* will be!) For one thing, her kids are not going on to college, at this point. She has said that if her daughter (for instance) were to decide to go to college, she feels she would be able to get her ready for it in about 1½ years. During that time she would be doing more academic, textbookish learning that would prepare her to enter the system. (And this is not necessary for all professions.) However, in many ways, her daughter is already *very* prepared for the system; it takes much more than mere academic expertise! There are "ability to learn" issues, love of learning, ability to research, relate, reason (tools of learning), firm establishment of personal and Christian values, understanding of godly world view, and a solid sense of life purpose. And all of these are being *very* well developed in her children. We have to have a vision of *God's* idea of "success in college" before we can head in the right direction. Too many kids go to college with mere academic prowess, and then fail in *God's* purposes for them being there.

~~~

Q: I know the books are wonderful and far superior to what a text book would contain. I guess I just have to wait until my kids are a little more mature?

A: Probably. Be sure to look closely at the ACS's and TAS's as what I have down there is not all book work! ☺

~~~

Q: When might they be ready for more serious study?

A: You'll be amazed at how much more they will be able and willing to grasp at age 17 and older! I told Tory I wish I'd had the 19-year-old model of him all the way
~~~

through high school! ☺ But we take – and love and work with – whatever we "have in hand", don't we? ☺ They don't start out mature, that's the whole purpose of the *process!* We just have to let God lead that process for *each* of us individually. We're *all* in the process of learning how to hear God's voice. We can expect to make some wrong turns and hear incompletely or imperfectly, but I'm *so* glad that my ability to hear God perfectly isn't the bottom line for pleasing God. His Word says that "without faith it is impossible to please God." So our *faith* – our leaning on God – is of much more concern to Him than our perfection. He is a *redeeming* God, so whatever we turn over to Him He will make the most of. Even if what's gone on in our lives was initiated by the enemy, we know that "what the enemy intended for evil, God will turn for good." *This* is the essence of what His redeeming power is all about! I'm *so* glad it applies to *anything* in my life! ☺

~~~

Q: Are there blank forms to fill out as TAS's or do I have to design my own from scratch?

A: Each form is so different, section and column wise, I felt it was pointless to make up a blank master. There was nothing universal enough. I'd make up a rough draft first, then do it on your computer (or by hand – which I did for years!!!) Be sure to leave plenty of blank spaces and lines for future additions to your "plan."

~~~

Q: When (or if) Tory applied to colleges, did you use his ACS/TAS forms as a type of course description in addition to his transcript or did you write up something else?

A: I would use the transcript, and any ACS forms only if more detailed info on particular classes was requested. Or if I did a portfolio, I might include them in there.

~~~

Q: In her book, *And What About College*, Cafi Cohen wrote a brief course description for her children to include in their college application packet. As I looked through your ACS/TAS sheets, I thought this may be a better way to give course description (more detailed) that would give a better picture of the "class" to a perspective college admissions board. But I was wondering if they may be *too* detailed for that, and if it might be considered over-done?

A: There's no "one way" or "best way" to do any of this documenting stuff. ☺ I made up the ACS forms for two main reasons, and neither of them was for a college admissions board as anyone on such a board would, as you suggest, most likely not need or desire that much detail! They have hundreds, if not thousands of applications to sift through, and at that point would probably not be needing that much detail. No, I did them for myself and for your homeschool moms – because I have been asked "What do (did) you (your kids) do, and how did you document it? – this non-traditional learning?" When I am trying to do something that no one has ever done before, I like as much detail as I can possibly get! This is what was *not* available to me when I wrote my book. I don't want a "blueprint", I just want to see how someone else organized and structured and documented it, what kinds of things they did, and from there I can take off and do my own thing! Just to give you another perspective, Cafi Cohen has told me she thinks I have more recording keeping stuff than is actually needed, in particular, to get to college. She is concerned that people will look at my stuff and assume they need to "do it all." I told her (don't worry, *kindly*) that I make it very clear all the way through my book that what I'm offering is a smorgasbord from which people can pick and choose what *they* need for *their* own purposes. They don't *have* to use it all. She countered that the mere presence of "so much" gives a "message." At this point I decided we were on two different "pages" in our thinking, and I didn't want for it to get tense. If I were to continue, what I would say is that I believe we simply have to recognize that there are "different strokes for different folks." What she presents in her book, which is great, would not have been nearly enough for *me.* (But it didn't come out until 4 years after mine anyway. ☺) The record keeping options she outlines for you (she doesn't provide masters, but explains how to do them) may very well be enough for you! And if so, go for it! But it wouldn't have been for *me.* Unfortunately, though, no one else had done what I wanted to do when I was first embarking on high school! That's what happens when you're a pioneer! ☺ (Diana McAllister's book *Homeschooling the Highschooler* was still in rough draft form (which she gave me a copy of!); Mary Schofield's book *The High School Handbook* was ⅓ the size it is now, and I didn't see it until I was already down the road a few miles. I knew a grand total of *three* other families who were homeschooling through high school, and they were *all* doing "the straight workbook thing." I knew I didn't want a "total diet" of that for my kids! We hadn't been homeschooling in freedom and creativity for nine years only to give it all up now!

~~~

Q: Which grade do you suggest I start using these?

A: I'd say no sooner than ninth grade. For junior high, use the format maybe to a simpler extent, if you really feel the need. Remember, though, most of those were "built" as we went. You don't want to just lock your child into what you want for them, though there's going to be a certain amount of that – which will vary from family to family and even child to child, depending on how self-motivated they are. You want to retain a lot of flexibility and room for their own interests to develop. To understand how all this comes into play, *please* read *Wisdom's Way!* This message will give you *vision!*

~~~

Q: A bad habit of mine is that I go through new books starting at the end! *Soooo* my frustration came at seeing some of the class content of Tory's schedules.
~~~

A: This is a good time to suspend that "bad habit"! ☺ And it's a prime example of why it's not only a good idea, but *imperative* that you read this book in order. ☺ Everything I say long *before* you ever get to the class plans will totally "color" what you see once you get to those. And just to clarify, they're not "schedules" at all, but are *class content plans* that were mostly done "after the fact"; which means, I did a lot of the "planning" as we went, and recorded *after* it was done! If you don't read *all* that comes *before* the forms, you will not get the "heart and soul" of my message, and *that's* what's going to free you up! This is *not* a binding message! It's very freeing, but you can't look at just the forms and get that impression from them.

~~~

Q: Doing reports on more than 10 different wars, the amount of reading, listening to tapes, watching videos and CD programs all in 60 hours!!! He just must be a whiz?!

A: Not at all. He's wonderful, and has a good mind, but, and this is not to discredit him in the *least,* but he's really not a "whiz" or "exceptional" academically *at all.* And he'd be the first to tell you that. However, he *is* exceptional in integrity, cuteness, in-love-with-Jesus-ness, *and* at basketball, *I* think! ☺ (And besides, the War Reports and most others just don't take all that long to do. They're mostly just one pagers.)

~~~

Q: I can't imagine that much work from my boys!

A: It's amazing what you can do a little at a time! Tory seldom did more than three hours of academic stuff in a day!!! And if he *did* do more, it was only because he *hadn't* done enough on other days and was catching up. I've heard from several other moms that they, too, were amazed at how all the bits and pieces add up! Take a closer look at many of the ACS's and you'll see that there are many "Read selected portions of [a book]. This means he did not read the whole book, but only portions of it, because that was all that I felt was necessary, or because he (or we) just wasn't "connecting" with it.

~~~

Q: Does he choose what he wants to do from all that and keep track of his hours spent?

A: What went onto the ACS's was what he actually *did.* However, he kept track of time, not on those, but on his "WAP" (Weekly Assignment Plan.)

~~~

Q: I am not giving up, mind you, but I do wish I had a partner nearby to do some planning with. I just don't like sitting alone with all this info and resources.

A: Two things on this: This is exactly why I wrote in such a conversational manner. I knew the feeling of being alone and "helpless" quite well, as there was *nothing* available when I first started! I remember sitting on the couch crying because I was so overwhelmed! I didn't know of anyone who was doing it the way I was thinking it ought to be done – with a lot of freedom and individuality! ...And also, you are *not* "partnerless" or alone, dear sister! Just like me, you have the Lord to help you. ☺ And if you aren't adept at hearing His voice (as I wasn't, and am still learning!), then *this* is as good an opportunity as any! In fact, this is how God teaches us – on the job! Hearing His voice is not a privilege reserved for only the "spiritually elite"! (whoever *they* are) It is one of your basic privileges as a child of God! "My sheep hear My voice." Are you one of His sheep? Then you get to hear His voice! What better place to apply and start learning this privilege!

~~~

Q: I believe I have enough books around here to cover the four years of college too! How will we ever use them all?!?! There's so much more good stuff to do than we'll *ever* have time to *do!!!*

A: Actually, that's exactly what we found too! There was no way we could do all that I hoped for in high school! So Tory is currently using the other "excess" resources now! You may not have such a "luxury" as this, so you'll need to ask God (your "Master Curriculum Counselor" remember?!) what your and His priorities should be. Tory graduated last year (1998), and this year is doing what we're calling "a year of in-depth discipleship and studies." He was free to do whatever he wanted, (work, local community college, go away to "whatever" training he desired, like Christ for the Nations, YWAM, etc. However sitting at home and doing nothing was *not* an option! ☺) But he came to us for counsel on what to do that following year. (Yay!) I didn't feel he was ready for the secular college scene, not even the local community college – your faith can get eaten up there as fast as in a larger university), and the funds weren't there for him to go away to a Christian college, so we set up a tailor-made plan for him. (See a sample of one week in Section 3) He is currently (as I write) doing this season, which is preparation for life, his calling (which we're all in the process of trying to hear God on), and possibly college next year. He's also working at a couple of part-time jobs and helping with the youth ministry. We believe God will show him (and us) the way as we progress. Many things are solidifying in him this year, spiritually, mentally and physically; and I – both Dave and I are – *really* pleased with what we're seeing in him!

~~~

Q: I am not easily overwhelmed by your stuff, but how do I deal with the fact that I *am* overwhelmed by all that I feel I need to do?

A: I'm glad you aren't overwhelmed by my stuff because it was designed to *ease* the burden! Not *add* to it! ☺ As for feeling overwhelmed by all there is to do, this is where we need to know that God does not overwhelm us. First, we need to get His vision for us from *Him.* And then once we have it – or start getting it – it is vital that we hear Him on what to do when; in what order. He doesn't expect us to implement the whole vision in one day!

Q: How few hours would you have for an area of study, or a "unit," and still record or consider this as its own separate class?

A: There are no "rules" (remember, we're *pioneers!),* but personally if he'd had only – but at least – 60 hours, for ½ credit, I would have recorded that, *unless* he needed the hours to make up the one "Spiritual Studies" class. So, again (as you've heard me say a thousand times), you make your system serve your and God's purposes; not you serve the system! ☺

Woops!

We are starting right out by changing things on you! I had said that the explanation would be on the left, and the form on the right. Well, nice idea, but because there are two pages of Algebra classes, (Algebra I and Algebra II) and two pages of the "Building Life Relationships" Explanation, this gets us completely "off" for a few pages! (And, of course, wouldn't it be the first few pages!) So the explanation for both Algebra I and II is on page (8), the ACS for Algebra I is on the page to the right (like it's s'posed to be) then and the ACS for Algebra II is on page 9, which puts the first page of the Building Life Relationships explanation on the wrong page! But it gets back on track right after that!

(And it gets even worse in spots! ☺ But you'll figure it out!)

 **ASSIGNMENT CHECK-OFF SHEET** ✓✓✓✓✓✓✓✓✓✓✓✓✓✓✓✓✓✓✓✓✓✓✓✓✓✓✓✓✓

Course: **Algebra I**	
Student: **Tory W. Shelton**	*Class Value:* **1 Credit = 120 Hours**

Directions: Complete, correct, and file each of the Daily Assignments in the textbook: Saxon Algebra 1/2: An Incremental Development, (which is equivalent to "Algebra I"). Check off the corresponding number at bottom of page when all the above is done for each.

Course Content Itemization

Whole number place value / Decimal numbers
Reading and writing whole numbers
Addition / The number ray / Ordering
Rounding whole numbers / Subtraction / Patterns
Multiplication / Division / The pattern
Addition and subtraction word problems
Multiplying and dividing by powers of 10
Dividing, ordering, and rounding decimal numbers
Points, lines, and rays / Angles / Perimeter
Divisibility / Word problems about equal groups
Prime numbers and composite numbers
Products of primes / Three-digit divisors
Multiplication word problems / Fractions
Expanding and reducing fractions
Fractions to decimals / Decimals to fractions
Rectangular area / Products of primes in cancellation
Multiplying fractions / Dividing fractions
Multiples / Average / Multiple fractional factors
English units / Unit multipliers / Subtraction
Whole number place value / Patterns
Reading and writing whole numbers / Addition
The number ray / Ordering
Rounding whole numbers
Multiplication / Division / The pattern
Addition and subtraction word problems
Multiplying and dividing by powers of 10
Dividing, ordering, and rounding decimal numbers
Points, lines, and rays / Angles / Perimeter
Divisibility / Decimal numbers
Word problems about equal groups
Prime numbers and composite numbers
Products of primes / Average in word problems
Three-digit divisors / Multiplication word problems
Fractions / Expanding and reducing fractions
Fractions to decimals / Decimals to fractions
Rectangular area / Products of primes in cancellation
Multiplying fractions / Dividing fractions / Multiples
Average / Multiple fractional factors
English units / Unit multipliers
Metric length conversions / Area as a difference

Areas of triangles / Least common multiple
Improper fractions and mixed numbers
Multiplying fractions and whole numbers
Fractional part of a number / Graphs
Adding fractions / Order of operations
Variables and evaluation / Multiple unit multipliers
Adding mixed numbers / Rate
Mixed number subtraction / Rate word problems
Equations; answers and solutions/ Equivalent equations / Addition-subtraction rule for equations
Reciprocals / Multiplication rule / Overall average
Parentheses / Division in order of operations
Multiplying and dividing mixed numbers
Division rule for equations / Exponents and roots
Volume / Order of operations with fractions
Evaluation of exponential expressions and radicals
Fractional part of a number / Fractional equations
Surface area / Brackets / Scientific notation
Decimal part of a number / Percent
Fractions and symbols of inclusion
Ratio and proportion - P^Q and $\sqrt[Q]{P}$
Decimals, fractions, and percent - Reference numbers
Equations w/ mixed numbers / Mixed no. problems
The distance problem / Proportions with fractions
Circles / Circumference & pi
Fractional-part word problems / Changing rates
Adding signed numbers
Using both rules to solve equations
Semicircles / Proportions with mixed numbers
Ratio word problems / Using ratios to compare
Forms of the percent equation / Percents under 100
Negative numbers / Absolute value
Visualizing signed / Powers and roots of fractions
Graphing inequalities / Right circular cylinders
Inserting parentheses / Order of addition
Implied ratios / Multiplication with scientific notation
Percents greater than 100 / Increases in percent
Multiplication and division of signed numbers
Algebraic addition - Using mental parentheses
Mode, median, and mean

Opposites / Evaluation with signed numbers
Rate problems as ratio problems
Formats for equations / Negative coefficients
Algebraic phrases / Surface area of a right solid
Trichotomy / Symbols of negation
Number word problems / Advanced ratio problems
Operations with signed numbers
Roots by cut and try / Order of division
Adding like terms / Variables on both sides
Two-step problems / Unequal quantities
Exponents and signed numbers
Multiplication of exponential expressions
Distributive property / Estimating higher-order roots
Powers with negative bases / Roots of negative numbers Negative exponents / Percent as a rate
Roman numerals / Percent word problems
Complex decimal numbers / Fractional percents
Simple interest / Compound interest
Markup - Markdown / Commission - Profit
Complementary and supplementary angles
Measuring angles / Copying angles / Construction
Triangles / Measures
Roman numerals greater than 50
Probability / Measuring lengths / Similar triangles
Probability: Independent events
Multiple-term equations / Polygons / Congruence
Parallelograms / Trapezoids
Perpendicular bisectors
Equations with X2 / Pythagorean theorem
Constructing triangles / Angle bisectors
English volume conversions / Base 2
Metric volume / Rectangular coordinates
Numerals and numbers / Larger than and greater than Pyramids, cones, and spheres
Advanced equations
Forming solids / Symmetry / Permutations
Adding in base 10 and base 2 / Equations of a line
Graphing a line
Appendix A Additional practice sets
Appendix B Glossary

Lesson Check-off Chart *(Check off lessons as they are completed.)*

1	2	3	4	5	6	7	8	9	10	11	12	13	14	15	16	17	18	19	20
21	22	23	24	25	26	27	28	29	30	31	32	33	34	35	36	37	38	39	40
41	42	43	44	45	46	47	48	49	50	51	52	53	54	55	56	57	58	59	60
61	62	63	64	65	66	67	68	69	70	71	72	73	74	75	76	77	78	79	80
81	82	83	84	85	86	87	88	89	90	91	92	93	94	95	96	97	98	99	100
101	102	103	104	105	106	107	108	109	110	111	112	113	114	115	116	117	118	119	120
121	122	123	124	125	126	127	128	129	130	131	132	133	134	135	136	137	APPENDIX A: Add'l Practice Sets		APPENDIX B: Glossary

[A.C.S.]
ALGEBRA I and II

I have to admit that this is about the *least* exciting of all my ACS's, content-wise and appearance-wise, so what a way to start. (Sorry, if any of you are math lovers and this is your favorite.) But I did have fun doing it, as I learned a new computer skill in the process!

For this one I got to use my Mom's scanner – which looks like a small copy machine. In fact, it *is!* ☺ Except that it's hooked up to a computer, so whatever you copy goes right into your computer! Yes, *very* spiffy! In this case, I scanned the table of contents at the beginning of the Saxon Math books, "Algebra1/2" and "Algebra 1."

SO WHY DID I NAME THEM ALGEBRA I & II?

Because in talking with the administrator of a home-school umbrella school, I found out that Saxon ½ is essentially the freshman year of Algebra, 1 credit worth; and Saxon Algebra 1 is the same as what they get in regu-lar schools the next year. Since "1/2" would give the impression that only ½ credit has been earned, I opted for a different numbering/titling system.

Hint from this administrator: She said the lessons should take *about* one hour each for the student to complete. If they take a lot longer, the student is probably in too difficult a book; if a lot shorter, it's too easy. But in the latter case, you should progress a ways through the book before deciding to move to a higher level book – or have the student take a few of the tests.

OK, TO FINISH WITH WHAT I SCANNED IN

On the ACS (Assignment Check-off Sheet) you see three columns, filled with all sorts of Algebraic terms. Here's how it got that way: After scanning in the table of contents, (by just laying the book face down on the glass of the scanner/copier), I "cleaned it up" – which you typically have to do when scanning anything. This just means you fix up anything that mistranslated in the process as the scanner sometimes (invariably) misreads some things. (Like it might read a "G" as a "6.")

Then I created my three-column table (chart) on the ACS, and set the font at a very small size. (I think this is maybe 5 point?) I then cut-and-pasted portions of the scanned table of contents into each of these three columns, pulling up terms from the line below that would fit on the next line up, just to make it all fit. That's why there are no individual chapter titles with the terms. They were there originally, but I deleted them. I felt it wasn't necessary for the purpose of this ACS to include what was being presented in which chapter; I just wanted what *all* the student was learning listed here in one place (page) so that if a math teacher down the road wanted to know what they learned, they could see it right here on one nice, neat page, and I wouldn't have to go hunt down my Saxon book from the neighbor I loaned it to who just loaned it to her homeschooling cousin who lives on the other side of the country and just moved. ☺ It's *all* right here! *(Note: I also saved several of Tory's work papers. He writes very small, sectioning the problems into their own little boxes as he goes.)*

LESSON CHECK-OFF CHART

OK, next look down at the very bottom of the page... There you see a chart with a bunch of boxes, starting with number 1. This is the *only* time you will see number 1 at the *beginning* of the chart, so don't get used to it! ☺ (I'll explain *why* when we get to it in the next ACS sheet.) Here I have just listed all the lessons of the book, from 1 to 137 plus Appendix A and B. So this is to check off lessons done, *not* time – as the rest of the ACS's will have, which is why all the rest of them are called "Hour Check-off Charts."

✓ ASSIGNMENT CHECK-OFF SHEET ✓✓✓✓✓✓✓✓✓✓✓✓✓✓✓✓✓✓✓✓✓✓✓✓✓✓✓✓✓✓✓✓✓✓✓✓✓✓

Course: **Algebra II**

Student: **Tory W. Shelton** | *Class Value:* **1 Credit = 120 Hours**

Resource/Directions: Complete, correct, and file each of the Daily Assignments in the textbook: Saxon Algebra 1: An Incremental Development. (equivalent to "Algebra II") Check off the corresponding number at bottom of page when all the above is done for each.

Course Content Itemization

Addition and subtraction of fractions
Lines and segments / Geometry review
Perimeter Area / Geometric shapes / Volume
Degree measure
Real numbers and the number line
Multiplication of fractions / Division of fractions
Unit multipliers / Sets / Surface area
Absolute value / Addition on the number line
Rules for addition / Definition of subtraction
Opposites and multiple signs
Rules for mult. of signed numbers
Inverse operations / Division by zero
Rules for multiplication & division
Exchange of factors in multiplication
Conversions of area and volume
Reciprocal and multiplicative inverse
Order of operations / Symbols of inclusion
Order of operations / Multiple symbols of inclusion
More on order of operations
Products of signed numbers
Evaluation of algebraic expressions
More complicated evaluations
Terms and the distributive property
Like terms; Addition of like terms
Exponents / Roots / Powers of negative numbers
Evaluation of powers / Product rule for exponents
Like terms with exponents / Statements & sentences
Conditional & Equivalent equations
Additive property of equality
Multiplicative property of equality
Solution of equations / More complicated equations
More on the distributive property
Simplifying decimal equations
Fractional parts of numbers / Negative exponents
Zero exponents / Decimal parts of a number
Volume conversions / Algebraic phrases
Equations w/ parentheses / Word problems
Products of prime factors / Greatest common factor
Statements about unequal quantities

Factoring the greatest common factor
Canceling / Multiplying fractions
Minus signs & negative exponents
Graphing inequalities / Ratio / Trichotomy axiom
Negated inequalities / Advanced ratio problems
Quotient rule for exponents / Distributive property
Negative exponents / 2-step problems
Like terms and negative exponents
Solving multivariable equations
Least Common Multiples / ...of algebraic expressions
Addition of rational expressions
Addition of abstract fractions
Conjunctions / Percents < 100
Polynomials; Addition of polynomials
Multiplication of polynomials
Percents greater than 100
Rectangular coordinates / Graphs of linear equations
Vertical and horizontal lines
Overall average / Addition of rational expressions
Power rule for exponents
Substitution axiom / Simultaneous equations
Dividing fractions / Rearranging before graphing
More on area and volume / Percent word problems
Rearranging before substitution / Subsets
Square roots / Evaluation using plus or minus
Product of square roots rule
Additive property of inequality
Addition of radical expressions
Weighted average / Domain
Simplification of radical expressions
Elimination / Set notation
More about complex fractions / Factoring trinomials
Trinomials with common factors
Subscripted variables / Factors that are sums
Factoring the difference of 2 squares
Scientific notation / Closure / Consecutive integers
Consecutive even and odd integers
Fraction and decimal word problems
Rational equations / Graphical solutions

Systems of equations with subscripted variables
Operations with scientific notation
Writing the equation of a line
Coin problems / Value problems
Multiplication of radicals / Division of polynomials
More on systems of equations
More on division of polynomials
Solution of quadratic equations by factoring
Slope-intercept method of graphing
Word problems with two statements of equality
Multiplicative property of inequality
Uniform motion problems about equal distances
Properties of the set of real numbers
Rational expressions / Pythagorean theorem
Uniform motion problems of the form $D_1 + D_2 = N$
Difference of two squares theorem
Distance between two points
Algebraic proofs / Rounding numbers
Uniform motion-unequal distances
Square roots of large numbers
Square root tables / Factorable denominators
Absolute value inequalities
Rational equations / Abstract rational equations
Equation of a line through two points
Functions / Functional notation
Line parallel to a given line
Equation of a line with a given slope
Radical equations / Slope formula
Consistent, inconsistent, dependent equations
More on conjunctions / Disjunctions
Linear inequalities / Probability
More on multiplication of radical expressions
Direct variation / Inverse variation
y^2 key - Exponential increases
Quotient rule for square roots
Advanced trinomal factoring / Factoring by grouping
Direct and inverse variation squared
Completing the square
The quadratic formula

Hour Check-off Chart *(Check off lessons as they are completed.)*

1	2	3	4	5	6	7	8	9	10	11	12	13	14	15	16	17	18	19	20
21	22	23	24	25	26	27	28	29	30	31	32	33	34	35	36	37	38	39	40
41	42	43	44	45	46	47	48	49	50	51	52	53	54	55	56	57	58	59	60
61	62	63	64	65	66	67	68	69	70	71	72	73	74	75	76	77	78	79	80
81	82	83	84	85	86	87	88	89	90	91	92	93	94	95	96	97	98	99	100
101	102	103	104	105	106	107	108	109	110	111	112	113	114	115	116	117	118	119	120
121	122	123	124	125	126	127	128	129	130	131	132	133	134	135	136	137	APPENDIX A: Add'l Practice Sets		APPENDIX B: Glossary

[A.C.S] Building Life Relationships

This is the class I created as a "platform" upon which to present the (my) ideals for pre-marriage, marriage and general life relationships. Because we are generally learning the ideals for marriage and life relationships in our home setting, there is less need for formal study in the area of "**Family, Marriage, and General Life Relationships**" *(see bottom section)* than there might be in a formal educational setting. But, because the world's idea of "**Pre-marriage Relationships and Attitudes**" is so foreign to God's idea, and because it takes a lot to come out of it when you are living daily in the midst of it, there is more "formal study" in that area, as you can tell by just a quick look at the form. ☺

SELF ESTEEM AND UNDERSTANDING

This was originally going to be its own separate class by the same title, ½ credit, but it turned out it just didn't take that long to do what I ended up wanting done. Starting at the top of the ACS, this is the first section you see. This is my "re-name" for "psychology," yet it's not *really* psychology in the world's sense of the word either. I just wanted to include a bit of study in the area of "self." Some might object to such a study on the grounds that this seems "self centered." I do not believe it is at all! – although someone could surely make it so. Understanding yourself – and how and why you think certain ways – is different from focusing on and serving (primarily) yourself.

Of course, this "study" can get carried away, as can any study, but I believe a few basics are vital to developing a good self image – which I do not see as being selfish or worldly – and developing healthy relationships with others. Jesus says "Love your neighbor as yourself." Loving myself includes understanding, at least to a certain degree, what makes me tick and also what aspect of "me" needs to die – as in "dying to self." It is this aspect that has yielded the most health in my life: "Unless a seed falls into the ground and dies, it cannot bear fruit."

PRE-MARRIAGE RELATIONSHIPS & ATTITUDES

Many of the resources here ended up being our "favorites" in this area. The idea behind having Tory do an "Essay on your own beliefs, hopes, intentions and goals in the area of relationships with the opposite sex" was to get him to *think* about this and to also *write*. It ended up that his beliefs are very strong – to the extent that he even lovingly confronts close friends who are compromising. So that, coupled with the fact that he did enough writing in other areas I hadn't planned on lead me to delete this assignment from his requirements. ☺ Again, flexibility is important!

In this section you'll see many resources. It took *many* books and much discussion to lay the foundation. In a society that sees dating as a God-given right, anyone who chooses to *not* be part of the dating scene is swimming upstream in a downstream world! And if you're trying to drag kids who'd *rather* be going downstream, it can be very difficult! I suggest you *not* wait until high school to start! Age 10 or 11 is not a moment too soon to begin a focused season of teaching and discipleship in this area!

But the most ideal timing is to get your own mind renewed in this area when your children are small – as in "in utero"! ☺ For this I recommend Pastor Reb Bradley's tape set *("Preparation for Courtship & Marriage")*, which is foundational! Even though it is directed to parents with small children, if the foundation has not yet been laid, *(which it hadn't with our family)* then *anytime* is appropriate for these tapes! – for teens to listen to as well. The "bigger picture" has *so* much more to do with the heart and purity than it does with any "method"! All the books I mentioned above deal with just that – the *heart* of the matter. I highly recommend Eric and Leslie Ludy's "Romance God's Way" seminar! They are *wonderful* and their seminar is funny, heart-warming, Jesus-centered. Their phone number is 303-702-0177. And their email address is: rgen@indra.com. *(Their ministry is "R Generation.")*

FAMILY, MARRIAGE, & GENERAL LIFE RELATIONSHIPS

In this section you'll see a variety of resources, all of which we are now carrying! *(including the ones mentioned above)* All three activities in the "Do" and "Plan/Lead" sections are right here in Form+U+la. (*Note: When You Can't Get Along* is out of print.)

This was a ½ credit class for us, but it could easily be a full credit. I wouldn't go so much by a goal, but by how long it takes to get to the place where you're satisfied with the results in your child's heart, which is totally unpredictable. It took us ½ credit to get there, though more time was probably actually spent than that.

It certainly doesn't work to cram things down your kids' throats. But at the same time too many parents give up on trying to disciple their children in this area, even though *they* are convinced of the rightness of it, simply because they don't feel they can change their children's minds. It's our *job* to change their minds – *and* hearts, but we have to follow the leading of the Holy Spirit if we want to do it His way and in His timing, and not get in His way!

✓ ASSIGNMENT CHECK-OFF SHEET ✓✓✓✓✓✓✓✓✓✓✓✓✓✓✓✓✓✓✓✓✓✓✓✓✓✓✓✓✓✓✓✓✓✓✓✓✓✓✓

Class: **Building Life Relationships**	
Student: **Tory W. Shelton**	Class Value: **½ Credit = 60 Hours**

BASIC REQUIREMENTS CHECK-OFF LIST

SELF ESTEEM AND UNDERSTANDING:	
READ:	☐ Life on the Edge by Dr. James Dobson ☐ Carpe Diem by Tony Campollo
TAPES:	☐ "New Attitude" by Joshua Harris ☐ "How to Set Values That Lead to Victory" by Dennis Peacock
PRE-MARRIAGE RELATIONSHIPS AND ATTITUDES:	
READ BOOKS:	☐ His Perfect Faithfulness and ☐ Romance God's Way (both) by Eric and Leslie Ludy by Eric & Leslie Ludy
	☐ Christian Courtship Vs. the Dating Game by Pas. Jim West ☐ Dating With Integrity by John Holzman
	☐ Finding the Love of Your Life by Neil Clark Warren, Ph.D. ☐ I Kissed Dating Goodbye by Joshua Harris
	☐ Of Knights and Fair Maidens by Jeff and Danielle Myers ☐ Passion and Purity by Elisabeth Elliot
READ BOOKLETS:	☐ "Establishing Biblical Standards of Courtship" (ATIA) ☐ "The Art of Choosing Your Mate" by Pastor Jim West
	☐ "Guidelines for Courtship and Preparation for Marriage" by Jeff Barth ☐ "Dating or Courting" by Allen Jones
READ ARTICLES:	☐ "A Love Story" by Matt and Jenny Chancey ☐ "Handling Committment & Responsibility"
	☐ "Dating Problems; Courtship Solutions" in New Attitude Magazine (Vol.1 No.2; 1993)
	☐ "Focus on the Family" mag article: "Courtship Makes a Comeback" by Jim and Annie Ryun
	☐ "Emotional Fornication" in New Attitude Magazine (Vol. 2 No. 3; Fall 1994)
LISTEN TO TAPES:	☐ "I Kissed Dating Goodbye" and... ☐ "A New Attitude for Courtship" by Joshua Harris
	☐ "Godly Courtship" 6-tape series by Denny Kennaston ☐ "Letting God Choose Your Mate"
	☐ "Preparation for Courtship & Marriage" 6-tape series by Pastor Reb Bradley
	☐ "Scriptural Romance" (Emotional, Physical & Spiritual Purity) 3-tape series by Jonathan Lindvall
ATTEND:	☐ Workshop: "Courtship Instead of Dating: A Better Way" by Gregg Harris
	☐ Attend "Romance God's Way Seminar" presented by Eric and Leslie Ludy
WRITE:	☐ Essay on your own vision, beliefs, hopes, intentions and goals in the area of relationships with the opposite sex
FAMILY, MARRIAGE, AND GENERAL LIFE RELATIONSHIPS:	
READ BOOKS:	☐ When You Can't Get Along by Gloria Chisholm ☐ Love is a Decision by Gary Smalley
	☐ Quick to Listen, Slow to Speak by Robert E. Fisher
WATCH: VIDEOS:	☐ "Making Love Work" seminar by Barbara De Angelis, Ph.D. ☐ "How to Handle Your Parents" by Dawson McAllister
	☐ "Love is a Decision: Hidden Keys to Loving Relationships" video-seminar by Gary Smalley"
LISTEN TO:	☐ Tape series: "Back to Patriarchy" - a "Conference for Men of Vision"
	☐ Tape: "Sports & Godly Families" by Jonathan Lindvall
ATTEND:	☐ Workshop: "Men and Godly Roles" by Stu Weber *(at conference by Gregg Harris and Chris Klicka)*
DO:	☐ ☐ 2 "Open Line 'n' Mind Evaluations" with siblings ☐ ☐ 2 "Home Reports" over a period of several mos.
PLAN/LEAD:	☐ ☐ 2 Family Meetings - using the "Family Meeting Guide"

HOUR CHECK-OFF CHART *(Check off hours as they are completed.)*

60	59	58	57	56	55	54	53	52	51	50	49	48	47	46	45	44	43	42	41
40	39	38	37	36	35	34	33	32	31	30	29	28	27	26	25	24	23	22	21
20	19	18	17	16	15	14	13	12	11	10	9	8	7	6	5	4	3	2	1

[A.C.S] BUSINESS & OFFICE SKILLS

I did not decide ahead that this would be a ½ credit class; I just had him do the various activities that were my ideas, recorded the ones he was already doing that were his ideas, and made the most of opportunities that came our way – no one's ideas, just "opportunity knocking." Sometimes one of these "knocks" would pre-empt another class's activity on the WAP (Weekly Assignment Plan). The key to "seizing the moment" is f-l-e-x-i-b-i-l-i-t-y. Other activities on here *were* able to be planned ahead.

I divided this class into several different areas of learning, skills, and/or experiences, as you can see listed in the left column. I did not start out with these areas in mind; I started with the experiences and skills and then just recently, when most of the "class content" had already taken place, broke everything down and gave it the various (left column) categories.

The first one, under "Seminars" – right at the top – is Gregg Harris' "Home & Family Business Workshop" which our family attended, and which I highly recommend! (Contact Christian Life Workshops at 503-667-3942 for info on that. They may have it available on audio tape!)

At first glance, this class looks very – uhhh, well "business-ish"! In actuality it *was*, but largely from a real-life perspective! Look closely at each of the items listed. Interspersed throughout the list you'll see a variety of items relating to our own family business which involves book producing (collating, comb-binding), photocopying, processing brochures and mailings, making calls, receiving and sending shipments and packages, and each of these activities are in their own different categories.

Tory has helped us set up our booth at curriculum fairs, helped man it, and was completely in charge of setting up a booth for Stil-Tees, a T-shirt company you may have seen at homeschool curriculum fairs.I divided this booth experience among "**Sales Experience & Skills**"; "**Mail Order Skills**"; "**Communication Skills**" ("Contact with Customers at Booth"); and "**Record Keeping**" (the first item listed). I *could* have just put all this under the single title of "Booth Experience," but because he actually learned a wide variety of skills in this one experience, I wanted the various skills enumerated. (Remember where I talked about "Dissect and Distribute"? – a method for creating a class? This is an application of that concept! I divided (dissected) this booth experience and distributed it among the various categories. An experience can typically be distributed among several different classes, or, as in this case, several different categories in one class. ☺ As usual, my ideas are ONLY to get your creative juices flowing!

What about "Organize Ski Trip for Youth Group at Church" under "**Organization Skills**"? Why include that in this class? Because the *skill* is the exact same thing he would do in an office setting! Whether the end result is a staff meeting, an appointment with the auditors, or a ski trip, all the same organizational skills are involved in putting the event together.

Also in the same category, as Captain of several different teams he was on over the years, Tory was involved in a variety of informal leadership roles and experiences. Since these require traits that are important to one involved in business -- tactfulness, working with others, integrity – I felt it was appropriate to include those experiences here.

Look under "**Photocopying Skills**" and see where it says "Have Color Copies Custom-made to our specifications." This was when we took some items that were going into Tory's Basketball Notebook down to get color copies of them. We needed them to be reduced to fit into the notebook, so we had experience with that, as well as with adjusting color -- which you will most likely have to be aware of when having a color copy made, or you could end up something that looks like the person has the flu.

Under "**Record Keeping**" it says "Make up Roster of Basketball Team..." Again, this was out of Tory's real life and something he did as "el capitán" of the team. He also got in some time under his "Computer" class for this same activity because he had to make up a "table" (chart) to do this roster. Now, let me ask you: Do you think he would have learned how to do charts any better – or any more "officially" – had he done it in a formal "Computer" class as opposed to doing it for something out of his real life? No! If you don't know anyone who could teach your child, then great, let a textbook have the honor, but which way do you think he's more likely to remember it better/longer? (Just trying to chalk up a few points for "real life learning" and pry a few white knuckles off a textbook. ☺)

Under the very last category of "**Business & Finance Skills,**" you'll see the two chapters out of "Applied Mathematics" which is an A Beka t-t-text-b-b-book!!! ☺ Yes, we actually used a t-t-textbook! But not in the "regular" way. Because Tory had all the Math (I felt) he needed in Saxon ½ and 1 (textbooks – which we *did* use the regular way), he did not need the math presented in this book. But I *did* want him to have the basic knowledge contained therein. Each chapter starts with a nice explanation of a particular concept. So we did two of the five units in the book for this class (the other three will appear elsewhere in these ACS's). I also included the concepts that were included under each of these two unit titles because I wanted to indicate here (on this ACS) that these concepts had been introduced in the class. They are not presented in depth in the A Beka book, but it does give a good overview and introduction.

For all activities in this "class"we recorded the time spent, but I must quickly add that we were *not* rigid or legalistic about it! "Roughly 4 hours for this" and "about 1 hour for that" sufficed. *(If anything, Tory was more exact than I felt necessary, but that's just him.)* Don't let yourself get bogged down being a slave to the clock! I'm sure we overestimated at times, and underestimated at other times. Some moms (kids) just round it to the nearest ½-hour, others to the nearest ¼-hour. "It all comes out in the wash." ☺

✓ **ASSIGNMENT CHECK-OFF SHEET** ✓✓✓

Class: **Business & Office Skills**	
Student: **Tory W. Shelton**	*Class Value:* **½ Credit = 60 Hours**

CATEGORIES:	BASIC REQUIREMENTS CHECK-OFF LIST
SEMINARS	☐ Attend "Home & Family Business Workshop" presented by Gregg Harris in Portland, OR
SALES EXPERIENCE & SKILLS	☐ Set up and man booth for Stil-Tee's T-shirts (at '97 W.H.O. Convention Curriculum Fair)
	☐ Help Set up and Man Our Booth at Curriculum Fair (at '95 W.H.O. Convention Curriculum Fair)
	☐ Write up Receipts for Individual Sales in receipt book
MAIL ORDER SKILLS	☐ Receive/Unload Truck Shipment of Books (from Tennessee: Sr. High Form+U+la and Lab Science)
	☐ Package Up Incoming Orders Applying Appropriate Postage (being sent via the Post Office)
	☐ Send Packages via UPS and ☐ Post Office (First Class and Special 4th Class)
ORGANIZATION SKILLS	☐ Organize Ski Trip for Youth Group at Church
	☐ ☐ ☐ Head or Co-Captain of Basketball & Soccer Teams (BB Head Capt. 1x, Co-capt. 1x; Soccer Co-capt. 1x)
COMMUNICATION SKILLS	☐ Make calls for Home-based Education Video-Correspondence Course
	☐ Write/Send Sponsor Request Letters for Basketball Funding ☐ Contact with customers at booths
BOOK PRODUCTION	☐ Collate Books (Senior High Form+U+la) and Packets ("Senior High Homeschooling Options & Resources")
	☐ Comb-bind Books (Field Trips, Booklet Building Book, Jumpstart Navigator)
RECORD KEEPING	☐ Keep records of sales and WA State taxes for Stil-Tee's T-shirts booth (at '97 W.H.O. Curr. Fair)
	☐ Make up Roster of Basketball Team (player's names, player number, parents, addresses, phone)
	☐ Set up a Student Record for Video-Correspondence Course ☐ Maintain (follow up on) Student Rec
	☐ Record Completed Student Worksheets for Course and Prepare Packg. for Certificate
	(See also the "Business & Finance Skills & Principles" Section below)
MISC. OFFICE WORK	☐ Process (Fold and Staple) Brochures According to Instructions
	☐ Prepare brochures for mailing (address, stamp)
PHOTOCOPYING SKILLS	☐ Copy Books (our self-published ones) and Handouts Using the Sorting Bins
	☐ Have Color Copies Custom-made to our specifications (size)
BUSINESS & FINANCE SKILLS & PRINCIPLES	☐ Record Deposit(s) on Deposit Record and Extend to Appropriate Columns
	☐ Read these topics in "Applied Mathematics": *(Do Exercises &/or further study of anything that looks helpful)* STEWARDSHIP (UNIT 1): ☐ Keeping Money Records (p.3) ☐ Cash Records (8) ☐ Assets (13) ☐ Liabilities (20) ☐ Capital (24) ☐ The Balance Sheet (28) ☐ Income, Cost of Goods Sold, and Operating Expensis (39) ☐ Profit or Loss (49) ☐ The Income Statement (55) ☐ Debits and Credits (67) ☐ The General Journal (78) ☐ The Ledger (92) MANAGING YOUR BUSINESS (UNIT 2): ☐ Ratio Analysis (119) ☐ Percent Analysis (128) ☐ Percent Analysis & Porportions (133) ☐ Average Collection Turnover (141) ☐ Inventory Turnover (150) ☐ Break-Even Point (154) ☐ Break-Even Point Using the Formula (162) ☐ Trade Discount (168) ☐ Trade Credit (174) ☐ Storage and Inventory (185) ☐ More English Measures (197) ☐ Metric Measures (204) ☐ Metric-English Conversions (208)

HOUR CHECK-OFF CHART *(Check off hours as they are completed.)*

60	59	58	57	56	55	54	53	52	51	50	49	48	47	46	45	44	43	42	41
40	39	38	37	36	35	34	33	32	31	30	29	28	27	26	25	24	23	22	21
20	19	18	17	16	15	14	13	12	11	10	9	8	7	6	5	4	3	2	1

[A.C.S]
CHRISTIAN ETHICS

This class was the one class that Tory attended at a Christian school. We didn't start out with him wanting to take a class there, or with us wanting him to do so, but rather it was required that he take one class in order to be able to participate in the school's – you guessed it – *basketball* program! ☺ He ended up doing the class the whole year, not just basketball season, and also participated in soccer during that season..

I *could* have just put "Attend classes at Columbia Heights Christian School" and had that be the entire content for this ACS, but that just would have looked way to sparse! Concise. Brief. *(Bleah!)* None of which I am familiar with. ☺ So I *had* to add some "meat" to the sheet. ☺ Thus we have the other requirements...

I have since taken off the requirement "Read Understanding the Times" because, realizing too late in the year what a huge and in-depth book it is, Tory is studying this book in-depth this year. (It's spring of 1999 as I write, and Tory is currently doing a year of in-depth discipleship and college-prep studies.) They didn't really use the book much in class, even though it and the workbook were the required texts, which disappointed me, but the class itself was good anyway.

Tory also took newspaper clippings he had cut out of the newspaper for "current events" – for which he got extra credit in his class. I have since added boxes across that line (box) – as many as he had done, which he checked off. So I did this "after the fact" (or "post-facto" as I say in my Lab Science book) as opposed to assigning a certain number of current events ahead of time. This means we just add a box every time he does one. ☺

This *also* means that this form (as is true for ALL my ACS forms) is actually a rough draft until the very end of high school. Now, if you are not doing workshops or writing books, you would probably have no reason to re-do all these rough drafts once all the work has been done, toward the very end of high school. Just leave your rough drafts as exactly that – *unless* you want to include them in a nice-looking portfolio, or have friends or relatives you want to impress. (I know this is challenging for you perfectionists – to not re-do the whole sheet – but I'm just trying to keep you from going crazy, and at least release you from feeling you *have* to re-do them once you're all done!)

I added "Take required quizzes / tests" just because it looks good and takes up another line. ☺ But seriously, he did a lot of study for and work on those, and since we did very little testing elsewhere, I thought it was good to put it here.

We had read *Where Do I Draw the Line* by Michael Farris some time ago (during high school), but because it fit under this title, I just added it here, to this class. It could just as easily have gone under "Social Studies" or "U.S. History & Government."

Just because I have a little extra space here, I want to take a moment to brag a bit. ☺ I actually have a purpose – *besides* getting to boast about my son – but first I'll share the incident... When Tory went to someone's home to deliver some books, he found a little post-a-note attached to the front door. The mom is a woman who had been on a mission trip to Mexico with Tory, then 18 (a whole group from our church went too), and she knew he was coming to deliver some books, but wasn't home when he came. So she left a note saying: "I'm so bummed I missed you! I brag about you to anyone who will listen. You are such an AWESOME young man!" With all the forms and charts and plans you see in this book, I want you to know that *this* is the kind of accomplishment and "test" I value most! "If I speak with tongues of men and of angels, and check off all my Assignment Check-off Sheets, but have not love, I am but a clanging cymbal." *(or something pretty close to that!?! ☺)* ❤

✓ *ASSIGNMENT CHECK-OFF SHEET* ✓✓

Course / Class / Unit: **Christian Ethics**	
Student: **Tory W. Shelton**	*Class Value:* **1 Credit = 120 Hours**

Resources / Instructions: Attend classes at Columbia Heights Christian School, read required book, and take tests. Bring in current events clippings for extra credit as desired.

BASIC REQUIREMENTS CHECK-OFF LIST	Done
Attend classes at Columbia Heights Christian School	
Read Understanding the Times	
Take current events clippings to class	
Take required quizzes / tests	
Read Where Do I Draw the Line? by Michael Farris	

HOUR CHECK-OFF CHART

(Check off hours as they are completed.)

120	119	118	117	116	115	114	113	112	111	110	109	108	107	106	105	104	103	102	101
100	99	98	97	96	95	94	93	92	91	90	89	88	87	86	85	84	83	82	81
80	79	78	77	76	75	74	73	72	71	70	69	68	67	66	65	64	63	62	61
60	59	58	57	56	55	54	53	52	51	50	49	48	47	46	45	44	43	42	41
40	39	38	37	36	35	34	33	32	31	30	29	28	27	26	25	24	23	22	21
20	19	18	17	16	15	14	13	12	11	10	9	8	7	6	5	4	3	2	1

[A.C.S]
Computer Skills Survey

Why the term "survey" in this title? I used this term because it indicates a broad spectrum of study areas, not just one major area. (You'll be getting a slight variation on this explanation in the "Lab Science Survey" class.) Rather our class included a wide variety of studies over the years that included areas that I'll be presenting momentarily. If my child were going into a computer profession, this would not have been enough. But it was for us.

This class turned out to be just ½ credit, but it can surely be just about *any* amount your student needs or determines according to interest. Had our son been more into "Computer Land," he would have had a whole credit of this class.

In the "**GENERAL KNOWLEDGE & BACKGROUND**" section (the top one), there are various books to be read. These just give a basic understanding of the computer, how it works, history, and background.

In the "**PRACTICAL APPLICATION**" section I have listed six different "T.A.S.'s" (Task Analysis Sheets) for the areas I wanted particular skills to be learned in, such as "Word Processing," "WordPerfect," "General Computer Literacy Skills," "Windows Literacy Skills," (if you have a Mac, you'd use this to make up a similar TAS for that), "America Online E-mail and Forum Operations," and "Internet Literacy Skills." Each one of these TAS's will directly follow this class ACS.

I don't think I need to go through those individually as they are pretty self explanatory. But I do need to explain why, for some of the tasks, there is a single box next to the item to be done/learned *(or sometimes more than one box)*, and why, for other skills, there are three columns headed at the top by "I / P / M." I explain this in the Q's and A's coming up right after this, on page 18. *(It's the first "Q" (question) in the left column.)*

On to the "**COMPUTER TERMS**" section of the ACS to the right... There are two items to do here, and both are on computer vocabulary. You and/or the student make these up according to the computer you have and programs you use.

The next section, "**SCANNER**," is simply how to operate a scanner. We didn't have one, so we went to my Mom's where she taught *me* first, and then I taught Tory!

Then in "**DISKETTE APPLICATIONS**" we have learning how to operate two particular programs, a card-maker program and a map-maker program.

In the "**CD-ROM APPLICATIONS & EXPERIENCE**" section I have listed all the CD-Rom programs I wanted our son to learn. Here's where your student can really rack up some hours, if needed or if so desired!

I know of parents who just assigned credits based on the student's expertise. This is essentially the same as "challenging a course" at a college. *(See pg. VI-B-6; left column.)* Again, the sky's the limit. Remember in doing all this that "all things are lawful, but not all things are profitable." Ask the Lord what is "profitable" and beneficial for *your* child. My heart and intent are not just to try to just "get by" with what I (or they) can; I want my children to *learn*, and I want "the hours of their day to be filled with good and pleasant (*and* valuable, *and* productive) things."

✓ ASSIGNMENT CHECK-OFF SHEET ✓✓✓✓✓✓✓✓✓✓✓✓✓✓✓✓✓✓✓✓✓✓✓✓✓✓✓✓✓✓✓✓✓✓✓✓✓✓

Class: **Computer Skills Survey**	
Student: **Tory W. Shelton**	*Class Value:* **½ Credit = 60 Hours**

BASIC REQUIREMENTS CHECK-OFF LIST

GENERAL KNOWLEDGE & BACKGROUND	☐ Read and do selected portions of Computer Applications workbook
	☐ Complete quiz on page 46 of Computer Applications workbook
	☐ Read Computer (an Usborne Young Scientist book) *(Use for Vocabulary Quiz reference)*
	☐ Read portions of Home Schooling and The Internet by Lynn Coleman *(Use for Vocab. Quiz ref.)*
	☐ Complete selected assignments in A Beka Using the Personal Computer *(Use for Voc. Quiz ref.)*
PRACTICAL APPLICATION: ♦ INTERNET ♦ EMAIL ♦ WINDOWS ♦ WORD PROCESSING	☐ Learn "***WordPerfect Operations***" up to indicated levels *(See Task Analysis Sheet = T.A.S.)*
	☐ Complete "***Word Processing Application Exercises***" up to indicated levels *(See T.A.S.)*
	☐ Learn "***General Computer Literacy Skills***" up to indicated levels *(See T.A.S.)*
	☐ Learn "***Windows Literacy Skills***" up to indicated levels *(See T.A.S.)*
	☐ Learn "***America Online E-mail & Forum Operations***" up to indicated levels *(See T.A.S.)*
	☐ Learn "***Internet Literacy Skills***" *(See T.A.S.)*
COMPUTER TERMS	☐ Make up one "Computer Terms" Vocabulary Quiz *(Do assigned and selected-by-you words)*
	☐ Complete 2 "Computer Terms" Vocabulary Quizzes *(Your own and Mom's)*
SCANNER	☐ Operate a Scanner and Program *(This is a scanner that is like a regular copy machine and is interfaced with a computer printer. Be able to copy item from the scanner into the program, edit ("clean it up"), print, save data)*
DISKETTE APPLICATIONS	☐ Operate a "Card-maker" program *(Grampa Shelton)*
	☐ Operate a "Map Maker" program *(Grampa Shelton, Gramma Edtl, or Merritts)*
CD-ROM APPLICATIONS AND EXPERIENCE	☐ Operate and Explore the Dorling Kindersley Biology CD-Rom: "The Ultimate Human Body"
	☐ Operate and Explore a CD-Rom Bible Study program *(Compton's)*
	☐ Operate and Explore a CD-Rom Encyclopedia *(Webster's)*
	☐ Operate and Explore a CD-Rom 3-D Atlas program
	☐ Operate and Explore a CD-Rom Interactive Spanish program *(Pro-One)*
	☐ Operate and Explore a CD-Rom Interactive Grammar program *(Pro-One)*
	☐ Operate and Explore a CD-Rom Interactive Vocabulary program *(Pro-One)*

HOUR CHECK-OFF CHART *(Check off hours as they are completed.)*

60	59	58	57	56	55	54	53	52	51	50	49	48	47	46	45	44	43	42	41
40	39	38	37	36	35	34	33	32	31	30	29	28	27	26	25	24	23	22	21
20	19	18	17	16	15	14	13	12	11	10	9	8	7	6	5	4	3	2	1

[T.A.S.]
COMPUTER
TASK ANALYSIS SHEETS

INTRO TO THESE TAS's

I'm not explaining every item on the sheets that follow because all they are is what I personally did, shown to you (a.k.a. "Homeschooling America") solely as an example of how to structure and organize whatever *you* are doing; to help you see how to get a learning experience down on paper. Modify these for your own purposes and needs, of course. So even if you don't understand everything on each one of these TAS's, it's not going to matter (generally speaking) because you don't need to be doing exactly what we're doing, and there is plenty of other stuff on them to give you ample ideas. Just edit for your own use and priorities.

I do not want people seeing my stuff as The Best Way to do a particular class. It worked for us. Period. I'm not an expert in this arena (or *any* arena, for that matter!); all I'm doing is hearing God for *my* kids, not anyone else's. Okay? ☺ I just don't want people following my examples *thinking* that I'm an expert, and then having them or their kids mad at me when they get into some class or take some test down the road, and find out they don't have what they needed. This can and will happen with *any* type of educational background, of course, but I just want to make sure you understand where I'm coming from, and the extent of my experience! ☺ (What you just heard was my "disclaimer.")

On the other hand, with that said, I know that some simply *want* to use my stuff and understand it as well as they can, because, with so much of the groundwork already laid out, they may as well take advantage of it, right?!? ☺ So here are a few Q's and A's... (I'll now revert back to the Q and A format used earlier in this section since that's how this explanation originated!)

Q What do the letters I, P & M stand for? I imagine these are the Skill Levels you refer to on the Basic Requirements Check-Off List. Are they Introduce, P (for ???), and M for Master? Give me a hint. ☺

A You're exactly on track! "I, P, and M" stand for Introduce, Practice (or Participate), and Master. *(I deal with this in more depth in my "Three Levels of Learning" article toward the end of Section 7, and mention it on page JP-2, in the second bullet down, so you can find out more detail about this in those places.)*

Q What's a Flash Session? (⅔ down in the left column of the "America Online..." TAS) Is that AOL specific?

A Yes. A Flash Session is just setting it up to go and retrieve the messages and/or send out ones you have ready to send. And yes, the term "Flash Session" *is* AOL specific, although other programs do the exact same thing either automatically (as does Eudora) or they may have another name for it.

~~~

Q What do you mean by the second-to-the-last skill on the Windows TAS which say: "REM - Be able to REM and UN-REM an application that is overloading the system"?

A I don't know what REM *actually* stands for, but I think of it as "Rapid Eye Movement" which has to do with sleep. *That*, in essence, is what you do to a program when you type in REM in front of it – you "put it to sleep" – and this is in a command line which I couldn't begin to tell you how to do. I'd have to call my computer-wiz uncle again to find out. I just know it helps unburden the system of active applications when it's getting overloaded. Since I now have a more powerful computer, I haven't needed to do this, and hopefully *won't* for a long time!

~~~

Q Do you have a booklist for this Computer Class? The titles you mention aren't enough for me to determine what type of book (library, textbook, etc.) or where to find it. Sometimes we can substitute books but I don't know what these books provide. Now, I realize that neither your addendum nor Form+U+La were designed to create complete courses, but you came so close for computers. Could you share those resources with us?

A We are now selling a few that I think are excellent and that I wish I'd had at the beginning when I was trying to learn all this stuff, but was totally overwhelmed and needed someone to take me by the hand. Well, that's exactly what these books do! They are *very* user-friendly and take you by the hand, step-by-step! *(And I still have far to go that these books will still help me with!)* The Dummies books are funny and absolutely fantastic! They offer just about every title you could imagine: *Word 97 for Dummies, WordPerfect for Dummies, Mac for Dummies, PC's for Dummies, Windows 95 (and 98 soon) for Dummies*, and more! (And they don't stop with computer!!! Oh no! That's just the beginning! There's *Herbal Remedies for Dummies*, *Fishing for Dummies*, *Piano for Dummies*, *Car Care for Dummies*, *Auto Repair for Dummies*, *Alternative Medicine for Dummies*, *Italian Cooking for Dummies*, etc. etc. etc. You name it, they've got it!!! Well, not everything, but a LOT!! Even *Buying a Car for Dummies*!! Prices and ordering info etc. are in my *Guide-a-log*.)

There's another book by the same company, *not* a "Dummies" book, but rather is in a "cartoon" form (this *darling* little diskette guy does all the "talking." It's called *Internet and the World Wide Web Simplified.* And another, like it, is *Computers Simplified.* One other, not the cartoon guy format, but in very clearly presented form with step-by-step instructions is *Teach Yourself Computers and the Internet Visually.*

Right after
the last of the five Computer T.A.S.'s
(which is the one for Windows)
since I have one page left over that I would
just hate to leave blank,
there's a little "bonus" for you!– and perhaps
for your child(ren) too!?
It's a short article called "Email 101"!
It can't help you with operating your particular
program, of course,
but it gives tips for writing
and setting them up, and those
cute little faces!

America Online E-mail & Forum Operations T.A.S.

Skill	I	P	M
BASICS: Access America Online			
Edit "Set-up" including Phone Access			
Use Keyboard to access areas			
Contact AOL Help about a problem/question			
Change own password			
E-MAIL FUNCTIONS: Check for E-mail			
Delete Unwanted E-mail			
Retrieve E-mail			
Access Favorite Places: File/Re-file E-mails			
Create New File Folders			
Re-name and Delete File Folders			
Create New E-mail			
Reply to E-mails			
Forward Emails			
Check Mail You've sent (online)			
Create Blind Carbon Copy			
Use Flash Session to Download E-mails			
Perform a Search (name, word, date) in File Manager			
Use Outgoing & Incoming Flash Mail bins			
Contact Aol Help with a problem or question			
LIVE FORUMS: Participate in a Live Forum			
Check "Who's Here" during Forum			
Access Interactive Online Game (Basketball Trivia)			
IM Someone During a Live Forum			
Access Online Interactive Game			
Contact "AOL Live" (Technical Support)			
LIBRARIES: Access/Peruse Online Library			
Download, View, Print Library Files			

Skill	I	P	M
FORUMS / BULLETIN BOARDS: Access 3:	☐	☐	☐
Access the Folders of each Bulletin Board			
Save a Forum Message by Creating new email and sending it to your own email address			
Access the chat area of any forum			
Access and Browse Forum Messages			
Respond to a Forum Message ("Add a Message")			
Create a new Forum Message ("Post message")			
Respond by email within a forum			
ADDRESS BOOK: Access Address Book			
Access Add. Book from E-mail and Use to Set up			
Add to and Edit Address Book			
IMING: I.M. Online; Check to see if there			
(Once responded to): Converse On-line			
Set up Buddy List			
PROFILES: Access Someone's Profile			
Create Your Own Profile			
Edit Your Own Profile			
Search Profiles for: City, specific interest			
EMAIL LISTS/LOOPS: Know what an e-mail loop is			
Understand the Diff. betw. a "List" and a "Loop"			
Be able to participate (at least briefly) in an E-mail List			
Subscribe to a List			
Unsubscribe from a List			
ONLINE NEWS AREAS: Access Online News			
Access Online Sports News			
SCREEN NAMES: Know How to Create & Delete			
AOL PREFERENCES: Be Able to Access & Edit			

General Computer Literacy Skills T.A.S.

SKILLS	I	P	M
Data Filing: Saving, Labeling, Retrieving			
Access Subdirectories			
Move Files			
Email: Know what it is and how to use, create, send, receive *(See separate "America Online Email & Forum Operations" T.A.S.)*			
Internet: Know what it is and how to access *(See separate "Internet Proficiency" T.A.S.)*			
Word Processing: Acquire proficiency for personal need/use *(See separate "WordPerfect Operations" T.A.S.)*			
Functions of Computers: Know various functions including: ☐ Word Processing ☐ Email ☐ Internet ☐ Data Spreadsheets			
Data Spreadsheets: Know what these are, how they are used, advantages			☐
DOS: Know what it is, basic fundamentals of operation			☐
Windows: Know what it is, and the basic fundamentals of operation *(See separate "Windows Literacy Skills" T.A.S.)*			☐
CD Roms: Understand basic functions and how to install (partially) and operate			☐
Drives: Understand the basics of A: B: C: D: and E:			☐
Printer: Know the various types and purposes, and how to use *yours* ☐ Dot Matrix ☐ Ink Jet ☐ Laser ☐ Color			
Scanners: Know the differences between earlier types and current (hand-held; copier)			☐
Computer Set-up: Know how to set up/connect Monitor, Unit, Keyboard			☐
Disk Formatting: Know how to format a disk from either DOS or Windows			☐
Mouse: Know what one is, differences (hand-held, ball, finger-operated), how to use *yours*			☐
Back-up Systems: Understand the basic types and pro's and con's of back-up systems (tape; zip)			☐
Install Diskette Application Program: Observe and/or do			☐

"Commune With Me" Guide

This got put here because I forgot to include it in the newest book-version of the Jumpstarter, (specifically the first section of this book) and there was literally NO other place in this entire book to put it but right here!

Remember, this is not a "To Do" list, but just several "heart starters"!

"COMMUNE WITH ME"

SIT IN JESUS' LAP and let him speak words of love to you. "Daddy God, I rest in your arms; Jesus just hold me against your heart; Spirit, enfold me within your wings; You are my place of contentment and peace." (To the tune of: "Jesus, How Lovely You Are" – or just make up your own! ☺)

GIVE THANKS - for specific blessings. "I will enter His gates with thanksgiving in my heart..."

PRAISE HIM - "...I will enter His courts with praise!" for who He is and everything He has done for you!

ASK FORGIVENESS - "Forgive us our sins..." Be specific as the Holy Spirit reminds you of what you need to ask forgiveness for. Consider this verse: "Who shall stand in God's holy place? He who has clean hands and a pure heart."

SET MY WILL TO FORGIVE - "...as *(meaning: "in the same manner as")* we forgive those who sin against us." Help me give grace and mercy to those I live with, "neighbors," the Body, slow or rude drivers, waitresses, *anyone* who hurts, offends, or sins against me.

WORSHIP - He doesn't care how I sound or look! He says (in Song of Sol.) "Let me see your face, let me hear your voice, for your voice is sweet and your face is lovely." Just worship God in Spirit and truth for who He is!

ASK FOR MY DAILY BREAD - "Give [me] this day [my] daily bread": which is whatever needs You see that I have spiritually, emotionally, physically, health-wise, time, etc.

PRAY - for family, friends (saved and unsaved) the Body of Christ, etc. as Jesus prayed in the garden: "Keep them from the evil one... May God's joy be fulfilled in them... Keep them in God's Name... Sanctify them in the Truth." Pray favorite Scripture verses over them.

PRAY A HEDGE OF PROTECTION around all the preceding and our home, health, "stuff," etc.

PUT ON MY ARMOR - as in Ephesians 6:10–17.

"EAT" THE WORD OF GOD - Read out of Psalms and/or Proverbs and/or the New Testament. Ask God to give me His "rema" out of His "Logos."

JOURNAL - Refer to journal for specific lists for/of any of the above. Write in journal out of my heart, during the good *and* the bad times. Listen for God to speak, and write down what He speaks to me.

ROTATE DEVOTIONAL MATERIAL: a good book, on-going prayer lists, further journaling, time in other areas of the Bible, communing with God... how ever He leads!

Internet Literacy Skills T.A.S.

Basic Skills	
Understand what the "World Wide Web" (WWW) is and how it basically functions	☐
Know the functions and differences of various online services (See "Online Mind Map")	☐
Understand the concept of "Surfing the Net"	☐
"Surf the Net" by Web Site Address (http://www...)	☐
"Surf the Net" by Category and Sub-category *(High School Basketball / Men's)*	☐
Try three different "web crawlers" *(and/or determine your preference and what you will use for what purposes)*	☐ ☐ ☐
Adopt "I will set no evil thing before my eyes" as a prerequisite to surfing the net *(Make private covenant with God)*	☐
Know how to create and send email from three different servers (AOL, Juno, Eudora)	☐ ☐ ☐
Know how to retrieve, answer, and file email on these servers	☐ ☐ ☐
Understand what "links" are and be able to use them. Do 5:	☐ ☐ ☐ ☐ ☐
Know how to answer an email from a web site. Do three:	☐ ☐ ☐
Be able to cut-and-paste text from a website, transfer it to email server, send to another computer	☐
Be able to retrieve above email in AOL for Windows, translate to ASCII file, retrieve in WordPerfect for DOS, and reformat	☐
Know how to access America Online and the various services it offers. Access 5 areas:	☐ ☐ ☐ ☐ ☐
Know what a "web site" is and how to access one. Access 10 web sites:	☐ ☐ ☐ ☐ ☐ ☐ ☐ ☐ ☐ ☐
Be able to follow instructions or work out a gliche with a technician over the phone *(3 times)*	☐ ☐ ☐
Participate in a live Internet Chat (Room)	☐
Access, create, and send an online postcard. Do three:	☐ ☐ ☐

Windows Literacy Skills T.A.S.

Basic Skills	I	P	M
Access and Use "FILE" Feature			
♦ New File (Create)			
♦ Open File			
♦ Save File			
♦ Save File as an ASCII File ("Save As")			
♦ Retrieve File saved as an ASCII File			
♦ Print			
♦ Print Set-up			
♦ Page Set-up			
♦ Change between "Portrait" & "Landscape"			
♦ Exit/Close a document			
Access "EDIT" Features (may be "Tools")			
♦ Undo			
♦ Cut			
♦ Copy			
♦ Paste			
Access "CHARACTER" Features			
♦ Bold			
♦ Italic			
♦ Underline			
♦ Superscript			
♦ Subscript			
♦ Change Font			
♦ Reduce Font (Using auto feature)			
♦ Enlarge Font (Using auto feature)			
♦ Reduce and Enlarge Font by number			
Access "PARAGRAPH" Features			
♦ Change Justification (Left, Right, Centered, Full)			
♦ Change Spacing			
♦ Change Indents			

Basic Skills	I	P	M
Access "CONTROL PANEL"			
♦ Color (Change the color combinations of the desktop)			
♦ Mouse Driver (know this is where mouse is connected/configured			
♦ Fonts (Be able to select different fonts and change sizes)			
♦ Time/Date (know how to change both)			
♦ Printer (understand that this is where printer is connected)			
♦ Desktop (Access/Change Screen saver, Wallpaper, Cursor Blink Rate)			
Access "FILE MANAGER"			
♦ Change location of doc's using the drag feature			
♦ Understand what it is and how it can be used			
♦ Know how to format a disk from here			
Access and Use "CALCULATOR"			
♦ Practice operations			
Access "CLOCK"			
♦ Change between "Analog" and "Digital"			
Access and Use "CHARACTER MAP"			
Access "DOCUMENT" Features			
♦ Change Margins			
♦ Change Tabs			
MAXIMIZE A WINDOW			
MINIMIZE A WINDOW			
MOVE A WINDOW using the bar across the top			
EXPAND / SHRINK A WINDOW, L & R, T & B			
ACCESS SEVERAL WINDOWS AT ONCE / SWITCH			
ARRANGE WINDOWS IN "TILE" OR "CASCADE"			
ACCESS / USE "HELP" Do 5 times: ☐ ☐ ☐ ☐ ☐			
WINDOWS TUTORIAL - Go through it. ☐			
REM - Be able to REM and UN-REM an application that is overloading the system			
WYSIWYG: Understand what this means ☐			

WordPerfect Operations T.A.S.

Skill	I	P	M	Skill	I	P	M
Start WP after Failure/Closure				Use Spell-check			
Page Formatting Options				Make Hard Hyphen (as to use in phone no.) (Home then "-")			
List, Find, Retrieve Files				Create Soft (adapting) Hyphen (Control + Hyphen)			
Bold feature				Re-set Tabs			
Underline feature				Use Arrow Keys to Navigate			
Italic feature				Search Forward & Backward			
Move Files (as from C: to A:)				Page-down and Up			
Delete Files				Home-Home Up/Down/L/R			
Fonts Choices - Set up				Use Alt Keys			
Font Sizes – Change				Use "Go to" to Find Page			
Tabbing & Re-setting tabs				Set up Page Numbers			
Top and Bottom Margins				Set up Headers & Footers			
Print Out a Document				Super/Subscript (3rd, H_2O)			
Print Options				View Doc – (4 sizes)			
Blocking				Use Thesaurus			
Move & Copy / Cut & paste				Switch Doc(ument)			
Center Text				Justification – L, R, Ctr.			
Change Page Directions: ▬ ▮				Widow/Orphan Options			
Close/Exit/Save Documents				Control-V Symbols			
Label and File Files				Search Feature			
Use Outline Feature				Create Line (w/ "graphic")			
Set up Columns				Redline and Strikeout			
Cancel/Undelete *(Use as copy)*				Double Underline			
Date Insert				Indent L and R			
Change Upper and Lower Cases				Create User Box			
Flush Right				Import Graphic into User Box			
Double Flush Right (.....) (as for Table of Contents)				Set up an Envelope			
Left & Right Margins				Set up Page for Labels			
Access and Use "Reveal Codes"				Create and Use Macros			
Create Tables; Change Columns				Create Footnotes			
Shade, Join, Split cells				Merge/Sort Documents			
Use Shadow and Outline				Search and Replace			
Set and Change Spacing				Use Shadow and Outline			

"EMAIL 101"

There are a couple of formats that are universally accepted ways of responding to emails.* (See note immediately following this section for the other one; I don't want to confuse you.) One is to put what the person wrote inside double arrows: <<like this>> ...and then respond to that. (I add the row of periods right after the last two arrows only because it's easier to spot where I start talking.) You don't *have* to use this format, it's just very very handy for the reader, so they (I) can know exactly what you're talking about and responding to. It's more "conversational" and is especially handy for those of us who handle hundreds of emails a week and have a short memory! ;-D People often come back, responding to an email from me, and say "Sure! That'd be great!" and since that was 48 emails ago, I haven't got a CLUE as to what they're talking about, so I have to go back and hunt through my emails (IF I haven't already deleted it!) and find out what in the world they were talking about! {:-/

* OK, the *other* format for "quoted text" (not to confuse you) is one that is automatically done by your email program, and that is a single arrow > at the beginning of each line of quoted text. This is fine too. My email program (on AOL) gives us both options, and I just happen to prefer the double-arrow format. If you have to do it manually, double arrows at the beginning and end are much faster than putting an arrow in front of each line.

There are lots of expressive faces that add humor and expression to a text-only note. Like these over in the next column, for example:

;-D = laughing/chuckling

;-) = winking

:-) = smiling

}-) = smiling with eyes closed; maybe blushing or sleeping with sweet dreams

{:-/ = puzzled

}-Q = sleeping and drooling

8-O = surprised

<g> = grin

<G> = big grin

There are ALL sorts of these little faces, and you can even make up your own! ;-)

Also, you might want to be aware that A CONTINUOUS USE OF ALL-CAPS is considered YELLING. So try to AVOID that unless you're REALLY wanting to be heard or make a STRONG point. ;-) If used for just one word, it's the same as bold; a little stronger.

And putting asterisks (*) around anything (usually just a word or short phrase) is the same as italicizing it – for emphasis.

A chuckle: ☺ I was already familiar with "Lifestyle of Learning" and commonly abbreviated it as "LOL" long before I started doing email, so the commonly-used email abbreviation of "LOL" was a bit confusing to me when I first saw it in an email, and I knew they weren't talking about Lifestyle of Learning! ☺ In email land, LOL means "Laughing Out Loud"! So when in "mixed company," like on forums, I use "L of L" instead of "LOL" to refer to Lifestyle of Learning. (Or just write it out.) And I usually do the laughing face: ;-D for "laughing out loud." FWIW! (= "for what it's worth) ;-)

And lastly, here's a "KEY" for several of the common email abbreviations that you'll see in emails, "chat rooms," BBS's (Bulletin Board Services), etc.:

♥♥♥♥♥

QUICK REFERENCE KEY TO COMMON EMAIL TERMS:

BFN = Bye For Now

BRB = Be Right Back

FWIW = For What It's Worth

FYI = For Your Information

IM(H)O = In My (Humble) Opinion

LOL = Laughing Out Loud

ROFL = Rolling On Floor Laughing

TIA or TA = Thanks In Advance

TIFN = That's It For Now

TTYL = Talk To You Later

[A.C.S]
EDUCATIONAL FOUNDATIONS

The motivating idea behind this class was that I didn't want my kids to just *be* homeschooled and not understand what was behind it. As they got older, I would hear comments that indicated to me that they really didn't understand what was behind what they were – and *weren't* – doing. It scared and saddened me to think that, after homeschooling for all their lives, they might not realize the need and reasons to homeschool their own children!

And how could I *expect* them to have an understanding of what real education was all about unless they too (like me) had gone through a season of renewing *their* minds! Even though they have always been homeschooled, that didn't mean they knew *why*. And since they have friends who do not homeschool, it would be easy for them to be swayed if they didn't have renewed minds in this area. Worse, they might not realize the importance of finding a mate that was totally committed to homeschooling.

Thus the class "Educational Foundations" was invented, (by yours truly) and not a moment too late! ☺ (By the way, "is committed to homeschooling our children" is now among the top priorities of all three of our children for what to look for in a mate!)

On to the form... The first thing you might notice about this class is that it is ¾ of a credit. Why the odd amount? This is because 90 hours is what we just happened to end up with after all that I wanted to be covered was covered. That's the only reason.

Directly under the class title it says in parentheses: "**COURSE: SOCIAL STUDIES.**" This is the general "course" I put this "class" under. (The difference between the two is explained in the glossary at the end of this book.) It's not necessary to organize/title it this way, but since we needed classes and credits for "Social Studies," I thought this was a very good way to fill a portion of it. History and Government studies (U.S. and Wash. State) and Current Events classes also counted toward the general "course" of Social Studies.

The first four sections on this sheet are **Books, Booklets, Articles,** and **Audio Tapes**. And then skipping one section, there's also a section for **Videos**; and skipping another section, there's a section called "**Attend**." All of these are things (materials, seminars, tapes, books) he "took in"; as in read, watched, attended, or listened to, in one form or another. It probably would have been better to have them placed together on the page, but oh well... ☺ I didn't think of it till it was all done.

These are all resources that I have collected from workshops, conventions, curriculum fairs, bookstores, mail order, etc. over the past sixteen years. There is no "magical formula" for this class; these are all just things that I wanted my son to have taken in and processed. So part of this class, not really listed anywhere, but assumed, is that we do a lot of talking. Big talks, comments here and there, whatever is needed at the time.

The section entitled "**Complete Worksheets, Papers, and Tape Reports**" is all stuff Tory "did," as was "**Tutoring**" and "**Practicum**" – which just means "putting into practice." (One of the two "Word Studies on Education" is actually supposed to be on Learning. And these are now in my Course Syllabus.)

One of the activities you'll see in "**Practicum**" is "Homeschool Time with Carlianne." This could be any of a variety of activities. Since Carlianne was (still is) seven years younger than Tory, and he's a boy and she's a girl, they didn't naturally gravitate to doing a lot of things together. (And Sharnessa is nine years older.) So this time was specifically to encourage them to do things together, like reading, crafts, cooking, games, math, a science experiment, etc.

There's a "template" for the cross-match Vocab Quizzes in the Homeschool Jumpstart Navigator for Younger Children. You just list the 5 to 10 words you want to define down the left column; then write out the definitions for each of them, in a different, non-matching order, down the right column. Label each of the definitions "A through E (for 5) or J (for 10), and put the letter of the correct definition next to the word it defines. Note: the one who makes up the quiz learns the most! That's why I started having the kids do them too! ☺

The practicum assignment "Design portions of Fill-in-blank Test on "Dictionary of the Human Body" is just something I made up for that particular book (by Dorling-Kindersly) because I felt it was a good basic foundation in human anatomy. Other than that (and Math) I never tested. ☺ They retained so much without it, I figured why introduce them to "test anxiety"? Sure, someday they'll need to take tests (and have at times,) but they'll be *ready* by then. ☺

✓ ASSIGNMENT CHECK-OFF SHEET ✓✓✓✓✓✓✓✓✓✓✓✓✓✓✓✓✓✓✓✓✓✓✓✓✓✓✓✓✓✓✓✓✓✓✓✓✓✓✓

Class: **Educational Foundations**

(Course: Social Studies)

Student: **Tory W. Shelton**	Class Value: **¾ Credit = 90 Hours**

BASIC REQUIREMENTS CHECK-OFF LIST

Read Books:	☐ For the Children's Sake by Susan Schaeffer MacAulay	☐ The Right Choice: Homeschooling by Chris Klicka
	☐ Home Grown Kids by Dr. Raymond Moore (selected ch's)	☐ Wisdom's Way of Learning (selected portions) by M. Howshall
	☐ Senior High: A Home-Designed Form+U+la (selected ch's)	☐ A Charlotte Mason Education by Catherine Levison
	☐ The Christian Home School by Gregg Harris (selected ch's)	☐ Booklet Building Book (Section 1) by Barbara Edtl Shelton
	☐ The Child Influencers by Dan Adams	☐ The Jumpstart Navigator by Barb Shelton (selected chapters)
	☐ Dumbing Us Down by John Taylor Gatto	☐ Onto the Yellow School Bus & Through the Gates of Hell - by Mary Hood, PhD
	☐ The Notebook Method by Marilyn Howshall	
Read Booklets:	☐ "Primer on Homeschooling" by Dr. Raymond Moore	☐ Homeschooling for Excellence by David & Micki Colfax
	☐ "A Two-fold Approach to Imparting Values" (in LOL Newsletter)	☐ "Home Education Myths" by Karl Reed
	☐ "Teacher's College Record" by Dr. Raymond Moore	☐ "Vision" (Pts 1 & 2 in L. of L Newsletter) by Marilyn Howshall
	☐ "Natural Learning" by Kathleen McCurdy	☐ "The Case for Homeschooling"
	Read "Handbook With An Answer": ☐ Original Booklet ☐ April and ☐ May Newsletters by same title	
Read Articles:	☐ "Education in Crisis" article in CWA (Christian Women of America)'s *Family Voice* Magazine - pages 4–15, and 17	
	☐ "Elijah Co. Catalog" by Chris & Ellyn Davis (selected portions)	☐ "Biblical S.A.T.'s of Learning" by Marilyn Howshall
Listen to audio tapes:	☐ "The Notebook & Unit of Life Method" by Marilyn Howshall	☐ "Relaxed High School" by Mary Hood, Ph.D.
	☐ "7 Natural Vital Signs of the Learning Process" by "	☐ "Nurturing Creativity in Children"
	☐ "Getting Ready for Achievement Testing" by Curt Bumcrot	☐ "Special Needs of a Child with ADHD" by Roger Nelson
	☐ "Fundamentals of Education" by Phillip Troutt	☐ "Alternatives to Textbooks" by Wendy Wartes
	☐ ☐ "Out of Fear & Burnout; Into Freedom, Focus & Refreshment" Tapes (Parts 1 and 2) by Barb Shelton	
	☐ "Homeschool Philosophy: Structure vs. Non-structure" Tape by Linda Krenelka	
Complete Work-sheets, Papers, & Tape Reports:	☐ Wisdom's Way of Learning (Worksheet)	☐ Home Grown Kids (Book Report)
	☐ "7 Natural Vital Signs of the Learning Process" (TR)	☐ "Learning & Teaching Philos." (from "Primer on Hsing" (W)
	☐ "Intro to a Biblical Lifestyle of Learning" (TR)	☐ "Handbook With An Answer" (W)
	☐ Write paper on "Downfalls of Public Education"	☐ "Teacher's College Record" (W)
	☐ Do Word Study on: "Education"	☐ Do Word Study on "Education"
Watch Videos:	☐ "Homeschooling: A Foundation for Excellence"	☐ "Why Homeschool Thru High School?" by Josh Harris
	☐ "Intro to a Lifestyle of Learning" by M. Howshall	☐ "Homeschooling 101" Video-Seminar by Barb Shelton
Tutoring:	☐ In your area of expertise and/or interest *(Basketball, Math, Science, etc.)*	
Attend:	☐ "Tools and Techniques" Workshop by Gregg Harris	☐ "The Right Choice: Home Schooling" by Chris Klicka
	☐ "Making the Transition from Elem. to Jr. & Sr. High School" Workshop by Gregg Harris	
Practicum:	☐ Make Cross-match Vocab Quizzes for High School and Elementary levels	
	☐ Do "Homeschool Times with Carlianne" in wide variety of subjects (See separate Task Analysis Sheet)	
	☐ Design portions of Fill-in-blank Test on Dictionary of the Human Body	

HOUR CHECK-OFF CHART

90	89	88	87	86	85	84	83	82	81	80	79	78	77	76	75	74	73
72	71	70	69	68	67	66	65	64	63	62	61	60	59	58	57	56	55
54	53	52	51	50	49	48	47	46	45	44	43	42	41	40	39	38	37
36	35	34	33	32	31	30	29	28	27	26	25	24	23	22	21	20	19
18	17	16	15	14	13	12	11	10	9	8	7	6	5	4	3	2	1

[A.C.S]
ENGLISH I

Let me first give you an overview of how I set up our three English classes...

Six areas of English – which I just chose because they sounded like a logical way to break down the larger area of English – are covered in our three credits of English. Two areas of the six comprise English I, which I'll be explaining here in this ACS, and then two other areas in English II, and the last two areas in English III. I will explain those last two English classes (and the four areas) in their own ACS's. The way I came up with the particular combinations (pairs) made absolutely *no* sense other than time-wise. I had recorded all the hours for all six areas as we went along, with no thought of how I would organize them later. Once we got to being "almost done," I could then see how it could be structured based on what had already been done, and what I felt was still needed to do. At that point I took into consideration how much of each area he had left to do (this was about half way through his senior year), and then combined the six components accordingly, in pairs. I had to do a bit of experimenting and shuffling to make it come out the way I wanted it to, ☺ but it wasn't hard.

OK, so why did I call these classes "English I, II and III"? ... Again, why not? No, there *was* a bit more thought and reasoning that went into it than that. However, there was no reason *sequence-wise* that I named the classes such. I *had* to give them some sort of title, and the Roman Numeral route made the most sense to me. I couldn't just call them all "English", yet "English 101," "102," etc. seemed too specific, hinting that they were sequentially related. Using the I, II, and III just gave them a title that enabled me to differentiate them from each other without insinuating that they were classes in a series, one a pre-requisite of – or related to – the other. This, at least to me, indicates that he took three classes of English. Since he didn't have whole credits in any of the 6 areas, it didn't work to have one class be "English Lit" and the other "English Comp." Even half credits didn't work for us as his classes were all pretty much roughly about ¾ credit of one of the two areas and ¼ credit of the other, time-wise. Or ⅔ and ⅓. But if your student had enough hours in each, you could certainly label them that way, being more specific and focused in each class.

This English class, "**Composition and Grammar**," is, as you see, divided into eight areas, all listed there in the left column. Seven of these areas are "Comp" alone, and the last (below the solid black line) at the bottom is just "Grammar." I came up with these areas just by looking at what he had done and deciding what to name them. (Kind of like how Adam looked at all the animals and decided what to name them just by looking at them. ☺) It wasn't hard, in fact, it was kinda fun. As you can see, they are all pretty obvious.

A few explanations... Under "**Writing Lessons**" at the top: When I say "selected portions" that means he did "some" of it, and either that was "enough" because he had learned "enough" out of that particular resource, or because he or I had "had enough" of that particular resource. ☺ But just because he didn't finish the book that didn't mean he couldn't get credit for what he *had* done. Even if he finished only a portion of a particular resource, he still learned *something* from it.

You will find many of the items and activities on this ACS in the ACS's for several other classes. For instance, the Field Trip Report on "Pioneer Farm Museum" was also a field trip under "Wash. State History." And the one on "David Manual Bronze Museum" *also* went under *both* "Fine Arts" and "Wash. State History." The Tape Report on "The Notebook Method" was also for "Educational Foundations"; the Lab Field Trip reports to Ocean Spray and Norpac were also for "Lab Science"; the book reports were all on books he read for "Spiritual Studies"; and the Concert/Performance Records were done for plays he saw that went under "Fine Arts." So basically then, I let the writing aspect of these other activities count for this class as well. But I would not count *all* the hours for *all* the possible classes.

However, I wanted Tory to do more compositional writing than just these tape, book, field trip, and concert reports, so that's what you see under "**Writing Application/Essays.**" I did not require a lot since it is not an area he enjoys. However, it was waaaay more than *he* would have felt was necessary ☺ – which I did not consider of great import since *he* felt writing wasn't necessary at *all* and *I* feel that *everyone* no matter what they do in life needs to know how to write – at least basically. (And I automatically win ☺ since I'm the mom, principal, and teacher.) What he did was, I felt, "enough" for a basic foundation in writing, and "enough" for him to take it further if he needs to. (He has more talent than he *thinks* he does – or wants! ☺)

Since I wrote up this ACS, he has also written an essay, even with footnotes and references, on "Abortion" for his "Christian Ethics" class at the Christian School And he did a very good job on it! So if you want to go ahead and write that in on this sheet, feel free! ☺

So what exactly will go on our transcript? The whole title goes on there, not just "English I," so that what we covered will be evident.

✓ **ASSIGNMENT CHECK-OFF SHEET** ✓✓✓✓✓✓✓✓✓✓✓✓✓✓✓✓✓✓✓✓✓✓✓✓✓✓✓✓✓✓✓✓✓✓✓✓✓

Class: **English I** (COMPOSITION & GRAMMAR)	
Student: Tory W. Shelton	*Class Value:* 1 Credit = 120 Hours

BASIC REQUIREMENTS CHECK-OFF LIST

Writing Lessons:	☐ Do selected Writer's Inc. exercises ☐ Do selected portions of Wordsmith
	☐ Read selected portions of The Writing Christian by Ron Brackin
	☐ Do selected portions of How to Write Clearly ☐ Listen to tape: "Report & Essay Writing" and take notes
Writing Application/ Essays:	☐ "America Is" essay ☐ "More Than a Carpenter" essay ☐ ☐ ☐ ☐ Write 4 essays for Scarlet Letter class
	☐ "Downfalls of Public Education" essay with Mom ☐ "Big Essay" for Scarlet Letter class
	Work on: ☐ "Venus" story ☐ "From Field to Freezer" ☐ Outline for "Hunting Guide Book"
	☐ Write "Basketball for Dummies" section of Basketball Notebook ☐ Write an acrostic poem
	☐ Write your own vision, beliefs, hopes, intentions and goals in the are of relationships with the opposite sex
Book Reports:	☐ Responsible Living in an Age of Excuses by Michael Farris ☐ Screwtape Letters by C. S. Lewis
	☐ How to Be Your Own Selfish Pig ☐ This Present Darkness by Frank Peretti
Tape Reports:	☐ Tape: "The Notebook Method / Unit of Life Approach" by Marilyn Howshall
	☐ Tape: "How to Set Values that Lead to Victory" ☐ Tape: "America's Godly Heritage" by David Barton
	☐ Tape: "Family Writing" by David Bragonier ☐ Tape: "Nurturing Creativity in Children"
	☐ Tape: "Why Homeschool Through High School?" by Josh Harris ☐ Tape: "New Attitude" by Josh Harris
	☐ Tape: "Report and Essay Writing"
Field Trip Reports:	☐ Field Trips to: "David Manuel Bronze Museum" ☐ Field Trip to: "Pioneer Farm Museum"
	☐ Lab Field Trip to: Ocean Spray Cranberry Lab ☐ Lab Field Trip to: NORPAC Lab
Journalism	☐ 3-day Journalism Homeschool Camp - Instructor: David Beard (at Goldendale, Wash.; by M.C.F.)
	☐ Create / Write / Design Basketball Notebook (with input from Mom)
Critiques & Concert/Per-form. Records	☐ Movie: "Civil War" Video ☐ Play: "M.A.S.H." ☐ Play: "The Fantastics"
	☐ Do Concert/Performance Record on "Piano Duet" and on... ☐ "Forever Plaid" Quartet Concert
Grammar:	☐ Read "How to Use Routine Activities to Teach the Parts of Speech" (ATIA resource)
	☐ "Easy Grammar" (selected portions) ☐ Test Preparation ☐ Do "Winston Grammar" (selected portions)
	☐ Edit Sr. High Video-Seminar syllabus

HOUR CHECK-OFF CHART *(Check off hours as they are completed.)*

120	119	118	117	116	115	114	113	112	111	110	109	108	107	106	105	104	103	102	101
100	99	98	97	96	95	94	93	92	91	90	89	88	87	86	85	84	83	82	81
80	79	78	77	76	75	74	73	72	71	70	69	68	67	66	65	64	63	62	61
60	59	58	57	56	55	54	53	52	51	50	49	48	47	46	45	44	43	42	41
40	39	38	37	36	35	34	33	32	31	30	29	28	27	26	25	24	23	22	21
20	19	18	17	16	15	14	13	12	11	10	9	8	7	6	5	4	3	2	1

[A.C.S]
English II
(Literature & Speech)

"English II" consists of "Literature and Speech." As I have previously explained, the only reason for this particular combination was the way the hours combined to form 120 hours.

All three English classes were done all the way through high school, at various times, not as a cohesive "class." So what you see here for English II was also done over all of Tory's high school years, in bits and pieces.

First, the "**Book Synopses**" section: There is a sheet entitled "Book Synopsis" in this book (see the Form Index to find it), and that's what all of these synopses are. I just created a form that contained "requests" for all the information I wanted him to find and think through on each synopsis. They are done on any of a variety of two-page synopses that can be found in the Great American Bathroom Books, also entitled "Compact Classics." (Same exact book; different, classier title. ☺)

There is a separate TAS (Task Analysis Sheet) for the complete list of all the synopses Tory did. I didn't feel it was necessary to include that here as this is totally elective. He just "happened" to do this many over the years; no particular reason for this number, just turned out to be that many. We stored all these in alphabetical order in his "English Notebook."

Who chose which ones Tory did? We *both* did; although I probably chose more than he did. It doesn't really matter; just make sure that the ones you feel are important are included.

The next section down is "**Videos of Books**," which is just what it says it is: the video re-makes of books.

"**Study Great Authors and Their Work**": I got this list from a friend of a friend, who majored in English and has a great appreciation of literature. I wrote as fast as I could as she shared all these ideas over the phone. I then added a few of my own choices to the list.

"**Study Shakespeare**": It would take much more than what I have here to do a complete, in-depth study of this subject area, but that was not my intent; only to give an intro to his work.

"**Study Poets**": I made up a Poet Synopsis sheet and that's what this is based on. (It's here in Form+U+la.) You might choose an entirely different assortment of poets than what I chose, and feel free to do so! ☺

"**Online Literature Class**": This was an eight-session (two classes per week) class lead by Cindy Marsch on America Online. There were about twelve other home-schooled high schoolers in the class, which met in a private chat room on AOL. Tory and I both enjoyed it! He was required to write several short papers on topics given by the teacher, as well as one large paper, which, for an extra (reasonable) fee, the teacher personally evaluated. (These all counted toward "English Composition" as well.) It was a very good experience!

"**Speech**": Tory attended a "Toastmasters" class, which I *highly* recommend! They are now charging for the class (which Tory just recently took again, this first one having been several years ago), and it is worth every penny! (It was $30, and came with a syllabus.) A sermon he did for youth group was also counted for this class. "Speech: Exploring Communication" is a textbook (gasp!) that my dear brother-in-law, an English major, gave us.

✓ ASSIGNMENT CHECK-OFF SHEET ✓✓✓✓✓✓✓✓✓✓✓✓✓✓✓✓✓✓✓✓✓✓✓✓✓✓✓✓✓✓✓✓✓✓✓✓✓✓

Class: **English II**
(LITERATURE & SPEECH)

Student: **Tory W. Shelton**	Class Value: **1 Credit = 120 Hours**

BASIC REQUIREMENTS CHECK-OFF LIST

LITERATURE:

Book Synopses

Do Book Synopses Reports on 84 book synopses. *(Many listed below; See Task Analysis Sheet for a list of all titles.)*

☐ ☐
☐ ☐
☐ ☐

Videos of Books

☐ "Lord of the Flies" ☐ "Romeo and Juliet" ☐ "The Crucible" ☐ "Grapes of Wrath" ☐ The Joy Luck Club"
☐ "To Kill a Mockingbird" ☐ "The Diary of Anne Frank" ☐ "Sense and Sensibility" ☐ The Scarlet Letter ☐ "The Firm"

Study Great Authors and Their Work

☐ Arthur Miller: The Crucible (S) ☐ Mark Twain: Short Stories (S) ☐ The Prince and the Pauper (S) ☐ ☐ 2 Stories (B)
☐ C.S. Lewis: Screwtape Letters (B) ☐ Nathaniel Hawthorne: The Scarlet Letter (B)
☐ Jane Austen: Sense and Sensibility (S) ☐ Charles Dickens: A Tale of Two Cities (S) ☐ Oliver Twist (S)
☐ Ernest Hemingway: The Old Man & the Sea (S) ☐ A Farewell to Arms (S) ☐ For Whom the Bell Tolls (S) ☐ The Sun Also Rises (S)
☐ John Grisham: The Firm (S) (V) ☐ Henry Wadsworth Longfellow: The Courtship of Miles Standish (S)
☐ Thornton Wilder: Our Town (S) ☐ O. Henry: Short Stories (S) ☐ The Gift of the Magi (S)
☐ Tom Clancy: Patriot Games (S) ☐ Thomas à Kempis: The Imitation of Christ ☐ Ayn Rand: The Fountainhead (S)
☐ Harper Lee: To Kill a Mockingbird (S)(V) ☐ Edgar Allen Poe: Telltale Heart ☐ The Fall of the House of Usher (S)
☐ George Elliot: Silas Marner (S) ☐ John Steinbeck: The Grapes of Wrath ☐ Of Mice and Men (S)
☐ Frank Peretti This Present Darkness (S) ☐ Agatha Christie: Murder on the Orient Express (S)
☐ Herman Melville: Moby Dick (S) ☐ Viktor Frankl: Man's Search for Meaning (S) ☐ Jack London: The Sea Wolf (S)
☐ Amy Tan: The Joy Luck Club (S)(V) ☐ Robert Lewis Stevenson: Dr. Jekyll and Mr. Hyde (S) ☐ Treasure Island (S)
☐ Emily Bronte: Wuthering Heights ☐ Leon Uris: Exodus (S) ☐ William Golding: The Lord of the Flies (S) (V)
☐ Charlotte Bronte: Jane Eyre (S) ☐ Victor Hugo: Hunchback of Notre Dame (S) ☐ Les Miserables

♦

KEY:
B = Book
S = Synopsis
V = Video

Study Shakespeare

☐ Henry VIII (S) ☐ Julius Caesar (S) ☐ Much Ado About Nothing (S) ☐ Synopsis of His Life (I-4-C1)
☐ Romeo and Juliet (S)(B) ☐ Watch Original Movie ☐ Watch New-version Movie ☐ Watch "West Side Story" ☐ Compare all 3

Study Poets

Do "Poet Synopsis" on each, and be familiarized with their poetry: ☐ Robert Frost (II-p.265) ☐ T.S. Elliot (II-p.275)
☐ Ogden Nash (II-279) ☐ Rudyard Kipling (II-p.261) ☐ Henry Wadsworth Longfellow (II-p.249) ☐ William Carlos Williams (II-p.267)

Online Literature Class:

Participate in 4-week, 8-session online literature class (on America Online) with Cindy Marsch, instructor, and 10 high school students: Class: ☐ 1 ☐ 2 ☐ 3 ☐ 4 ☐ 5 ☐ 6 ☐ 7 ☐ 8
Write 4 weekly essays on assigned topics: ☐ Essay #1 ☐ Essay #2 ☐ Essay #3 ☐ Essay #4
☐ Write "big essay" for class (1,000–1,500 words / Yours ended up _______ words.)

SPEECH:

Speech Components

Toastmasters 9-week Series of Classes (13 hrs) Week: ☐ 1 ☐ 2 ☐ 3 ☐ 4 ☐ 5 ☐ 6 ☐ 7 ☐ 8 ☐ 9
☐ Prepare and present sermon for Youth Group (Father's House) (1 hour)
☐ Read selected portions of Speech: Exploring Communication *(See Task Analysis Sheet for specifics.)*

HOUR CHECK-OFF CHART *(Check off hours as they are completed.)*

120	119	118	117	116	115	114	113	112	111	110	109	108	107	106	105	104	103	102	101
100	99	98	97	96	95	94	93	92	91	90	89	88	87	86	85	84	83	82	81
80	79	78	77	76	75	74	73	72	71	70	69	68	67	66	65	64	63	62	61
60	59	58	57	56	55	54	53	52	51	50	49	48	47	46	45	44	43	42	41
40	39	38	37	36	35	34	33	32	31	30	29	28	27	26	25	24	23	22	21
20	19	18	17	16	15	14	13	12	11	10	9	8	7	6	5	4	3	2	1

[A.C.S.] ENGLISH III (STUDY & RESEARCH SKILLS & VOCABULARY)

English III is comprised of "Study and Research Skills" and "Vocabulary"; again, only because it worked out that way time-wise. (hours)

Tory also counted writing a number of definitions for the "**Glossary**" of his "Basketball Notebook." (I will someday have a book out on that.)

VOCABULARY

English From the Roots up is an excellent resource by Joegel Lundquist. It gives a basic foundation in the Latin and Greek roots of words, which are"keys" that unlock much of the English language. I have heard of kids as young as 10 years old doing it, but think that junior high and high school will get the most out of it.

The "**America the Beautiful**" word definition was something I thought of because I didn't understand most of that song, ☺ and (therefore) thought it would be a good exercise for my kids. It is here in Form+U+la. (See the Form Index.)

The **Reader's Digest** "**Word Power**" quizzes are an excellent vocabulary resource!

The "**God's View of Science**" word studies can be found in my book "**Lab Science: The How, Why, What, Who, 'n' Where Book.**"

The "**Word Alive**" study sheet is what I made up as a format for studying a particular word. These also counted for "Spiritual/Biblical Studies."

The **Vocabulary Quizzes** were already explained in English II. They counted both making them up and taking them, their own, mine, and each others'.

The **Pro-One CD-Rom** was rather dry, but had some good material in it. Plus it was "something different" for English. ☺ (It could have counted for "Computer Skills Survey" too.)

RESEARCH & STUDY SKILLS

We got the "**How to be a Superstar Student**" video from The Teaching Company. They have a lot of great videos, but be careful, you can easily get hooked into thinking that you need *all* of them and that you can't possibly do as good a job. Just keep in mind that it's mostly about how to do well within the traditional school system – some of which they can certainly use now, but most of which will be helpful if they go back into the system, like in college. I had hoped it would have more "heart" stuff, but it's mostly externals. But it's still good. Remember, we aren't trying to emulate the traditional school system; God is doing a *new* work. We can utilize resources, but He has something better for us, which is going to involve our kids doing more of the learning for themselves! (To contact this company, and request a free catalog from them. call 1-800-832-2412.)

I found the "**SAT's Got You Freaked?**" video at a garage sale. Tory enjoyed this and said that anyone who was planning to take this test would greatly benefit from this video. Check at video stores for it.

The **research** that Tory did for his "**Downfalls of Public Education**" and "**The Scarlet Letter**" paper he got to count for this area of English, as his research for a class he took from a homeschooling dad.

The rest of what's listed here are just resources that we had found – and that he did – over the years and counted here in this class.

While it *appears* that this girl is raising her hand in a classroom, in reality she is in the middle of a delight-directed study project, and is saying...

✓ ASSIGNMENT CHECK-OFF SHEET ✓✓✓✓✓✓✓✓✓✓✓✓✓✓✓✓✓✓✓✓✓✓✓✓✓✓✓✓✓✓✓✓✓✓✓✓✓✓

Class: **English III**

(STUDY & RESEARCH SKILLS / VOCABULARY)

Student: **Tory W. Shelton**	Class Value: **1 Credit = 120 Hours**

BASIC REQUIREMENTS CHECK-OFF LIST

VOCABULARY

☐ "English From the Roots Up"

☐ Do "America the Beautiful" word definition study and re-write

☐ Do *Reader's Digest* "Word Power" Vocabulary Quizzes (10) ☐☐☐☐☐☐☐☐☐☐

☐ Do "God's View of Science" word study on: ☐ "Chemistry" ☐ "Biology" ☐ "Physics"

☐ Read The Story of American English (selected portions)

Do 12 Word Alive Studies: ☐☐☐☐☐☐☐☐☐☐☐☐

☐ Crossword Puzzles

☐ Define words during reading (gave a blanket 5 hrs. for this)

Make up Vocabulary Quizzes (8): ☐☐☐☐☐☐☐☐

Take Vocab Quizzes: ☐☐☐☐☐☐☐☐

☐ Work with "Pro One" CD-Rom: "Vocabulary

☐ Write definitions for Basketball Notebook Glossary

RESEARCH & STUDY SKILLS

Watch 8-part Video Series: "How to Be a Super Star Student" video: ☐☐☐☐☐☐☐☐

☐ Watch Video: "SAT's Got You Freaked?"

☐ Research for essays: "Downfalls of Public Education," "The Scarlet Letter") ☐☐

☐ Do Research for U.S. History Class (taught by Keith Merritt)

☐ Read Book: A.C.H.I.E.V.E.M.E.N.T. by Robert B. Blackmer

☐ Familiarize yourself with Avery FactFinders Quick Writing Guide

☐ Familiarize yourself with "Webster's Concise Encyclopedia" CD-Rom

☐ Read selected portions of The Art of Being A Successful Student by Helen M. Thompson

☐ Read selected portions of The Sat I FOR Dummies

HOUR CHECK-OFF CHART

120	119	118	117	116	115	114	113	112	111	110	109	108	107	106	105	104	103	102	101
100	99	98	97	96	95	94	93	92	91	90	89	88	87	86	85	84	83	82	81
80	79	78	77	76	75	74	73	72	71	70	69	68	67	66	65	64	63	62	61
60	59	58	57	56	55	54	53	52	51	50	49	48	47	46	45	44	43	42	41
40	39	38	37	36	35	34	33	32	31	30	29	28	27	26	25	24	23	22	21
20	19	18	17	16	15	14	13	12	11	10	9	8	7	6	5	4	3	2	1

[A.C.S.]
FINE ARTS SURVEY

The activities on this sheet are divided into various types of "artsy" categories. The more general category of "Drama" is actually divided into the top three sections of this sheet (though not labeled "Drama"): "**Attend Concerts and Performances**," "**Drama (Participatory)** *(meaning they took part in it)*, and "**Attend Plays**." If your student has experienced very few of these types of activities, they could all go under the single category of "Drama."

As our son, Tory, embarked on high school, I knew there were certain areas of "**Fine Arts**" that I wanted him to experience, both observationally and actively. I wanted his experience of music, dramas, and performances to extend beyond what is currently available in the "For Christian Teens" realm. And I knew that, unless community-type concerts were on his list of requirements, he would never choose to attend one of his own volition. ☺ *(I'm happy for you if your child wouldn't dream of any other type of musical entertainment, but that's not the case for us. I enjoy many of those types of things, but my husband never has, and Tory has taken more after him.)* After having attended several such events, Tory now admits that he didn't hate them nearly as much as he *thought* he would, and even says he "almost slightly enjoyed some of them and actually *did* quite a enjoy a few of them"! (In all fairness, the ones he "almost slightly enjoyed" I just "vaguely enjoyed" myself! ☺)

As we set out to do high school, I began to get a feel for the general direction I wanted this class to take. However, I had no way of knowing what opportunities would actually come our way, or what Tory would be interested in doing on his own. This is another example of why it's very difficult, if not impossible, to have your classes completely planned at the beginning! At least this is true if you want to be free to take advantage of opportunities that God brings your way all along the way. However, you could still go pretty far with a generic plan. For example, instead of writing the names of the actual concerts as I did here on this sheet (which you would have no way of knowing four years ahead anyway), just write: "**Concert**:_______________" leaving a blank space or line to fill in whatever your student ends up attending.

Concerts generally last two hours, so you can do some figuring of hours (if you're using hours in your system; and I'm using hours *and* content) based on that figure for at least that portion of your credit/class. And if you are having your student do "Concert/Performance Records," those each take about 15 minutes to do.

So do you have your student do a "**Concert/Performance Record**" for every concert or performance they attend? I hope not, unless they really *enjoy* doing them and *want* such a record of every event they've attended. Otherwise it will very likely cast a wet blanket over something that you're wanting him/her to learn to enjoy.

The "**Concert/Performance Record**," the "**Plot Wholesomeness Evaluation**," and the "**Play or Movie Critique & Analysis**" are all in this book. *(See the Form Index.)*

Because I just recently came across a wonderful resource, I have now added a new category to my ACS (not included on this form) which I'm calling "**Fine Arts Literacy**." I got that word ("literacy") from Diana McAlister, one of the co-authors of the book *Homeschooling the High Schooler*. *(A quick digression: I highly recommend that anyone designing/coordinating their own high school courses have this book in their personal library! Diana, a dear sister in Jesus, was the one God used to launch me into high school and inject into my heart the fire, into my guts the nerve, and into my head the gall to think I could actually do this!)* Anyway, into this new category I have put two require-

(Text continued after the Fine Arts Survey ACS ☞)

Sharnessa's Heart and Vision for the Arts

(excerpted from a letter she wrote)

"...For a few years now, I have been excited about the move of God to bring "the Arts" back into the church – and the church back into the Arts. I believe He desires to do this in such a dynamic way that the world will be touched by Him through this medium. God is calling Christians to excellence, which means that what we produce in the area of the Arts should be at least as good as what the world produces, if not better! The world sees most of what the Church in general has produced (as far as the Arts go) as being "low quality" – not usually attractive to the unsaved. It's not meant to be that way! The Arts were *God's* arena to begin with! What we produce should "blow the world away." The world should be drawn to Jesus by what we do. Not that it's supposed to be alluring in a sensual sense, but by their seeing the purity, truth, and love of Jesus in it.

I know that God wants to use me in the Arts, particularly singing, dancing, and acting. But I don't want to just "perform" or "get a name for myself"; I want to draw people to *God* through what I do. I'm SO excited about being part of all of this, and about being on the "front lines"! (Hey, my mom doesn't call me her "little Joan of Arc" for nothing!)

God has already used me in this area to a degree in several performances and productions that I have either been involved in or that I myself have choreographed, directed, and produced. But I now feel God is leading me to go deeper with it all as He has opened the door for me to take this to the next level! I have been accepted at the School of Creative Ministries in London, which is under IBIOL – the International Bible Institute of London. This is a one-year program. [*And it's what she is completing as this book is in the final stages of being printed in June of 99!*] Part of the school's focus is to give the students a broad perspective of the Arts, from theater to the entertainment industry. But most importantly they focus on the *relationship* with God, not on religion or legalism, but on the *reality* of Jesus, for *real* people in a *real* world with *real* needs. The combination of God and the Arts at this school is so rare and so incredible!..."

✓ ASSIGNMENT CHECK-OFF SHEET ✓✓✓✓✓✓✓✓✓✓✓✓✓✓✓✓✓✓✓✓✓✓✓✓✓✓✓✓✓✓✓✓✓✓✓✓✓✓

Class: Fine Arts Survey

Student: Tory W. Shelton | **Class Value:** 1 Credit = 120 Hours

Basic Requirements Check-off List

Attend Concerts and Performances:	"Violin Concert" by Scott Yoo (Community Concert)	
	"Nevada Dance Theatre"	
	"Act I" Live Performance at The Old Fashioned Christmas Bazaar (@ Puyallup Fairgrounds)	
	"Piano Duet" (Community Concert)	
	"Pianist John Bayless" (Community Concert)	
	"Side Street Strutters" (Community Concert)	
	Concert with Columbian Chorale, Quintesce, and the Bell Choir (Jean Merritt in it)	
	"Swingin' on Broadway" LCC Concert (Grampa Lee Edtl in it)	
	"Forever Plaid" Male Quartet Concert/Performance (Community Concert)	
Drama (Participatory)	Participate in "Celebrating Jesus" Production *(produced by Sharnessa)*	
	Participate in the play "Treasures of the Snow" *(with "School Biz" Homeschool Group)*	
	Take Drama Class @ Columbia Heights Christian School *(Mr. Jim Mortland)*	
	Do Radio Drama *(own spontaneous creation)*	
Attend Plays:	Easter Play at Lv. Church of the Nazarene (Gramma & Grampa Shelton and Fountains in it)	
	Play: "M.A.S.H." (Pete and Julie Ouellette and Feena Spence in it)	
	Play: "Give 'Em Hell, Harry!" by Kevin McArthy (Solo Performance)	
	"The Fantastics" Live Play (starring Pete Ouellette)	
Tour Art(istic) Displays:	The David Manuel Bronze & Nez Perce Indian Museum in Joseph, Oregon	
	Art and Craft displays (at Puyallup Fairgrounds; Old Fashioned Christmas)	
Evaluating Music:	Watch Video: "Rock's Primal Scream"	
	Do 5 "Music Evaluations"	☐☐☐☐☐
	Do 3 "Concert/Performance Records"	☐☐☐
	Do 3 "Plot Wholesomeness Evaluations"	☐☐☐
	Do 3 "Play or Movie Critique & Analysis"	☐☐☐

Hour Check-off Chart

(Check off hours as they are completed.)

120	119	118	117	116	115	114	113	112	111	110	109	108	107	106	105	104	103	102	101
100	99	98	97	96	95	94	93	92	91	90	89	88	87	86	85	84	83	82	81
80	79	78	77	76	75	74	73	72	71	70	69	68	67	66	65	64	63	62	61
60	59	58	57	56	55	54	53	52	51	50	49	48	47	46	45	44	43	42	41
40	39	38	37	36	35	34	33	32	31	30	29	28	27	26	25	24	23	22	21
20	19	18	17	16	15	14	13	12	11	10	9	8	7	6	5	4	3	2	1

ments, one of which is: Read "Renaissance Art" Notebook. This is a 3-ring notebook full of photos of many works of art: sculptures, paintings, and buildings of the Renaissance. The other new requirement under the new **"Fine Arts Literacy"** category is: In the **"Dictionary of Cultural Literacy"** read selected portions of the chapter entitled "Fine Arts" (pages 158–194.) I will tell you more about this wonderful resource in a separate article at the very end of this section entitled "**Explanation of Cultural Literacy Resources.**" It's really about much *more* than an article!

AND NOW... *SHARNESSA'S* FINE ARTS CLASSES

The previous "Fine Arts Survey" ACS is just *one* way to create a Fine Arts class. There is nothing "official" or "best" about what I have put together. And there are SO many different ways you could create one!

Sharnessa, who ended up with four credits in Fine Arts, had several different types of classes. Since one of her credits is structured essentially the same as Tory's (on previous page), I'm not including it here. But I thought you might like to see how I took her areas of more heavy interest and created classes out of them! Keep in mind, though, that these were done in an earlier format which is different at the top and does not have the "Hour Check-off Chart" at the bottom.

A student in "regular school" might attend a choir or band class with 40 other students for a full year and earn their Fine Arts Credit that way. In that light, do you think my (or your) homeschoolers have "missed out" on a *thing!?* I'm not putting down such classes, but you're not going to convince me that our kids who get to develop their own interests and take advantage of a variety of activities (of their own choosing) are deficient in how they earned that credit!

"ART & CRAFTS" A.C.S

Directly to the right is the "**Art & Crafts**" ACS. As you can see at the bottom, she had 60 hours of activities, which, for the most part, were of Sharnessa's own choosing. Everything except for "Personal Drawing/Sketching" and "Decorate Corners and Creative Niches of own Bedroom" took a particular amount of time that she had recorded as she went along. Two of these, **"Personal Drawing/Sketching"** and "**Decorate Corners...**" were things that Sharnessa actually did much more of than is indicated here. We added in what I felt was a fair "ceiling" on the hours and what was needed to bring the hours up to 60 for the ½ credit. She probably had enough hours between the two to get her another ¼ credit, and I would have felt fine doing that. But, since she wasn't exactly in need of hours for Fine Arts, I just kept it at that. ☺

"CHOREOGRAPHY & PRODUCTION" A.C.S.

Looking at the following page, you see the ACS for Sharnessa's "Choreography and Production" class. I just collected all the various activities that Sharnessa had done in this area – her delight, in case you hadn't figured that out yet – and added up the hours for all of them. With a little help – more trimming off than anything – the hours came out to 180, which is 1½ credits within our system.

The way I would do it if I were creating this class now would be to add the 60-box Hours Check-off Chart at the bottom of the "Art & Crafts" ACS, a 180-hour box at the bottom of "Choreography & Production", and, of course I would re-do the top area to look like my newer design (for Class title, Student's name, and Class Value.) Even though I have deleted the "Resources/Instructions" section (near the top) from most of my ACS's, this is one class I decided to still keep it for, as it gave all the disjointed activities some cohesiveness.

"ART OF LIVING" T.A.S.

On the *following* page is the T.A.S. for what was to be the "Art of Living" segment of Sharnessa's main Fine Arts class. I say "was to be" because this was something I planned *ahead* that didn't end up coming to fulfillment. She did several of the activities there – which is what got me going on making up this sheet – but the rest just didn't pan out. I've included this form here because some might find it helpful in creating such a class – or *portion* of a class – for or with your own daughter. *Or* you might want to use it with a Home Ec. (or "Home & Life Management Skills") class.

Student: Sharnessa Shelton	☐ Full year (1 Credit) ☐ ___/___ Credit

Resources: Lessons, Homeschool Moms, "How Great Thou Art" Art Course Book, Aunt Jean, School Biz, Personal Creativity, Mom, Reference Books

Instructions: Complete the crafts, assignments, quilt lesson, displays, costumes, and art course as follows:

BASIC REQUIREMENTS CHECK-OFF LIST	TIME (in hours)	DONE
Quilt-making Weekend Seminar with Aunt Jean	10	
Craft Classes at School Biz	19	
Personal Drawing/Sketching	12	
Costume Design for Performances	5	
View Art & Craft Displays	4	
Fundamentals of Art Course: "How Great Thou Art"	8	
Decorate Corners and "Creative Niches" of own Bedroom	2	
Total Hours =	60	

✓ *ASSIGNMENT CHECK-OFF SHEET* ✓ . . . **CHOREOGRAPHY & PRODUCTION**

Student: Sharnessa Shelton	☐ Full year (1 Credit) ☐ ___/___ Credit
Resources: Own Gifting, Dance Instructor and Classes, Mom, "Studying the Experts," Dance Clinics	
Instructions: Choreograph, instruct, practice, produce, direct, and/or perform these dance/drama performances.	

BASIC REQUIREMENTS CHECK-OFF LIST	**NUMBER OF HOURS**	Done
Dance & Assist w/ Choreography of Shekinah "Angel Dance" (1991)	48	
Dance & Assist w/ Choreography of Shekinah "Angel Dance" (1992)	31	
Choreograph, direct, and produce "Celebrating Jesus" Drama/Dance Performance with 25 teenagers (1993) ■ "50's" Dance (poodle skirts) to "Celebrating Jesus" by Carman ■ "Walk with Jesus, Turning Away, Warfare, Victory, Return" Drama/Dance to "Ashton" by Michael W. Smith ■ "Angel Worship Dance" to Michael W. Smith song:	36	
Choreograph, direct, and produce "Wings of Grace" and "Scream" Dance for Youth Alive *(with Kindra, Julie and Stacey)* (1995)	31.5	
■ Costume design and production for the above Choreograph and instruct "Dance of Rebellion" for Mechowlah Company (1995)	26	
■ Costume design and production for the above (")	2	
Choreograph Little-Girl's Hip-hop Dance *(for Carlianne, Bailey Sample, and Jessica Wafer)*	2	
Choreograph Little-Girl's Angel Dance *(for Grace Kempf and Carlianne)*	2	
Teach Line-Dancing at Jeanne Nelson's Birthday Party	1	
Teach Line Dancing and Aisle Dancing to 50 people at own 18th Birthday Party	2	
Total =	180 hrs. = 1½ Credits	

Art of Living

Task/Activity	I	P	M
Hostessing:			
Plan and put on a Progressive Dinner (Five stages: Appetizers, Bread/Soup, Salad, Main Dish, Dessert)			
Attend Victorian Tea for mothers and daughters at Columbia Hts. Assembly of God			
Plan and put on a Buffet for a Party			
Plan and put on a Reception			
Plan and put on a Baby or Wedding Shower			
Study different types of invitations and prices			
Plan and hostess a Video Car Rally			
Plan and hostess a Limousine Excursion			
Help plan, prepare, cook a Thanksgiving meal			
Help plan and do a child's Birthday Party			
Help clean and prepare home for company			
Personal Nurturing:			
Set up "Personal File" (Goals, Rituals, Thots, Verses)			
Read *Living a Beautiful Life* by Alexandra Stoddard			
Read *The Hidden art of Homemaking* by Edith Schaeffer			
Write your daily, weekly, monthly, annual rituals			
Creating Memories:			
List the traditions you want to do with your own family someday – current and new ideas			
Establish "Grace cards" system and add to Personal File			
Help Carlianne with her "Family Traditions" Notebook			

Task/Activity	I	P	M
Dishes and Tables:			
Identify and know uses for various types of dishes (China, Porcelain, Crystal, Corningware, Stoneware, etc.)			
Learn creative table settings and napkin folding (See "Decorative Napkin Folding" booklet from Current)			
Learn creative ways to make centerpieces			
Decide on your own dishes for hope chest			
Environment Enhancement:			
Study different types of decor (country, modern, etc.)			
Re-decorate our main bathroom			
Design your ideal guest room (colors, decor, amenities)			
Study use of color in rooms – moods created			
Study window treatment for different effects			
Study use of space, decor; simplicity & busy-ness			
Heirlooms, Family Treasures, & Keepsakes:			
Take photos of the pretty corners of your room			
Mount above photos in pretty booklet or journal			
Know how to identify heirlooms and antiques			
Create your own "Heirlooms Album"			
Set up and maintain photo albums			
Organize/file your own packets of photos			
Set up Keepsake Boxes and files for your own keepsakes and treasures – label			
Safely and appropriately store long-term keepsake bags in storage shed to keep from spoiling			

[A.C.S.]
HEALTH & SAFETY

I had originally intended this class to be one credit, but "when all was said and done" and Tory had done everything that I felt was necessary for this class and, lo and behold, it only came out to ¾ of a credit, I said to myself "Hey, why not?" *(See digression in right column.)*

As you can see, this ACS for Health is divided into several categories, with specific assignments in each one. I should mention that, having gone through the Weigh Down Workshop, I have since deleted "Calculate Appropriate Number of Fat Grams (per day) for your body" from the list because I no longer feel this is necessary to be concerned about if a person embraces a lifestyle of moderate eating. (waiting until hungry to eat, stopping when satisfied.)

The "**Word Alive Study**" and "**Scripture Verse Study**" mentioned in the "Study" section are done using the forms by those titles that are here in this book. (See Form Index.)

My husband gets The "**Welles Newsletter**" via his job. It's informative and fun and encouraging to read.

Notice down at the bottom of the "Basic Requirements Check-off List" is a whole section of "**Traffic Safety.**" This was the Driver's Ed Class Tory took through the public school district (for $150, which is what the public school kids had to pay too.) This class is 60 hours of his health credit only because that's where it's put in the schools. It didn't matter to me where it went, so this was as good a place as any. It could have even been a class of its own – "Traffic Safety" – ½ credit worth. I didn't do that because it would have left only ¼ of a credit of Health.

If you are a "Weigh Downer" you may be wondering why I have all the "health-supporting" resources as part of this class. For one thing, Tory did all this *before* I went to Weigh Down. *(Actually, he went for a session of Weigh Down too just to get an understanding of the concepts. He didn't need it weight-wise!)* However, this is the *one* point I disagree with Gwen on – her view of "healthy food," so I would still have had him do most of it even if I'd been creating his class *after* going to Weigh Down. ☺ *(Someday I will explain all this in greater detail in an article I am writing. ☺ I agree with 99.5% of what Gwen teaches, so the .5 % I don't agree with is not something I want to make an issue of. Weigh Down is absolutely excellent!!!)*

The "**Wisdom's 7 Pillars for a Healthy Lifestyle**" goes along with a workshop I've done on this topic, where I liken 7 pillars for physical health to 7 pillars for spiritual health, and then to 7 pillars for a healthy homeschool. Basically these pillars (which I just came up with myself, so they're nothing "official") are: Food, Water, Sleep/Rest, Sunshine (not overdoing it or tanning), Fresh Air, Exercise, and Elimination. All these translate and co-relate physically, spiritually, and homeschool-wise.

Hey, Why Not?!?

DIGRESSION: *As many of you more perceptive ones are probably picking up on, I am saying "Hey, why not?" quite often these days. ☺ There is actually a very good reason for this. ... If my child knows and has experienced all that I feel is necessary in an area, and has done so in ¾ of a credit, then I see no reason to <u>not</u> stop there, rather than trying to force a full credit. "But what about how "¾ credit" will look on the transcript?!?" you understandably ask. Well, I have come to the point where I am realizing a couple of things. One is that our transcript is a record of what we have done, not a determiner of what we will do. For example, many states don't even require Health at all. But I did, because I felt it was valuable and necessary for life. And we covered a whole lot of stuff that the schools wouldn't dream of covering; and we <u>didn't</u> cover a whole lot of stuff that the schools cover that <u>I</u> wouldn't dream of covering. So, since many of our classes don't look at all like the schools', it's all just a "records game" anyway.*

The other thing I realized is that God is perfectly capable of opening doors for my (and your) kids no matter what's on their transcript. If a school is going to reject a homeschooled student because they have only ¾ of a credit instead of 1 in Health, then I'm sorry, but I just don't believe they were supposed to be at that school. ☺ My son is a "picture of health," has a good understanding of and is living a healthy lifestyle – including the spiritual aspects which are absolutely indispensable to being a healthy person, and which the schools are totally missing! I'm sorry, but the more I see of the system, the less I feel the need to bow down to it as my standard. God will open THE ONE door that He wants to open for my (our) kids. I know this because I have been faithful to do what I believe God wanted my children's education to consist of (not perfectly, but that's where grace comes in), and have followed, to the best of my ability, what I felt He was leading. I am confident that God will honor this. "Without faith it is impossible to please God," which means that with faith it IS possible to please Him. As you look through all our ACS's and TAS's it is evident that we have not "slacked" or "faked" our way through high school. But whether or not what we have done is "state approved" or acceptable at several colleges, I just plain <u>do not care</u>. We are on <u>God's</u> path. All that needs to open is <u>one</u> door at a time! And He IS leading us! I refuse to operate out of fear and miss what God has for us! ❧

✓ ASSIGNMENT CHECK-OFF SHEET ✓✓✓✓✓✓✓✓✓✓✓✓✓✓✓✓✓✓✓✓✓✓✓✓✓✓✓✓✓✓✓✓✓✓✓✓✓✓

Class: **Health & Safety**	
Student: **Tory W. Shelton**	Class Value: **¾ Credit = 90 Hours**

	BASIC REQUIREMENTS CHECK-OFF LIST	**DONE**
Books to Read:	Eat More; Weigh Less by Dean Ornish, M.D. - *Read selected portions*	
	Home Made Health by Dr. Raymond Moore - *Read selected portions*	
	The McDougall Plan by John McDougall - *Read selected portions*	
	Fit or Fat by Covert Bailey - *Read all*	
	Fit For Life II by Harvey and Marilyn Diamond - *Read selected chapters*	
Videos to Watch:	Watch "Fit or Fat" Video by Covert Bailey *(Twice)*	
	Watch "Fit For Life" Video by Harvey and Marilyn Diamond	
Cooking:	Make 6 health-supporting dishes: *(See our health-supporting cookbooks and/or sections of books: The McDougall Plan Cookbook, Fit For Life. 1 soup; 1 main dish; 1 salad; 1 dessert; 1 snack; 1 option)*	☐ ☐ ☐ ☐ ☐ ☐
Do:	Calculate Appropriate Number of Fat Grams (per day) for your body	
Study:	Complete "Wisdom's 7 Pillars for a Healthy Lifestyle" *(First, do generic "Wisdom's 7 Pillars" lesson with Mom)*	
	Do "Word Alive" studies on: ☐ "Food" ☐ "Strength" ☐ "Health" ☐ ________ ☐ ________	
	Read articles: ☐ Hypothermia ☐ Walking ☐ Cholesterol ☐ Fruit ☐ Beta Carotene ☐ ________	
	Read 25 issues of "Wellness Newsletter" *(Health, Safety, Nutrition, Weight Control, Stress Reduction, Fitness, Exercise, Healthful Living)* ☐	
	Complete a "Scripture Verse Study" on: II Timothy 2:5: *"If anyone competes as an athlete, he does not win the prize unless he competes according to the rules."*	
	Complete a "Scripture Verse Study" on: I Timothy 4: 7–8: *"...Discipline yourself for the purpose of godliness, for bodily discipline is only of a little profit, but godliness is profitable for all things, since it holds promise for the present life, and also for the life to come."*	
	Complete a "Scripture Verse Study" on: Philippians 3:13–14: *"...Forgetting what lies behind and looking forward to what lies ahead, I press on toward the goad for the prize of the upward call of God in Christ Jesus."*	
	Devise own "Personal Health Planning Sheet" - (Twice: First time early in high school; Again at end, after further study)	☐ ☐
Traffic Safety	Complete 6-week Driver's Ed (Traffic Safety) Course through Longview School District (30 classes @ 2 hours per class = 60 hours)	Check off hours in top section of chart below ↓

HOUR CHECK-OFF CHART

(Upper section is 60 hours of Traffic Safety; Lower section is 30 hours of Health)

90	89	88	87	86	85	84	83	82	81	80	79	78	77	76
75	74	73	72	71	70	69	68	67	66	65	64	63	62	61
60	59	58	57	56	55	54	53	52	51	50	49	48	47	46
45	44	43	42	41	40	39	38	37	36	35	34	33	32	31
30	29	28	27	26	25	24	23	22	21	20	19	18	17	16
15	14	13	12	11	10	9	8	7	6	5	4	3	2	1

[A.C.S.]
HOME & LIFE MANAGEMENT SKILLS

You can probably tell by looking at this ACS that this class was formerly "Home Ec." That previous title just didn't encompass all that I wanted my class to encompass, as you'll soon see.

This class is really a collection of a wide variety of skills and areas that I felt were necessary to effectively managing a home and life. ☺ Thus the name of the class. You will probably come up with other components that I've left out, that are important to you. Let's start at the top of the list...

PERSONAL ORGANIZATION - The first item in this section at the top was a seminar by Gregg Harris. Tory , 18 at the time he attended, very much enjoyed it, and also got the Noble Planner! ... The other portion of this section is to complete the "Personal Organization T.A.S." (There is a much simpler one for girls in Section Four under "Home Ec." – the "precursor" to this one.) Just the one for Tory (boys) is here in this section, but it wouldn't take much to adapt it to a girl.

HOME MAINTENANCE & REPAIR - This TAS is in this section, hopefully in the same order that I'm listing all the sections in. ☺ (It's getting to be pretty confusing.) These are the skills I feel are necessary for one who is going to live on their own. (Some will not be necessary for living in an apartment, but they are all listed here anyway.)

COOKING - The "Basic Cooking Skills" TAS and the "Gourmet Cooking" TAS are next. (See comments on this directly below.) However, the "Baking & Cooking Terminology" is under "Home Ec" in Section Four. (only because it started out there, before I actually and formally invented "TAS's" and it's too messy to move it.) I think it's just as important for boys to learn this stuff as for girls. And *no*, I'm *not* a "women's libber"! Men will very likely live on their own for a time before they marry, even if they're at home they can still help the family by preparing meals once in a while. Once married, there will be ample opportunities to make dinner, either regularly taking turns if they enjoy cooking (as my husband does!), or when Mom is sick, very pregnant, shopping, or *whatever!*

GOURMET COOKING - This class is not one of my requirements. But this is a suggested line-up of dishes for such a class. (This would be a portion of Home Ec. hours. Or if a student was interested in becoming a chef, it could go under Occupational Skills.) This is one class for which I made up the requirements without input from my kids because if left to them, they'd include such complex, intriguing items as: Hot Dogs Ordinairre,* Fruit Cocktail de la Cán, Hamburger Helper Extraordinnaire (with parsley added), and Pudding-in-a-Cup a la Cool Whíp (pronounced "coo-ale hweep"). Perhaps my standards are too high, but I'm going to stick to my guns on this. If, however, along the way they do come up with something requiring a bit more talánt and brain-power, I'll certainly consider substitutions.

(*Mary Schofield, one of my original "Homeschool Parent Testers" and author of *The High School Handbook* adds: "We thought "Hot Dogs Ordinairre" *was* gourmet!"

VEHICLE MAINTENANCE - All three of these TAS's, "Bicycle Maintenance," "Car/Truck Maintenance," and "Lawnmower Maintenance," are also in the Packet. As for girls doing any or all of these? I have heard of several who felt it was important for them too! Know thy child and hear thy God! ;-)

CARPENTRY - My husband and my dad, both skilled carpenters, helped me come up with the "Carpentry Skills TAS" which is in this section. One mom wrote to me with an excellent idea: "My husband has assigned our sons the task of assembling their own sets of tools or a "needs" list to gradually collect enough adequate tools (like a "hope chest" for girls) so that by the time they leave home they'll have their own and won't have to be borrowing Dad's! We check this list when grandparents want to know what they "need" for Christmas or Birthdays."... Isn't this an *excellent* idea?!?!?

HOUSEWORK SKILLS: The "Housework Skills" TAS is included here in this section (coming right up, in fact) It's just a 2-page list I compiled from a variety of sources that gives all the individual housework skills I feel are important to have, at least to some extent – not necessarily "mastery" but at least being familiar with them. (Mastery will come with practice!) Ideas for how to *do* them in the most efficient ways are in books like "Is There Life After Housework?" (by Don Aslett) or any housecleaning type book. I've always had "Job Charts" for the kids through the years that I've changed and added to as they got older.

The "Family Jobs" is a one-page list of *all* the jobs that need to be done "today" – deeper cleaning jobs that aren't on the regular job lists. I put a circle and a line wide enough to put an initial on to the left of each job. So the list would start out like this:

O___ Sweep front porch

O___ Clean front entry (sweep, hand-mop)

O___ Clean microwave, inside and out

O___ Clean dining room French doors

O___ Clean windowsills

O___ Vacuum whole house

O___ And lots more!

This is something I have a "master" for in my computer, and then I tailor it for whatever needs to be done today. I put it on the table with a pen, and then people choose a job, put their initial next to it, go do it, then, when

✓ ASSIGNMENT CHECK-OFF SHEET ✓✓✓✓✓✓✓✓✓✓✓✓✓✓✓✓✓✓✓✓✓✓✓✓✓✓✓✓✓✓✓✓✓✓✓✓✓✓

Class: Home and Life Management Skills

Student: Tory W. Shelton	*Class Value:* 1 Credit = 120 Hours

BASIC REQUIREMENTS CHECK-OFF LIST

Personal Organization	☐ Attend "Managing Life's Priorities" presented by Gregg Harris
	☐ Complete: "*Personal Organization T.A.S.*"
Home Maintenance	☐ Complete: "*Home Maintenance and Repair T.A.S.*"
	☐ Take care of the home when Dad or Dad and Mom are gone
Cooking	☐ Complete: "*Basic Cooking Skills T.A.S.*" ☐ Complete: "*Gourmet Cooking T.A.S.*"
	☐ Complete: "*Baking & Cooking Terminology T.A.S.*"
Vehicle Maintenance	☐ Complete: "*Bicycle Maintenance T.A.S.*" ☐ Complete: "*Car/Truck Maintenance T.A.S.*"
	☐ Complete: "*Lawnmower Maintenance T.A.S.*"
Carpentry	☐ Complete: "*Carpentry Skills T.A.S.*"
Housework Skills	☐ Complete: "*Housework Skills T.A.S.*"
	☐ Skim and familiarize yourself with Don Aslett's book: Is There Life After Housework?
	☐ "Job Chart" Jobs *(through the years)* ☐ Do "Family Jobs" *(with bigger, deeper-cleaning housework jobs)*
Financial Skills and Wisdom	☐ Read "Budgeting" and do applicable exercises ☐ Do "Living Expenses Comparison Study"
	☐ Read booklet: "A Biblical Perspective on Debt-free & Prosperous Living"
	☐ ☐ ☐ Listen to "Father's House" sermons on "Tithing" ☐ Read Vineyard brochure on "Tithing"
	Read these sections of Steck-Vaughn "Consumer Math": ☐ **DRIVING A CAR** (Unit 5: pages 104 – 124) ☐ **OWNING A HOME** (Unit 6: pages 126 – 146)
	Read these sections of Abeka "Applied Mathematics": *(First read selected portions of Units 1 and 3 in the "Social Studies" ACS under "Economics")* **INCOME & TAXES (UNIT 4)** ☐ Earning a Living (341) ☐ Deductions from Gross Pay (348) ☐ Social Security Tax (351) ☐ Income Tax Return 1040EZ (356) ☐ ...1040A (359) ☐ ...1040 (361) ☐ State Income Taxes ☐ Sales Tax (374) ☐ Property Tax (380) ☐ Corporate Income Tax (385) **BANKING (UNIT 5)** ☐ Checking Records (393) ☐ Checks and the Register (398) ☐ Proving (balancing) the Checkbook Balance (411) ☐ Electronic Banking (417)
Hunting Skills	☐ Go Elk, Deer, and Grouse Hunting with Dad
	☐ Learn Proper Techniques for Dressing (gutting out) the animal, Skinning, and Hanging
	Learn Techniques for: ☐ Butchering ☐ Wrapping ☐ Smoking ☐ Freezing (the above Meat)

HOUR CHECK-OFF CHART

120	119	118	117	116	115	114	113	112	111	110	109	108	107	106	105	104	103	102	101
100	99	98	97	96	95	94	93	92	91	90	89	88	87	86	85	84	83	82	81
80	79	78	77	76	75	74	73	72	71	70	69	68	67	66	65	64	63	62	61
60	59	58	57	56	55	54	53	52	51	50	49	48	47	46	45	44	43	42	41
40	39	38	37	36	35	34	33	32	31	30	29	28	27	26	25	24	23	22	21
20	19	18	17	16	15	14	13	12	11	10	9	8	7	6	5	4	3	2	1

done, come back and check it off (in the circle) and choose another job. Sometimes they prefer to choose them all at once, taking turns, then they know what's expected and can "look forward" (??) to it. ☺

FINANCIAL SKILLS AND WISDOM: "Budgeting" was just a workbook I found at a curriculum sale. The "Living Expenses Comparison Study" is here in this section, with the other T.A.S.'s for this class. (It's right before "Lab Science Survey's ACS.) The other items are things I have found "along life's way" at churches we've visited, our church, conventions, people, etc.

HUNTING SKILLS: This is how I broke down Tory's hunting experiences. He is now able to take an animal from field to freezer, (with a little help from his dad)! He's gotten nine! By the way, any of you with hunters might want to have them make a Hunting Notebook! The title could be "From Field to Freezer." I had wanted Tory to make such a notebook, and even got all the pictures for it, but he ended up doing a Basketball Notebook instead. They are both very large projects – at least the way we would have done them. An *one* was plenty. ☺ By the way, there is a "Grading Criteria and Course Assessment" form for "Hunting." *(Even if you're not in the least interesting in hunting, you might want to take a look at it anyway; it's probably not at all what you would have expected.* ☺) Did you look?! Well, are you beginning to see how real-life all my stuff is!?!) And about the "Hunting" Grading Criteria sheet... This could *possibly* be a whole class, depending on your lifestyle and priorities, but it might be just a "unit." (a term I use loosely) As with everything in the Grading section, they're just ideas to use with grading the stuff, not forming a class; that's what *this* section is for!

"PERSONAL ORGANIZATION" T.A.S.

(For Boys)

We put this T.A.S. under "Home & Life Management Skills," or you could put it under "Life Skills" or wherever else you think it fits. I organized all the stuff I wanted Tory to be adept at into the categories of...

ORGANIZING PERSONAL SPACE:

This includes organizing the bedroom (broken down into all the specific areas) and his car, including "equipping" it. (Yes, he worked to pay for his own car. ☺) There are many books that will help with all of this if you are not naturally inclined that way. He did not count every cleaning or organizing session! Only the organizing/set-up aspect.

TIME MANAGEMENT:

This includes setting up his daily planner, writing up long-term and short-term goals, the other three activities listed there. (By the way, Gregg Harris offers a great seminar on personal planning! He has created the "Noble Planner" which is an excellent life and daily planner for men.)

ORGANIZING YOUR STUFF:

Here we have "organizing photos" (pictures and negatives), setting up a workout bag, keepsakes box (which is added to as time goes by), and a flat keepsakes folder, which is just a notebook with pocketed dividers to hold flat items that would otherwise get crumpled in on their own with all the larger keepsakes.

PERSONAL RECORD KEEPING:

The "Articles & Ideas From Mom" notebook idea was started by *my* mom who does this for all her children. Whenever she finds articles that she thinks we'd find interesting, she makes six copies of them and sends them to all of us! She already set up a notebook for each of us, including several categories with tabbed dividers so that we are ending up with a very good and handy reference book, written by others, but from Mom's heart to us!

Setting up and keeping an address book is pretty basic, but it's a good idea for your kids to get one set up *before* they leave home, and include all their parents' friends as well as their own friends.

(The negatives part should actually be in the "Organizing Your Stuff" section above.)

The Warrantees Notebook consists of an extra-wide notebook filled with plastic sheet protectors that open at the top. As we purchase items, we slip the warrantee, instructions, receipt, etc. into a plastic sheet protector, and stick a tab onto that, in alphabetical order. *So* handy when something breaks down!

PERSONAL ORGANIZATION T.A.S.

ORGANIZING PERSONAL SPACE:

ORGANIZE BEDROOM *(including floor, shelf over bed, space behind bed, walls, entertainment center, drawers, and the following major areas:)*	
Organize Closet *(incl. hanging clothes, floor, shelves, with boxes)*	
Organize Desk, Shelves over Desk, and Work Area	
Organize Your C.D.'s and Audio Tapes *(at home)*	

ORGANIZE AND "EQUIP" YOUR CAR: ☐ Glove Compartment ☐ Trunk ☐ Music ☐ C.D.'s "to go" ☺

TIME MANAGEMENT:

Set up and Use Daily Planner *(Style and amount of detail are completely your choice)*	
Write up Long and Short-term Personal Goals: *(in major areas of life, including: Relationship w/ God, Family/Friends, Church/Ministry, Education, Job, Talents, Personal Edification)*	
Use "Focus Affixer" for one month *(and adapt to own needs each week)*	
Keep log of all activities for one week *(using "Day of Accountability" sheet)*	
Start personal journal *(use store-bought journal or set up 3-ring notebook)*	

ORGANIZING YOUR STUFF

Organize Photos *(in whatever way is most appropriate for you: Album, Files, Packages)*	
Set up Workout Bag *(with whatever contents you need)*	
Set up Keepsakes Box	
Set up Flat Keepsakes Folder *(To go in Keepsakes Box)*	

PERSONAL RECORD KEEPING

Set up "Articles & Ideas From Mom" Notebook *(Mom will help with categories, and will send you input for this throughout the rest of her lifetime)*	
Set up "Personal Organizing" File or Section of the above notebook	
Set up own Address Book or Rollo-dex *(Transfer any addresses you want for your own use, once living out on your own, from the Family's address book)*	
Set up Negatives Organizing System *(Or have system set up and ready to go when you start taking own pictures – first roll!)*	
Set up Warrantees Notebook or File *(Include all items that you have already; Do in Alphabetical order for easiest accessibility)*	

Home Maintenance and Repair T.A.S.

TASK/ACTIVITY	DONE	TASK/ACTIVITY	DONE
GENERAL:		**INDOORS:**	
Install a lock		Properly hang items on walls	
Be familiar with referring to Consumer Guides to evaluate appliances and all home products		Know uses of: various types of screws mounting braces, and their uses, i.e. wood, metal, sheetrock, etc.	
Know where your fuse box is and how to replace a fuse or...		Be able to paint walls	
Know where breaker box is and how to re-set a thrown breaker		Know the helpful techniques in painting (i.e. brush into the wet paint; not allowing paint to get all the way to the top of brush; uses for dirrerent kinds of brushes; when to use a roller (foam or fuzzy), pad or brush.)	
Know how to choose interior and exterior paint		Repair wall holes with putty	
Know the diff. betw. latex, stain, varnish, etc.		Be able to hang/set up shelves	
Use weather and all-purpose caulking		Be able to start a fire and keep it going	
Know what to look for in home appliances		Be able to hang wallpaper	
Be able to use basic tools such as hammer, screwdriver, tape measure, etc.		Be able to mount moulding	
OUTDOORS:		**PLUMBING:**	
Paint house		Remove toilet (to get item out) & replace donut	
Paint trims and/or planters (and know how to be neat)		Be able to install new faucet fixture	
Know which tools/supplies to have on hand for own needs		Unplug overflowing toilet with plunger	
Have a place to store these properly *(where they won't rust, and are accessible)*		Be able to set up washer and connect to proper water hoses	
Clean out gutters		Unclog sink or tub pipe using chemicals	
Cut and stack wood, chain saw or or axe (safely and appropriately)		Be able to set up dryer and vent to outside	
Know how to dig a hole and set a post for a fence		Be able to change water filter	
Know where your natural gas shut-off is located and how to operate it		**CITY:** Locate main water shut-off to your home ...OR... **COUNTRY:** Have some knowledge of your well	
Know how to pour cement for simple jobs		Shop for and install or replace a garbage disposal	
BASIC VEHICLE CARE:		**BASIC FLOWER/PLAN GARDEN & YARD CARE**	
(For complete listing of Vehicle Maintenance & Repair tasks, see separate "Vehicle Maintenance & Repair T.A.S." sheet)		Mow lawn and know how to operate electric, gas powered, or hand-pushed mower	
Clean out inside of vehicle *(clean out garbage, litter bag, fold blankets, straighten & re-organize organizer caddy & tapes)*		Know optimum frequency and height for grass, considering your weather conditions	
Know how to wash outside of vehicle *(proper equipment and techniques, in different types of weather)*		Know how to plant a garden w/ plant/flowers that are appropriate to your area, your sunlight/ shading conditions, to each other (height, compatibility), and to your level of dedication to maintain	
Clean/polish dashboard and instrument panel			
Vacuum vehicle carpet		Water lawn (freqency for your area; have appropriate sprinklers)	
Clean (wash) vehicle carpet		Water garden (and have appropriate supplies)	
Wash car windows, inside and out		Know how to weed	
Be able to winterize your vehicle; antifreeze, etc.		Be able to administer fertilizer to garden/lawn	
BASIC ELECTRICAL SKILLS:		Know how to do (and practice!) basic pruning	
Change the plug on an electric cord		Be able to identify common pests and how to eradicate	
Be able to operate voltage tester		Know when/how to prune ornamental trees & shrubbery	
Be able to operate continuity tester		Be able to choose appropriate fertilizer(s) for lawn and garden – organic and chemical	
Be able to use needlenose pliers to strip wire			
Be able to install new receptacle on existing circuit			
Be able to replace ceiling light fixture		Be familiar with your hot water tank	
Be able to install a dimmer switch		Know your central heating system, ducts, controls	
Be able to rewire incandescent lamp		Know your air conditioning system	
Be able to replace ballast on flourescent lamp		General wiring skills for telephones	
Be able to get correct light bulb for flashlight		General wiring skills for computer set-up	

Basic Cooking Skills T.A.S.

Skills	Specific Item	Recipe / Source	✓
Rolled Biscuits			
Muffins			
Rolls *(yeast, raised)*			
Yeast Bread			
Quick Bread *(banana, pumpkin, etc.)*			
Cornbread			
Bread Machine Bread			
Garlic Bread			
Whipped Cream			
Coffee *(pot of fresh-ground from beans)*			
Cookies - rolled *(use cookie cutter)*			
Cookies - drop			
Bar Cookies			
Brownies			
Cinnamon Rolls			
Donuts			
Cream Pie			
Lemon Merengue Pie			
Baked Fruit Pie			
Pumpkin Pie			
Crisp *(Apple, Cherry, Peach, etc.)*			
Pudding *(box, cooked)*			
Cake - from scratch			
Coffee Cake with Topping			
Frosting - from scratch			
Candy - cooked			
Scrambled Eggs			
Fried Eggs *(hard, medium, soft)*			
Boiled Eggs			
Poached Eggs			
Egg Salad (Sandwiches)			
Deviled Eggs			
Hot Cereal *(oatmeal, crm. of wheat, etc.)*			
Gravy *(in pan - from scratch)*			
White Sauce *(cheese, creamed tuna)*			
White Rice - slow cooking			
White Rice - instant			
Rice - *(Mom's chicken or beef flavored)*			
Rice-a-roni			
Brown Rice			
Baked Potatoes			
Raspberry Cordial			
Nachos			
Mashed Potatoes - from scratch			
Mashed Potatoes - instant			

(BASIC COOKING SKILLS T.A.S. *continued*)

SKILLS (CONT'D.)	SPECIFIC ITEM	RECIPE / SOURCE	✓
Macaroni and Cheese *(homemade)*			
Chicken Noodle Soup *(homemade)*			
Vegetable/Bean/Barley Soup			
Potato Soup			
Spaghetti (from scratch)			
Bean Enchiladas			
Fettucini with Alfredo Sauce			
Lasagne			
Spaghetti Pie			
Chicken and Dumplings			
Pizza			
Chili (con Carne)			
Jam or Jelly - fresh/freezer			
Can Fruit *(water bath)*			
Green Salad			
Pasta Salad			
Fruit Salad			
Caesar Salad			
Potato Salad			
Jello/Fruit Salad			
Waffles			
Pancakes			
Maple Syrup *(Homemade)*			
Brown Sugar *(from molasses & white sugar)*			
Roast Beef			
Baked Pork Chops			
Baked Chicken			
Baked Ham			
Swiss Steak (elk)			
Sweet 'n' Sour Meatballs			
Porcupine Meatballs			
Meatloaf			
Fish - Baked or Broiled			
Deep-fried Fish			
Roast Turkey (whole)			
Stuffing - in Turkey			
Dip - for vegetables or chips			
Cook Frozen Vegetables			
Steam Fresh Vegetables			
Corn on the Cob			

GOURMET COOKING T.A.S.

BASIC REQUIREMENTS CHECK-OFF LIST

ITEM	DATE	RECIPE SOURCE	Done	RESULTS	Time
Seafood Thermador					
Stuffed Flank Steak					
Cantonese Meatballs*					
Decorated Cake*					
Lattice Top Pie*					
Cherries Jubilee					
Shrimp Creole					
Chocolate Torte					
Marzipan					
Mint Truffles					
Dipped Candy*					
Custard Pudding*					
Quiche Lorraine*					
Turkey (or Chicken) Divan*					
Shiskebobs*					
Hors d'oeurvres*					
Soufflé*					
Petit Fours					
Fondue*					
Crescent Rolls*					
English Toffee*					
Peaches Flambé					
Student's Choice:					
Student's Choice:					
Student's Choice:					
Student's Choice:					
Student's Choice:					
Student's Choice:					
				Total Time =	

Vehicle (Car, Truck) Maintenance T.A.S.

TASK/PROCEDURE	I	P	M
Know basic functions/operation of the vehicle			
Know how to fill with gas - what type for optimum performance			
Be able to check the oil			
Be able to change the oil			
Put antifreeze in radiator			
Know your vehicle's antifreeze type and percentage w/ water			
Be able to change a flat tire			
Know when to have studded tires put on and off			
Know how to deal with overheating engine			
Take part in replacing break pads			
Be able to change fuses			
Be able to change dome lights			
Be able to jump-start a vehicle (w/o killing anyone)			
Be able to fill tires with air (w/o popping them)			
Be able to select and change wiper blades			
Be able to add window-washer fluid			
Be able to clean eroded battery terminals			
Know what's involved in engine tune-up			
Be able to change headlights			
Be able to check transmission fluid			
Be able to check brake fluid			
Be able to check and fill steering fluid			
Know how to check radiator fluid and fill			
Know differences and effects of front, rear and 4-wheel drive			
Know how and when to turn on hazard lights			
Be able to change air filter			
Be able to check and change spark plugs			
Be able to change brake and back-up lights			
Know safety precautions for driving in snow			

Bicycle Maintenance T.A.S.

TASK/PROCEDURE:	I	P	M
Know basic functions/operation			
Basic repair of bicycle			
Lubricate a bike chain			
Re-apply a bike chain			
Adjust bicycle seat			
Be able to pump up a tire			
Clean bicycle properly			
Be able to true a wheel			
Adjust the brakes			
Lubricate all moving parts *(shifters, cables)*			
Be able to fix a flat tire			

Riding Lawnmower T.A.S.

TASK/PROCEDURE	I	P	M
Know basic functions/operation of the mower			
Fill mower with gas			
Check the oil			
Be able to start mower			
Safely operate mower			
Safely park mower			
Clean fouled plug			
Gap the spark plug			
Remove blade and take in to be sharpened			
Check linkage on transmission			
Clean air filter			
Tighten brakes			

Carpentry Skills T.A.S.

TASK/ACTIVITY	I	P	M
PRELIMINARY:			
Gather basic carpentry reference books			
Set up organizer drawers for paraphernalia			
Set up own tool box with basic supplies			
Know different types of wood and purposes			
Understand lumber grading & numbering system			
SHOP SAFETY:			
Know proper way to store paint			
Know how to safely handle hand and power tools *(as listed in section below and to the right)*			
FAMILIARITY WITH BASIC TOOLS:			
Hammer			
Screwdrivers			
Pliers			
Wrenches			
Level			
Handsaws (Rip, Crosscut, Pruning, Coping, Tree)			
Tape measure			
Chisel			
Squares (L, triangle, adjustable/butterfly)			
Plumbline (and chalkline)			
Fasteners: *(Have a knowledge of each; which shapes and lengths are needed for which purposes)*			
♦ Nails			
♦ Screws			
♦ Bolts, nuts, etc.			

TASK/ACTIVITY	I	P	M
OPERATION OF THESE POWER TOOLS:			
Electric sander			
Electric screwdriver			
Electric drill			
Table saw			
Power saw - hand-held			
Mitre saw *(esp. for cutting 45° angles, as for corners)*			
Router			
Jigsaw			
CONSTRUCTION EXPERIENCE:			
Be familiar with / Know how to:			
Using a pattern (for furniture, toys, etc.)			
Drawing/Layout of project			
Job Planning (which tools, lumber & supplies you'll need)			
Trimming (windows, doors, etc.)			
Use of putty (to "cover a multitude of sins"!)			
Nailing (without leaving dents)			
Hanging doors			
Wood Finishing:			
♦ Sanding			
♦ Routing and edging			
♦ Painting			
♦ Staining			

Housework Skills T.A.S.

TASK/PROCEDURE	I	P	M
CLOTHING CARE:			
Separate and sort dirty laundry			
Wash laundry (& know appropriate temp's, leave 4 hrs. max., etc)			
Simple Mending (rips, patches, buttons, etc.)			
Straighten laundry room			
Basic spot removal (oil, coffee, blood, tea, soda, etc.)			
Fold, sort, and put away clean laundry			
Iron Clothing (be aware of heat diff's for various fabrics)			
Take clothes for dry cleaning			
Waterproof shoes and boots			
Polish shoes and boots			
Hand wash clothing (nylons, delicates) & hang to dry			
KITCHEN UPKEEP & CLEANING:			
Clear table & put away leftovers (label, date)			
Set table; basic and formal			
Rinse, load, & put away dishes in dishwasher			
Wash & dry dishes & pots by hand			
Clean counters (and under items)			
Clean out refridgerator (old food, wipe inside, sides, top)			
Clean stove top and control panel			
Clean under stove burners			
Clean oven			
Wipe surfaces of appliances and cupboards			
Clean out under sink			
Arrange cupboards in efficient manner			
Defrost freezer			
Operate appliances *(stove, ovens, dishwasher, blender, mixer, bread mach)*			
Set and adjust temperatures in fridge / freezer			
BATHROOM:			
Be able to properly: Clean toilet, inside and out			
Clean sink and counter			
Clean cupboard and drawer surfaces			
Clean bathtub			

TASK/PROCEDURE	I	P	M
GENERAL:			
Dust furniture			
Use vacuum cleaner, attachments			
Shake and machine wash area rugs			
Scrub and spot clean walls and doors			
Mop and scrub floors			
Change vacuum cleaner bag and belt			
Polish wood furniture			
Empty wastebaskets			
Wash windows, inside and out			
Wash window screens			
Clean mini-blinds			
Shampoo carpet			
Clean runners under sliding doors			
Clean combs and brushes			
Clean light fixtures			
Use broom and dustpan			
Fold blankets and sheets neatly			
Replace light bulbs & understand wattage			
Change bed sheets			
Clean fireplace			
Wash mirrors, and French doors			
Clean window sills			
Clean cobwebs (corners, pictures, shelves)			
Put away groceries			
Clean a messy room (using the "5 Sack System)			
Damp-wipe kitchen chairs (all over) and table			
Check and clean/change furnace filter			
Vacuum and clean air and eating vents			
Vacuum upholstery and curtains			
Use vac. cleaner & attachments / put away			
Strip and re-wax linoleum floors			
Strip and re-wax hardwood floors			

(HOUSEWORK SKILLS T.A.S. continued)

TASK/PROCEDURE	I	P	M
PET CARE:			
Feed pets			
Set up kitty litter box, change			
Clean pet living area & food/water bowls			
Give pet a bath			
Purchase and apply flea collar and powder			
Be able to take pet in for shots – know price			
Be able to have pet spayed – know price			
YARD/OUTDOOR SKILLS:			
Wash vehicles			
Water grass and garden			
Put away hose properly			
Sweep porches, patios, walks			
Clean out between wood slats			
Burn garbage and fire safety			
Correctly discard garbage / take to dump			
Pick up trash/junk around yard			
Trim trees and bushes			
TELEPHONE SKILLS:			
Write down phone messages			
Answer phone appropriately & take verbal mssgs			
Create answering machine message			
Set up wiring for answering machine			
Place long distance calls using card			
Place collect calls			

TASK/PROCEDURE	I	P	M
ORGANIZING SKILLS:			
Set up a medicine cabinet			
Set up kitchen cupboards and drawers			
Set up cupboard under sink			
Set up hall closet			
Set up linen closet			
Set up bathroom cabinets and drawers			
Set up closets and storage areas			
Set up personal drawers, shelves, & closet			

LIVING EXPENSES COMPARISON STUDY

(Not all of the following will necessarily be expenses that every person will want or feel a need for! Inclusion of a specific item in no way implies its necessity!)

Expense	Least Expensive ("Cheap," "chintzy," near poverty level)	Mid-range (Average, "OK," "socially acceptable")	Expensive ("Posh," "ritzy," extravagant, *nice*)
Rent			
Food			
Car Insurance			
Life Insurance			
Medical/Health Insurance			
Eye Care			
Dental Care			
Utilities/Electricity			
Water / Sewer / Garbage			
Phone			
Cable (TV)			
Gas			
Vehicle Maintenance			
Car Payment			
Entertainment			
Clothes			
Gifts / Cards			
Visa / Mastercard			
Newspaper			
Home Maintenance & Improvement			
Toiletries			
Education / Books			

[A.C.S.]
LAB SCIENCE SURVEY / BASIC SCIENTIFIC LITERACY

This class started out being "Science" (general) and "Lab Science," but as we went along, it became difficult to distinguish between the two, so I just lumped it all into one class that I named the above title. This is why there are two credits of this "one" class. I realize the title is long, but since science is so important, and this is two credits' worth, I wanted a comprehensive title for our transcript.

Why the term "survey"? Because it indicates a broad spectrum of study areas, not just one, as biology or chemistry would be. Rather our class included a wide variety of studies over the years that included such areas as biology (the majority), chemistry (a little), geology and astronomy (3-day camps on each of those), creation science, and many lab experiences. If my child were going into a scientific occupation, this would not have been enough, but I felt it was ample for our purposes, which are actually stated in the class title, thus part of my reason for naming it that. ☺ Let's go through each section on the ACS:

PRELIMINARY / SET UP: The first step was to set up a Lab Science Notebook. Step-by-step instructions for doing this, including the reproducible masters, are in my book "Lab Science: The How, Why, What, Who 'n' Where Book," (which shall hereafter be more succinctly referred to as LSB for Lab Science Book.) ... The "Why in the World Would I Want or Need to Study Lab Science" worksheet is also in LSB, and you can probably guess from the title why I might have felt the need to make up such a sheet. ;-)

SCRIPTURE STUDY: The "God's View of Science" study sheet, which can be used for any branch of science, is in the LSB. The "Word Alive" study is in this book..

SCIENCE SEMINARS / CAMPS: All three of these camps were offered to homeschoolers and taught by people who were experts in their fields. The "Unlocking the Mysteries of Creation" seminar offered by Dennis Petersen is excellent, and is also in book and audio tape form, both of which we sell. (He may even have a video out now.) He is fascinating to listen to! These are appropriate for junior high through adult. Another great resource for Creation Science that I didn't get onto this ACS is "The Evolution of a Creationist"! It's by a man who started out a confirmed evolutionist, but because of some persistent students in a college class he was teaching, eventually "evolved" (by God's design) into a confirmed creationist! *(We carry that too!)*

FIELD TRIPS: I wanted our kids to see how science is applied in real-life work settings, thus we went on five field trips. The other two, not listed here, were to the lab of a pediatric clinic and the lab of our local water filtration plant. You'd be surprised at how many places have labs! Would you have thought that an Ocean Spray Cranberry harvesting factory would have one?! Or a paper manufacturing plant?

CD-ROM'ing: Just a CD-Rom program. ☺

READING: I had my kids read the first 2 sections of my LSB because these are designed to open eyes to the need for and wonder of science. The other books listed are ones I found from a variety of sources and felt were valuable.

LAB ACTIVITIES: The masters for "Nature Drawings"; "Microscope Drawings"; "Loupe Drawings" *(based on the "Private Eye" approach and book)*; "Science Clippings"; "Dissection Drawings"; "Lab Experiment Record"; and "Scientist Interview Sheet" are all in my LSB. We got cow eyeballs from a butcher for free and had a friend who was attending nursing school *(taking anatomy)* help us with that dissection. ... We got the fetal elk via a road kill that my husband came upon at work. *(He's a police officer.)* The norm would be to do a fetal pig, which you'd get as a preserved specimen, but getting it this way (fresh) was far superior, as the texture and color were more natural. *(dead, but natural* ☺) If you can get a fetus from a butcher, that's the best, but if that's not available, the preserved specimens would still be very interesting and educational. I chose the frog, starfish, crayfish, and earthworm dissections because they were included in a preserved specimen kit we got from Carolina Biological Supply. ... We also bought dissection tool kits and the dissection pad/cover, which were not expensive at all and made our work very do-able! The preserved specimens came with a very good, small but concise, how-to manual. We also got much use out of Kathleen Julicher's materials, including "Experiences in Biology."... I got the experiments out of several science books we had on hand. They were not textbooks, thus I made up the "Lab Experiment Sheet" (in LSB) which we could use for any experiment we found. ... The "Scientist Interview" should actually *not* be under "Lab Activities," but in its own category. And this could count under English too, under "Study & Research Skills" as interviewing is a key skill in research. ... One lab activity I left out was butchering a deer, which we got pictures of, from start to finish. In fact, *lots* of pictures are included throughout the Lab Science Notebook! (It's quite a treasure, at least to *me!* ☺) I should actually have included a section on this form for "Creation Science." I don't know why I didn't, but it *is* inter-woven throughout all our studies. I didn't hesitate to use materials that were not specifically Christian because it's all just knowledge of the world God designed. But there were a few resources that were too pointedly evolutionistic, so we stopped using them; too much to weed out.

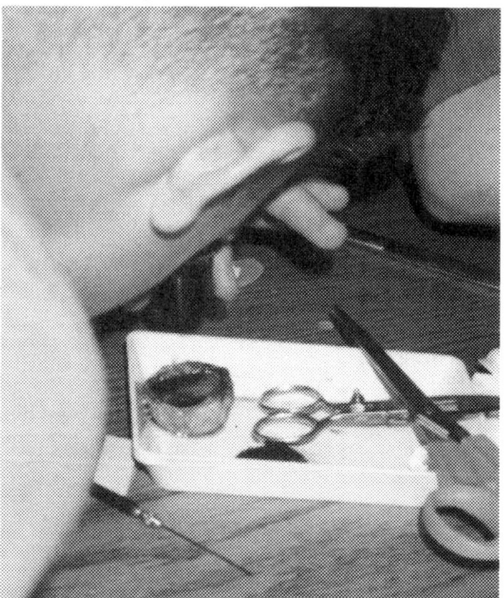

Tory examines a cow eyeball (obtained from a butcher); the retina, to be exact, through a (jeweler's) loupe. Fascinating!!!

✓ ASSIGNMENT CHECK-OFF SHEET ✓✓✓✓✓✓✓✓✓✓✓✓✓✓✓✓✓✓✓✓✓✓✓✓✓✓✓✓✓✓✓✓✓✓✓✓✓

Class: **Lab Science Survey / Basic Scientific Literacy**

Student: **Tory W. Shelton** | *Class Value:* **2 Credits = 240 Hours**

Basic Requirements Check-off List

Preliminary / Set Up: ☐ Set up Lab Science Notebook *(with divider tabs, copies of forms, photos, etc.)*

☐ Complete "Why in the World Would I Want or Need to Study Lab Science" Worksheet

Scripture Study: Complete a "God's View of Science" for: ☐ "Biology" ☐ Chemistry

☐ Complete a "Word Alive Study" for the word "Science"

Science Seminars/Camps: ☐ Attend 3-day Astronomy Camp - Taught by Dr. Robert McClung

☐ Attend 3-day Crime Lab (Criminology) Camp - Taught by Prof. James R. Goodwin

☐ Attend 3-day Geology Camp - Taught by Terry Tollefson

☐ Attend "Unlocking the Mysteries of Creation" seminar - presented by Dennis Peterson

Field Trips: ☐ Attend Lab Field Trip to Interox Hydrogen Peroxide Plant

☐ Attend Lab Field Trip to Norpac Paper Manufacturing Lab

☐ Attend by proxy the Lab Field Trip to Ocean Spray Cranberry Testing Lab

CD-Rom'ing: ☐ Explore the Dorling Kindersley Biology program: "The Ultimate Human Body"

Reading: ☐ ☐ Read sections 1 and 2 of Lab Science: The How, Why, What, Who, 'n' Where Book

☐ Read From Field to Lab and do assigned activities throughout the reading

Read these Usborne Books: ☐ Essential Biology ☐ Essential Chemistry ☐ Essential Physics

☐ Read book: Johannes Kepler (Godly Heroes Series) ☐ Read Selections from Visual Dictionary of the Human Body

☐ Read: Structures and Materials ☐ Forces in Action ☐ Plastics ☐ Radiation ☐ Oils

Lab Activities: Complete 2 Nature Drawings *(including: Yellow-shafted Flicker (bird); Centipede upper body)* ☐ ☐

Do 6 Microscope Drawings *(incl: Fat Cell Process; Diatoms; Algae; Eudorina Protozoa; Spirogyra Algae; Butterfly Wing)* ☐ ☐ ☐ ☐ ☐ ☐

Draw 7 Loupe Drawings *(including: Insect Wing, Moss, Pumice, Decomposing Leaf, Dusty Miller Leaf, Twig)* ☐ ☐ ☐ ☐ ☐ ☐

Collect 10 Science Clippings from newspapers and science/health magazines ☐ ☐ ☐ ☐ ☐ ☐ ☐ ☐ ☐ ☐

Do 6 assigned Dissections: ☐ Cow Eyeball ☐ Fetal Elk ☐ Frog ☐ Starfish ☐ Crayfish ☐ Earthworm

Do the 5 assigned Lab Experiments, including a "Lab Experiment Record" for each one ☐ ☐ ☐ ☐ ☐

☐ Interview 1 scientist using the "Scientist Interview Sheet"

Hour Check-off Chart

120	119	118	117	116	115	114	113	112	111	110	109	108	107	106	105	104	103	102	101
100	99	98	97	96	95	94	93	92	91	90	89	88	87	86	85	84	83	82	81
80	79	78	77	76	75	74	73	72	71	70	69	68	67	66	65	64	63	62	61
60	59	58	57	56	55	54	53	52	51	50	49	48	47	46	45	44	43	42	41
40	39	38	37	36	35	34	33	32	31	30	29	28	27	26	25	24	23	22	21
20	19	18	17	16	15	14	13	12	11	10	9	8	7	6	5	4	3	2	1

120	119	118	117	116	115	114	113	112	111	110	109	108	107	106	105	104	103	102	101
100	99	98	97	96	95	94	93	92	91	90	89	88	87	86	85	84	83	82	81
80	79	78	77	76	75	74	73	72	71	70	69	68	67	66	65	64	63	62	61
60	59	58	57	56	55	54	53	52	51	50	49	48	47	46	45	44	43	42	41
40	39	38	37	36	35	34	33	32	31	30	29	28	27	26	25	24	23	22	21
20	19	18	17	16	15	14	13	12	11	10	9	8	7	6	5	4	3	2	1

[A.C.S.]
OCCUPATIONAL EDUCATION

You may notice that this class is for Sharnessa, not Tory, as the rest of the classes are. This is because Tory never really did enough in this to form a whole or partial credit, whereas Sharnessa did, so I just used her form.

There should actually be boxes for 120 hours at the bottom there in the Hour Check-off Chart. Woops! ☺

The activities for this class were all unique and didn't really fit into categories, therefore everything is just listed item by item. As you'll see upon closer inspection, almost all of these activities were just part of Sharnessa's "real life." I didn't think up most of this class; it just unfolded, and I just grabbed what I saw happening (most of it "after the fact"), and funneled it into this class title. Do you think that a student who took an "Occupational Ed." class out of a textbook and listening to a teacher lecture had a richer, more comprehensive experience than what follows?

SECRETARIAL AND OFFICE SKILLS: This was a variety of skills that Sharnessa learned and used working for our home business. It involved orders, packaging, phone skills; copying, collating, and comb-binding books; working with a variety of office machines, people skills, working under a boss, following through on assigned tasks, and a host of other activities and skills.

ATTEND HOME & FAMILY BUSINESS WORKSHOP: This is a workshop by Gregg Harris our family attended in Portland. It's also on audio tape and is very good! (Of course, I love *anything* by Gregg!)

RECORDING STUDIO EXPERIENCE: On several occasions Sharnessa had the opportunity to record in a recording studio (singing). *(She is now, as of this writing in April '99 about to use this skill on a professional basis, as she was offered a job in London to be part of an evangelistic "pop band" that will tour both the U.S. and the U.K., and will produce 5 CD's through Sparrow Records!)*

TEACH DANCE LESSONS: At various times, Sharnessa taught a variety of dance lessons to girls her sister Carlianne's age (9 years younger), and to a group of younger girls at church. She was paid for doing so (by us too), and put on little recitals. At the graduation reception for Sharnessa, Carlianne (our then-10-year-old) and a friend performed the dance Sharnessa had taught them! (something about "Leaning on Jesus.") ... *And by the way, Sharnessa was recently asked to choreograph the production that is produced at the end of the year by the School of Creative Ministries that she is attending as a student! (This is at I.B.I.O.L. – International Bible Institute of London.)*

LEARN TO OPERATE: Computer; Printer, Copy Machine; Comb Binder, Adding Machine, Answering Machine (and record Homeschool Info Line and Business Line messages.) These became common utilities to Sharnessa as she worked for us.

INTERVIEW AN ON-DUTY POLICE OFFICER: Sharnessa rode with him in his police car for part of a shift. She took along a clipboard with the "Occupational Interview" sheet on it (page IV-77), and asked away! The form opened up other topics of conversation as well. This officer was a friend of my husband's, who is also a police officer. But you don't have to know one to be able to do this. (There may be a minimum age requirement for riding; 16 for our dept.)

TELEPHONE SKILLS: Sharnessa learned how to screen calls for me and make calls for the homeschool video-correspondence course we offered – both of which were invaluable to me and I missed sorely when she left home! Not only did they help me, but they also increased her skills of dealing with people, as many of the calls were very "stretching" for her. ☺ We had many talks, before and after calls, about how to approach the situation, and what could have been done differently.

HOME CONSTRUCTION: When we turned our garage into a bedroom, storage room, and office, Sharnessa and Tory both helped Dave, not just "gophering" but helping in every stage of the process.

REMODELING: Sharnessa has helped with wallpapering and painting jobs, indoors and out, and I also let her totally decorate the main bathroom including deciding a theme and *doing* it all herself!

FINANCE SKILLS: Here I just listed the various skills that were part of Sharnessa's job working for us, skills that had to do with finances.

BOOK PRODUCTION: These are the skills she did for us that pertained to producing our books. (We self-publish all the books I have written.)

LITERATURE PRODUCTION: There have been a variety of packets, handouts, and flyers that we have produced for our business, and Sharnessa has been involved in the production of all of them. As she got older, I was even able to give her a complicated job (involving various colors of sheets, stapled items within a packet, machine-collating, etc.) and send her down to the copy machine alone to do it all!

VIDEOTAPING: Sharnessa went with me to several of the places that I did interviews for my "Senior High Homeschooling Options & Resources" video-seminar and did the video-taping for me, plus she video-taped the seminar portion of it in our home (at our kitchen table).

PHOTOGRAPHY: The first two activities, Read/Do "Ace Photography" book was my idea, but she was interested in it; the field trip to the photographer (who did our wedding and was also a homeschooler) was my idea; taking sunrise photos was hers.

✓ **ASSIGNMENT CHECK-OFF SHEET** ✓✓

Class: **Occupational Education**	
Student: **Sharnessa Shelton**	*Class Value:* **1 Credit = 120 Hours**

BASIC REQUIREMENTS CHECK-OFF LIST	Check Here When Done
Secretarial and Office Skills *(See Task Analysis Sheet for specifics)*	
Attend "Home and Family Business" Workshop *(presented in Portland, OR by Gregg Harris)*	
Recording Studio Experience: Record songs and Learn about equipment *(with and in the recording studios of Larry Harrold and Felix Moore; and with Chuck Tilton)*	
Teach Dance Lessons *(Hip-hop, Lyrical, Tap to Carlianne, Gracie Kempf, Jessica Wafer)*	
Learn to operate: Computer *(Separate class and Task Analysis Sheets)*	
Copy Machine *(See separate Task Analysis Sheet for individual skills)*	
Comb Binder *(to comb-bind the books Mom has written)*	
Interview an on-duty Police Officer *(Officer Darr Kirk, on patrol in Kelso, Wash.)*	
Telephone Skills: KZOE Share-a-thon phone manning; Make calls for Course; Screen calls for M.	
Home Construction *(New bedroom, Storage area, Office)*	
Remodeling: Wallpaper, Paint, Choose Color Scheme, Decorate *(Main bathroom)*	
Finance Skills: Record deposit entries, Make deposits, Record and itemize Course payments, Talley totals on Deposit Records	
Book Production *(Copy, Collate, Comb-bind, 3-hole punch, Assemble Clear-vue covers)*	
Literature Production *(Packets, Brochures, Flyers, Copying and Collating)*	
Videotaping *(of Senior High Video – Seminar portion at home; Interviews around the state)*	
Photography: Read/Do "Ace Photography" book	
Field Trip to Photograper Arne Loren's Studio	
Take sunrise photos and make booklet or cards	

HOUR CHECK-OFF CHART *(Check off hours as they are completed.)*

60	59	58	57	56	55	54	53	52	51	50	49	48	47	46	45	44	43	42	41
40	39	38	37	36	35	34	33	32	31	30	29	28	27	26	25	24	23	22	21
20	19	18	17	16	15	14	13	12	11	10	9	8	7	6	5	4	3	2	1

[A.C.S.]
P.E.

P.E. (GENERAL)

Under "Sports Areas" I have listed all the various types of sports Tory has participated in. I didn't feel I needed to get extremely detailed on this as I can't imagine anyone at a college (or wherever) really needing to know any more than this.

Why do I have the time spent at each activity on this ACS but not on most of the others? My original plan, when I first wrote the book, was to do this for *all* the classes – put the amount of time for each activity of each class. That got too cumbersome, and there really wasn't any need to be that specific about the time. So I pulled in the "Hour Check-off Chart" at the bottom which has (is) a place to check off the hours in one place (chart). However, at that point I had already kept track of his hours in these individual areas of P.E. so thought "Why not?"

He received 1¼ credits for this class since the hours came to 150. He had even more, actually, but this was close to an "even" number. I could have put all the hours together in one chart at the bottom, and did so on another ACS somewhere else, but this was just what came to me at the time. ☺ (Don't they call this "creative license"? I use mine a lot!)

P.E. / SOCCER

This is the same idea as for basketball, which I'll explain next. The thing I wanted to point out here was that he came up eight hours short of what he needed for one credit, so we carried over – "borrowed" – eight hours from basketball, which he went way over in, but we didn't document. *(This is no big deal, but I wanted to show you how it was done.)*

P.E. / BASKETBALL

Tory played basketball at a Christian school for three years. His time through all three years came to (at least) 2 credits. I could probably elaborated more on the "Time includes" part under "Description." I put "Time includes practices, free gym time, and games." The "free gym time" was when he went to another school that opened its doors to anyone wanting to come and practice at 6 a.m. every day! One thing I forgot to include was the 4-day "Point Guard Basketball College" he attended. (It's really more like "camp" but that's what they call it, probably because it's more serious and focused. They had actual classes too in addition to gym time that included "tutoring.") And another thing I forgot was Tory's participation in a "Son League" Basketball and a day camp. If he was applying to some big basketball event or for a basketball scholarship, those would be appropriate places to include details on all that.

Here's the Table of Contents of Tory's Basketball Notebook: (much of which went under other classes.)

"BASKETBALL FOR DUMMIES" SECTION

- How the Game is Played
- The History of Basketball
- Positions
- Court Layout
- Defense
- Teamwork
- Game Timing Information
- Ways to Score
- Foul Information
- Glossary and Pictorial Glossary

> LAST MINUTE BONUS!
>
> After Tory's three P.E. ACS's is a photo of our family involved in doing "Homestyle P.E."; and then there's a P.E. ACS for Sharnessa that I found at the last minute when looking for something else! ☺

PHOTO SECTION: Photos of Tory and team in action.

NEWSPAPER CLIPPINGS: Clippings from the newspaper on the team's and Tory's accomplishments. (Tory wanted both "the good and the bad" included!)

BASKETBALL HALL OF FAME: When Dave and I were back in Springfield, Mass., Dave discovered this museum and went there and took pictures for Tory! We had no idea that this town is the "birthplace of basketball" until we got there. *(We went there because I was speaking at a homeschool convention; Dave visited the museum while I was in the hotel room organizing and preparing for my workshop.)*

SCRIPTURE STUDIES: "Word Alive Studies" and "Scripture Verse Studies" that can be found in here. These are on various words and verses relating to sports.

BASKETBALL TIPS & METHODS (all these were found on the Internet)

- "Basketball Coaching Philosophy"
- "Basketball Program Philosophy" (of a High School)
- "10 Commandments of Basketball Parents"

KEEPSAKES & SPECIAL MEMORIES

- Picture of Tory with his "Most Inspirational" Trophy
- Basketball photo card
- The "Players Magazine" that featured a story on him!
- Color copies of all 3 years' teams
- Basketball Awards Dinners
- Son League Luncheon and "Hustlers" award T-shirt
- Honoring our Coaches
- The "Top Players of the League" roster (that Tory was on)

✓ ASSIGNMENT CHECK-OFF SHEET ✓✓✓✓✓✓✓✓✓✓✓✓✓✓✓✓✓✓✓✓✓✓✓✓✓✓✓✓✓✓✓✓✓✓✓

Class: **P.E. (GENERAL)**	
Student: **Tory W. Shelton**	Class Value: **1¼ Credit = 150 Hours**

SPORTS AREAS:	TIME	DONE
Skiing - At White Pass (in Washington), and at Mount Hood (in Oregon)	24 hours	
Broom/Gym Hockey - At Calvary Community Church; Longview, WA	25 hours	
Workout/Weights - At YMCA & Physical Impact (workout studio in Longview)	15 hours	
Ice Skating - West Edmonton, Canada, and Clackamas Town Center Rink	3 hours	
Jogging - Around Columbia Heights and Lake Sacajawea (Longview)	10 hours	
Trampoline - On full-size trampoline at home	5 hours	
Sledding - On sled run at home	20 hours	
Rollerskating - At Skate World (in Kelso, WA)	10 hours	
Softball - For Father's House team with Church League	20 hours	
Raquetball - Two sessions at Mint Valley Raquetball Complex (Longview)	2 hours	
Bicycling - "Tour de Blast" at Spirit Lake Highway near Mt. St. Helens, S.H.A.R.C. (St. Helen's All-terrain Riders' Club) Autumn Bike Tour, General bike riding and prep for the Bike Tours	16 hours	
TOTAL →	150 hours = 1¼ credits	

HOUR CHECK-OFF CHART - One Credit:

(Check off hours as they are completed.)

120	119	118	117	116	115	114	113	112	111	110	109	108	107	106	105	104	103	102	101
100	99	98	97	96	95	94	93	92	91	90	89	88	87	86	85	84	83	82	81
80	79	78	77	76	75	74	73	72	71	70	69	68	67	66	65	64	63	62	61
60	59	58	57	56	55	54	53	52	51	50	49	48	47	46	45	44	43	42	41
40	39	38	37	36	35	34	33	32	31	30	29	28	27	26	25	24	23	22	21
20	19	18	17	16	15	14	13	12	11	10	9	8	7	6	5	4	3	2	1

REMAINING ¼ CREDIT:

30	29	28	27	26	25	24	23	22	21	20	19	18	17	16	15	14	13	12	11	10	9	8	7	6	5	4	3	2	1

✓ ASSIGNMENT CHECK-OFF SHEET ✓✓✓✓✓✓✓✓✓✓✓✓✓✓✓✓✓✓✓✓✓✓✓✓✓✓✓✓✓✓✓✓✓✓✓✓✓✓✓

Class: P.E. / Basketball	
Student: **Tory W. Shelton**	Class Value: **2 Credits = 240 Hours**

DESCRIPTION:	TIME	DONE
Participate on the Basketball team at Columbia Heights Christian School for the 1995, 1996 and 1997 seasons; Team Co-Captain for the 1996 and 1997 seasons. Time includes practices, free gym time, and games.	240 hours	
TOTAL →	240 hours = 2 credits	

HOUR CHECK-OFF CHART

(Check off hours as they are completed.)

FIRST CREDIT

120	119	118	117	116	115	114	113	112	111	110	109	108	107	106	105	104	103	102	101
100	99	98	97	96	95	94	93	92	91	90	89	88	87	86	85	84	83	82	81
80	79	78	77	76	75	74	73	72	71	70	69	68	67	66	65	64	63	62	61
60	59	58	57	56	55	54	53	52	51	50	49	48	47	46	45	44	43	42	41
40	39	38	37	36	35	34	33	32	31	30	29	28	27	26	25	24	23	22	21
20	19	18	17	16	15	14	13	12	11	10	9	8	7	6	5	4	3	2	1

SECOND CREDIT

120	119	118	117	116	115	114	113	112	111	110	109	108	107	106	105	104	103	102	101
100	99	98	97	96	95	94	93	92	91	90	89	88	87	86	85	84	83	82	81
80	79	78	77	76	75	74	73	72	71	70	69	68	67	66	65	64	63	62	61
60	59	58	57	56	55	54	53	52	51	50	49	48	47	46	45	44	43	42	41
40	39	38	37	36	35	34	33	32	31	30	29	28	27	26	25	24	23	22	21
20	19	18	17	16	15	14	13	12	11	10	9	8	7	6	5	4	3	2	1

✓ *ASSIGNMENT CHECK-OFF SHEET* ✓✓✓✓✓✓✓✓✓✓✓✓✓✓✓✓✓✓✓✓✓✓✓✓✓✓✓✓✓✓✓✓✓✓✓✓✓✓

Class: **P.E. / Soccer**	
Student: **Tory W. Shelton**	*Class Value:* **1 Credit = 120 Hours**

AREA OF FOCUS:	**TIME**	**DONE**
Soccer - Participate on soccer team at Columbia Heights Christian School for the 1995, 1996 and 1997 seasons; Team Captain for the 1997 season. Time includes all practices and games.	112 hours	
(Excess hours carried over from Basketball)	*8 hours*	
TOTAL →	120 hours	

HOUR CHECK-OFF CHART

(Check off hours as they are completed.)

Last 8 hours are carried over from Basketball

120	119	118	117	116	115	114	113	112	111	110	109	108	107	106	105	104	103	102	101
100	99	98	97	96	95	94	93	92	91	90	89	88	87	86	85	84	83	82	81
80	79	78	77	76	75	74	73	72	71	70	69	68	67	66	65	64	63	62	61
60	59	58	57	56	55	54	53	52	51	50	49	48	47	46	45	44	43	42	41
40	39	38	37	36	35	34	33	32	31	30	29	28	27	26	25	24	23	22	21
20	19	18	17	16	15	14	13	12	11	10	9	8	7	6	5	4	3	2	1

Used with permission

Photo by ***White Water Photos by Thomas Davis***

“Homestyle P.E.”!

Otherwise known as “White Water Rafting”! ☺
Here we’re on the Deschutes River in Oregon in 1996, and what a blast!!!

From left to right, back to front:
Dave, Diane, Nicole, and Ryan Tweet;
And then middle-back around to front:
Dave, Barb, Sharnessa, Tory, and Carlianne Shelton

✓ *ASSIGNMENT CHECK-OFF SHEET* ✓✓✓✓✓✓✓✓✓✓✓✓✓✓✓✓✓✓✓✓✓✓✓✓✓✓✓✓✓

Class: **P.E. (GENERAL)**	
Student: **Sharnessa Shelton**	*Class Value:* **1 credit = 120 hours**

BASIC REQUIREMENTS CHECK-OFF LIST	Done
Complete assignment studying frequency, duration, of exercise (from *Fit or Fat* by Covert Bailey)	
Read selected readings on heartrate information	
Complete heartrate data sheet	
Dance activity:	
Tap Dance	
Jazz Dance	
Lyrical Jazz Dance	
Ballet/Jazz worship	
Miscellaneous (50's, Line Dancing)	
Personal Exercise Workouts	
Skiing	
Walking / Jogging	
Specialized Weight Training	
Stretch and Flexibility	
White Water Rafting *(See opposite page!!!)*	
TOTAL HOURS:	

HOUR CHECK-OFF CHART - ONE CREDIT:

(Check off hours as they are completed.)

120	119	118	117	116	115	114	113	112	111	110	109	108	107	106	105	104	103	102	101
100	99	98	97	96	95	94	93	92	91	90	89	88	87	86	85	84	83	82	81
80	79	78	77	76	75	74	73	72	71	70	69	68	67	66	65	64	63	62	61
60	59	58	57	56	55	54	53	52	51	50	49	48	47	46	45	44	43	42	41
40	39	38	37	36	35	34	33	32	31	30	29	28	27	26	25	24	23	22	21
20	19	18	17	16	15	14	13	12	11	10	9	8	7	6	5	4	3	2	1

[A.C.S.]
SOCIAL STUDIES

This is really a "catch-all" class of smaller areas of study that can be classified as "Social Studies." Any one of these sections could actually be a whole class in itself, if there were enough hours accrued for it. The number of hours that all these studies put together amounted to came to 90, thus the ¾ credit. ☺

I should point out that this is a good example of starting with what I wanted my kids to know, and then finding a place to "pigeon-hole" it academically. So I did not start with the areas that I'll list in a moment, and then try to figure out how to fill them. I started with what I wanted them to know, as well as what *they* wanted to know, and then found a place to put it. Let's take each of the areas one at a time:

GENERAL SOCIAL STUDIES:

"**Psychology vs. Christianity**," an audio taped seminar by Jonathan Lindvall, is an excellent comparison of these two areas. By the way, he has produced many tapes on a wide variety of topics, not just "courtship and betrothal" for which is well known.

Marshall Foster produces the **Mayflower Journal** which comes out monthly, and always has a very timely and interesting article. This assignment was to read any five of these.

The other three items are resources I have collected over the years. In particular, "**13 Things You Gotta Know to Make it as a Christian**" by Josh McDowell is excellent, especially for younger teens. (And we do carry it!)

CONTEMPORARY SOCIAL STUDIES:

There was a TV special on "**Martin Luther King,**" his life and his cause, that Tory watched, that was very good.

We have some **booklets by Karl Reed on Abortion** that actually cover more than just abortion, but also aspects of government and civil disobedience. These are booklets that will bring about some good discussion with your student; I would definitely read them *with* my child, not just assign it. These matters need to be discussed and hashed over; convictions and attitudes formed about them. It's vital that this happen *with* a parent.

I found "**Understanding the New Age**" at a garage sale (it's from a Christian perspective.)

Tory had to write an essay on a topic of his choosing for the Christian Ethics class he took at a Christian school (which is what qualified him to play basketball there; he had to take one class.) So he chose to do his **essay on "Abortion."** It was good for him to write an essay for someone besides me. ☺

For that same class he would frequently take articles he had clipped from the newspaper on a **current event** topic of his choice, and the class would discuss it.

PHILOSOPHY OF RELIGION:

This all started with a book I'd had for years called "**So What's the Difference?**" It was actually a bit too heavily detailed for the high school level; details that, unless you really need to know them, just aren't that relevant or important for a teenager to know; or even an adult. So it was a bit "much," but somewhat interesting. We could have quit reading it, as I feel free about doing that if it's just not what we expected or hoped, but the kids decided to just finish it.

At the local community college we attended a **debate called "is There Enough Evidence to Prove the Existence of the Judeo-Christian God?"** Not a good place for anyone who already suffers from high blood pressure, it was very interesting! We were not accustomed to that level of anti-God thinking! The Christian who argued for God was very good at what he did, but he basically set himself right into the lion's den and was pretty much eaten alive. The non-Christians – which were in the majority – were not there to learn or listen; they obviously just wanted to ridicule and win. But it was a good experience for all of us. (I and our two teens went together. They were both hoping I wouldn't "contribute" when they opened the mic for input; and I didn't disappoint them!) We hoped that seeds were planted anyway: "My word goes forth, and shall not return to me void..."

CIVICS:

Sen. Linda Smith (our senator from Wash. state) came to town and did a **town hall meeting** that we attended. We were quite impressed with how she fielded difficult questions! And, of course, we got pictures with her. Another time she did a **debate** with her contender that was another blood-pressure-elevator! Very eye-opening!

The "**God and Caesar**" book is one I'd had for many years and was actually beyond Tory as a senior, so we never did it. (He'd do great with it now, and may add it to his studies this year.)

ECONOMICS:

Marshall Foster has a 3-tape series on "**Finances.**" And "**Debt-free Living**" was a booklet that came in the mail. The Investment study was a chapter **in"Applied Mathematics**," a t-t-textbook by A Beka. ☺ It gave an overview of each of these topics here.

✓ ASSIGNMENT CHECK-OFF SHEET ✓✓✓✓✓✓✓✓✓✓✓✓✓✓✓✓✓✓✓✓✓✓✓✓✓✓✓✓✓✓✓✓✓✓✓✓✓

Course: **Social Studies**	
Student: **Tory W. Shelton**	*Class Value:* **¾ Credit = 90 Hours**

BASIC REQUIREMENTS CHECK-OFF LIST

General Social Studies	☐ Listen to Tape: "Psychology vs. Christianity" by Jonathan Lindvall
	☐ ☐ ☐ ☐ ☐ Read 5 Articles by Marshall Foster in the Mayflower Journal
	☐ Read book: Marco Polo ☐ Listen to tape: "Teaching World Views" by Cathy Duffy
	☐ Read: 13 Things You Gotta Know to Make it As a Christian by Josh McDowell
Contemporary Social Issues	☐ Watch video: "Martin Luther King" (TV Special)
	Read and discuss the 3 "Abortion Letters" booklets: ☐ Part I: "Your Honor" ☐ Part II: "An Analysis of Five Ways Civil Disobedience Violates Scripture" ☐ Part III: "God, Government, and My Neighbor"
	☐ Read selected chapters of Understanding the New Age by Russell Chandler *(Christian perspective)*
	☐ Write essay on "Abortion" using at least three reference sources
	Current Events: Choose articles of interest in newspaper or magazine and take to Christian Ethics Class at Columbia Heights Christian School ☐☐☐☐☐☐ ☐☐☐☐☐☐
Philosophy of Religion	☐ Read and discuss: So What's the Difference? by Fritz Ridenour (Phil. of Rel.)
	☐ Attend: Debate at LCC: "Is There Enough Evidence to Prove the Existence of the Judeo-Christian God?"
	☐ Read and discuss: "Comparison of Religious Theory in the Late 20th Century" *(a concise, humorous-but-accurate parody which is a good synopsis of each of the major religions of the world)*
	Interview: 3 Christians about the four Christian terms: Evangelical, Pentecostal Charismatic, and Fundamental, and write up composite ☐☐☐☐
Civics	☐ Attend: Town Hall Meeting at Kelso City Hall with Senator Linda Smith
	☐ Attend: Debate between Brian Baird & Linda Smith, candidates for U.S. Senate 1996 Election
	☐ Read: God and Caesar – Christian Faith and Political Action by John Eidsmoe
Economics	Listen to Tapes: ☐ Marshall Foster tape series on "Finances" ☐ Debt-Free Living
	Read booklet: ☐ "A Biblical Perspective on Debt-free and Prosperous Living"
	Read books: ☐ Whatever Happened to Penny Candy? ☐ The Totally Awesome Money Book
	Read these topics in "Applied Mathematics": *(Do Exercises &/or further study of anything that looks good)* **INVESTMENT (UNIT 3)** ☐ Reading the Stock Exchange Table (223) ☐ The Stock Market (244) ☐ Buying and Selling Stock (254) ☐ The Stock Market Game (264) ☐ Certificate of Deposit (269) ☐ The Savings Acct. (276) ☐ Finding the Principal in a Savings Acct. (287) ☐ Real Estate (291) ☐ Corporate Bonds (308) ☐ Mutual Funds (318) ☐ Compound Interest (323) ☐ Real Return on an Investment (328) *For reading on "Income and Taxes" and "Banking," see Units 4 and 5 in same textbook.)*

HOUR CHECK-OFF CHART

90	89	88	87	86	85	84	83	82	81	80	79	78	77	76	75	74	73
72	71	70	69	68	67	66	65	64	63	62	61	60	59	58	57	56	55
54	53	52	51	50	49	48	47	46	45	44	43	42	41	40	39	38	37
36	35	34	33	32	31	30	29	28	27	26	25	24	23	22	21	20	19
18	17	16	15	14	13	12	11	10	9	8	7	6	5	4	3	2	1

[A.C.S.]
SPANISH

Tory's Spanish class had a combination of resources. I'll go over them one by one...

SPANISH CLASS: VAMOS A SABER

This class was taught by the grandmother of a homeschooled boy we knew. She and her husband (and this boy and his parents as well for a few years) had been missionaries in Bolivia, so she knew the language well! She organized the material she wanted to teach into a notebook she called "Vamos a Saber" (which I think means "We will speak), and also made audio tapes of herself going through the material (selected portions) so the students could practice at home. It was well done, and Tory enjoyed it and learned a lot!

POWER GLIDE SPANISH PROGRAM

This program consists of several audio tapes and a handbook. Tory used and liked this, but didn't do the whole thing. I've heard of homeschoolers who found it very helpful. They have them for several other languages as well.

FIELD EXPERIENCE IN USING SPANISH

Tory went down to Mexico with a group from our church on an outreach trip (twice). Of course, while there, he got to put his Spanish into practice, but found that they talk a lot faster than a Spanish teacher or tape do! ☺ He was able to do a little bit of communicating though!

PRO-ONE CD-ROM SPANISH PROGRAM

This is a good way to get some basics, but if you are looking for something colorful and "fun," this isn't it. But there's something to be said for "dry learning" when you really want to get in and be serious about learning the basics without having to deal with a lot "bells and whistles" which can be very distracting and time consuming.

PRACTICE SPEAKING SPANISH WITH MOM

This is nothing extensive, but I felt that all the very light practice of Spanish that Tory and I did (still do) just around home ought to count for something. Things like "Dondé está el teléphono?" (We have a cell phone that spends more time being lost than found.) Or "Pasa me la sal." ☺ Not very impressive to a native Spanish speaker, but we're always kind of tickled with ourselves! ;-)

TAKE GENERAL SPANISH QUIZ

This should actually be deleted because we never did it, but I left it on the form in case any of you might want to use the idea. I started out making up a general Spanish quiz, but by the time he had taken the class and used Power Glide, I felt it just wasn't necessary.

SPANISH NOTEBOOK *(not on the form; a bonus!)*

The following is from "Kelli" on the Senior High Form+U+la list who contributed her (and her son's) ideas for a Spanish Notebook! (I added just a few ideas.)

1. Cognates - These are words related in spelling and meaning in both English and Spanish, but are pronounced differently. There are a lot of cognates, and learning them is a quick way to learn vocabulary. For example: Many words that end in "ción" in Spanish are changed to "tion" in English. So, invitación is invitation, adoración is adoration, salvación is salvation, etc...

2. Popular Spanish names for boy and girls [From Barb: I am still called "Burbujas" my nickname of "Bubbles" from high school days!]

3. Family names (i.e.. grandmother, grandfather, mom, dad, brother, sister, aunt, uncle, cousin, etc...)

4. Parts of the body

5. Store and Grocery store words (tools, car parts, animals (pets), food, etc...)

6. Names for the rooms in a home

7. Verbs used in Christianity (to save, to redeem, to deliver, to pray, to fast, etc...) Or anywhere!

8. Spanish names for Jesus, like "El Salvador" = "the Savior"; "El Rey de los Reyes" = "the King of Kings"; "El Pan de Vida" = "the Bread of Life"; "La Vid" = "the Vine"

9. Books of the Bible

10. Greetings *(there are dozens of ways to greet someone in Spanish)*

11. Useful expressions, like "How are you?" *(Como estas?)*

SPANISH FOR GRINGOS

This book isn't listed since I discovered it later, but it's a *great* resource for teaching basic conversational Spanish!

✓ ASSIGNMENT CHECK-OFF SHEET ✓✓✓✓✓✓✓✓✓✓✓✓✓✓✓✓✓✓✓✓✓✓✓✓✓✓✓✓✓✓✓✓✓✓✓✓✓✓

Class: **Spanish**	
Student: **Tory W. Shelton**	*Class Value:* **½ Credit = 60 Hours**

BASIC REQUIREMENTS CHECK-OFF LIST	Done
Take Spanish Class, "Vamos a Saber," through Mrs. Elaina Kempf (30 hours)	
"Power Glide" Spanish Program	
Field Experience in using Spanish (5 hours)	
Use the "Pro-One" CD-Rom Program for Spanish	
Practice speaking Spanish (informally) with Mom	
Take general Spanish quiz	

HOUR CHECK-OFF CHART

(Check off hours as they are completed.)

60	59	58	57	56	55	54	53	52	51	50	49	48	47	46	45	44	43	42	41
40	39	38	37	36	35	34	33	32	31	30	29	28	27	26	25	24	23	22	21
20	19	18	17	16	15	14	13	12	11	10	9	8	7	6	5	4	3	2	1

[A.C.S.]
SPIRITUAL CONFERENCES & MEETINGS

This class was originally a segment of the "Spiritual Studies" class, but because it grew (and grew and *grew*) to be so large, I just made it into its own class! In fact, not only did it end up being it's own credit, but one and a half credits! We didn't plan that, by the way. It just turned out that way. Again, there were a few "variables" that I adjusted the amount of time we counted to make it come out even.

The only reason I have the hours of each event on this one (and not on most of the others) is because they were readily available. Already all recorded, so (again) *Why not?!?* ☺ I had started out recording all the hours for all the individual activities and sections on all the ACS's. But later I realized that getting *that* detailed – hour-wise – was just not necessary, not to mention cumbersome. But since these hours for the conferences and such were *already* recorded, since we kept track of each event, I decided to include them here for "just in case" someone ever wanted to know more specifically. (and also as an example for those who *do* want to record hours this extensively. I will warn you though – it can and *will* get cumbersome! So just be careful of that, or you'll burn out! – even on a good thing!)

In case you're wondering, under **"Youth Alive Conference,"** the **"Stretch Team"** was something Tory was asked to be on at the Youth Alive Conferences which were headed by Bob Smith (who wrote *Raising Teens (Almost) Hassle Free.*" I don't think that's quite exactly the title, but close.) Teens who were known in their churches as being "spiritual minded" and "leaders" and "servant-hearted" (those three tend to go hand-in-hand, don't they?!?!) were hand-picked by their youth leaders. They served in various ways throughout the conference, and met for extra fellowship and Bible Study throughout it, between the activities and their duties. The kids on this "stretch team" were definitely s-t-r-e-t-c-h-e-d!

If you will look down at the bottom entry **"Youth Group meetings w/ Shekinah** *(our church; it means "the radiant presence and glory of God."*☺), **Grace Fellowship, etc...**" you'll see that the number of hours for that is 15. Right to the left of that, you'll see it says "**ceiling** →" – pointing over to the 15 hours. This is one of those "variables" I mentioned a moment ago. Tory actually spent *many* more hours than this at youth group. I just decided on 15 hours as a ceiling (limit) pretty arbitrarily. He already had many other hours for both this and the Spiritual Studies class, so he didn't *need* any more hours. I could easily have counted 40 or 60 or 100 hours over the course of the years! And if you want to do so for your child, feel free! You'll definitely want to consider how much "spiritual" stuff is actually happening at these meetings, and if there's ungodly interaction between the genders, which will require some discussion with your teen. These meetings that Tory attended were pretty spiritually focused for the most part; he just didn't happen to need the hours.

In this photo, Tory may not look like he's doing anything particularly "spiritual," and, indeed, he's not at a "Spiritual Conference or Meeting." But he is working on a "Word Alive!" study! ☺

(Besides, I just had to include at least ONE "studious-looking" picture somewhere in this book just to prove they actually studied!)

✔ ASSIGNMENT CHECK-OFF SHEET ✔✔

Class: **Spiritual Conferences & Meetings**	
Student: **Tory W. Shelton**	*Class Value:* **1½ Credits = 180 Hours**

BASIC REQUIREMENTS CHECK-OFF LIST	**TIME**	**Done**
Attend: "Youth Alive" Conference in 1993 (Bob Smith leader/coordinator)	21 hrs.	
Attend: "Youth Alive" Conference and be on "Stretch Team" in 1994 (")	31 hrs.	
Attend: "Youth Alive" Conference and be on "Stretch Team" in 1995 (")	31 hrs.	
Attend: "Plumbline" (taught and facilitated by Jerry Praetzel)	21 hrs.	
Attend: Ski Retreat with Grace Fellowship (Dan Anderson leader)	3 hrs.	
Attend: "Youth Alive Weekend" (Bob Smith leader)	8 hrs.	
Attend: Ski Retreat with Shekinah (Dan Anderson leader)	3 hrs.	
Attend: "Basic Life Principles" Conference (taught by Bill Gothard)	32 hrs.	
Attend: Vineyard "Fan the Flame" Seminar and "Clinic" (led by Wes Campbell)	5hrs.	
Attend: "Impact 2000" Weekend (Bob Smith leader)	10 hrs.	
Youth Group meetings w/ Shekinah, Grace Fellowship, and Father's House *(ceiling →)*	15 hrs.	

HOUR CHECK-OFF CHART

(Check off hours as they are completed.)

180	179	178	177	176	175	174	173	172	171	170	169	168	167	166	165	164	163	162	161
160	159	158	157	156	155	154	153	152	151	150	149	148	147	146	145	144	143	142	141
140	139	138	137	136	135	134	133	132	131	130	129	128	127	126	125	124	123	122	121
120	119	118	117	116	115	114	113	112	111	110	109	108	107	106	105	104	103	102	101
100	99	98	97	96	95	94	93	92	91	90	89	88	87	86	85	84	83	82	81
80	79	78	77	76	75	74	73	72	71	70	69	68	67	66	65	64	63	62	61
60	59	58	57	56	55	54	53	52	51	50	49	48	47	46	45	44	43	42	41
40	39	38	37	36	35	34	33	32	31	30	29	28	27	26	25	24	23	22	21
20	19	18	17	16	15	14	13	12	11	10	9	8	7	6	5	4	3	2	1

[A.C.S.]
Spiritual Studies

This class ended up being a whopping 300 hours, when all was said and done, *and* read, *and* watched, and written and listened to. ☺ It was easiest to categorize this material into how it was presented (book, tape, video, etc.) rather than trying to figure out labels for the content, especially since much of this could fall into many different categories, and one book or tape may cover many aspects. It doesn't matter; this is just a way to organize it; that's all. Let's take it a section at a time...

WORD & SCRIPTURE VERSE STUDIES:

The "Word Alive" studies also counted for English, as you may remember. Both the "Word Alive Study" and the "Scripture Verse Study" masters are here in this book. (See the Form Index.)

BOOKS TO READ:

These are all books that Tory read, some chosen by him, more chosen by me since he didn't really know what to read in the earlier years. (Later on he started taking more of an interest in choosing.)

READ SELECTED PORTIONS OF THESE:

This just means that, for one reason or another, he didn't finish these books. It might have just not been interesting to him, or above his head, or "too" something. ***Foxe's Book of Martyrs*** was pretty gory, which didn't disturb him too much (like it did me), but after a while we found it to be rather repetitious. So it seems to me that it's meant to be more a documentation than "good reading." Either way, it's very impactful!

My Utmost for His Highest by Oswald Chambers was a little deep for him back then, but he reads it daily now and loves it! (So do I! It's a "refining fire"!)

I read ***Embraced by the Light*** waaaay back before anyone had any doubts about it. I know there are a few things in it that are "off," but not knowing that at the time, I got a lot out of it!!! "To the holy, all things are holy." The Holy Spirit used it *greatly* in my life for some very specific purposes, and spoke to me profoundly through this book. But it would probably not be wise to blanketly recommend it. It might be good for a mature-in-the-Lord parent to read it *with* their child, *after* they have read it themselves. (I'm not going to argue about the pro's and con's of this book; God used it profoundly, and there is still much truth in it, so no need to write to me about this. ☺)

ARTICLES / BOOKLETS:

The first two booklets mentioned here, **"A Time to Seek God"** and **"Discerning of Spirits,"** are two of my favorites, and I have included them in my recommendations for a Season of Re-education! The other three are from ATIA., Advanced Training Institute of America' *(aka "Basic Youth" or "Basic Life Principles" headed by Bill Gothard.)*

FORMAL & INFORMAL STUDIES TO DO:

The "**Religion Types**" interview was not in the original edition of Form+U+la, but is in this one.

"**Survival Kit for Christians**" is a Bible study that comes in several different age levels. (It's not one that I can get; look at a Christian bookstore.)

The "**God's Actions**" study is here in this book. (See the Form Index for locating *all* forms.)

"**Bible Challenge**" and "**Choices**" are games we purchased several years ago from a Christian bookstore.

VIDEOS TO WATCH:

These are ones we borrowed from friends or checked out at our local Christian bookstore. As for "**The Hiding Place,**" I'd recommend reading the book first, or a biography about her, before watching the video. Just a personal preference. I think that part of the benefit of reading is that it challenges the reader (or listener) to have to create the scenes in their mind. If a video has been watched first, much less "creating" will need to be done in the thought processes as characters and places have already been seen visually. No harm done, but just less challenging. Also many times the movie version is not accurate, so I've heard, so if reading the book after having seen the film, there my be inaccuracies that will "color" or "taint" the true story. No hard-and-fast rules, though; just do as you feel so inclined or lead. ☺

TAPES TO LISTEN TO:

These are tapes I have been collecting over the years. I didn't get them specifically for this class; I just knew I wanted my kids to listen to them. They are *all* good, but I particularly want to point out that *any* tapes by Pastor Reb Bradley are excellent!!! (We carry several different titles by him now, on a variety of topics including anger, lust, child training, courtship, homeschooling, TV, etc.)

✓ ASSIGNMENT CHECK-OFF SHEET ✓✓✓✓✓✓✓✓✓✓✓✓✓✓✓✓✓✓✓✓✓✓✓✓✓✓✓✓✓✓✓✓✓✓✓✓✓✓

Class: **Spiritual Studies**	
Student: **Tory W. Shelton**	*Class Value:* **2½ Credits = 300 Hours**

BASIC REQUIREMENTS CHECK-OFF LIST

Word and Scripture Verse Studies:	**Word Alive Studies** on: ☐ "Redeem" ☐ "Build" ☐ "Counsel" ☐ "Diligence" ☐ "Dedication" ☐ "Die" ☐ "Harmony" ☐ "Glorify" ☐ "Abide" **Scripture Verse Studies** on: ☐ II Pet.1:5 ☐ ________ ☐ ________ ☐ ________
Books to Read:	☐ In the Presence of Mine Enemies by Howard Rutledge ☐ Winning God's Way by Loren Cunningham ☐ Carpe Diem by Tony Campolo ☐ The Release of the Spirit by Watchman Nee ☐ Vanya by Myrna Grant ☐ Twice Pardoned by Harrold Morris ☐ Just Do It! by Buddy Harrison ☐ Screwtape Letters by C.S.Lewis ☐ Winning God's Way Loren Cunningham ☐ Life After Life by ??? ☐ Beyond and Back by Ralph Wilkerson ☐ This Present Darkness, Piercing the Darkness, The Prophet by Frank Peretti ☐ Hearing God by Peter Lord ☐ Hind's Feet on High Places Hannah Hurnard ☐ Experiencing the Depths of Jesus Christ by Jeanne Guyon ☐ How to be Your Own Selfish Pig Susan Schaeffer MacAulay ☐ To Hell and Back Maurice Rawlings M.D. ☐ Love, Acceptance and Forgiveness by Jerry Cook ☐ In His Steps Charles Sheldon ☐ Comeback and... ☐ When You Can't Come Back by Dave Dravecky ☐ The Three Battlegrounds, ☐ The Divine Antidote, ☐ The Place ofImmunity, and ☐ Holiness, Truth, and the Presence of God by Francis Frangipane ☐ The Mark of a Christian Francis Schaeffer ☐ The Christian's Secret of a Happy Life Hannah Whitall Smith
Read Selected Portions of these:	☐ My Utmost for His Highest by Oswald Chambers ☐ Called to Suffer, Called to Triumph ☐ Embraced by the Light by Betty Eaddie ☐ Foxe's Book of Martyrs by John Foxe
Articles/Booklets:	☐ "A Time to Seek God" and ☐ "Discerning of Spirits" by Francis Frangipane ☐ "The Eagle Story" ☐ "Applying Basic Principles" and ☐ "The Overlooked Requirements for Riches, Honor & Life (ATIA)
Formal & Informal Studies to Do:	☐ Do "Religion Types" interviews and worksheet ☐ "Survivial Kit for Christians" ☐ "God's Actions" study ☐ Play "Bible Challenge" with Mom ☐ Play "Choices" with the family ☐ Own Quiet Time with God
Videos to Watch:	☐ "Twice Pardoned" (Harrold Morris) ☐ "God's Generals" ☐ "The Hiding Place" (re: Corrie Ten Boom)
Tapes to Listen to:	☐ "Vision" (Casey Treat) ☐ "Hearing God" (John Paul Jackson) ☐ "God, TV, and Your Family" (Reb Bradley) ☐ "Why You Were Created," ☐ "How to Set Values That Lead to Victory," ☐ "What I Wish I Knew in High School," ☐ "10 Lies That Can Destroy Your Life" from "Straight Talk to Teenagers by Dennis Peacock ☐ "Husbands, Fathers, Protectors" by Pastor Reb Bradley ☐ ________

HOUR CHECK-OFF CHART - FIRST CREDIT

120	119	118	117	116	115	114	113	112	111	110	109	108	107	106	105	104	103	102	101
100	99	98	97	96	95	94	93	92	91	90	89	88	87	86	85	84	83	82	81
80	79	78	77	76	75	74	73	72	71	70	69	68	67	66	65	64	63	62	61
60	59	58	57	56	55	54	53	52	51	50	49	48	47	46	45	44	43	42	41
40	39	38	37	36	35	34	33	32	31	30	29	28	27	26	25	24	23	22	21
20	19	18	17	16	15	14	13	12	11	10	9	8	7	6	5	4	3	2	1

SECOND CREDIT

120	119	118	117	116	115	114	113	112	111	110	109	108	107	106	105	104	103	102	101
100	99	98	97	96	95	94	93	92	91	90	89	88	87	86	85	84	83	82	81
80	79	78	77	76	75	74	73	72	71	70	69	68	67	66	65	64	63	62	61
60	59	58	57	56	55	54	53	52	51	50	49	48	47	46	45	44	43	42	41
40	39	38	37	36	35	34	33	32	31	30	29	28	27	26	25	24	23	22	21
20	19	18	17	16	15	14	13	12	11	10	9	8	7	6	5	4	3	2	1

LAST ½ CREDIT

60	59	58	57	56	55	54	53	52	51	50	49	48	47	46	45	44	43	42	41
40	39	38	37	36	35	34	33	32	31	30	29	28	27	26	25	24	23	22	21
20	19	18	17	16	15	14	13	12	11	10	9	8	7	6	5	4	3	2	1

[A.C.S.]
TYPING / KEYBOARDING

This is one of the simpler classes to document. As you can see, I just broke down the various components of the class into specific areas that were to be checked off when done. You really don't need any more for this class than a good computer program for teaching typing.

My personal feeling about the class of Typing, whether or not it is needed for a high schooler, is that, in today's world, it is absolutely indispensable. "Hunt and peck" is a handicap, and I don't say that just because I'm a writer and do a lot of typing. There is almost no occupation or calling for which a person would not need or want the skill of typing. It only enhances a resumé and one's ability to advance and expand both in work and home skills. You don't have to be fast (probably 40 to 50 WPM is ample), but you do want the basic skill. The more you use it, the better and faster you'll get! And the higher your WPM, the better it will look on your resumé!

We are totally sold on "**Mavis Beacon Teaches Typing**" having used the very first version (on 3" disk that we had to boot up!), the middle version (better), and now the CD-rom which is absolutely fantastic! So, since I have some space here, I'm going to spend some time extolling the virtues of this program...

My favorite game– woops! Did I say "game"?! I mean *serious typing practice exercise* is "the races"!!! (where a bug splats on the windshield for every mistake you make! I can really get that windshield covered with 'em!) You "compete" with the car next to you, which is your old time. Your WPM (Words Per Minute) at any given moment is shown in the MPH (Miles Per Hour) gauge. ☺ If you start going slower than your old time, the other car starts passing you in the lane next to you; you can see it in your windshield! ☺ (If you go really slow or stop, it disappears on the road way ahead of you, just like a real car would.) If you start passing your old time, you can see the other car in your rear-view mirror! It is just *too* clever and fun! That's just *one* of many clever – and very effective – games! One of Carlianne's favorites is the grocery store check-out aisle!

Mavis Beacon also has one really helpful feature where the keyboard is shown with two moving transparent hands over it! As you type a different letter, the hand and fingers move to that key to show you which finger should be pressing that key! Of course, you won't need this once you're a pro, but it's excellent for beginners!

As for the WPM, you and the student can determine your own goal(s), which is going to affect how long it takes to finish the course. For Tory, it turned out to be a ½ credit course because that's how long it took to get to the WPM he wanted to reach and do all the exercises. However, at college-level he is now working on it again, wanting to improve even more! Totally *his* idea!

This program is very inexpensive for what you're getting... only $29.95! There are many typing programs available, and I'm sure the others are just fine. We just happen to ALL really enjoy this. (Our 12-year-old too!) It is very thorough, personal (keeps a record of everything you do and your progress and weak areas), is totally user-friendly, and is easy and fun to use!

The "**12 Practice Exercises**" were in the "Owner's Manual" for one of the Mavis Beacon programs we had. (One comes with each program, but they're different with each new version, so the page numbers on my form will probably not match what you'll find in the new version.) These were interactive with the program, set up to work together, so the manual wouldn't do you any good without the program.

Any typing they do outside of working in your particular typing program (which could even be a workbook) can also be counted as part of your class.

Sharnessa, my wonderful secretary for five years, doing office work. (Sorry she's not typing, in keeping with the theme of this page, but there wasn't room for this picture on the "Occupational Education" write-up page where it actually belonged, and I couldn't find a picture of her typing. ... I have to be honest: My office was much neater when Sharnessa worked in it! Come to think of it, so was the house!)

✔ ASSIGNMENT CHECK-OFF SHEET ✔✔

Class: **Typing/Keyboarding**	
Student: **Tory W. Shelton**	*Class Value:* **½ Credit = 60 Hours**

BASIC REQUIREMENTS CHECK-OFF LIST	Check Here When Done
Learn and practice correct typing form according to instructions for proper finger placement in Mavis Beacon program.	
Use the various typing games and exercises in program interchangeably, as you see fit, to improve weaker areas.	
Read the "Owner's Manual" instruction booklet for Mavis Beacon Teaches Typing.	
Achieve the WPM *(Word Per Minute)* proficiency of your choice: ______WPM	
Practice typing/keyboarding skills by typing up own compositions and letters.	
Type a business letter of your own (or for Mom) using one of the four correct business letter formats *(as shown on pg 38, 42–45 of the Owner's Manual)*	
Do the 12 Practice Exercises on pages 61 – 72 *(Use chart here:)* #1: ○ #2: ○ #3: ○ #4: ○ #5: ○ #6: ○ #7: ○ #8: ○ #9: ○ #10: ○ #11: ○ #12: ○	

HOUR CHECK-OFF CHART

(Check off hours as they are completed)

60	59	58	57	56	55	54	53	52	51	50	49	48	47	46	45	44	43	42	41
40	39	38	37	36	35	34	33	32	31	30	29	28	27	26	25	24	23	22	21
20	19	18	17	16	15	14	13	12	11	10	9	8	7	6	5	4	3	2	1

[A.C.S.]
U.S. HISTORY, GOVERNMENT, & GEOGRAPHY

As you can see, this class is divided into four aspects of history, as well as U.S. Government and Geography...

EARLY AMERICAN HISTORY

Under this area there are several excellent resources. I have to say that *anything* by Marshall Foster (especially the audios and videos where you can *hear* him) is excellent. He also does in-person seminars which I cannot recommend highly enough! He gives a biblical perspective on *true* history and I doubt there is one detail of world or U.S. history that he doesn't know!

Two excellent resource companies for this area (and many others) are "Beautiful Feet Books" and "Bluestocking Press," both of which are listed in Section 8.

CIVIL WAR

This was actually covered more in the previous section, but these resources here were *solely* on the Civil War; more specific. The "Movie Critique" mentioned is here in Form+U+la. (See Form Index.)

GENERAL AMERICAN HISTORY

A homeschool dad, the husband of a friend, taught this class for several sessions to his own two high schoolers, my two, and one or two others.

Richard "Little Bear" Wheeler does excellent seminars in person, and even brings (and wears) authentic costumes of the various periods!

STUDY FAMOUS AMERICANS

The book "Great American Statesmen and Heroes" gives a brief bio on many great Americans, including quotes, which are sometimes difficult to decipher due to the difference in language and spelling, but worth the effort! (An aside: I asked myself one day, what if our country was being lead by people such as these in this book? Their zeal and integrity are amazing! And then I realized that the problem is not so much the leaders, because there *are* people – leaders – of this caliber today, but the problem is just as much the followers! We live in a country that is full of people whose values have greatly disintegrated; people who have spurned righteousness and purity for pleasure and lust. There *are* leaders today, people of integrity, but they are disdained, labeled "narrow minded," and ridiculed. I fear for our nation. How much longer will God wink at the lack of integrity and morality? ... Stepping down from soapbox now...) There are probably hundreds of biographies you can read. "Heroes of the Faith" is one series; and "Sower Series." (And we carry these too!)

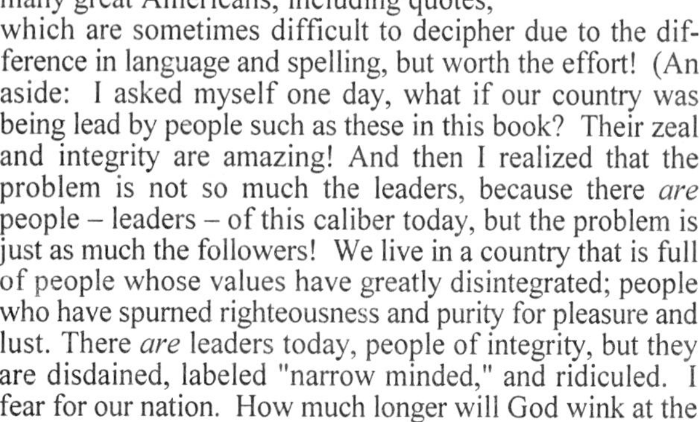

U.S. GOVERNMENT

All the resources here are books that present various aspects of government that all contribute to a bigger picture. The first two are booklets that give an overview of the Constitution and the Bill of Rights in language that even *I* can understand. *The Land of Fair Play* is appropriate for younger high schoolers; even junior high. But, though the presentation is "youngish," the content gives a good basic foundation in more complex government issues.

GEOGRAPHY

Any map or puzzle can be used to learn the location of the 50 states. There are more and more good CD-Roms (and books still work quite well too! ☺) to become familiar with the U.S.'s major cities, industries, climates, topography and geography. (The "Facts 50" computer game mentioned here is an *old* one on a 5" floppy disk that I doubt they even make anymore!)

DISCLAIMER

What I have come up with for this class may be *far* different from what you (or a history buff) might! So please don't just take my word for it, or follow what I have done and think that it's as comprehensive as you might hope or want yours to be! These are ideas for some good resources and for how you could structure a class, but there are *many* excellent resources out there now, and someone with a greater knowledge of this area might recommend a different course of study.

✓ ASSIGNMENT CHECK-OFF SHEET ✓✓✓✓✓✓✓✓✓✓✓✓✓✓✓✓✓✓✓✓✓✓✓✓✓✓✓✓✓✓✓✓✓✓✓✓✓✓

Class: **U.S. History, Government & Geography**	
Student: **Tory W. Shelton**	Class Value: **1 Credit = 120 Hours**

BASIC REQUIREMENTS CHECK-OFF LIST

Early American History:	☐ Read: The American Covenant by Marshall Foster & Mary-Elaine Swanson
	☐ Listen to "The American Covenant" Tapes Series by Marshall Foster
	☐ Read: The Light and the Glory by Peter Marshall and David Manuel
	☐ Listen to Tape: "America's Godly Heritage" by David Barton
	☐ Read: America's Providential History by Mark A. Beliles & Stephen K. McDowell
	☐ Attend Seminar by Marshall Foster: "Battle for the 21st Century"
	☐ Read portions of George Washington edited by W.B. Allen
	☐ Read portions of The Making of George Washington ☐ Read Liberty
Civil War:	☐ Read: The Diary of Clotee, A Slave Girl by "Clotee"
	☐ Read: Go Free or Die (About the life of Harriet Tubman)
	☐ Read: The Underground Railroad *(Life on the Road to Freedom)* edited by Ellen Hansen
	Watch movies: ☐ "The Civil War" and ☐ "Roots" and ↓
	☐ Do Movie Critique on one of the above: ____________ *(which one?)*
General American History:	☐ Attend U.S. History Class taught by Keith Merritt
	☐ Read selected portions of The Rewriting of America's History by Catherine Millard
	☐ Listen to Audio Tapes by Richard "Little Bear" Wheeler
Study Famous Americans:	☐ Read: Great American Statesmen and Heroes by Catherine Millard
	Read Biographies: ☐ Francis Scott Key ☐ Johnny Appleseed ☐ Abe Lincoln ☐ Johannes Kepler
	☐ Read: 50 Famous Americans by Orville Webster III
U.S. Government:	Read Booklets: ☐ "U.S. Constitution" ☐ "The Bill of Rights"
	☐ Read Book: Land of Fair Play ☐ Take part in Street Law Class taught by Jean Merritt
	☐ Read Book: Constitutional Law for Christian Students by Michael Farris
	☐ Read: Evaluating Books: What Would Thomas Jefferson Think About This? by Richard J. Maybury
	☐ Read Book: Are You Liberal? Conservative? Or Confused? by Richard J. Maybury
	☐ Read Book: Whatever Happened to Justice? by Richard J. Maybury
Geography:	☐ Be familiar with the locations of each of the 50 states ☐ Take Write-in Test on the 50 States
	☐ Use 3-D Atlas CD-Rom to become familiar with the U.S.'s major cities, industries, climates, topography, geography
	☐ Use/Practice "Facts 50" Computer (Game) Lessons to become familiar with the capitals and industries of the United States

HOUR CHECK-OFF CHART *(Check off hours as they are completed.)*

120	119	118	117	116	115	114	113	112	111	110	109	108	107	106	105	104	103	102	101
100	99	98	97	96	95	94	93	92	91	90	89	88	87	86	85	84	83	82	81
80	79	78	77	76	75	74	73	72	71	70	69	68	67	66	65	64	63	62	61
60	59	58	57	56	55	54	53	52	51	50	49	48	47	46	45	44	43	42	41
40	39	38	37	36	35	34	33	32	31	30	29	28	27	26	25	24	23	22	21
20	19	18	17	16	15	14	13	12	11	10	9	8	7	6	5	4	3	2	1

[A.C.S.]
(WASHINGTON)
STATE HISTORY & GOVERNMENT

I call this the "non-desk potato approach to state history"! ☺ It's ideal for families who love to travel, and a good impetus for those who don't!

For this class we had the privilege of having input from my dad – who *loves* history! I never appreciated it as a child, but I have now grown to love it! – even though I still cannot remember dates and facts! I don't think I ever will! But history is very enriching! What's that saying: "Those who do not know the past (history) are bound to repeat its mistakes"?

If you don't have a "Grampa" to help, then be sure to take advantage of the "docents" at museums who are happy to come in and take you on a tour. You'll usually need to arrange to schedule one ahead of time; and from what I have heard, they do this free of charge! They love to talk about their favorite topic, and you'll love to listen! You will get *much* more out of a museum this way! The stories they share make the history come alive!

By the way, it's a good a idea to talk with your docent *before* the tour so you can fill them in on the level your kids are at and how much to expect attention-span-wise. It's possible to get a particularly long-winded guide (you'll probably be able to tell from your phone conversation) who "loses" the kids – *and* maybe even the adults – so be aware of perhaps "steering" them in a different, shorter version. This can be awkward, so pray for wisdom. More often than not, though, they are adept at keeping their audiences!

This was a ½ credit class for us only because that's all that's required in Washington State, and I used that as my guideline. But as we got into the field trips more, I realized it could *easily* have been a full credit! And we could have made it part of our "Social Studies" credits, or even just an "elective." But it was too late at that point. (It doesn't really matter, as I'm sure you know by now if you've read much of my stuff; you get to do what you want anyway! ☺)

Even though this class is Washington specific, there are several ideas you can still glean and use from what I have laid out! My book, *Washington State History & Government: The Search-it-out Route*, has been used by homeschoolers in other states as well; only about 20% of it would be unusable by out-of-staters.

And do not hesitate to go beyond your state borders. For example, the David Manual Bronze Museum we visited is actually in Joseph, *Oregon*, (not Washington) as are the Columbia River Maritime Museum and Fort Clatsop. I have absolutely *no* problem counting visits to these next-door state places as part of our "Wash. State History" study because the history of our – or *any* – state involves the entire area, its history does not just take place within the confines of our state borders! The state borders were established long after the history of the area took place, so the history of adjacent states is really not "isolatable." For instance, the Oregon Trail, even though it cuts through several states and ends in Oregon, was vital to anyone living in Washington! The Columbia River borders northern Oregon and southern Washington, so the museum in Astoria, Oregon is as much a part of Washington's history as it is of Oregon's! Additionally, life in 1750 in what is now mid-Oregon was remarkably similar to life in northern Washington, just as you would find today!

SET UP NOTEBOOK:

The first step for this class is to set up a (Washington) State Notebook into which all the other studies (on paper) for this class will be placed. Complete directions for this are in my book.

THE STUDY OF OUR HISTORY:

All the forms mentioned here as assignments are in the book (not *this* book; the Wash. St. History & Gov't. book). I also give many ideas for doing "a history project of your choice."

THE STUDY OF OUR GOVERNMENT:

Again, all forms *and* the literature for each are in the (WA) book. As for the court and legislative information, this is very similar from state to state; there would be only minor discrepancies as "court is court" and "lawmaking is lawmaking"; it's just not going to be *a lot* different from state to state. I'm sure the new budding legislatures figured: "Why re-invent the wheel?"

(The only state, I discovered, that does not have both the House of Representatives *and* a Senate is Louisiana! I wonder why!? If anyone finds out, please let me know! BEShelton@aol.com or BEShelton@kalama.com)

The (WA) book also includes forms for taking notes at a city council meeting, interviewing anyone working in the law/court system, tips for writing a letter to the editor and your legislator.

THE STUDY OF OUR CULTURE & GEOGRAPHY:

This can cover a wide variety of field trips and studies! And if we'd had more time, we would have done far more than just five of those field trips! *(But Tory was involved in working and sports, which keep you pretty bound to practices; something you should seriously consider, and something I was not happy about. It was a hard "compromise" for me.)*

✓ ASSIGNMENT CHECK-OFF SHEET ✓✓✓✓✓✓✓✓✓✓✓✓✓✓✓✓✓✓✓✓✓✓✓✓✓✓✓✓✓✓✓✓✓✓✓✓✓✓✓

Course: **Washington State History & Government**

Student: **Tory W. Shelton** | **Class Value:** **½ Credit**

Resources: The Book/Manual: Washington State History & Government: The Search-It-Out Route; Museums and historical sites around Wash. State; Literature from these places, (Grampa) Leland Edtl

BASIC REQUIREMENTS CHECK-OFF LIST

Basic Requirements	Check When Done
Set up a Student Notebook for Study of Washington State History & Government *(See Section 6 for instructions)*	
THE STUDY OF OUR HISTORY:	
Complete "Purposes for Studying History" sheet, casually interviewing at least one historian, researching as well *(pg. II-15)*	
Interview one "Elder" using the "Our Elders Interview Questionnaire"	
Visit required museums and historical sites, and complete SIO sheets on each one *(See separate T.A.S.)*	
Complete "Location Data Sheet" (with photo(s) for each museum and site visited	
Do one history project of your choice. *(Types of projects are on pgs. II-8 – 9; areas are on pages II-10 – 13)*	
Collect and compile "Decade at a Glance" sheets for the 20 decades of WA's history *(keep in Notebook)*	
Practicum with younger child(ren): CHOICES: Unit of History using time line or map ... Take on F.T. to Museum ... Project	

THE STUDY OF OUR GOVERNMENT:

Read the literature for and complete the following SIO (Search-It-Out) sheets:				"Washington's Constitution"	
"Symbols of Wash. State"		"Washington Statehood"		"Our National Heritage"	
"The Legislative Process"		"Types of Court Cases"		"Legislative Terms"	
"How to Testify at the WA St Legislature"		"Superior Court"		"How a Bill Becomes a Law"	
"Supreme Court / Temple of Justice"		"Voting"		"Court Terms"	
"Elected Officials & Their Functions"		"Legislative Building"		"District and Municipal Courts"	

Write a letter to the Editor or a Legislator using the tips on pgs IV-84 – 89

FIELD TRIPS: ☐ Attend a Town Hall Meeting with a State or U.S. Legislator
☐ Attend one City Council Meeting and take notes using the guide on page IV-97
☐ Tour the County Jail ☐ Attend a Court Hearing ☐ Tour the Court House with a Bailiff

Interview a Police Officer while on duty patrolling: ____________ *(Name)* ____________ *(City)*

Interview any TWO: ☐ Judge ☐ Bailiff ☐ Records Clerk ☐ Attorney ☐ Police Chief ☐ Candidate ☐ Mayor ☐ City Manager

THE STUDY OF OUR CULTURE & GEOGRAPHY:

Requirement	Check When Done
Create a "Washington: From My Viewpoint" Booklet *(a photographic look at WA; Use K.Howshall's word/drawing of Washington for title)*	
Watch "Mt. St. Helens: Explosive Evidence for Catastrophe" Video - by Geologist Steve Austin, Ph.D. (published by ICR)	
Draw a map of Washington, including the major cities and terrain features (included in WA Notebook)	

FIELD TRIPS: to 5 cultural sites *(Underlined ones are required)*: ☐ Mt. St. Helens ☐ NW Trek ☐ Art ☐ Ethnic ☐ Space Needle ☐ Maritime ☐ Flight/Air ☐ Nuclear ☐ Industry ☐ Railroad ☐ Aquarium ☐ Military ☐ Armed Srvc. ☐ Stonehenge ☐ Rain Forest ☐ Science ☐ Conservatory

HOUR CHECK-OFF CHART

60	59	58	57	56	55	54	53	52	51	50	49	48	47	46	45	44	43	42	41
40	39	38	37	36	35	34	33	32	31	30	29	28	27	26	25	24	23	22	21
20	19	18	17	16	15	14	13	12	11	10	9	8	7	6	5	4	3	2	1

[A.C.S.] WORLD HISTORY, PROBLEMS, LEADERS, & GEOGRAPHY

I renamed this class to be a more appropriate title for what I wanted to include. I believe we can learn *much* from the leaders of the past, therefore there is a whole section on just that area, and named it... *(as we now jump into the sections of this ACS...)*

PEOPLE WHO HAVE AFFECTED THE WORLD'S THINKING, TECHNOLOGY, & DIRECTION:

The Great American Bathroom Books *("GABB" for short, which we also used for Literature, and which are also called "Compact Classics"– a little more dignified)* contain brief but ample backgrounds on each of the people listed in this section. They can be found under the same headings as I have listed here: "Pioneers in Science"; "Leaders in Crisis"; "Inventors/Innovators"; "Artists & Philosophers"; and "Leaders of the Spirit." Of course, there are hundreds more, but these are more commonly known people, and a good starting point. *(There wouldn't be time to read all about all of them in even a 10-credit class!)*

Something that would have been *excellent* to add to this section is *America's God and Country,* compiled by William Federer. *(We carry it.)* It's a *thick* book!

GENERAL HISTORY:

The first resource in this section is one I didn't discover until late in Tory's high school experience: Diana Waring's "What in the World's Going on Here?" tape sets, Volumes I and II. These are excellent! And she has *more* out now too! (And yes, we carry them!) She is *very* lively, *very* knowledgeable, and *very* interesting. I dare anyone to try to fall asleep listening to her! (Or Marshall Foster!) She gives a good overview of history over many centuries.

The whole next section starting with "□ Read Scientists (from Archimedes to Einstein)" is based on the Usborne book by that exact title. It covers *all* that you see here! Of course, it's not Christian, but you'll get that perspective from Diana and Marshall! This is just good "basic knowledge" but with the Christian parts left out.

We also read a series of five booklets (early high school level) called "The History of Western Civilization." It is very brief (sketchy), but a good overall intro.

PREJUDICE & RACISM ISSUES

This included two major "racism" issues: slavery and the holocaust. The lecture by Thomas Blatt was excellent! The rest is pretty self-explanatory. The "Movie Critique" mentioned is here in this book.

Digression: After reading and watching these movies and books, I have to admit I had some *major* forgiveness issues to deal with in my heart. You may want to watch for these in yourself or your children. I found myself aching for all the suffering "we" – Americans, whites, Germans (one of my ancestral nationalities: "Edtl"), etc. inflicted upon so many innocent people! How could such atrocities have been inflicted by man upon his fellow man?!?!? One day God really dealt with me heavily on this, and brought me to a place of: # 1 - forgiving them and, in a sense, myself, because I'm capable of the very same thing! ("There but for Jesus go I"!) And # 2 - giving the horrible burden of it to Jesus, who died on the cross for those very things. He showed me that *He* took that burden upon himself, and I am not to carry it. But I still feel like apologizing to every such person I see! {:-) "I'm *so* sorry for all that your people have suffered! *Please* forgive us!"

"MAJOR WARS":

Somewhere in this book (uh-huh, see the Form Index) is the "War Report" that you make copies of to use in this study. (And you might check out the pages before and after that for more ideas for studying State History.)

GEOGRAPHY:

For this we just used the "3-D Atlas" CD-Rom. There are now *many* great resources out there though! (In fact, I'd love to hear from anyone who's found a good one that has worked for them! *(Email me at BEShelton@aol.com or BEShelton@kalama.com.)*

"CULTURAL LITERACY":

I put out a request to my SHF-L list some time ago asking for input on what people feel are important topics for anyone to be at least somewhat aware of and informed about, and this is what I compiled! Several of these are defined (in a paragraph, not just a quick dictionary definition) in the Dictionary of Cultural Literacy, a 600+ page book that covers just about everything in science, history, fine arts, medicine, etc. that you could possibly think of!!! (And, yes, we carry that too!)

What I would suggest you do, as far as this list here on the ACS goes, is *discuss* these topics with your student! Get Dad, Mom, grandparents, friends *(not necessarily all at one time)*, and ask *them* for their insights and opinions on these areas. You'll get a variety of perspectives – which is *exactly* the idea! To become knowledgeable of the various ways to look at any topic. Truth is truth, but there are many areas that are neither black nor white. It's not only fascinating, but helpful to gain insight as to *why* people think the way they do. I believe helps us formulate our own opinions and views, and, once formed, to be more able and gracious in leading others to truth and right thinking.

✓ ASSIGNMENT CHECK-OFF SHEET ✓✓✓✓✓✓✓✓✓✓✓✓✓✓✓✓✓✓✓✓✓✓✓✓✓✓✓✓✓✓✓✓✓✓✓✓✓✓

Class: **World History, Problems, Leaders, & Geography**

Student: **Tory W. Shelton** | Class Value: **1 Credit = 120 Hours**

BASIC REQUIREMENTS CHECK-OFF LIST

People Who Have Affected the World's Thinking, Technology and Direction *(See GABB #1)*	Pioneers in Science: ☐ Euclid ☐ Isaac Newton ☐ Galileo Galilei ☐ Nicolaus Copernicus ☐ Albert Einstein
	Leaders in Crisis: ☐ Winston Churchill ☐ Ben Franklin ☐ Abe Lincoln ☐ George Washington ☐ Thomas Jefferson
	Inventors/Innovators: ☐ Thomas Edison ☐ Marie Curie ☐ Orville & Wilbur Wright ☐ Louis Pasteur ☐ Antoine Lavosier
	Artists & Philosophers: ☐ Leonardo da Vinci ☐ Ludwig van Beethoven ☐ Michelangelo ☐ Aristotle
	Leaders of the Spirit: ☐ Martin Luther King, Jr. ☐ Mahatma Gandhi ☐ Florence Nightingale ☐ Confucius
	☐ Read: <u>Vanya</u> by Myrna Grant ☐ <u>Christopher Columbus</u> ☐ <u>Susanna Wesley</u> ☐ Christopher Columbus
General History	Overview of History: Listen to Audio Tapes: ☐ "What in the World's Going on Here?" - Volume I ☐ Volume II
	☐ Read: <u>Scientists (From Archimedes to Einstein)</u> (Usborne Book) ☐ What is a Scientist? / How Science Began / Expanding Frontiers ☐ Early Scientific Ideas ☐ Science in Ancient Greece ☐ Medicine in the Ancient World ☐ Islamic Science ☐ Science in Medieval Europe ☐ The Earth and the Sun ☐ Understanding the Human Body ☐ Science During the Renaissance ☐ Scientific Societies ☐ The Rise of Experiment ☐ Classifying the Natural World ☐ The Age of the Earth *(Secular view)* ☐ Evolution *(Overview)* ☐ The Growth of Modern Chemistry ☐ Studying Electricity ☐ The Fight Against Disease ☐ Waves & Radiation ☐ The Science of Life ☐ Opening Up the Atom ☐ The Origin of the Universe *(Secular view)* ☐ Women in Science ☐ Key Dates in Scientific Discovery
	Read/Do "The History of Western Civilization" Series: ☐ Book I ☐ II ☐ III ☐ IV ☐ V
Prejudice and Racism Issues	☐ Attend "Escape From Sobibor" Lecture by Thomas (Toivi) Blatt, Escapee from a Nazi death camp ☐ Watch "Holocaust" video ☐ Watch "Escape From Sobibor" (the movie about the author's experience)
	☐ Watch "Amistad" (the revolt of slaves on a slave ship), and ☐ Do Movie Critique
Major Wars:	**Do War Reports on:** ☐ French-Indian War ☐ Independence War ☐ War of 1812 ☐ War of Mexico ☐ Civil War ☐ Indian Wars ☐ Span-American War ☐ World War I ☐ W.W. II ☐ Korean War ☐ Viet Nam ☐ Desert Storm
Geography:	☐ Use "3-D Atlas" CD-Rom to and the globe to become familiar with the World's major countries and cities, climates, topography, geography
Cultural Literacy:	**Research and write synopses on:** ☐ Abortion ☐ Affirmative Action ☐ AIDS ☐ Alternative Health Care ☐ Alternative Power Sources ☐ Assisted Suicide ☐ Child Abuse ☐ Christian Persecution ☐ Cloning ☐ Computer Glitch of the Year 2000 ☐ Constitutional Rights ☐ Date Rape ☐ Discrimination in the Workforce ☐ Economic Interdependence *(ie. why the European crash lead to the 1930 crash here; will Asian crash will lead to same?)* ☐ Environment/Ecology ☐ Euthanasia ☐ Evolution vs. Creation ☐ Excessive Government Power ☐ Food Source & Clean water ☐ Global Warming / Ozone Layer Holes / El Niño ☐ Gay Rights ☐ Germ Warfare ☐ Goals 2000/Outcome-Based Education ☐ Gold Standard ☐ Government Provided/Regulated Child Care ☐ Gun & Ammunition Control ☐ Healthcare ☐ Home Education ☐ Immunization Issues & Dangers *(both ways)* ☐ Internet / World Wide Web ☐ Ku Klux Klan ☐ Medicinal Marijuana ☐ The "Millenium Bug" (Y2K problem) ☐ Modern Medicine *(drugs, invasive tests, surgeries)* ☐ National Deficit ☐ Nuclear Power & Disarmament ☐ Parental Rights ☐ Political Correctness ☐ Polution ☐ Population Control *(and the atrosities in China of taking this too far)* ☐ Pornography ☐ Rights of the Child ☐ Romance God's Way vs. Dating *(AKA "Courtship" or just a matter of "Purity")* ☐ School Violence ☐ Second Coming of Christ *(Millenium, the Tribulation, 666, the Rapture)* ☐ Sexual Discrimination ☐ Space Program ☐ Substance Abuse ☐ Survival & Preparedness Skills ☐ Tax Credits for Alternative Education ☐ Transition from Industrial Age to Information Age ☐ Values Clarification ☐ Waste Disposal Problems ☐ Watergate ☐ Women in the Military

HOUR CHECK-OFF CHART *(Check off hours as they are completed.)*

120	119	118	117	116	115	114	113	112	111	110	109	108	107	106	105	104	103	102	101
100	99	98	97	96	95	94	93	92	91	90	89	88	87	86	85	84	83	82	81
80	79	78	77	76	75	74	73	72	71	70	69	68	67	66	65	64	63	62	61
60	59	58	57	56	55	54	53	52	51	50	49	48	47	46	45	44	43	42	41
40	39	38	37	36	35	34	33	32	31	30	29	28	27	26	25	24	23	22	21
20	19	18	17	16	15	14	13	12	11	10	9	8	7	6	5	4	3	2	1

CULTURAL LITERACY RESOURCES

(Not ASC's or TAS's; just mentioned on some of them)

Because these resources (especially "The Dictionary of Cultural Literacy,") can be used in so many different classes, I felt it would be the most efficient use of space to explain them just once, and in one place. I discovered these wonderful resources toward the end of Tory's high school experience. I wish I had known of it sooner as this is what I had been looking for, but didn't know existed!

I first want to address the issue of the Cultural Literacy movement. I'm not *real* familiar with it, just mostly with these books, but I do know that there are opinions about this movement *(which I found at a website. Just type in "Cultural Literacy" and you'll get address for several sites, both "pro" and "con.")* that express concern about its use in schools and by educators. One concern is that it is focused only on filling young minds with too many facts. This is a valid concern in the school setting, or in homes that follow a rigid "school at home" approach. But I don't have as much concern about it within the homeschool setting as long as the resources and concepts are used as tools and parents implant this into the larger picture of the child's *total education.* I explain that more in-depth in Section 1 of this book, and how to utilize this resource right here...

Individual sections of the "Dict. of Cultural Lit." (D.O.C.L) can be used as "starting points" and/or "fillers" for many classes. With 619 pages including the index, this book is huge! It has chapters on a wide range of topics that I'll be listing and detailing in just a moment. In each chapter is listed, in alphabetical order, many terms, concepts, people, and events, and concise definitions for each.

For instance, under "Life Sciences" here are the words listed and defined on one page that I chose randomly: "Anatomy, Animal Kingdom, Arthropods, Asexual Reproduction, Australopithecus, Bacilli, Bacteria, Balance of Nature, Basal Metabolism, Biochemical Pathways, Biochemistry, Biodegradable, Biology, Biomass, Biophysics, and Biosphere." The average explanation/definition is about 3 to 6 lines long; with 2 columns per page.

What I do is go through the section in the D.O.C.L. that I am going to use and pencil in a very faint circle next to each of the items I want my child *(which will now be Carlianne)* to read. (It's a good idea to read much of this *with* your child!) There are some, though not many, that I don't feel are all that important for them to know. How easy it will be for a person to make such determinations will vary according your forté, but if you truly feel unequipped, and if the Lord deems it important for you (your child), then He'll provide someone who can do this for you.

Here are all the chapters of the book. Next to each one I've written the class or classes that it could fit into. If you have another idea, go for it! *(Remember, I am not an expert; just someone with a lot of ideas who has been asked to share them. That's all.* ☺) Note: Some of these may be for just portions of the stated classes; not the whole class.

THE BIBLE - "Spiritual & Biblical Studies"; "Philosophy of Religion"; "Social Studies"

MYTHOLOGY AND FOLKLORE - "Social Studies"; "Philosophy Of Religion" (Note: For religious reasons, some may have a problem with mythology. I personally believe that if mythology is studied without going into too much depth and with an explanation of its nature, it will only increase the student's understanding of the world they live in; I don't feel it will confuse or taint them. But at the same time I wouldn't want to spend a lot of time on it.)

PROVERBS - "Life Skills"; "Philosophy. Of Religion"

IDIOMS - "English": Vocabulary – finding out what these sayings mean; and Composition – knowing what to avoid in good writing!

FINE ARTS - "Fine Arts Survey"

WORLD LITERATURE, PHILOSOPHY & RELIGION - "English (Literature)"; "Philosophy of Religion"; "Social Studies"

LITERATURE IN ENGLISH - "English (Lit)"

CONVENTIONS OF WRITTEN ENGLISH - "English Composition"

WORLD HISTORY TO 1550 - "World History, Problems, Leaders, and Geography"

WORLD HISTORY FROM 1550 - "World History, Problems, Leaders, & Geography"

AMERICAN HISTORY TO 1865 [and "FROM 1865] - "U.S. History And Government"

WORLD POLITICS - "World History, Problems, Leaders, and Geography"

WORLD GEOGRAPHY - "World History, Problems, Leaders, and Geography"

AMERICAN GEOGRAPHY - "U.S. History, Government, and Geography"

ANTHROPOLOGY, PSYCHOLOGY, AND SOCIOLOGY - "Social Studies"

BUSINESS AND ECONOMICS - "Business & Office Skills"; "Social Studies"; "Economics"

PHYSICAL SCIENCES & MATH - "Math"; "Lab Science Survey / Basic Scientific Literacy"

EARTH SCIENCES - "Lab Science Survey / Basic Scientific Literacy"

LIFE SCIENCES - "Lab Science Survey / Basic Scientific Literacy"

MEDICINE AND HEALTH - "Health"; "Lab Science Survey / Basic Scientific Literacy"

TECHNOLOGY - "Lab Science Survey / Basic Scientific Literacy"; "Social Studies"

(Whenever I add any of these areas onto an ACS, I include the page number so Tory doesn't have to hunt for it.)

Why would I be so excited about a resource of this nature? Because I see it as a tool for the accomplishment of a couple of goals. For one thing, I think that having an understanding of terms, definitions, and specific pieces of history will enable us to more easily get the "big picture" because there are so many smaller pieces of the puzzle that must be present in one's understanding! Maybe it's a question similar to "which came first, the chicken or the egg?" But to me, either way, this book (DOCL) is going to unlock some doors and expand my understanding, and hopefully do the same for our children as well.

The other benefit from this book is that it will give the reader more "common ground" with the unsaved! It will help you build a knowledge base that will provide you with more things to talk about, and more to be "enough aware of" that we can at least then ask intelligent, sincere questions. This has opened many doors for me with people.

Just to clear up the confusion about the cultural literacy books... The book I was referring to in any of my ACS's is called "The Dictionary of Cultural Literacy" (The ISBN number is 0-395-65597-8) and is by E.D. Hirsch Jr, Joseph F. Kett, and James Trefil. This is the huge 600+ page book with all the terms and definitions.

There is another book called *Cultural Literacy*, and it is also by (just) E.D. Hirsch, Jr. (The ISBN number for this book is 0-395-43095-X.) Interestingly, the subtitle for *both* books is: "What Every American Needs to Know." Much smaller, 251 pages, this book is the "heart and soul" – the philosophy *behind* the need to be culturally literate. It has a very large 60-page section which is a list of all the things one should be literate about; not the definitions, just the words. It appears to me that the bigger one, the Dictionary of Cultural Literacy, is the organized *answer* to the question people would naturally ask when looking over that list: "OK, so what *are* these things?!?"

Just for the record, I personally think *both* books are equally important.

Linda Bullock (*Listen My Son)* explained the books well, so I'll reprint her comments here, adding my own *lengthier* comments: (I've put her words in bold.)

"There are actually several cultural literacy books. The original book "Cultural Literacy" is fascinating to read, and convinced me that as Christians, we need to be aware of cultural literacy."

[Barb:] I think it was probably the Holy Spirit who convinced Linda of this, as the book is secular in nature. ☺ Just didn't want anyone expecting it to be a great *Christian* work! However, as Linda has indicated, and I explained, it is a great book – or *tool – for* Christians!...

"Paul (in the Bible) was culturally literate. He knew about other religions and world views, and so could refute their beliefs."

[Barb:] This is such a good point! I think, too, that it's important to have enough awareness about the world we live in to be able to identify with people. We can get "so heavenly minded we are of no earthly good." This is *not* an excuse to be worldly minded. But the Word admonishes us to be "in the world, but not *of* it. We should probably also be aware of the fact that not everyone is called to be a great "defender of the faith" as Paul was, and many are today. Not just well-known leaders either. But I know many wonderful Christians who will *never* be used in this way. That's just not their function in the Body of Christ. And they should feel ***no*** guilt if that's not their calling.

But no matter what ministry we are called to, and no matter what our spiritual giftings, or our strengths, I feel that a basic foundation of knowledge in this area is still important – for our own growth, depth and understanding, and it will also make us more available to be used in this area even if in just "small" ways – with our own family and friends, who make up our "circle of influence." Just my opinion. *(Back to Linda...)*

The Cultural Dictionary for adults is awesome, but the philosophy behind all this says that to be really effective, cultural literacy is something that must be learned during the elementary years so that it becomes part of our thinking, not just information floating in our brains. So there's a smaller dictionary for the elementary years. And of course, for those who have younger children, there is the series of books: *What Your First (Second, Third, etc) Grader Needs to Know.* Actual books to read to your children.

[Barb:] I had seen these books but always steered away from them because they felt "heavy" – like they could too-easily put me (or anyone) into bondage. It felt like I would be yoking myself to the world's idea(s) of what my children should be learning and losing my freedom to hear God. In fact, His ways and His truth *do* still need to be our top priority. "A man cannot serve two masters."

But I'm seeing something different now. As I have gotten into this Dictionary, I'm seeing it all in the light of this resource being a useful *tool*, as opposed to how I had been viewing it – as a *master*. This shift in perspective makes all the difference in the world! I also acknowledge and should mention here that having walked in freedom from the world's requirements for the past several years makes it quite a bit easier for me to quickly get a healthy perspective toward these resources. I pray for that freedom to be yours soon, as well!

And before anyone makes a single purchase or visits any of these web sites, I highly recommend you get this "right" between you and God – i.e. that you see it as *He* sees it! This is similar to what I have said here in my book about how we view and use Student Learning Objectives. That's where I get into the "Intimidometer." ☺ You might even go back and read that again, this time inserting "cultural literacy resources" for the word "Student Learning Objectives." *(See pages IV-2 and 3. But I actually deal with the fear issues in much more depth in Sections 1 and 2.)*

The bottom line is that, as with *any* resource, these books must be regarded as our "servants," not us as servants to them! They are tools to help us accomplish God's work in our children. Period. No more, no less. Use them if you feel so led, but don't let the idea rule you by making you think that, without this particular mass of knowledge, your child will be ill-prepared or unusable by God!

The Dictionary of Cultural Literacy
(the BIG 600-page book)
ISBN 039-543-7482 - $24.95

Cultural Literacy ISBN 039-475-8439 - $12

(And we do carry these books.)

Testimony Tidbits from a Recovering Homeschool Mom

When the following homeschooling mom , Leslie Gamell in Osteen, Florida, first looked at all my completed class plans (the ones that are now in section 6-D) she panicked, wondering how in the world my son could do so much! (You may recall her questions.) She sent out a plea for help on the SHF-L list** (see page VIII-7 for the write-up on SHF-L), **and several moms on the list and myself came to her rescue with encouragement and clarification. Now, here's her response to all that...

"After I sent that post, I sat down to really study the class plans again. After studying them, I found that Tory had only read *portions* of some of the readings and not complete books in all cases. Even just that made me feel better. Since then, the Lord has really opened my eyes and shown me so many things that they (my kids – separately) and we (together) have read on many topics as we did "unit studies" on various historical eras. The Lord began to show me, using your class plan models, that we have some classes complete, or near complete, if I only lay them out in a similar way!

"About two years ago, my son did a study on wilderness survival just because he felt like it, while I (in my old wineskin) was trying to get him to "put that stuff down and get your *school work* done." This study took him about six months. The other day I had him go to the library and list all the books he used for this study, whether he read them in entirety or portions. He filled a page! The material is a permanent record in his brain because it was such a passion of his. He learned many things that are covered in a biology book I have no idea how many hours he spent but it was *many, many!* (I remember vividly because I was *always* on his back to put it aside! ... Hindsight is 20/20, isn't it?)"

I had told her: "You'll be amazed at how much more they will be able – and willing – to grasp at age 17 and older! And I had told Tory I wish I'd had the 19-year-old model of him all the way through high school! ☺ But we take – and love and work with – whatever we 'have in hand,' don't we? ☺ They (we) don't start out mature, that's the whole purpose of the process! We just have to let God lead that process for <u>each</u> of us individually." ... And to that she answered: "Thank you for this, Barb. The Lord also showed me your wisdom in having the kids read books on homeschooling. *(See "Educational Foundations in 6-D.)* I always thought of that stuff as MY reading material and never thought of having the kids read it. <u>Thank you</u> for the wonderful prompt. But I will wait for them to mature a little more. From my vantage point this looks like never-never land, but in my heart, I know it will come all too soon."

I had also shared: "Even if what's gone on in our lives was initiated by the enemy, we know that "what the enemy intended for evil, God will turn for good." THIS is the essence of what His redeeming power is all about!!! I'm SO glad it applies to absolutely ANYthing in my life!" To which she answered: ..... "This encouragement was like a sweet, cool shower on a hot day, pouring over my weary self." ***... Thank you for sharing so much of your heart, Leslie!***

Section Seven:

LIFESTYLE OF LEARNING APPLICATIONS

This section will help usher you into freedom and excellence in true education, and will help in acquiring a life-long love of learning – IF you are willing to let go of some old ways and old thinking. Please *please* do N☺T skip this very *vital* section❣

❥ The Heart of This Message ❥

Marilyn Howshall

My message is not about how to homeschool, but about being a Christian. This does not "look like" the model seen currently in today's mainstream Christian culture. If it was, all you would need to do is continue doing school at home, and you could get the same kind of fruit that you see everyone around you producing. If you want godly fruit, you have to know God and do things His way. "For as the heavens are higher than the earth, so are My ways higher than your ways, and My thoughts than your thoughts." (Isaiah 55:8, 9) You won't learn what His way is until you decide you want to know what it is and surrender your will to become completely teachable of the Holy Spirit.

This message is about parents who truly know how to walk a surrendered life, by faith—with the cross—becoming empty of self and filled with the love of God that imitates who He is for their children. So filled with the love of God that they are able to love their children as unique individuals instead of enforcing generic, artificial substitutes in exchange for what God has planned for the child's unique purpose. This can only occur if the parent is willing to take steps of obedience in faith instead of acting out of the fear of their hearts, obeying what the world says instead of allowing God to be their fear and dread. "Let God be your fear, and let Him be your dread, and He will be your sanctuary." (Isaiah 8:11)

Our Christian faith seems to get put on the shelf for homeschooling, for "ministry," and for many other worthwhile purposes. The Christian faith to be truly the Christian faith is going to "fly in the face" of every preconceived idea we might have been conditioned to believe. True faith in God will require a true faith walk, step by step, not necessarily knowing what the steps are, but following only the one the Holy Spirit says to do next. Most decisions we make are based upon assumption that we are supposed to do something a certain way. Why is that? We are not listening to the Holy Spirit!

If God is the one who called you to homeschool, then He has the plan and the right to be the director of it! Be still for awhile and know that He is God and you will begin to hear His quiet voice directing you. He will not lay it all out for you. He will tell you one or two things to begin working on now. As you obey He will give you another step. This is called dying to self, becoming still, empty, and willing to ...

Continued on last page of this section...

LIFESTYLE OF LEARNING APPLICATIONS

INTRODUCTION OF THIS SECTION

I (Barb)imagine you're curious to find out how on earth this system I'm presenting in this book is going to mesh with a highly natural lifestyle of learning. Actually, the two don't really "mesh"; they are separate approaches, with similar hearts and elements, but different means of getting there. If I had to put it in a nutshell, which is almost impossible to do accurately (or for *me* to do, period!), I'd have to say that *Lifestyle of Learning* will help you get an understanding of the *heart* of true education for yourself first, and then for your children as an outflow of your own walk with the Lord. My materials will serve in two ways: 1) as a catalyst in pointing you toward putting yourself through a "season of re-education and renewing of the mind," which will include many of Marilyn Howshall's writings, and those of others, and also 2) as a "bridge" to help you transition from being *bound* to the "world's ways" to being *free* to hear God for *your* family.

I strongly encourage you to read this whole section carefully, even if you haven't understood or thought much about the concept of "lifestyle of learning" up to this point. I think you're in for a pleasant surprise. I suggest that you *begin* with the message by Marilyn that begins to the left and ends on the very last page of this section. It's something in her words that will serve to "hem" this section at the beginning and end. ☺

God is doing a *new thing* in His people, and is using homeschooling as an "avenue" through which He wants to bring His people closer to Him – which, in turn, involves turning the hearts of the fathers – *and* mothers – toward their children, and children's hearts toward their parents.

Questions and Answers

Q *What is a Biblical Lifestyle of Learning?*

A The whole panoramic painting of *lifestyle of learning,* though some brush-strokes are similar to traditional methods of teaching and learning, is essentially completely different in composition. Imagine a painting of a field of red and white tulips and then a painting of the U.S. flag flying in the breeze: *Both* have red and white, both have curvy, flowing strokes, but they're not at all the same painting! Let's take a look at some of the differences:

➤ Whereas traditional education is a one-size-fits-all, cookie-cutter, run-em-all-through-the-same-mill approach, with generic requirements and generic applications, lifestyle of learning is highly *individual.*

➤ Traditional schoolwork is geared toward producing a *product* that looks good; a lifestyle of learning focus is on the *process* of learning, this means that the product may come more slowly, being less impressive to the casual observer in the earlier stages of learning, but which will become deeper and more meaningful, resulting in a superior product *later,* as it is allowed to unfold naturally.

➤ In traditional education, the child is motivated externally by rewards and competition. In a lifestyle of learning he is – or eventually *becomes* – motivated from *within.* The student may not *start out* with motivation to learn, but as the parent sets the example and provides a learning-condusive environment, he inevitably catches it. Please note: *The parent sets the example.* This is not a "Do as I say, not as I do" approach! In fact, if this has *not* been true in the past – if Mom has *not* been a learner or setting that example – then even her going through a "season of re-education and renewing of the mind" can be a starting-point opportunity for her to set such an example. Of course, though, it will ultimately involve much more than that, as she begins to seek and discover delights and interests within herself that she senses God wants to develop – but in *His* timing and in *His* way.

➤ Traditional school is directed and prescribed by the compartmentalized curriculum set forth by the school system and has nothing to do with the *real life* of the child. In a lifestyle of learning, a child's learning is totally *relevant* to his life, with all the skills he needs for life being taught in relation to the life he is experiencing at any given time. Thus the *love of learning* is developed in the child (and kept alive and flourishing) rather than merely the *love of accomplishing academic goals* and *acquiring "knowledge that puffeth up."*

➧ Material learned in traditional school is typically forgotten within two weeks after the test because the student hasn't learned it with *delight,* or it wasn't relational to his life in any significant way. What a child learns in a delight-based lifestyle of learning pursuit, having been nurtured and given the time and freedom to take root and develop, carries the student of *life* through the rest of his life!

➧ Traditional school is something you *do* six hours a day; lifestyle of learning is a way you *live* twenty-four hours a day.

♥♥♥♥♥

***Q** Why do so few people pursue the freedom of a lifestyle of learning if it's so excellent? None of my friends are interested!*

A First, there are probably a lot more people doing it than you think; they're either just not in your area, or they're not very "visible." ☺ But on the other hand, as appealing as the lifestyle of learning concept is to most people, the greatest problem, surprisingly, lies in the freedom it brings. Liberty requires more responsibility on the part of the parent than the security found in prepackaged curriculums.

Another reason people might shy away from it is due to the way "natural learning" has been promoted and represented in homeschool circles, often with little or no structure involved in their approach to it. People intuitively – and correctly – feel the need for structure in order to meet their goals, and consider it foolish or even dangerous to pursue education without it.

Yet another reason for not pursuing it is that some might equate "self-directed" with "self-*centered.*" They get the impression that delight-directed study simply feeds the flesh with no concern for discipline – which, "for the moment, seems painful rather than pleasant." In fact it might seem, at first glance, that the two concepts of *delight* and *discipline* cannot work together. In reality they are a necessary and integral part of each other. Just because an activity is delight-directed doesn't mean no discipline is involved, and just because there's discipline, doesn't mean there's no delight. Notice that the verse says the discipline *seems* painful, and then it also goes on to say "but later it yields the peaceful fruit of righteousness to those who have been trained by it." As you read the remainder of this section (7), the truth will begin to unfold. *(Hang in there! It'll be worth it!)* Marilyn Howshall shows in *Wisdom's Way of Learning (the four-book set that presents the Lifestyle of Learning message)* and also her subsequent books *(that continue to develop the message in more focused areas,)* that there is a tremendous difference between "natural learning" as it has been taught and a *"Biblical* lifestyle of learning." These books will open parents' eyes and free them to pursue a more godly, delight-directed education for their families, enabling them to see their children, their roles, their *lives* in a much fuller and meaningful way, and equipping them to implement what they see (their vision) into their family's lives!

♥♥♥♥♥

***Q** Your system (in this manual) appears to have quite a bit of structure to it as indicated by your forms, charts and measuring credit hours. I thought a lifestyle of learning was unstructured. Won't this method cramp a newly developing natural lifestyle of learning? How can something be natural and structured at the same time?*

A On the surface it might *appear* that a lifestyle of learning is unstructured. In reality it is not. Learning *must* involve some sort of structure, or there would be no order in it, no direction, no ability to build precept upon precept, no way to determine if goals were being met. *And there will be a wide range of structure types and degrees from family to family.*

Natural should not be equated with *lack of structure,* making the education sloppy and haphazard. Without structure, the entire universe would fall apart. Our bodies have a skeletal structure that holds them together and makes it possible for them to function. Likewise, a building has an underlying structure. Without structure, even natural learning would have little purpose. There must be goals, direction, and a means by which to determine if they have been met.

The *right* kind of structure gives you freedom to move forward with your plans, without which your goals would be only a vain vision. The elements of my system, (as with the elements of the *Lifestyle of Learning* framework,) are designed to be used with a great deal of individuality and creativity *based upon the goals of your family and your own unique lifestyle and interests.* Far from *cramping* your lifestyle, if you press into hearing God on how much of what kinds and degrees of structure to implement, you will have what you need that will allow you to operate *freely.* He will lead you! The only thing that will hinder you is your individual level of creative thought and application, which the Lord can expand if you will let Him, and if you will let Him remove the chains that are keeping you from walking in the freedom He offers.

***Q** Won't the system presented in this manual just keep us bound to worldly standards that we're trying to get set free from?*

A To answer that, we have to first define more clearly *what* we are trying to get set free *from*. There's nothing inherently wrong or even worldly with a system of measuring credit hours to determine a student's performance. Nor is there anything worldly about establishing a set of requirements for the student to meet during his course of instruction and upon completion of his course to demonstrate what he has learned.

Rather, what we're trying to get set free from is meeting *unbiblical* requirements – *not* from having *any* requirements at all. The requirements themselves will reflect the worldly *or* the Christian values of the parents. Without *some* set of requirements, how can you know what your goals are, or what you are moving toward? Credits, credit hours, charts, academic requirements, forms... these are all *neutral tools*. Just as the homey-ness of a house is not determined by its structure, but by its *inhabitants*, likewise, what you fill your educational framework with is what will determine its value and whose standards are reflected – yours or the world's. Of course, even if the framework you choose is "the best," it *is* possible to err on the side of either being too dependent on the system, or making the system too rigid and binding. Follow the leading of the Holy Spirit in how much, and even *how*.

It is important to remember that most homeschoolers are connected to our cultural system in one way or another, and in varying degrees at different times. This doesn't mean they have to be enslaved to it, but it *is* common for one family to have an older child who has been a part of the public (or private) school system as well as younger children who have never entered it and probably never will. Each child will require an *individual* plan based upon where he is at the time. For this reason, the parent may need to operate with varying degrees of freedom for their different children. Transitions are difficult for children, too. It is more important to minister to the child's *heart* than it is to switch systems. God takes *us* one step at a time. Likewise, we need to take our *children* one step at a time even in our desire to get them – or ourselves – set free.

I am simply utilizing, modifying, and personalizing a familiar structure to help parents in the transition ("bridge") between traditional learning and a freeing approach to education during the homeschool years.

♥♥♥♥♥

***Q** How would you say your and Marilyn's messages mesh, compare, dovetail, connect, and/or differ?*

A The emphasis of Marilyn's message is different from mine. Her focus is in the area of applying biblical principles to the learning, living, lifestyle training, discipleship, and equipping processes of education. All of her materials present principles and models for you to apply uniquely to your situation. As described in *Wisdom's Way of Learning,* her focus is primarily on discipling and preparing her children for their individual purpose as it unfolds to their understanding. All methods and curriculum she uses support this goal. She uses traditional curriculum only in the area of mathematics. The academic emphasis is on developing the learning tools instead of on filling subjects.

Marilyn extends many of her children's interests into the application of learning tools and the development of her children's products that truly reflect their own learning process. The student's advanced learning skills are then extended into the development of entrepreneurial and career plans to equip them with a well-rounded education in addition to establishing them in fulfilling life purpose. This approach requires a generous amount of applied faith, discipleship, and creativity. Her family's personal goal structure requires a minimum amount of record keeping. She collects her children's projects and analyzes the content, skill level, etc. of each project, but not according to worldly "rulers." She does not try to make sure that all subjects are addressed in a project (as is typical of the "unit study" approach), and feels this is unnecessary for her family. The subjects get covered automatically as their lifestyle of learning progresses and unfolds. She keeps a portfolio of documentation of her children's work in case the future calls for some form of transcript and would be able to translate her records accordingly. However, she does not believe this will be necessary because she and her husband are helping their children get established in various forms of self-employment before late adolescence. (She does not at this time have a generic version of her record keeping system available for others. This would actually be difficult to do since each child's interests and skill levels vary widely.)

Many moms may not at first be able to pursue a lifestyle of learning with the same degree of freedom that Marilyn's family does. This is where I believe my materials can come in. My method is different from Marilyn's, and the two are not necessarily interchangeable, yet mine will free you to leave behind many of your traditional and binding methods of homeschooling. (Just don't use it as a crutch to keep you from truly following the Lord's leading.)

My focus is not on just filling subjects either, though it started out that way – or so I *thought*. And it might appear that way at first glance. But as my understanding of education deepened, I realized that the filling of subjects is the "external form"; but that I was really much more concerned about the *hearts* of my children and their walks with God than about simply filling credits. So that developed as we proceeded and began to be reflected more in my framework – the *meat* of the framework.

Though our approaches *are* very different, it would be possible to "translate" the work and experiences a child had/did/produced in a lifestyle of learning approach into the format I present here in this book, for record-keeping purposes. Generally speaking, however, there would not necessarily be any *true* need to do this, unless it was perhaps for a skeptical husband who is not sure that "enough" is being done, or if it were requested from a college or institution the student wanted to apply to. The student would not necessarily even realize which structure was being utilized, unless he was asked to keep track of time. But Mom would most likely have a general idea of that, which is really all that would be needed.

My focus is essentially the same as Marilyn's, but with a different framework. The content of my kids' education includes a wide range of hand-picked experiences and pursuits, real books, some traditional activities, and lots of interaction. All of this is then pigeon-holed into subjects. This approach gives my family our own unique "lifestyle of learning" – unlike the Howshalls,' but that's not the intent. Marilyn does not expect everyone (or *anyone)* to be clones of her. She presents principles and truths that we are free to apply as we feel led by the Lord.

The content, approach, and framework we choose are all *tools* through which *true education* is taking place. This book presents, among many other things, a method of record keeping that will help you to tally up hours and assign credits for the learning activities that you pigeon-hole into the various academic subjects. It will help you set up a "Master Framework Plan" within which you'll work for the duration of the high school years. (By the way, I do *not* recommend that you go to this kind of a structure any earlier than high school; you *might* start it at the junior high level *only* if the student is really ready.)

As you have probably noticed, my method entails more paper work than Marilyn's does. This can provide security and organizational help for moms who need the structure to free them to use more of real life in the learning process.

In utilizing one or the other's methods, you will definitely be approaching the learning process from different places. Any and all aspects of learning, traditional and non-traditional, formal and informal, can be included in your framework; I utilize a system only to organize and structure it all into some sort of a framework. The system serves *my* purposes; not the other way around!

Whatever you do in developing your own family's lifestyle of learning, I hope that you will be able to utilize any of our resources as freely as possible, utilizing any structure only to the degree you need for your own peace of mind. However, and please hear me on this: I do not recommend that you embrace my framework and approach without getting the heart of Marilyn's message. Marilyn's message is the "meat," or the "heart and soul" of truly educating your children; mine points you to Jesus, enlightens you, and points you to Marilyn's message – which points you to Jesus. My system, at the same time, can give those who feel they need it a traditional-feeling framework that they can fill with non-traditional, real-life learning, or whatever mixture of traditional and non-traditional they feel so lead to use.

♥♥♥♥♥

Q ***What about graduation?"***

A Your student may be so freed up in his attitudes about graduation that he won't even be interested in it. There is absolutely nothing wrong with this. If my child felt this way, I would feel no urge to instill in him the "graduation bug." This will probably be tied in with the particular friends he has and how "tight" he is with them and how highly he esteems "graduation."

It is possible, though, for a high schooler educated in a lifestyle of learning manner to have good friends with different educational values than his, and still enjoy the idea of taking part in a graduation. It all depends on how you view graduation. As Diana McAlister has pointed out, most cultures have some sort of "coming of age" ceremony or a standard to meet. Many Indian cultures have elaborate rituals that can extend over several days or weeks during which rigorous skills or feats must be successfully completed or attained for a "boy" to be deemed a "man." In the Jewish religion, the Bar Mitzvah ceremony is when a boy, about thirteen, is recognized as a "son of divine law," a highly significant passing from one level of maturity into a greater one.

In our American culture there is no such marking point of passing from one level of maturity into another. This is probably why high school graduation has such an elevated significance in our country. *College* graduation probably actually has more meaning as the subject matter and level of intensity are generally much greater at the college level – especially since it was the *student's* choice rather than the choice of a *parent* or a compulsory attendance law. (Plus they're four years older.) But, even so, high school graduation is generally viewed as the more cherished event in one's memory. So I did not view Sharnessa or Tory's desire to graduate as being based on insecurity or false standards. Tradition can be a good thing as long as it does not take the place of God, His values and goals, or His leading in our lives.

Q So what kind of "requirements for graduation" could you even have for a high schooler who has been living in a lifestyle of learning environment?

A Let's say your child wants to "graduate." You will probably want to devise *some* sort of standards for him to fulfill, in order for "graduation" to hold any meaning. I believe there is an additional benefit though. I don't know about you, but when I have a list of stuff I need to do in a day, I feel a real sense of satisfaction in having completed them. By giving your child a "program" or set of goals, you provide him with the opportunity to experience the "thrill of victory" – a sense of accomplishment – once they are completed. And, *to me,* the thrill of this form of "victory" is much more delightful and significant than any victory felt by an Olympic athlete. (I don't know if an Olympic athlete would agree with that though!)

What could your standards, criteria, or goals consist of? How might you go about determining them? This will depend completely on three things: your student's interests, level of initiative, and your own set of standards for "acceptability" or "excellence." The mature, "bonafide" lifestyle-of-learning student *will* be operating within a high level of initiative. This student will not *need* your guidelines to be based upon time in any way, even indirectly. But you will want to take his documented, filled requirements and translate them into some sort of framework for the sake of assigning credits. Otherwise why would you even be bothering with credits at all? Several samples are given in Section 6-A on "Formulating Your Framework."

If your teen will stay on task, continuing to pursue areas of interest unpushed by you throughout the season of high school, he is already in a place of "higher learning"! You'll get to observe the inevitable process of transition as he continues "graduating" himself to more mature subject material and allows his interests to overflow into academic areas that have previously received less attention in his life. If this is the case, then your requirements could be as simple as *"Continue what you're doing for the next four years."*

This is one point at which *Lifestyle of Learning* differs from the current teachings in the area of "natural learning." A love of learning does not ensure a *biblical* education, only a *delight-directed* one! In a Christian home, the parents do the guiding in the learning process, if their ultimate goal is their student's discipleship to Jesus Christ. Their degree of involvement will depend upon the level of self-motivation of the student.

The direction the parent gives a self-motivated learner can be as simple as a written outline of the course requirements based upon what the parent deems valuable and necessary for a well-rounded biblical education. This will become clearer as you seek the Lord more seriously and heart-full-y, ☺ and find practical help, guidance and applications of biblical principles in all of Marilyn's writings. *All* of her books answer the many questions she has received as homeschool moms begin implementing the truth they come upon. It will *grow* as *you* do! There should be requirements even in the lifestyle-of-learning family no matter how much the child is motivated. The difference is that the child will enjoy doing whatever you set before him to do! What a blessing! But it involves *training the heart.* More on this in Marilyn's books!

♥♥♥♥♥

Q What are some examples of Lifestyle of Learning-style requirements for High School?

A There are so many diverse possibilities, so many ways to implement them, that I hesitate to present *any* for fear that your ability to apply them to your own life will be restricted. But trusting that you'll let these ideas here serve to *broaden* your "requiremental" horizons rather than shrink them, here are a few suggestions to get you going:

➤ Begin with your child's delights. If your child has an interest in a particular area of science, you may want one of your requirements or guidelines to be completion of a project already thought of, started and developed by your child. If history is a love, you could say that a certain amount of time – something *general* like weeks or months, not hours – be spent on a particular project. Whatever it is, it needs to encourage and enhance what he is already involved in doing, not heap a pile of heavy, unnecessary requirements on a student who is already fairly well motivated.

➤ You might want to have some additional, minor areas that you require, like, for instance, American History. Emphasize points that you feel constitute a good *foundation* of knowledge in this area. There isn't time to learn it *all!*

➤ If your child is involved in a word study, such as what Marilyn describes in *Wisdom's Way Book #3* as being a season of in-depth study in a particular area of study, you may want to have your requirement be to take that word study to the "third layer" of definitions. Then have the student organize and categorize his material, putting it into a neat, readable, polished, finished form such as a notebook or portfolio. Come up with whatever you and your child agree would be acceptable standards.

➤ There are skills in the area of "tact and grace" in dealing and working with people and relating to them that I want to instill in my kids. I feel that this subject is more important – or at least more *relevant* to their effectiveness in bringing others to the Lord – than history is, yet it won't be found in any state's line-up of credit requirements. I am not giving them workbooks for these areas, but rather their main avenue of learning is two-pronged: I have searched for interesting beneficial resources that we read or listen to together; plus, we make the most of the *real life* experiences that God orchestrates especially for my family! He's doing it all the time in *every* family! You just have to become *tuned in* to what God is doing and place a higher value on it – perhaps even asking God to help you *recognize* it when it comes along.

♥♥♥♥♥

Q What if my student wants to have basically such goals, but would like to feel he has reached a certain level of competency in a few specific skills?

A Goals *are* fun to reach, and if they are realistic and relational to the student's *real life,* they can be *valuable* as well as fun. If this is your student's desire, and you see it as being a desire that's healthy for and beneficial to him, you might do the following:

Devise a set of standards or accomplishments in area(s) of interest such as Computer, Cooking, Astronomy, Carpentry, just as a few ideas, seeking much input from him on the direction *he* wants to take his study so that you do not inadvertently end up taking over and squelching his self-directedness. (E-a-s-y to do!)

Just be careful, especially you perfectionists, to not demand *full maturity* as your standard. This needs to be something *measurable* so that it can be set and then accomplished. *You, as a parent are not required to produce the results;* God only expects you to *cooperate with Him* in his training plan for your child. Then your job for that season of his life is done. As he matures into adulthood, you will relinquish one role and take on a new one.

♥♥♥♥♥

Q How would you start a high schooler who has been exposed only to traditional means of education in a lifestyle of learning? How will my teen respond to the switch-over from his familiar system to this?

A Getting down to specifics, perhaps you have a teenage child you've been homeschooling who, as he or she is approaching the high school years, is getting antsy and dissatisfied. You're now hearing comments like, "Do I *really* have to *homeschool* all the way through high school?!"

Or perhaps you now want to take your teen out of school, and his heels are firmly anchored in the hallowed ground of the traditional school setting. Or you have *already* taken him out and the constant clashes are getting both of you nowhere *fast!*

One other possible scenario (people in many situations will be reading this book) is that you have been reading about the *Lifestyle of Learning* message and are convinced that this is God's design and plan for your family. However, your teen, upon hearing your thoughts about it so far, is extremely leery, putting it mildly. *"But, MOM! I don't want to be* ***dumb****! (or "weird") I want to learn what my friends are learning!"* Ah, yes... Herein lies the great frustration for us as parents! *We* know that if they follow our leading – as *we* follow the leading of the Holy Spirit – they will *far surpass* their friends in wisdom *and* in knowledge *and* in the fear of the Lord! But *they* don't know this yet! So what do we do in the midst of this gap?!

I wish there was one right answer someone could give you. But we're dealing with living, active, unique beings here, not to mention a Living, Active, Unique God who knows and feels our frustrations – those of both parent and child. The *real* question is *"What will work in harmony with what God is doing in the heart of this child?"* Your plan needs to come from God – for each individual child. *(You thought I was going to give you "Step 1, 2 and 3"? Sorry that doesn't work here!)*

If you can identify with one of these scenarios, you couldn't be in a more exciting, life-changing place! Or a scarier one, either! Scary, because leaving old familiar ways involves moving into uncharted territory – *alone*. And "uncharted" and "alone" are not comfortable places to be. But they are the two places in the universe where you're the most vulnerable and open to receive the blessings and direction God has in store for you! (Read *Secrets of Solitude.)*

So how do you get started at the high school level? No one can really give a complete answer here, but here's a little direction:

■ **Put yourself through a season of educating yourself** in and about a lifestyle of learning. This has much more to do with the *heart* than it does any forms, frameworks, or methods. *All* of Marilyn's books, explain the whole picture in greater detail.

■ **Limit or eliminate the TV.** If you haven't done this already, you'll be surprised at the shroud television places over one's creativity. If you don't feel the need to get rid of it altogether, you might at least consider having a rule that it stays off during the day. This will generate a void that anyone with even the *lowest* level of creative brain cells will eventually *have* to find ways to fill with *something.*

■ **Then give your child some space.** Some time and space may need to be given for the void to surface, but it *will* surface. During this "de-programming time, "let the student do pretty much nothing except responsibilities he may have in the home. Don't worry about "getting behind" or "losing ground." You're on different terrain now, remember. Terrain with no worldly-restrictive corrals. "He maketh me lie down (and *rest)* in green pastures (that haveth no fences)."

■ **Talk with your teen about his interests.** Find out what he likes, and together pursue getting materials to work on or books to read. If he comes up on the short end for ideas, take him down to a Christian bookstore (or your own book shelves) or the library, and just start going through some possibilities! (A few ideas are coming up in just a bit!) Look at a book on science experiments or an Audubon Society book of birds, field flowers, trees, animals. Don't worry about its possibilities; just let that aspect fall in line.

■ **Entice his spiritual taste buds.** It can be challenging to figure out how to do this, but it will be well worth your effort. You must first do a lot of *listening.* I found *backrubs* to be one of the most "spiritual" things I could do with Sharnessa, as she would talk about *anything* and *everything* to keep me doing it! ☺ You don't have to direct the conversation into "spiritual" things, unless the Lord (really) directs that way. Just listen, listen, listen. Understand. Empathize. Every so often, give advice. Some philosophies today discourage the giving of advice. That's foolishness! Not God's way at all! That would negate the wisdom spoken to youths in Proverbs to heed the advice of their parents. When I slip into my "Advisor" mode with my kids, I tell them: "This is in my job description as your parent. If I didn't *care,* or if I only wanted you to *like* me, I'd just keep my mouth shut. But I can't; I have to do the job God has given me! *Because I love you that much!"*

But that's not the "enticing" part. *That* involves sharing out of my life. Not just my successes or the parts that look good, but the *pits* and *valleys* as well. The failures. The embarrassments. The tears. The humbling moments. They need to see it *all.* Well, as much as God leads you to share with them or let them see; you don't want to lay inappropriate stuff on them, or stuff that's too heavy, but that's probably a lot less than we think. When my kids walk out our door, I want them to have seen firsthand how a Christian handles rejection, flops, humiliation, conflict with others – life in general, which isn't always pretty. In fact, come to think of it, it's very *seldom* pretty. It's downright messy, hard, and painful! But with Jesus it's *good.* "What the *enemy* intends for evil, God will turn for *good.*"

One more idea. I have my kids do "Word Alive Studies" and "Scripture Verse Studies." Of course, these could be done on any ol' topic and still be fruitful because wherever God sends forth the Word, it does not come back to Him void. But if you listen to your child for "spiritual cues," you can use these studies (in the Potpourri section) to whet their spiritual appetite.

I wanted Sharnessa to do one such study, so I asked her if she had any words or verses that were intriguing to her. She said she didn't. Later, in the course of the day, we did a lot of talking about some exciting and "heavy" things that had been happening in her life. At one point she asked me what "spiritual warfare" was. She had sensed some "odd feelings" directly before she had been involved in an experience where one of her friends, for whom she had been praying for two months, turned from sin and hardness of heart and came to God, with tears and total repentance. She wondered if the feelings she had felt had anything to do with spiritual warfare. This had happened the day before, and she was mulling over the amazing course of events that had taken place. She also told me (the same day) that a leader at one of their youth meetings had told her God would use her to "break down walls in people's lives." This is exactly what had happened in the spirit the day before!

What I saw were two prime opportunities for studies that would be needs in *her* and would relate directly to her life! Not that everything that happens to our kids needs to be made into a study! That would get really old! But she had a *desire* and a *need* to dig into the word and find out what *God* had to say about this. So I suggested she do a Scripture Verse Study on a verse having to do with spiritual warfare. We could look up the verse on "we wrestle not with flesh and blood, but with principalities" (Eph. 6:12) She could do a word study on the word "wall." But to make it more appealing and relevant, I told her that if she was going to be involved *with God* in breaking down walls, she would need to be acquainted with *what these walls were!* I opened up the concordance to *walls* and found four exciting scriptures that I shared with her. Here they are, as well as the questions I asked her to get her thinking about it more in-depth...

"...I will break down the *wall of...*" *"Sharnessa, to participate in breaking down walls, you'll need to know **what** those walls **are**!"*

"...we grope for the *wall* like the..." *"What does this tell you about a wall? If a person needs to **grope for a wall**, what state must they be in? ...Darkness!"*

"...every fenced *wall...*" *"Is this fence keeping protection in or out? When is a fence good and when is it not?"*

"Behold I walked by the field of the sluggard and its walls were broken down..." *"Hmmm. Doesn't* *sound like these walls were supposed to be broken down, does it? How do you build up walls, and how do you know which walls should be built, and which should be broken?"*

Since this study *applied directly to Sharnessa's life,* it had an entirely different meaning and application than if I had merely assigned a word to her. In fact, it *wouldn't* have had *any* application that way, period! So be aware of that as you consider Section 4, the Potpourri of Learning Guides! They will be best used when you see a need for them arise.

♥♥♥♥♥

***Q* What if my child has a less-ripened appetite for learning? In fact, he doesn't seem to be interested in ANYthing!"**

A In order to answer this, let's first take a look at how the child arrived at this unhappy place in the first place. In traditional methods of education, our children are fed *head knowledge* year after year during the season of their life when their character is being formed and their life-long attitudes toward learning are being developed. At this critical point, the type of education they get will either "make" them or "break" them. Traditional education typically "breaks" them because it drains the delight of learning right out of their hearts. Also, the level of skill development hasn't had a chance to become what it should be because of the chopping up and pre-digesting of knowledge into bits and pieces of unrelated, subjects that are totally segregated from their real lives.

The result? By the time they reach the high school years, the age when they should be gaining momentum and becoming willing to embrace a deeper, stronger, broader education, *they're burned out!* Their love of learning has shriveled up! So, in the early years, the skills and character should be developed and the love of learning thoroughly *allowed.* You can't *teach* it, though you can *model* it. The only way this can happen, especially if it *didn't* happen in the early years, is to ***quit force-feeding academic knowledge,*** fearing they won't grow, learn enough, or "meet requirements."

As this begins to happen, you may want to set up some goals and requirements that you know will involve a certain amount of time for each activity without your having to say "spend two hours doing such-and-such." There's something overwhelming and joy-squelching about being required to fill a certain block of time. Fulfilling a specific *goal* is more enjoyable; and making learning *enjoyable* for a burnt-out child is our goal at this point! The bulk of the student's learning activity may need to remain in traditional curriculum if this is what he is most comfortable with for the time being. Include at least one delight-directed study pursued in a non-traditional way so that your child can begin to experience the difference, and eventually experience the pleasure of learning. *This is a godly pleasure! He designed it to be this way!* Your child may then be more willing to wean off of the dry stuff in favor of more stimulating pursuits. In the meantime...

Get excited about something *yourself* and bring your child into it. Find a few resources that you'd enjoy reading, or projects you'd enjoy doing, and bring your child into it.

Watch him/her while he/she isn't looking. ☺ Everyone is interested in something. Even if your child has said he is interested in *nothing,* he actually *is.* Just give him a little time for the traditional school dust to settle, and then *watch and listen closely.*

Pray a lot! A line of one of my favorite songs says "nothing, nothing, *absolutely nothing;* nothing is too difficult for You!" Since there are no "exception clauses" in the Word, we can only assume that God is fully able to do a work in *your child* as well. Entrust your child – and your anxious heart in the meantime – to Him. Be watching for His answer, but not with a troubled heart, or his words will get lost somewhere between his mouth and your heart. "Be still and know that I am God."

When he *does* express or start forming an interest, don't judge it. Dr. Raymond Moore tells the story of one boy who was in this very situation, "Absolutely no interests of any kind" his mother lamented to Dorothy (Mrs. Raymond) Moore. She encouraged the mother that there *must* be *something* the boy was interested in. "No, nothing," she frustratedly answered. Later in the course of the conversation, it came out that the boy had a passion for motorcycles! Since the mother didn't consider this to be of value or "studious" she hadn't even counted it as an interest! But this is exactly what Dorothy used to help bring the boy out of his lethargy. They were able to take this interest many directions and get him interested in learning – *right where **he** was!* So, unless the interest is ungodly (like rock music) let your child be who he is and start where he's at.

Start dabbling. After a season of laying low, start bringing in books and resources that you think would sound interesting to your student. Insist on a reading time each day. Perhaps one hour of sitting together on the couch with a cup of hot chocolate, and maybe three 20-minute sessions in each of three books, or however long you *want* to go, but at *least* that as a discipline. If you have chosen good books, *something will spark a fire somewhere.* (A few ideas coming right up!)

***Q** Can you recommend a few good resources that might help light that "spark" of interest in a disinterested child?*

A First of all, the main benefit coming from these resources will be to get the parent and child reading out loud together in a cozy seating arrangement. *(Like on the couch, even touching, if they'll go for it.)* The books themselves are good, but the most important aspect of reading is the *shared experience*, and the conversational interaction that will be the result. That "spark of interest" will be ignited by you, the parent, *much* more than by the books themselves. Although that *will* happen too. Here are a few ideas, along with some comments from a few people who have read them and *enjoyed* them!...

➧ ***We Thought We Heard the Angels Sing*** by James C. Whittaker – A captivating story of several men whose plane went down in the ocean for *(I can't remember how)* many days *(but it was an awfully long time!)*, how the ordeal affected each one individually and how they survived. We all loved it!

➧ ***Bruchko*** by Bruce Olson – The first five chapters are a little slow. In fact, you'll wonder if you got the right book. But persevere! It's *well* worth it! This book is an amazing story of a "regular guy" who wanted to take Jesus to the Motilone Indians in an "irregular" but practical way, but a way that was so effective it changed entire villages at a time! *And* the "approach" of many missionaries since!

➧ ***Vanya*** by Myrna Grant - **Sharnessa at 16 said:** *"This made me really thankful for our country where we have freedom of religion. Sure, America needs help, but I think people take for granted the great amount of freedom we have. It also challenged me in my walk with Jesus. It made me less fearful of dying for Him."* **(B❤:)** Me, too! In fact, God used it to do a deep work in me in the area of fear of the end times and persecution. The hotter the "frying pan," the hotter and more glorious and real (to us) the presence of the Lord becomes! Vanya's visit with an angel to the "outskirts of Heaven is absolutely amazing! If it's *that* good on the outside, how *awesome* it must be *inside!*

➧ ***Within Heaven's Gates*** by Rebecca Springer - This gives an intimate, detailed portrayal of Heaven in ways that only someone to whom God has given a glimpse of it could portray! Utterly breathtaking, surprising, and amazing, it will make Heaven more real to you!

➧ ***China Cry*** by Nora Lam – An inspiring story of God's providence and miraculous provision in extremely difficult circumstances.

➧ ***Story of Liberty*** by C. Coffin – This book presents the Christian *truth* behind the events surrounding the founding of our country. And, by the way, once the Christian "aspects" of our history are deleted, we are left with not just "incomplete" stories, but outright "lies." If someone were to retell the story of your life, but leave (as in *take)* out all parts having anything to do with God, what would be left? The *truth?*

➧ ***Beyond and Back*** by Rev. Ralph Wilkerson - **Sharnessa said:** *"This book creates an exciting anticipation for heaven! I'm pumped to get there – **when** my time comes! It opens up your eyes to how unconditional and awesome God's love really is." (This book is not easy to find; there are book searches, either businesses (in shops) or on the Internet that can help.) B❤: And another common thread for many who have experienced this is a tremendous awareness of how completely perfect and deep God's love is for <u>them</u>, and in turn, love for God and others becomes an all-consuming passion in their lives. So the "fruit" is good!*

➧ ***Pilgrim's Progress*** by John Bunyan – This book is both adventurous and spiritual. It can help to open your kids' spiritual eyes and help you (Mom) know how to talk with them about the basic truths of the gospel. Sometimes we have so much "churchianity" that we miss the simplicity of how to follow Jesus."

➧ ***Roots*** by Alex Haley — **Jamie Cosby** (S's friend) **said:** *"This book lets you see how slave traders and some slave owners **truly** treated the slaves. It was very surprising on both the good and bad sides. It gave me a lot of insight on what the whole slave issue was all about."*

The above list is, of course, very limited, just a place to start for ideas, since you asked! ☺ (If you are unable to locate any, we do carry most of them.)

You might also want to go to a Christian bookstore and find some good "how to be your best" books having to do with success in life, setting and reaching goals, getting vision; or short, inspirational books that would motivate your teen to think along those lines.

If you have a high schooler who needs a new way to think about learning, here are a few suggestions:

✎ Take a look at all the resources in my "Educational Foundations" class and mark any that you want your student to read. These will all help him change his thinking about "learning." Otherwise, because he won't have any understanding, he could balk at your new ideas for pursuing education.

✎ After he's read them, (or while he's reading them), spend some time discussing what he's read with him! This needs to be done without pushing or insisting on your own way.

✎Through all of this, allow him to express his doubts, ponderings, and questions. In love and grace, and with understanding and respect (ground rules for your home!), hash it over together! ☺

♥♥♥♥♥

***Q** If I'm getting the newsletters (or am on the WWOL list), do I really need to get the books?*

A You will glean a lot from the newsletters and the WWOL email list, but please don't think that what you gain from either can possibly take the place of what Marilyn has already written in *Wisdom's Way* and her books! You will spend less on *all* of her books put together than on *one* text/workbook program for *one* child! And you *and* your family will gain *infinitely* more from these than that one child would gain from the snazziest math program. If necessary, just lay down one subject for a year! It won't hurt a thing! Your family needs to get onto the right track *much* more than your child needs to "keep up with his agemates." Believe it or not, your child can catch up in a month, if need be. Just review this year what he/she did last year if you *really* can't live without it.

♥♥♥♥♥

***Q** I love the heart and the basic ideas of lifestyle of learning, but do your materials help break down the ideas a little better – into bite-sized, more "do-able" pieces?*

A Eventually, as traditional ideas fall away from your thinking, and you begin to better understand the heart of true education, you will realize that you don't really *need* as much "breaking down" as you *thought* you did, because you'll see the broader, deeper picture more clearly. To help you "get there," Marilyn has written several smaller books that are *very* practical helps in getting the vision for your family, and practically applying it in all the needed areas of life! My materials help you organize what you're doing in a little more traditional-seeming manner *(retaining your creative freedom!)*, put it into some sort of a framework if you feel you need that, or if it would help your spouse feel better about it, and gives you many bouncing-off points. But you don't (***really***) need me to help you break anything down into bite- sized pieces. Anyone – including you – can do that. One page at a time; one picture at a time; one sentence at a time. Today, for example, we'll set up a notebook for each child. The following day Mom could have mini-conferences with each child to figure out a favorite topic for each child – which you might get out of Marilyn's notebook ideas, my Booklet Building Book (BBB), or right out of your child's own interests – which is the "*Ultimate* Idea" anyway! Then you can copy the masters (in the book) for whichever booklets are chosen. The next step is to gather drawings, information, supplies, or photos for the pages – or go *take* some photos – and start *doing* the project. That's all my BBB is – just "starter ideas" for a myriad of projects! It's helpful for those who *think* their children don't have any interests because they have been so stifled by the wet blanket of "too-much-school-at-home."

It might be a little "messier" this way, but, know what? Real life *is* messy! ☺ Are you running a *home* full of *children*, or an *army* full of *soldiers*?

Another step is to find a place where your notebooks will be safe, and easy to find and pull out when needed. Remember, they're done just one page at a time! It might take months or *years* for one booklet to actually get done. The *process* is what counts; not the *product.* ☺

♥♥♥♥♥

***Q** I've tried reading Wisdom's Way, but I'm just not getting it; it just doesn't make sense to me.*

A I urge you to *continue* reading them anyway. Ask the Lord to open your eyes as you read them over and over – until it starts sinking in. All of her books will help you tremendously as you attempt to shed "the system" and don your own, tailor-made lifestyle of learning. Each one of these booklets was written in response to hundreds of specific questions Marilyn has received from those attempting to understand and develop the message in their own lives, as God leads each one individually. Be encouraged. This isn't nearly as hard as it looks. It's hard only if you're looking at it from a "school at home" perspective.

♥♥♥♥♥

Just for Moms with Several Children

Even if you do not have several children, you might want to read this anyway, as I'm sure you'll glean some good things, but if not, just skip on ahead over to the bottom of page 11.

***Q** How can those of us of us with large families make this way of 'schooling' work*

with so many? I haven't seen Marilyn deal with this anywhere in her writings, so I'm asking you because it seems your stuff is based a lot on Marilyn's work.

A Yes it is, but having only three children, I'm not knowledgeable on doing it with more than that. ☺ As far as Marilyn not addressing this matter, actually she *has,* but you just didn't realize it. ☺ She cannot tell anyone how to implement this with *many* children any more than she can tell a person with just one or two children how to implement it with their one or two. She shares what she did/does with her own family only as an example of how to apply the principles of this message; not as your *guide*.

Most importantly, she sends you to *God* for your answers. God is totally practical and specific! You will find *your* answers for your family, what*ever* size it is, only as the Holy Spirit reveals it to you, which you aren't going to hear until you: **1)** lay down your old ideas – and let them die; what's of God will come back anyway; **2)** quit arguing with God about it and realize He really does have an answer for you and *can* and *wants* to lead you to and in it; **3)** listen and watch for His answers; and finally, **4)** just DO it.☺ This is also called "obey" in God's Word. ☺

This is why you need to be reading her materials *not* with eyes that are looking for her "methods" at this point, but with eyes to see the *heart* of her message! Before any of it will make much sense, you need new eyes! New *understanding!* New *vision!* This is what the Lifestyle of Learning material will help you gain! You'll get insights, encouragement, hope, examples, tools, how-to's, and ideas. But *you* have to take all this to the Lord and ask Him to do His work in *you;* to renew your mind in the area of education and learning. *Then* the rest will begin to fall into place and *happen*.

You can't expect a plant to grow where you haven't planted seeds. Your old ideas and securities need to be sown into the ground and die! The *old* will die a natural death and simply fall away as the *new* grows bigger in you. "Unless a seed is sown into the ground, it cannot..." – cannot *what?* "...it cannot bear fruit." Or another translation says "...it cannot bring forth *life*"! We can't look for fruit if we haven't first sown the seed. This is a step-by-step process. So start there, sowing the seed, and "in due season you will reap a harvest of blessing," and will start seeing your children differently. *And* education! *And* handling, managing, and facilitating education for "multiple children." ☺

♥♥♥♥♥

Q I know of one mom with six children who was so discouraged that they went back to using a full curriculum because she couldn't imagine having six notebooks with six children, etc.

A I don't mean to be contrary, ☺ but I can't imagine going back to a pile of work/textbooks with six children, thinking *that* will be any easier, and trying to make them *do* all that! Notebooks are a *much* more manageable – and lighter – burden than curriculum!!!

Instead of having several different notebooks for each child, you could have *one* notebook per child, broken down into a few favorite topics. Then once the child is older, they could take out the content of one or more and make them into separate books.

I talked with one very *desperate* mom at the Mass. Homeschool Convention (*several* times; she kept following me around from workshop to workshop, and then hung out at our booth), who had six children she was homeSCHOOLing, and she *and* her children were absolutely miserable. They lived a "lifestyle of *yearning*" – for joy and *LIFE!* I order to get that many kids to do all that *stuff*, she had to run her home militaristically. She asked me the same question you just did: "How do you do a lifestyle of learning with so many kids?!?" I told her she had the wrong idea of what a "lifestyle of learning" was. She saw it as just another system that would be similar to the other, maybe a little more fun, and that would replace the other system, but would just be a little less structured and whole lot more messy. This spelled C-H-A-O-S to her. And it *would* have been if that's where she had stayed stuck. I emphasized her need to *renew her mind* as to what true education was before trying to pour "new wine" into her "old wineskin." Her old wineskin would burst if she didn't do this! In fact, to be honest, that old wineskin doesn't need *refilling*; it needs to be *replaced entirely!*

(End of Q's and A's)

♥♥♥♥♥♥♥♥♥♥♥♥♥♥♥♥♥♥♥♥♥♥

WHERE TO GO FROM HERE

If what you've read in this section rings true to your heart, I recommend you start by seeking God for what to do next. He has some very specific and good plans for your child, whether you've been into doing a *Lifestyle of Learning* for some time, or are just beginning the trek – or should I say "quest"?!?

NEWSLETTER

I have mentioned the *Lifestyle of Learning* newsletter several times. The Howshalls have been producing this since the beginning of 1997. The newsletter is designed to give parents pursuing this form of education continued insight and inspiration to keep seeking God and His way for your family as well as creative ideas to implement the concepts on a practical level. It is *not,* however, intended to take the place of *Wisdom's Way of Learning.*

This newsletter, which has actually been the "precursor" of many of Marilyn's new books, continues to develop the *Lifestyle of Learning* message on an ongoing, monthly basis. They have graciously offered it free of charge in the past, and it is still available for free to those who honestly cannot afford the newly-suggested (as of April 1999) donation of $15 per year. However, I urge you to really search your heart regarding this matter of being "able to afford" it.

WHICH REMINDS ME...

[DIGRESSION:] I have seen a tendency within the body of Christ to think that ministries "owe them something." This tendency, sadly, seems to be even greater within the homeschool movement. A few well-meaning, kind-hearted homeschool leaders have actually perpetrated this thinking, with some saying they wouldn't *dream* of charging homeschoolers for their books/services. I have to ask *why?* Just because it's a ministry? Just because they're parents who have chosen a noble path? (The path that, really, probably pretty much all parents *ought* to be taking anyway.) That's wonderful, but I believe "The servant is worthy of his hire."

Those who are providing ministry, services, and products to the homeschooling community are doing so at great personal sacrifice and cost, in time and financial investment. There is no reason that they (we) need to feel they (we) should not receive financial remuneration for their services and products.

Rather than having an expectant, "receiver's attitude, we need to ask God for a "cheerful giver's" heart! A heart to keep those who are serving us in business! They appreciate compliments and letters, but it takes *cold hard cash* to pay the high costs involved in serving homeschoolers. If we bless those who are serving homeschoolers, we will not only be blessed in return, but will be helping to make it possible (and desirable) to continue in their service to us rather than feeling they are being taken advantage of. *(stepping down from soapbox now.* ☺) Okay, no guilt! ☺ Just re-think this matter, and if the Lord reveals such an attitude in your heart, open it up to the "adjusting" of the Holy Spirit. ☺ And, just for the record, I have not said all this on behalf of the Howshalls, or *any* ministry, for that matter; it's just something I've personally encountered many times in many arenas over the past sixteen years of dealing with homeschoolers, and I wanted to bring it to your attention. ☺ [END OF DIGRESSION]

CONTACTING THE HOWSHALLS

If you want to subscribe to the newsletter or request a free product catalog, write to the Howshalls at: Lifestyle of Learning / P.O. Box 1750 / Eatonville, WA 98328-1750.

WISDOM'S WAY EMAIL LIST

You can subscribe to the online WWOL *(Wisdom's Way of Learning)* email list by sending the simple message: "subscribe wwol-list" to: **hub@xc.org**. *(This is as of April 99. Changes may have occurred by the time you read this, if so, please try contacting Sherri Rimmer at: sdrimmer@xc.org for updated info.)*

MASTER PLAN FRAMEWORK IDEAS

See Section 6-A, "Setting up Your Framework," for an overview of *several* frameworks for high school, including one for how you might set up a high school plan within the context of a lifestyle of learning, and then organize the record-keeping aspect of *one* project of that plan.

THREE LEVELS OF LEARNING

On the following four pages I have included an article I wrote about five years ago, *after* I had written the first edition of this book. I have revised it since then, but it's the same basic idea.

If you aren't sure of the quality of your child's *(or your own)* education, this is a good place to start evaluating it, and understanding the varying qualities in learning activities.

To the same end, you'll find "Seven Natural Vital Signs of the Learning Process," a segment in *Wisdom's Way of Learning,* eye-openingly helpful.

THREE LEVELS OF LEARNING

When some pretty radical ideas about learning first started touching my life through Marilyn's writings, my whole concept of *learning* began to get stretched in some very uncomfortable ways. As my prior and quite well established ideas about education were shaken, I started getting defensive: "My kids have been *learning*! We haven't been doing just a bunch of junk for the last eleven years!"

But when I let the Holy Spirit give me a closer look at what we had actually done in our homeschool, the first thing I realized was that I did not have God's perspective on education or child raising, and so I began to allow Him to change and deepen and refine and purify that. I share on that in other places in my writing, but the focus of this article is on the quality of the learning itself...

LOOKING CLOSER

I began noticing that there were different "qualities" or "aspects" of learning. While it was true that I had never gone the traditional route of using workbooks, textbooks, and pre-packaged curriculum, our learning did not look entirely like what Marilyn was presenting. And while I used a lot of real-life stuff, I had still done most of the "directing" and "creating."

I also examined and dissected the learning I myself had done over the past almost forty years (at the time; I'm now 46) – including my school years up through the then-present – and I found that all of it could be filed into one or more of three levels. By learning, I mean any and every experience, formal and informal, that has been part of forming my mind, my spirit, my life, my ideas, my skills, and my values.

OVERVIEW OF THE THREE LEVELS

I will delve into each one separately, but first I want to present the whole picture so you can see where each fits in. Starting right at the foundation, I saw these three levels to be: "Observe," "Practice," and "Master." (See the chart to the right ☞)

Back in grade school when I had watched Sister Laura up at the chalkboard diagraming a sentence, I was operating in the *observation* level of learning. If I could have avoided going any further in absorbing this stuff, you can be sure I would have! But, alas, opening up to page 43, there was a whole page of sentences that I was expected to *practice* and *participate* in diagraming myself! It was not until later, much later when I began writing (stuff that I *wanted* to write because I actually had something to *say*) that I took this diagraming stuff seriously because, **A)** now I did not have to do it; **B)** it was actually kind of fun, but what made it worth pursuing was that: **C)** it was actually helpful in my writing, and **D)** it increased my self esteem (superficially, at least) to find I was good at something that most people found difficult and distasteful.

MASTERY LEVEL

Granted, reason "C" was the most honorable reason, but *mastery* is not always pursued for purely honorable reasons. A gold medal has motivated many an athlete into mastery of the ice, a mile, or a balance beam; and the desire and hope of world fame has spurred many beyond a mere *knack* into an enviable *proficiency*. In other words, pride, fame, money, and self-promotion have played a large part in prodding many up to the mastery level of their interest. But more on that in a moment.

Whatever the motivation, true and pure *mastery* must be a *choice*. I had been out of school for many years before I made the choice to write. I had to get de-programmed and freed up from the intimidation that the words "study" and "practice" triggered in me before I was free to start.

Full mastery in the arena of writing has yet to come (obviously), but with a few books behind me now and several articles published, I am on my way – and am having a ball! No one has to urge me to sit down and edit a manuscript. Boredom is not even in my vocabulary anymore, even in the midst of prolonged exposure to writing. It is as natural to me as eating; it is enjoyable and fulfilling; and I *own* it in my heart! (Now if I can say all this about housework someday, I will be in Heaven! Probably *literally!)*

I want to add that the income that now springs from my writing endeavors is "nice," but that definitely takes a back seat to the satisfaction of knowing

I am doing what God called and created me to do. He has several purposes for my life, one major one – to know and love Him – and a few minor ones (the reasons He gave me the particular talents and personality He did), and it is a total thrill to be walking in at least a measure of it!

APPLYING THIS TO HOMESCHOOLING

What I did not understand as I first read the *Lifestyle of Learning* message was what in the world my kids had been doing all those first eleven years of our homeschooling? I knew it simply was not true that they had not learned *anything*. It was evident that *some* degree of learning had taken place in many areas because I saw good stuff in their lives and character. They possessed a variety of skills and poise. They could actually talk and write with ample correctness, and they could communicate with adults, and speak in front of a group of people. They could even pray for people (*with* them) and had been known to correct adults (graciously and humbly) on trivial and not-so-trivial points of knowledge. So it was not like their heads or hearts were empty. But then, what *was* it like?!

That's when the three levels of learning started to materialize. I began to see that the bulk of what they had done was rooted in the first two levels.

OBSERVATION LEVEL (1)

We had been on more field trips than I care to disclose, but I began seeing that those field trips comprised learning that was primarily at the *observing* level, or, as Mary Schofield calls it, the *exposure* level. *(In other sections of this book I call it the "Introduce," along with "Practice" and "Master," but for different purposes. For my purposes here, I call it "observe.")* Had they actually *learned* anything? I mean *besides* how to be quiet while they observed something, or while someone pumped information into them, and how to behave correctly in a group setting?

Sure, they learned some good though generic stuff about the world they lived in. They learned how a bakery works, what it looks like inside an ambulance, how a college science lab works, what happens to river water at the water treatment plant, and how hydrogen peroxide is made.

Was all or any of the above "learning"? Certainly! But most of it was not usable in or relevant to the world they lived in on a day-to-day basis. Nor was it pursued out of their own interest or desire.

At this level, motivation to act comes from *without*, from *external* forces such as parents, peer pressure, hope of reward (like getting a star on a paper or being granted a privilege), or from the threat of penalty or losing a privilege. On a "Good / Better / Best" scale, *observation* usually earns only the meager rating of "Good." That's not a *law*, and it won't *always* be so, but it just depends on the situation at hand, the focus, the purpose, and/or the need.

PRACTICE OR PARTICIPATORY LEVEL (2)

Other activities our kids were involved in took them a bit deeper into the learning process because they *practiced* things and/or *participated* in them. Like visiting a museum, taking an art class, presenting a report at Homeschool Kids' Day, making a favorite recipe over and over, learning the multiplication tables, or drawing a map of a state.

Again, there was nothing wrong with any of these activities. Many had a positive influence on them, gave them valuable knowledge, deepened relationships with us, family, or their friends, grandparents, and adults, and taught them how to behave in public and interact with and in front of other people better. On our "Good-to-Best" scale, these *participatory* or *practicing* activities scored a "Better."

The inward motivation of the thrill that stems from completing a goal, or the satisfaction of keeping up with a scope-and-sequence lesson plan also falls into this category because *this* variety of joy flows out of *accomplishment,* not out of the *activity* itself.

MISSING SOMETHING

Observing and Practicing or Participating are very valuable aspects of learning. But, at the same time, if they comprise the *full extent* of the learning process, the resulting education will be shallow, lacking in maturity and depth.

If we never really *own* anything or if we never become so delighted with or immersed in a skill or activity that the thought of getting to do it actually propels us out of a warm bed on a chilly morning, then we are not only missing out on one of life's highest joys, we are missing out on true, real, deep, permanent learning.

We can liken going only to the second level to observing someone make a lovely loaf of bread; then getting in and rolling up our sleeves ourselves and mixing, kneading, forming, and popping one in the oven; but then we leave before getting to smell the mouth-watering aroma as it bakes, and then sinking our teeth into a warm, buttery slice.

We can also compare the levels of learning to a rose. Observing is like a tight bud, and Practicing and Participating opens the rose petals a bit. But Mastery brings the rose into full blossom; full beauty, full essence of fragrance, full maturity of purpose. Mastery scores a rating of "Best." Mastery is true education in full bloom.

BACK TO MASTERY

So how and when does mastery take place? There are no cut-and-dry guidelines, no hard-and-fast rules, no sure-fire tests. Mastery simply happens when it is

THREE LEVELS OF LEARNING

LEVEL OF LEARNING	STUDENT'S SOURCE OF MOTIVATION	PROGRESSIVE EXAMPLES	OUTWARD SOURCE OF MOTIVATION	TIME REQUIRED TO LEARN	LEVEL OF INTEREST W/IN STUDENT	FRUIT IN STUDENT'S LIFE	STUDENT'S RESPONSE IF FORCED
OBSERVE (dabble, exposure) (Consumptive)	**External** (Done to or for the student)	A. Read about museums B. Read about birds & feathers C. Read grammar & writing D. Read about or observe tap dance	**Parents** *and/or* **Student Learning Objectives**	Brief Exposure	None or Whim	Student is aware the knowledge exists or experience is happening Only consumes	At Best: Tolerance Worst: Resistance (Displays a general lack of interest and/or cooperation)
PRACTICE or PARTICIPATE (Basic experience or study)	**Either External *or* Internal** (Done with or by the student)	A. Visit museum B. Collect feathers & learn about or care for birds C. Do some writing referring to rules D. Learn to tap dance	**Student Learning Objectives** *and/or* **Parents** *and*/or **Self**	Longer exposure *or* Repeated activity	Interest and passion begins to develop	Gains a feel for the thing	At Best: Boredom Tolerance Worst: Resentful
MASTER **(Delight directed pursuit)** **(Productive)**	**Internal** (Done by the student on own initiative)	A. Able to relate details about the museum from memory B. Study birds & feathers, mount in book, adding details C. Skillfully apply grammar skills in own writing D. Perform or choreograph tap dance	**Self**	Very Prolonged Exposure *or* Repetition of activity Freedom to continue until satisfied and/or mastered	Passion	Owns it in the heart Enjoyable and fulfilling Natural part of life Knows without having to look it up or be shown again Knows enough to be useful to the student and others Can produce something	At Best: Resentful Worst: Rebellion and/or Burn-out

allowed to unfold. It has to be given time, opportunity, and value. One can easily spot a flower garden that is in full-bloom; it is a joy to listen to one who has mastered a challenging piano piece (beyond "Chopsticks" or "Heart and Soul"); and you would have no problem eating a piece of homemade toffee that a child-baker has tried over and over to the point that he has it perfected.

Mastery is not hard to spot; it is just hard to *allow* or *facilitate* when we cling to old, traditional, binding methods and ideas. We have to be willing to let our "old wineskin"go and receive a new one if we want God to pour in His new wine!

APPLICATION TO OUR WALK WITH GOD

The three levels of learning have a useful application to our walk with God. In fact, the following example describes the three levels more effectively than any other illustration could, probably because it is actually the ultimate goal of education...

The "generic Christian walk" can take place on any one of three levels, even fluctuating between them. We can observe Christianity in the lives of other Christians: our friends, the pastor giving a ser mon, a television evangelist, a radio talk-show teacher.

At this level we merely dabble in it or are exposed to it – or observe it. As we sit in church we can let everything from the sermon to the praying to the worship be done to or for us. We are aware that something is happening, and it may even appear to the casual observer that it is happening to us, or that we are even doing it! Yet we are not really entering into the activities ourselves. It is all happening outside of us. If any activity is forced upon us – or our kids – without the heart being involved or addressed, we can expect tolerance at best or resistance and lack of cooperation. This is not always bad – no one usually wants to clean a toilet, fold the laundry, or do the dishes because the job looks appealing. Yet these need to be done. Once the heart is dealt with, they can be done out of a motivation to please the Lord and serve each other though.

We can go deeper and choose to participate in the Christian walk, and practice what we – or the preachers -- preach. Though self-discipline may be involved at this level, we find ourselves gaining interest and momentum. This comes from the hand of God. It is *not* of the flesh because the flesh resists the ways of God.

But when God sends forth His word it does not return to Him void, so if we continue to participate in His life and practice His ways, His word will bear fruit in our lives in due season. Not by our own merit, but because of His word and His promise.

It is only in continued and focused participation – as our whole heart is "sold out" to Him-that we attain *mastery.* We spend time with God and read and study His Word of our own choosing. We repent because we know we need to and we *want* to. We attend seminars, read books, and listen to tapes, because we are hungry for truth and change within our lives. At this point no one is pushing us to do any of the above; we *are free* to continue on to mastery.

"Mastery" in our walk with God might be considered to occur in having walked with Him for an in depth season, having totally surrendered everything to Him that He graciously revealed to us – not in having attended the most seminars, read the most books (or the most *spiritual* books) or spent the most hours in prayer.

"Mastery" in this realm is a matter of the *heart,* not of accomplishment. The books, church services and seminars are, at best, tools in God's hands to mold our hearts. We do not attain to *mastery* <u>it</u>self, but to the *Master* <u>Him</u>self. As we do so, there will be evidences that He is doing a work within us, and there will be a wonderful outworking and culmination of it as well...

First, the evidence of mastery will be seen in two main areas. One is how much of Jesus others see in us. The glory is not ours to desire or to have. "We Christians... can be like mirrors that brightly reflect the glory of the Lord. And as the Spirit of the Lord works within us we become more and more like Him" (II Corinthians 3:18).

The second evidence is how drawn people are to Jesus in our presence. "If I be lifted up, I will draw all men unto me." Do we try to draw them to ourselves? Or to *Jesus*?

The *culmination* of mastery lies in the joyful discovery that we have become "useful to the Master, prepared for every good work..." not because we have now attained perfection, but because we have become yielded and pliable, "having (had) our faculties trained." What joy to realize that He is actually using us, even with our limitations and imperfections!

At this level we are propelled from within, the delight of the Lord being all the motivation we need. *"The one thing I want from God, the thing I seek most of all is the privilege of meditating in His temple, delighting in His incomparable glories and perfection..."* Self discipline is not even particularly necessary here, except for perhaps the polishing or completion of an aspect of a skill or a product. But on the whole, wild horses could not keep us away because we have tasted and seen that the Lord is good. At this point "the Lord himself (has become) my inheritance, my prize; He is my food and drink, my highest joy."

This, I submit, is the sole purpose and the supreme goal of learning and ultimately mastering anything: to find ourselves at a place where we are completely yielded to, satisfied by, immersed in, transformed by, and (therefore) useful to The *Master* Himself. ❧

❤ ❤

A prayer... ...for you:

May the Lord BLeSS and guide you
and free you from every encumbrance
as you seek His perfect will for your family,
a plan that will not over-burden or burn out.
May you position your heart and soul to
receive His peace, joy, and vision,
and fall in love with the One
who IS the Truth
and the
Way!

On the next two pages
are two letters from the same mom,
glimpses of her heart at two different stages
of the process the Lord took her through
(to which she yielded, I might add! ☺)
Her thoughts are very encouraging
to anyone who may be a bit skeptical toward
or confused about the Lifestyle of Learning message.
Even if you're not (confused or skeptical)
I know you'll still find them most
edifying and delightful!

And don't miss the last half of
Marilyn Howshall's thoughts on the very
last page of this section!

Skeptic Turned Believer

I began reading *Wisdom's Way* with much scepticism. We had been doing our own unit studies for 7 years, and I had found a literary based unit study just a year ago, that "I" was *very* happy with, so I just kind of breezed through the book shrugging my shoulders and saying in secret to Marilyn, "Okay...so get to the point". But when I finished it, something tugged at my heart and I decided to take Marilyn's advice and enter a season of study. I opened to the Lord for His speaking.

Some have said things like: "Marilyn shares what God showed her, how she implemented it, and calls it "biblical." That seems to suggest that she sees this methodology as the one true way and that her book will now show *us* the way. So is it Marilyn or Jesus we are to follow?" My realization, when I read the book, is that Marilyn is just pointing the WAY. Through an intense season of study and search for what the Lord considers true learning, Marilyn has made many exciting discoveries, and rather than hoard these riches to herself, she has shared this wisdom with us. She points out that all of the treasures of wisdom and knowledge are hidden in Christ. So she went to Him, as should we, through the Word and through prayer, and He met her there. And that is where He has been meeting me too. And I have to testify that what she shares as being "biblical" truly *is.*

I would encourage anyone with the same doubts to put Marilyn's words to the test. Go to the Lord in areas that you have question marks in. He is so faithful. Marilyn graciously shares verses on the side panels of her book that are full of light. When I prayerfully read the verses and other related verses that the Lord takes me to, the Lord gives me strong confirmations.

For example, I took Marilyn's suggestion and chose to do a study on the word "knowledge." I was beside myself with what I found, and it totally matched many of the things that Marilyn shared in her book regarding truth and the importance of training our children's hearts that they might have a love of the truth which produces a love of learning.

When I came to the five Learning Tools presented in *Wisdom's Way,* I didn't see it the first time around. So I challenged the Lord and I was brought to tears when He showed me that this is absolutely biblical. Again, Marilyn provided verses that demonstrated the principles involved and I looked up the verses and many others so that I too could receive light. They were as follows:

Research - 2 Tim. 2:15
Record - Luke 1:1-4
Reason - Eccles. 7:24, 25
Relate - Prov. 22:17
Rhetoric - 1 Peter 3:15

Now I think the challenge will be to keep the vision that the Lord has mercifully given me, and not run off on my own tangent and turn this into another system or "methodology". It is instead, as Marilyn points out, a framework, a biblical framework or set of principles, in which God is free to move according to the needs of each child.

When I read *Wisdom's Way* and the newsletters, I had the feeling that Marilyn was saying, "Here is what the Lord showed me, and here is what we did in our home with this vision. Now you go to the Lord, and He will show you what to do in your home with this vision."

With love from a former skeptic who's now *SOLD!*
Pam Heath; Homeschooling Mom of 2 boys for 10 years; Georgia

Finding Him As I Seek

This beautiful letter was sent to the Wisdom's Way list. Read and be blessed!

My dear sisters,

I have to tell you! Our God is *soooooooooo* faithful! I say that all the time, but it's *soooooooooo* true! Recently, I have been floundering around quite a bit trying to work through some confusion and questions about how the Lord would have me go on with homeschooling, and feeling quite desperate. I've had lots of questions about preparing my kids for college, because my husband believes it is a must. I've been filled with the fear of leaving educational gaps and just doing the academic part wrong. I wondered, if my children go to college, how would a love of learning be useful there? College is like public school. Won't they have even more difficulty entering college when they haven't been educated by the same system? I bombed out in college and I went through the system. Was my inability to do well in college because I failed to study as diligently as I should have in my earlier years? Maybe I just wasn't smart enough. How can I equip my *kids* when I couldn't even do well *myself,* and am not even sure what the cause of my failure was? I know that this is the fear that Marilyn warns us about and that this is a classic case of what Barb calls "gapaphobia." I've wanted very much to heed their warnings, but I've realized that these nuisances have to be removed from me by the *Lord*, because I cannot do it myself. But praise God from whom all blessings come! He is addressing my concerns, calming my fears, and giving me answers in such a timely fashion. He is SO good to me. Lord, what am I that You are mindful of me?!

I am still getting educated on what REAL learning is and it is such a labor for me because I have been so "conditioned"! It is difficult for me to understand what learning REALLY is! Isn't that incredible?! I am reading Barb Shelton's book *Homeschool Jumpstart Navigator* and she has an article in there by Marilyn called "The Biblical S.A.T.'s for a Lifestyle of Learning" which is a summary of what a lifestyle of learning is all about. A quote, "We need to renew our minds to God's truth about learning, for it is only with a new mindset that we will be able to carry out God's plan freely and confidently." My poor mind is still trying to sort through all this truth that Marilyn has presented to us and I can't believe how very foreign it is to my understanding. How VERY sad!! My heart grieves for the children in the system. :-[

I realize now that I have to read and reread this stuff again and again and again for it to sink in because the truth is, I don't know HOW to learn. All my life I have been told WHAT to learn. The basis of my educational experience was memorization. I passed tests and classes by memorizing the material, but I never processed it. I have never been any good at memorization either. I went to college trying to recall the information I memorized and tried to memorize the new material that was given in the textbooks. Because the volume of information was so great, there was *no way!* I failed miserably and ultimately gave up, settling for a vocational school and choosing from their limited selection of "careers." Now another quote, "We need to be aware of the SERIOUS EDUCATIONAL GAP that is created when even one of these (Biblical S.A.T.'s, which are the ***S****cience,* ***A****rt,* and ***T****ools* of learning) is not present in a child's learning process." I was never taught HOW TO THINK, hence the *real* gap!!! I'm telling you, processing information is a real struggle for me. I now see that if we are not taught how to reason, to think, and to reach our own conclusions in a search for truth, it is that much more difficult for the Lord to bring us to the point of being hungry and thirsty for Him!! The Lord said, "Look at the birds of the air...", "Consider the lilies...", "Consider the ravens...", "And do you consider this, O man...?

The Lord has revealed so much to me about the devil's strategy in my life and the lives of the masses of children who are in the public school SYSTEM. I have been, and they are now being *desensitized!!* Death to sensation! Numb! Lifeless! The system tells us what to think, and we like cattle are herded into the ways of this world. Is it any wonder that society is in the state that it is in now?! Oh Lord Jesus!!! I am now seeing that if my boys learn how to think for themselves, and come to love finding out about the things of life that interest them and that are relative to their own lives, *they will be ABLE to learn ANYTHING they need to learn,* when they need to learn it! IF the Lord directs them to college, they will be able to process information because they will be able to reason for themselves. Oh hallelujah!! Thank you Lord for sparing my children! Thank you for showing their mother the WAY to enliven their thought processes! Thank you for rescuing them out of the hands of the evil one! Thank you for your unfailing LOVE for ME!!! and your endless mercy! You are so RICH in mercy! and You are so RICH to all who call upon Your name! You are so worthy of our praise!!!!!!

Growing in His life,
Pam Heath; Homeschooling Mom of 2 boys for 10 years; GA

Continued from first page of this section...

be broken of ideas, methods, agendas, beliefs, etc. His way will take you upstream from everyone else. His way will produce the rich godly fruit of true, healthy relationships within your family; children that grow to be mature, thoughtful young adults instead of selfish teenagers; children that are strong in character and able to embrace the will of God for their learning process; children who are able to pursue their own development of character, purpose, and quality relationship with God; children who are taught of the Lord not just taught of their parents. The grace-based training of children in obedience, relationships, and purpose will produce all the godly fruit you could ever hope for. It will look completely different from what you currently see in Christians today. If the goal of your instruction is love from a pure heart (which will only come with an emptying of self), then you will provide a strong foundation of character in your children that will enable them, with your help, to acquire a strong and unique, God-designed education which will include creative vocational purpose.

The fruit of rebellion, boredom, loss of soul life, loss of purpose, weak selfish character and bad relationships are all a result of how children were treated and provided for. Are their true needs being met? If mom is not dead, their true needs are being neglected, what's more mom isn't even aware of what they are! Most of mom's decisions are based upon her need to feed her own fears!

Children are schooled, churched, socialized, culturalized, virtualized and parentified from a young age until they have no uniqueness or creativity, no character or true fellowship in healthy relationships, no maturity or love to give, and most of all they become void of the very Life we think we are trying to give them.

Academics is not what education is about! Nor is it about indulging the child's every whim in the name of education along the lines of extra-curricular activities outside the home just for the sake of doing them – just like doing academics for their sake alone. Children need to interact with their environment in order to develop. They need constant instruction in every aspect of life and learning from a godly, loving source of wisdom. This foundation will make the academic aspect of education far simpler to achieve. If you faithfully work on your foundation, the academic aspect of education will begin to make much more sense to you. Give this time and work on your foundation now. I love you but God loves you more and wants to become intimately involved in all the details. ~Marilyn Howshall ❤

"Lord, shine a light of truth into the dark areas of our soul
and strike our hearts with conviction to listen to You."

Section Eight:

RESOURCES, REVIEWS AND GLOSSARY

Here you'll find some of my favorite homeschool resources, reviews and comments on this book (in case you want something to pass on to others or put in your homeschool newsletter), info on my Lab Science Book, plus info on the Sr. High Form+U+la Email List, and a Glossary of terms in this book.

What Curriculum Do You Use?

***Q** I stumbled across your address while researching about homeschool. We're trying to decide what program to use this year. What curriculum do you use, and would you recommend it?*

A The easiest and fastest way for me to answer your questions is to send you a brochure on all the stuff we carry – which is really more of an *approach* than just "products" or "program." (And this offer is extended to those of you reading this book.) We carry all the books that I have written as well as those of many other authors. At the time of this writing, the Howshalls are pulling back from all wholesaling, so you'll need to get Marilyn's books directly from her. *(P.O. Box 1750 / Eatonville, WA 98328-1750)*

By the time you get done reading *Form+U+la,* you'll have a pretty good idea of what I think of as "curriculum." I use some traditional stuff, but mostly a combination of real or "living" books, life experiences, *whatever life (and God) brings our way and what we are hungry to know.* Though it's a much bigger picture than that! (As you can see!)

The one piece of advice I'd give you at this point (and it's even free ☺) is to set aside your hunt for curriculum and *first* put yourself through a season of educating yourself as to what *true education* is. I assure you, it's very different from how the world defines and does it, and what you hunt for will be very different "before" and "after" this season! Until we discover what *real* education is, and what God has for us, we will spend many wasted dollars on trying to imitate the world's system, which is actually what *anyone* whose mind is not yet renewed *is* doing at this point. You can't avoid it. This is nothing to feel bad about; it's very natural considering the society we live in. But the traditional school system and methods are not producing the godly fruit in our children that we so desire, so why imitate them/it? God has a "burden" that is LIGHT; the world's idea of education is *heavy; God's* yoke is easy and right-fitting, the *world's* is *not.* If you will re-focus the thrust of your search onto renewing your mind in this area before continuing to amass knowledge about curriculum, your priorities and what you seek after will shift greatly.

I hope I'm not blowing you away or appearing rude. ☺ The Lord has me in a ministry of helping new homeschoolers get onto a better path than following the world's ways, and it does tend to make one a bit uncomfortable and insecure (sometimes even angry!) at first. ☺ But I assure you that it is well-worth the temporary discomfort!!! *Many* moms have written to me and thanked me for having steered them onto this path so that they could avoid the pitfalls that many of their friends are falling into. We have been homeschooling for sixteen years and graduated our oldest in 1996 and our son last year (98). We homeschooled both "from scratch"! ☺ And we still have a 12-year-old daughter we are homeschooling. We are very pleased with the fruit in our children and wouldn't trade homeschooling for any other form of education, even if we received scholarships to the "best" private school in the world. ☺

I wish that when I'd started 16 years ago *(as of May of 1999)*, I'd had the resources available that you (we) have now! We had *two whole books!* And oh how I searched for *more!* But many home educators – pioneers – were still in the process of being discipled by the Lord and learning it all for themselves, to begin with! Now it's available to you much more readily. (For more info on my own suggestions for a "Season of Reeducation and Renewing of the Mind" please see my *Homeschool Guide-a-log,* available only through us, which fully explains my suggestion plan.) – B❤

RESOURCES, REVIEWS, AND GLOSSARY

WHERE'D EVERYTHING GO?!?

If you have seen the Resource Section of either of the previous editions of this book, you may be wondering where all the resources went that I *used to* have in this section. ... I decided to not include them anymore because it's just impossible to stay up to date. I commend all the homeschool authors who *do* keep on top of which books are still available for how much and through whom, but that is just not part of what I feel called to do. And for those of you who don't realize it, it's quite a task! *(It took me about 3½ hours and several calls just to update the resources listed on pages 2-3!)* I regularly receive requests from several of them for updates on my materials.

Also, back when I first wrote the book, high school homeschooling was much less heard of, and therefore it was much more difficult to find resources. In fact, there were far fewer to be found! But since then, authors, how-to books, curriculum and resource companies, and conventions are *much* easier to locate.

Cathy Duffy's Curriculum Manual for Jr./Sr. High is a great resource for finding out such information, and much more! *(She also has a younger-age edition.)* And Ted Wade's *Home School Manual* is well-done too. I'll let them have that job! So please do not let me know if you are unable a company with the info I've given, or ask me where to find them. ☺ My focus is the how-to and inspirational stuff, and I don't even *want* to keep up with the other!

WHAT'S IN THIS SECTION *(NOW)*?

I'm listing and explaining what you'll find in this section here in one spot to save space. It's mostly a hodge-podge of stuff that I thought you'd find useful and/or interesting:

✎ **RESOURCE COMPANIES** – With the above in mind, all I have kept of the old section is a list of resource companies (on the next page) as well as several high school related resources and a few resources relating to legal matters. If I were you, I'd contact each of them that sound interesting and ask for their most recent catalog.

If you are getting this book directly from us, hopefully there was a catalog included with it. Obviously, companies with their own catalogs are not going to include ours in with it. If you didn't get one, we'd be happy to send you one if you just call or email us with an address – postal (also known as "snail mail" in Email Land.) Our main email address at the time of this writing is BEShelton@aol.com. But since our future with AOL (America On-Line) is unsure at this point because of local access number deletions, we may be off of that before too long. ☹ Try it first, though, *(as I use it most)*, and if you don't get a response from me within a couple of weeks, then try me at my other e-mail address, which is BEShelton@kalama.com. Wish I could be a little more certain about it. ☺

✎ **REPRODUCIBLE CATALOG REQUEST FORM** To make it even easier to request catalogs, I have provided a reproducible catalog request form you can use to request catalogs! First fill in the forms with your own name and address, then make several copies of this sheet, and then insert the names of the addressees at the top of each form. Or you could just stick on address labels. I know that you know how to write letters, but I thought this might save you a little time. I've included my name in the letter just so they'll know I'm recommending them. ☺

✎ **REVIEWS OF FORM+U+LA** – Coming right after those, you'll find a couple of reviews of this book. I include them here for three reasons:

1) **to help those who may not know if this book is for them or not.** You may be looking at this without the benefit of someone who can really tell you anything about it, and I guarantee that a quick-look through it will *not* give you the same impression that these "reviews" and comments give!

2) **for use in getting the word out on this approach to high school.** These reviews are good to share with friends to whom you're not quite sure what to say, or put into homeschool newsletters. We just ask that, for newsletters, you state the source – this book – and say "Used with permission from Barb Shelton."

3) **to broaden the horizons of those who might be looking at this book a bit too narrowly.** Some are so highly focused on the technical end – record keeping, making a transcript or portfolio, etc. – that they don't see the rose for the rosebush. Yes, this book is about the records, but I hope you'll see that it is much more!

✎ **INFO ON THE SHF-L EMAIL LIST** – This is something that those of you who have access to email will probably want to look into! It's a free email "list" (also called a "loop" or "list-serve") that's based on this book. I get on there occasionally, but am not "regular" – which I say only to not disappoint those who might join thinking they are going to have immediate and frequent access to me. It's not that I'm so exclusive or elite; it's just that I have only so much time – like everyone else – and a home life I have to guard if I'm going to actually be a mom while I have kids at home. The time we have with them is *so* short; as I found out too quickly with Sharnessa who's now been gone since she was 20. I was glad to have a couple more years with her than many do, but the time was still too short.

At this point in time, I have no plans to discontinue the list. I have a wonderful assistant, Donna Heck, who makes it possible for me to be on it by taking care of a lot of behind-the-scenes stuff, as well as questions out on the list – things that would put me "over the edge" in my ability to cope with the hassles involved in such an undertaking. Also, to Larry Wilson, who made this list possible, a wonderful brother in the Lord and *excellent* person to work with, I owe a great debt of gratitude. If at some point down the road you find that you are unable to get onto the list, it'll probably be because we discontinued it. ☹ Time has a way of changing priorities, and it may be that it just wasn't working out anymore. Unfortunately, we make revisions on this book as seldom as possible. If so, my sincere apologies.

✎**GLOSSARY –** Don't miss the Glossary at the very end of this section! It's not your typical glossary, and is actually very informative and helpful! (Not to mention funny!) So you probably *really* don't want to miss it. ☺

CHRISTIAN RESOURCES ONLY?

Just for the record, I do not avoid non-Christian publications unless they are downright or forthrightly anti-Christian, or contain too many abrasive references. If they are what I would consider "benign" (they just don't know any better), I feel comfortable using them. Besides the educational value offered in these books, I also use them to teach my kids how to sift through material and sort the chaff from the wheat. I don't want their first contact with non-Christian material to be after they've left my home. I want to have walked through it *with* them long before that day!

A prime example of this is the Usborne Books. I have found them to be excellent teaching aids. Colorful, and packed with understandable, attractive illustrations. Others share my view, but some have found their occasional "graphicness" to be offensive. I simply get out my trusty black marker and "correct" the picture.

HOMESCHOOL RESOURCE COMPANIES:

Beautiful Feet Books - 139 Main St., Sandwich, MA 02563, 1-800-889-1978

Builder Books - P.O. Box 5789 / Lynnwood, WA 98046-5789 / (425)778-4526 1-800-260-5461

Bluestocking Press - P.O. Box 1014 / Placerville, CA 95667-1014 / (530)621-1123

Christian Life Workshops - P.O. Box 2250 / Gresham, OR 97030 / (503)667-3942

Elijah Company - 1053 Eldridge Loop / Crossville, TN 38555 / (931)456-6284

Farm Country General Store - Rt. 2 • Box 412 / Metamora, IL 61548

Hewitt Educational Resources - P.O. Box 9 / Washougal, WA 98671-0009

Home School Legal Defense Assn. - P.O. Box 159 / Paeonian Springs, VA 22129 (540)338-5600

How Great Thou ART - P.O. Box 211, Dept. HST-95, Bishopville, MD 21813

In His Steps - 1618 Kendolph / Denton, TX 76205 ... Website: http://www.inhissteps.com

Institute in Basic Life Principles - Box 1 / Oak Brook, IL 60522-3001

Institute for Creation Research - P.O. Box 2667 / El Cajon, CA 92021 / (619)448-0900

Lifetime Books & Gifts - 3900 Chalet Suzanne Drive / Lake Wales, FL 33853

Lifestyle of Learning - P.O. Box 1750 / Eatonville, WA 98328-1750

Math-U-See - 1-800-475-8585

Moore Foundation - run by Dr. & Mrs. Raymond Moore - Box 1 / Camas, WA 98607

R.O.C.K Solid - 3039 Burkit Lane / Jacksonville, FL 32226 (904)751-3569

LOTS MORE COMPANIES!

There are many more companies that sell homeschool resources. The fact that any are not listed above means nothing. I would have included many more, but just didn't have the space. I'm sure I've even omitted some that carry my book, this is only because there are over fifty, and I just couldn't include them all. My sincere apologies to anyone who feels left out.

HIGH SCHOOL RELATED RESOURCES

Companies:

Educational Support Foundation - Diplomas, Transcripts, Report Cards, Stationery Letterhead! Custom-made for your *own* homeschool! National speaker and homeschool leader Hank Tate, a homeschool dad himself, operates this home-based business, and his products are of *professional* quality! For info, samples, and prices, contact: (713) 870-9194 *(Voice and Fax)* E.S.F. / 1523 Moritz / Houston, TX 77055

Josten's - offers diplomas (in leatherette cases), caps and gowns and tassels, announcements, etc. Contact the national U.S. number at 1-800-854-7464, and they will refer you to an area representative.

❤❤❤❤❤

Helpful Books for Homeschooling Through High School:

Homeschooling the High Schooler by Diana McAlister and Candice Oneschak (Contact: Family Academy / P.O. Box 66839 / Seattle, WA 98166 / (206)246-9227)

The High School Handbook by Mary Schofield (Contact: C.H.E.A. (Christian Home Educators Assn. of CA, Inc.) P.O. Box 2009 / Norwalk, CA 90651-2009)

(The) Christian Home Educator's Guide to Junior & Senior High by Cathy Duffy (Published by: Home Run Enterprises / 16172 Huxley Circle / Westminster, CA 92683)

Lab Science: The How, Why, What, Who, 'n' Where Book by Barb Shelton. *(See comments coming up, and postal and email addresses on publisher page of this book.)*

LEGAL RESOURCES

"**Home School Legal Defense Association**" - A membership organization established to provide a low-cost method of obtaining quality legal defense for homeschooling families. For $100 a year *(as of 1999, and it has been this price for many years)*, a family is provided with representation by a lawyer at every stage of a potential legal proceeding. For a brochure, write to them at *H.S.L.D.A. / P.O. Box 3000 / Purcellville, VA 20134 (540-338-5600)*

"**Rutherford Institute**" - A non-profit legal and educational organization committed to the rights of home educators and others dealing with issues of religious rights. The Institute does not offer a legal insurance program, but will use its best efforts to defend homeschoolers' rights through legal assistance, limited only by time and/or financial constraints. Ask for their free resource listing. You may subscribe to the Home School Brief for $25/year which will also give you access to free legal information. For further information or advice, call their Home Rights Hotline at (804)978-3888, or write: *P.O. Box 7482 / Charlottesville, VA 22906-7482.*

Dear ____,

We've heard from Barb Shelton that your company offers a brochure or catalog with several resources she highly recommends for homeschooling through the high school years. Would you please send a copy to:

____

____

____

Appreciatively,

Dear ____,

We've heard from Barb Shelton that your company offers a brochure or catalog with several resources she highly recommends for homeschooling through the high school years. Would you please send a copy to:

____

____

____

Appreciatively,

Dear ____,

We've heard from Barb Shelton that your company offers a brochure or catalog with several resources she highly recommends for homeschooling through the high school years. Would you please send a copy to:

____

____

____

Appreciatively,

Dear ____,

We've heard from Barb Shelton that your company offers a brochure or catalog with several resources she highly recommends for homeschooling through the high school years. Would you please send a copy to:

____

____

____

Appreciatively,

Here's a letter I received on this book...

August 19, 1996

Dear Barb,

I want to thank you for "Senior High: A Home-Designed Form+U+la". I don't know whether it's better as a devotional or as a high school help, but it really "fits the bill" for both. Normally, I try to do a bit of devotional-type reading in bed before I go to sleep, and I knew I had a special book on high school, (a special book – *period*) when my morning or afternoon reading of "A Home Designed Form+U+la" ended up being extended to include my bedtime reading slot. I looked *forward* to reading it at that special time, and I don't feel that about many books.

The best thing I liked about the book was feeling I was in the Holy of Holies. Your constant encouragement of what you have received there – personally, and for life's education for yourself *and* your children – made me feel like "I've *got* to pray myself!" But it was not out of the too-usual duty, obligation, and dreary necessity, but that deep inner necessity of heart's desire and eager anticipation of living water for myself, our family's education in life, and the living presence of Christ. And your direction to "go ask Him for what He has for you" is *gold.*

Your book has living water. It also helped me feel free-er to be a pioneer and not a sheep-fold sheep – either to follow the traditional school way or to do what "all the other" homeschool families are doing.

The second thing I like best about "Form+U+la" is your heart coming out all over. Your (contagious) joy – especially when mine is ebbing or gone -- your jokes, your "snide" thought-comments (on How-one-feels-under-pressure-and-attack-but-responds-with-a-gracious-tongue-and-a-growing-heart-of-compassion 101), your tears - in short, *you* shared *yourself.* Wow, what a gift and treasure.

The third thing I like best is the information and help about forming the content of classes, alternate ways of teaching subjects (like real-life stuff that everybody likes, but who'd have thought it would ever "count," and how would I record it, anyhow?), and your forms. The forms are why I actually bought the book originally. (Mary Schofield in "The High School Handbook spoke of and showed some, and I said, "I *need* those!") Your comments about "school-at-home" burnout were so true, and your alternative of true homeschooling (*life* with learning and books) is a light in a dark place. (I have six wonderful, active, live-wire kids!)

The fourth thing I like best (and I would say this is not too far down on my list) is your way of writing, so that by parentheses, different size fonts, quips, and all your other means, I feel like I'm sitting across the table from a welcome friend drinking tea (the fourth cup), and chatting about all this school stuff and the Lord.

Love, Sue Howe; (California)

A ❥Review❥ of
SENIOR HIGH: A HOME-DESIGNED FORM+U+LA

by Julie Lofurno

Homeschooling mom of four, for 12 years

If you are homeschooling a high schooler (or intend to do so soon), I would like to recommend a wonderful resource called Senior High: A Home-Designed Form+U+la. Barbara Edtl Shelton has written a wonderfully detailed manual on how to plan the high school years in a non-traditional manner - and keep track of it all! Interspersed throughout the book is Barb's unbeatable humor which will keep you in stitches while you become educated, motivated, and encouraged.

Last year, as we embarked on our umpteenth year of homeschooling (which included a ninth grader), we realized that there would be a type of transition in the way we were teaching our oldest child. We began to ponder the possibility that he would want to attend college and, with that in mind, knew we had to "buckle down" and structure his course of study toward that end. We began to think about things like high school "credits" and transcripts, neither of which we were familiar with. We wondered, How can we use the events of everyday life - like changing the oil, babysitting, meal preparation - and fit them into his academic schedule? Well, this book was our answer.

This book will teach you how to set up the FRAMEWORK SYSTEM. It does not tell you what books to use; rather, how to document what you devise. This system accommodates ALL methods of teaching - traditional or non-traditional:

- If you like the pure textbook route, this system will work for that.
- If you are like us and prefer a mixture of textbooks and real learning experiences, (including apprenticeships), this record keeping system is perfect!
- If you consider yourself "unorganized," this manual will help you "get it together"!
- If you are an organized person by nature, you will LOVE this book!

Barb takes the reader step-by-step in figuring what subjects to teach over the high school years, how many credits to assign, and even how much time should be spent daily, hourly, weekly and yearly to achieve the goals you and your student set. She helps you understand that biology need not be learned all in one year and shows how easy it is to document biology-related activities over a longer period of time if you desire to stretch the subject out a bit. For example, last year we took advantage of oceanography workshops sponsored by ODU. We documented six biology hours, even though we did not begin an "official" biology course until this year.

The record keeping forms are well designed and are all accompanied by complete instructions and examples... I photocopy all the forms needed for the entire year and organize them into a note-book by subject. (Directions for all this are in the book, too.) Our high schooler is then in charge of maintaining them week to week. This has totally changed our son's approach to his education! He is completely in charge of it!

The Learning Guides cover a vast variety of subject areas and are great "springboards" for other learning activities. These are not your typical fill-in-the-blank forms. They require "thought" and analysis. You can put the report in your child's portfolio as is, or let your student us it as an outline for a report.

Senior High: A Home-Designed Form+U+la is well worth the price. It will not only motivate and encourage you, but it will motivate your high schooler to take control of his education and, if college is on the horizon, equip him for the task ahead.

SHF-L Email List

I wanted to somehow introduce you internet users to the email list that is based on my book, *Senior High: A Home-Designed Form+U+la*. So I went back through a few posts I've saved over the months and pulled the following excerpts out. (I have not included any last names to protect the privacy of those who prefer to remain private.) By the way, in Email Land, we use these <<...>> arrows to enclose anything from a previous email that we are quoting. Or sometimes there will be a single arrow at the beginning of each line of the quoted material. *– Barb*

♥♥♥♥♥♥♥♥♥♥♥♥♥♥♥♥♥♥♥♥♥♥♥♥♥♥♥♥♥♥♥♥♥♥♥♥

A FEW E-MAIL SAMPLES♥

♥♥♥♥♥♥♥♥♥♥♥♥♥♥♥♥♥♥♥♥♥♥♥♥♥♥♥♥♥♥♥♥♥♥♥♥

Dear Barb,

I attended the homeschool conference you spoke at in June and purchased your "Form+U+La" Book. I have been reading the e-mails from the Form+U+la list. I have looked up your catalog on your web page at the "Our Home Our School" site. After reading your book I was so blessed. I'm on my second time through and am trying to get what ever I need found and copied and set in order. My 14 year old daughter and 16 year old son are in the process of reading your book as well. God is my greatest counselor. In your book you talk about a season of drawing close to God for the purpose of letting Him do a work in my heart and that the season of drawing close may feel a wilderness. I thought that my wilderness season was coming to an end, but when I heard your "Choosing Your Own Sr. High Framework..." seminar and read your book, I guess the Lord is now telling me I've only scratched the surface. From all the discussion I've read on the SHF-L list, and the bit of info in your catalog, I think I need to read Wisdom's Way of Learning next. If I understand correctly, it doesn't apply only to homeschooling but to every aspect of our life as we walk and grow with the Lord. (YES!) I cannot pass on to my children what does not reside in me. Meaning, if it's not in my heart and hanging as fruit on my tree then I sure can't expect to see it in my children. Have I got it? (Yes, dear sister! You've GOT it!)

Colleen Houger; WA (houger@televar.com)

♥♥♥♥♥♥♥♥♥♥♥♥♥♥♥♥♥♥♥♥♥♥♥♥♥♥♥♥♥♥♥♥♥♥♥♥

Dearest Barb,

I have learned sooo much from you already with your Lab Science book, Form+U+la, and SHF-L. You have no idea how much more confidence I feel now to continue in the way I felt led to mentor my many blessings from the Lord (a.k.a. "children"!) I had *great fears* that I would have to resort to textbook teaching (gasp) for the high school years in order to make it work with our very own learning lifestyle. Besides, the lessons I've learned from you are applicable to all my children at any level. – not just high school! I must thank you again for dedicating yourself to this ministry. For those homeschooling with an eternal perspective, your books are such a blessing.

In Christian love, LML

♥♥♥♥♥♥♥♥♥♥♥♥♥♥♥♥♥♥♥♥♥♥♥♥♥♥♥♥♥♥♥♥♥♥♥♥

How to Subscribe to the List

To subscribe to SHF-L (the Sr. High Form+U+la List), just send the message "join shf-l" to: Lwilson@xc.org (It doesn't matter what you put in the subject line. Normally you'd send it to a hub address, but the list server is undergoing some management transitions, so Larry Wilson has graciously consented to intercepting all subscriptions and handling them for us.) That's it! Shortly thereafter you'll receive a "Welcome Letter" from me that I wrote at the onset of the list when it first started in June of 97, and it is automatically sent out to everyone who subscribes. It's long ;-) *(← winking face)* **and thoroughly explains my heart, the list's purpose, guidelines, my participation (its extent), and how it all works. *Please read it all!* :-) Anytime you want to quit – even right away ☹ you can do so quickly and easily; simple instructions are at the bottom of each post you receive on the list, so you don't have to remember a thing!.**

~Barb Shelton

♥♥♥♥♥♥♥♥♥♥♥♥♥♥♥♥♥♥♥♥♥♥♥♥♥♥♥♥♥♥♥♥♥♥♥♥

Dear Barb,

Hello... I unsubscribed from all my lists for the summer except yours!!! I love Senior High Form+U+la! When I go camping I take several books along to read and yours has traveled with me all summer!! Thanks so much.

Love in Christ, Donna J.

♥♥♥♥♥♥♥♥♥♥♥♥♥♥♥♥♥♥♥♥♥♥♥♥♥♥♥♥♥♥♥♥♥♥♥♥

What to do about Lab Science?!?!

First here's the write-up on it that appears in our catalog (and several others who carry it.)

CATALOG DESCRIPTION: Lab Science can put off even the most seasoned homeschoolers, and has probably intimidated more people out of homeschooling through graduation than any other! Yet it has the potential to be one of the most enjoyable, life-enriching areas of study! This book, with the insight and encouragement of experts in science and homeschooling, will help you view Lab Science in a different light, and shed God's light on the subject as well! In the "Excerpts from the Experts" section, one half of the book, Barb has gathered the wisdom of over 30 science and homeschool experts to give you rich "food for thought" and many practical hands-on ideas to use in formulating your own lab science class. In another section there are several sample frameworks and fundamentals for a variety of Lab Science courses, from "natural" to "traditional" approaches. Plus the book has many helpful record-keeping and lab science notebook forms to help you organize your materials and activities, and coordinate (or even create from scratch) your own course of study, with and without college on the horizon. There are *many* wonderful (non-textbook) books "out there" available for use with science, but the tendency at the high school level is to just toss all those aside and head for textbooks. What a waste of excellent resources! This book will help you utilize what you already have and determine what you really need to get! There's a whole section listing many traditional and non-traditional resources – books, audio tapes, videos, and lab science supplies – from which you can draw. This book is guaranteed to replace the "jitters" with delight and practical help and direction! Worth its weight in Au! (the Periodic Table name for gold.) Softbound / 160 pages / $11.95

Comments from several readers:

Because of how intimidating Lab Science is to most homeschoolers, and because of how much potential I believe this subject has, (but many get so intimidated, they miss it!), I'm dedicating two pages of this book to saying a few things about it; letting others say a few things, actually... (Don't worry, you aren't paying a cent extra for me to include this in with this book.) So here we go!

"I have just finished reading Barb's Lab Science book and it is *fantastic*. It has more information than I thought was possible on this subject! I had just ordered Dr. Wile's *Exploring Creation with Biology* and really liked it but was looking for a little better explanation on how to do the Lab Notebook. Well I got plenty from Barb's book!! Wow did she ever simplify that for me!! I am such a perfectionist and I just couldn't figure out how these lab pages should look. Barb did all the work for me. All I have to do is copy out the pages I think we will need and "wha-la!" – a nice neat Lab Science notebook that fits my perfectionist mind perfectly! Thanks Barb!!!! You've done it again! Sorry I waited so long to order it!" *~Barbara O.; Homeschooling Mom*

"What a great looking book! I get questions on what to do about high school and need more good resources to recommend. This book will certainly be one of them!" *~Dr. Ruth Beechick (Author of many books for homeschoolers and teachers)*

"I wanted to thank you for writing the lab science book. I read your book on Senior High last year (several times) and have planned out the high school years for our children. (In fact, your book on Senior High freed me from the fear of the high school years and measuring credits, etc. and it is because of this book that I can face the high school years with confidence and excitement.) However, there was one class that I was dreading, lab science – any lab science. I know our family well enough to know that we do NOT do well with textbooks. But at the same time I didn't feel confident enough to put together my own course in lab science as I would wonder what to exclude and include. However, after reading in your book what all of the various "experts" had said, and then looking at the various frameworks, I have come to realize that I can design a lab science course that will meet the needs of my kids and not feel tied to textbooks." ... The best thing about your lab science book is that it now has me excited about teaching lab science (except doing the dissections) because I understand the reasons for studying it. Your book gave our upcoming studies a sense of "purpose" and I realized that our study of the sciences would not be "meaningless". The courses that I had been dreading the most are now ones that *almost* excite me the most (ok – so I really prefer history). ... Thanks for writing your book. Not only did it help me to design a plan for our studies, but it helped me get excited about them also!" *~Peggy Flint; Homeschooling Mom, Anchorage, AK*

"Barb, your book is great! It will help me in selecting research books. You honor me by making my books part of such a wonderful collection of science materials." *~ Janice VanCleave (Author of 26 Science Experiment Books)*

"Where was this book 3 years ago?! I sure could have used it! My children are now 18 and 16 and have always been homeschooled. Lab Science presented a challenge to me. We ended up choosing two different courses of study that didn't fit our homeschool philosophy – which is to be self-directed and independent in all our studies. This book would have given us the freedom and the help we needed to design an autonomous course of study; had I had it, I could have easily designed my own course. It's very practical and well-rounded. It offers several different perspectives and choices, so you can pick and choose components of any or all that will work best for *your* family." *~Janice Hedin; Homeschooling Mom for 15 years; High School Speaker; Support Group Leader*

"This book gets my "must-have-if-you-are-homeschooling-a-high-schooler" seal! It offers much more than what to teach for Lab Science; it also has lots of creative ideas! An abundance of excellent information packed into one book!" *~Robin Scarlata; Author and Homeschool Curriculum Consultant*

"This book changed my mind from feeling the need to send my daughter to a community college for Lab Science. We can just use this book and figure out a good lab course on our own! And we are both relieved and excited about that! ... This book is an excellent resource to try to help high schoolers with lab science – the most intimidating subject for homeschoolers!" *~Vickie Smith; Homeschooling Mom*

"What I love about this book is that Barb has found all these great resources, excerpts, and insights, compiled them all into one book, and along with her own material given us a smorgasbord of ideas and concepts to choose from!" *~ Teresa Kempf; Homeschooling mom for 8 years* *(at the time, back in 1996)*

Lab Science Testimony of a Homeschooling Mom

"I bought Barb's Lab Science book a couple of years ago after attending one of Barb's seminars. I have to say that, until then, I had totally misunderstood what the book was about! I just figured it was a "lab" book. But I also knew that I liked Barb's ideas and figured that if she wrote the book, then it was something I probably needed!! ☺ I didn't get around to reading it for many months as "Form+U+La" totally engrossed me during that time.

It was during the fevered time when the frenzy of buying and selling curriculum begins that I began thinking about my younger son's science requirements. I bought a physical science textbook, a biology text and lab book, chemistry text and some intro physics stuff – all of which I ended up having to re-sell!)

I was really struggling with a feeling of incompetence and also just not wanting to do some of the subjects since I didn't like them myself! Of course, this started the old concern about educational gaps, and pretty soon I was completely overwhelmed! As I began planning the courses and actually getting input from my son and husband, I began to see that I was getting caught up in the "school at home" thing again.

Well, first thing I did was pray for direction! Then I got out Barb's *Lab Science* book and began reading. I finished it in a couple of hours and started over again. I talked with my husband some more and prayed some more. After skimming the book again I was exhausted and went to bed. Sometime in the night I awoke with the idea in my head that I needed to do something different in science for my son. I couldn't sleep until I got up and figured it out. I picked up the Lab Science book again and it seemed to open right to the page of the class that Barb had designed for her own kids called "Lab Science Survey Framework." I thought "Eureka!" *("I found it!")* This is what I need to do for my son! He is not interested in a college degree that requires heavy science, but he does like machines and physical science type stuff and I felt there were some things he just needed exposure to in order to be a well-rounded person.

The next morning my husband and I wrote an outline for our son based on the concept of "large to minute." (That's mīnūte – as in "small") We began with the big picture (astronomy), to earth science (including weather and ecology), to biology, to microbiology, to chemistry and physics. Using Dorling Kindersley's "Science Encyclopedia" as the main resource, we then filled in with 15 to 20 other books from Usborne, TOPS, Dorling Kindersley, Kathleen Julicher, biology/anatomy coloring books, etc. I have many books on rocks, weather, oceans, volcanos, birds, etc. Whatever I had on hand, we tried to use as a resource even if it was just to read a page or two from it. I have to give a big thumbs up to *The Private Eye!* This concept is being used in all of our studies – not just science! *(This is the intent of the author, who helps you apply it widely)* The only textbook we will use is for biology in order to plan the experiments and labs that my husband thinks are important. He is a microbiology major so this is important to him! Plus, we have an excellent microscope and many, many slides already, so we figured, *why not?* We are sharing these resources with several homeschool families from our church. I am writing a schedule of dissections and experiments so they can plan their studies to coincide with our lab times.

I shared all of this to say that Barb's book allowed me to see that lab science doesn't have to be intimidating or taught only from a textbook by a certified teacher! Her *Lab Science* book is full of ideas from people who have already "been there, done that"! It is sort of like going to the grocery store and picking fruit to make a big fruit salad! If you made it completely out of apples and oranges it wouldn't be nearly as good as if you added peaches, watermelon, cantaloupe, cherries, passion fruit, mango... and the list goes on! Our science course is composed of pieces from many of the ideas presented in Barb's *Lab Science Book* – and I must say *It looks delicious!*

As a starting point for anyone considering teaching a lab science, I highly recommend *Lab Science: The How, Why, What, Who, 'n' Where Book.* It is *not* a lab science textbook but rather a collection of experiences, ideas, resources, and writings of over 30 home education and science teachers, authors and suppliers, for example: Cathy Duffy, Janice VanCleve, Diana McAlister, Candice Oneschak, Mary Schofield, Marilyn Howshall, TOPS, Tobin's Lab, Mike Bolinsky *(Science Labs-in-a-Box)*, just to name a few. No, I don't have any "stake" in pushing Barb's book but I know from experience how intimidating the thought of lab science can be. My boys are in 9th and 12th grades and plan to go to college, one into the medical field. I just feel this book helped me to get past my feelings of total inadequacy and the idea that I had to teach with a textbook. It's not a miracle worker, *(nor is it intended to be!)*, but it sure helped me!" *~Donna Heck; OR*

THE HOMESCHOOL GUIDE-A-LOG
(FOR REAL-LIFE HOME EDUCATION)

Written and Compiled by Barb Shelton

Affectionately called just "the Guide-a-log," this unique book contains just for starters, all the products that are dearest to my heart – all the ones I have written plus a variety of other favorites. Oh, but that's only 10% of the book! There are also many articles to help in some vital areas! It was written for those who are burned out on "school at home" and for veteran homeschoolers who are beginning to realize that God has something different and better for them, and are wanting to RE-start.

It includes articles on: socialization, homeschooling through high school; taking a child out of public school; transitioning to home; the purpose of education; goals of child training; legal and testing issues; romance and relationships God's way *(purity, integrity, and abstinence, but from a heart to please God, not from a legalistic or rule-based angle)*; "Decide to Be a Homeschooled Mom"; "Hearing God 101"; "The Diploma Quest(ion)"; and "Biblical S.A.T's of Learning," just to name a few.

One chapter presents an in-depth description of my suggestions for a "Season of Re-education & Renewing of the Mind " and a few reviews on my books that are detailed enough to help you decide if they are really what you want. It includes these sections: "For Those Just Starting or RE-starting Out"; "Season of Re-Education and Renewing of the Mind"; "For Parents with Children of All Ages"; "Intro-only to the Jumpstart Plan for Younger Kids"; "For Senior/ Jr. High School Age"; "To Enhance Relationship with and Discipline of Younger Children "; "For Mom's Personal R & R (&R)"; "Spiritual Edification and Deepening Your Walk With God"; "Romance & Relationships God's Way"; "Legal Issues and Testing." Here's a break-down of the contents of the Guide-a-log: *(It can be ordered only through us.)*

- Product pages - 18 pages (only about 10% of the whole book)
- Explanations on various concepts - 17 pages
- Articles written by me (yours truly, Barb) - 32 pages
- Articles by several other authors I admire - 32 pages
- Reviews; Comments, Letters from Readers - 22 pages
- Jumpstart Plan Suggestions for Sr./Jr. High - 22 pages *(which, as of the new revision of Form+U+la (that you are now reading) is no longer really needed.* ☺)
- Intro to the Jumpstart Plan for Younger Kids *(7 pgs. of "Smorgasbord of Real-life Learning activity ideas)* - 10 pages total
- Order Forms - 8 pages *(including two sets of the 2-page "Season of Re- education" outline for future reference)*

❤ *Comments From a Few "Satisfied Readers"* ❤

"What's the Guide-A-Log??!! Only the best home-school resource I've seen in 8 years of homeschooling! ☺ It's like explaining chocolate – it's in a category by itself and compares to nothing!"
~ Joan V. in WI

"I just wanted to write and commend you on the fabulous job you did on the G.A.L. I was blessed by reading it. ... Your writing style is suited to my "reading" style, I suppose, because I found your stuff *very* easy to read.....while still getting tons of helpful info. I enjoyed everyone else's articles too!!!!"
~ Dana, a homeschool mom in Alabama

"Today I join the ranks of the tea-sippin', tear-drippin', head-noddin' homeschoolers reading the Guide-A-Log. I have only skimmed it yet but I find myself getting excited all over again at all the varied opportunities. Barb, I wanted to publicly thank you for making your views known in such a gracious, yet straightforward manner... In our "anti" society, Christians have a (sometimes, unfortunately, deserved) reputation for being strident and illogical and .... you know all the adjectives! On the contrary, your tone is completely honest and firm, but at the same time gentle and edifying, challenging the reader (oops, I almost said listener, because it's almost like talking with you to read your books!) to further study and application of the Word. (For those of you who have not seen the GAL, I am referring to the "Legal Issues" section and a few others.) Anyway, just a heart-felt "thanks" to Barb (and Dave!!!!)
~ Amy Beckel in WA

GLOSSARY OF TERMS

USED IN THIS MANUAL

A.C.S. - "Assignment Check-off Sheet " - also known as a "Completed Class Plan." These are described in detail with many examples given in Section 6-D. They are actually the completed class plans for all of our high school classes for Tory, and a couple for Sharnessa. (I practiced on Sharnessa, and then mastered the art (IMHO) with/for Tory! ☺) As an example, please see the ones for "Home & Life Management Skills" and "Educational Foundations" in that section.)

class - For purposes in this book, a *class* is a specialized area of study, possibly done with other students, but just as likely done on an individual basis or with a parent. Two *classes* under the *course* of Math would be Algebra and Geometry. A few *classes* under the *course* of Social Studies would be U.S. History, Economics, and Current Events. The *course* of P.E. could be made up of the *classes* of aerobics, basketball, ballet, track, and weight lifting.

collage - the cover that your student may or may not want to make for his high school notebook.

college - the place your student may or may not want to attend after high school graduation.

course - a generalized area of study like English or Occupational Skills that could cover a wide spectrum of more specific classes (see above) or areas of study or experience.

credit - If using the *time-based approach,* one credit equals a certain amount of time spent in a particular area of study. If using the *content* or *project-based approach,* one credit is whatever you deem to be an "adequate" amount or content of knowledge.

dangling preposition - any word such as "of," "to," or "for" left at the end of the sentence. You'll find many of these in this book because I write like I talk. I just can't bring myself to say "Take a look at that with which you have to work." Sorry, too weird. I prefer the improper "Take a look at what you have to work with."

delight directed learning or study - See the three article/excerpts on this on the last three pages of Section 6-B.

education - According to Webster's 1828 Dictionary, education is: "The bringing up, as of a child; instruction; formation of manners. Education comprehends all that series of instruction and discipline which is intended to enlighten the understanding, correct the temper, and form the manners and habits of youth, and fit (prepare) them for usefulness in their future stations. To give children a good education in manners, arts and science is important; to give them a religious education is indispensable; and an immense responsibility rests on parents and guardians who neglect these duties." Read *Wisdom's Way of Learning* for a complete definition (view, understanding, vision) of this widely little understood, misunderstood and misapplied word.

Guide-a-log, The Homeschool - The book I wrote that I suggest as a starting point for *anyone* considering homeschooling *or* wanting to RE-start from having burned out on "school at home." See write-up on it on previous page. Among many other things (it's 170 pages!), it contains my suggestions for a "season of re-education and renewing of the mind."

him/he - the grammatically correct writing term for male *or* female Are we hu**man** beings or "hu**woman**beings"? Color me old fashioned or "sexist," but I'm using "him" to refer to *both* genders. "He" *does* accurately refer to both male and female; "she" (sorry) does *not.*

homeschool pioneer - any of those who started homeschooling when only Dr. Raymond Moore and John Holt were talking about it, and started doing it because we knew it was right, but were going to get a lot of flack from everyone we knew. And we did! At the time we were called "radicals" and "fools." We had no intentions of being "pioneers" but do like the sound of that term much better! ☺

kid - originally referring to a baby goat, this term is now almost universally accepted as being an alternative term for "children." Some think it is disrespectful. I, on the other hand, do not. ☺ I just don't. Respect comes through in my relationship with them. The word "kids" is just faster and more concise, which, as you can tell from the rest of this book, I *need!*

lab science - the subject that's infamously responsible for freaking out homeschool parents and keeping them from thinking they can home school all the way through to graduation. In reality it is just studying how God did what He did in creating and ordering His creation, and has the potential of bringing us closer to Him as it opens our eyes to the amazing THOUGHT and DESIGN that went into every aspect and every molecule of His creation! Only if a student is going on to seriously pursue a study of science do I feel they need to tackle the *(To gain freedom and insight in this area, and get help and insight to tackle – and enjoy – this subject, read my book Lab Science: The How, Why, What, Who, 'n' Where Book. Comments from readers earlier in this section.)*

learning guides - the sheets you'll find in Section Four that *look* like typical, standard worksheets at first glance, but are actually dissimilar. Typical worksheets provide *consumptive learning* in that the student simply consumes pre-digested, spoon-fed material – which is OK in limited quantities, but who wants to live on TV dinners? These learning guides, on the other hand, (which you will very likely start making up yourself based on priorities *you* have for your children) encourage *productive learning* by directing a student to search out knowledge in the real world he lives in, and glean more valuable learning out of events, books and activities.

lifestyle of learning - DATA OVERLOAD!!! . . . DATA OVERLOAD!!! Woops! Sorry, but evidently the definition of this one is too expansive to fit into the data bank here. ☺ See Sec. 7 for a brief intro; *Wisdom's Way of Learning* for a *full* explanation.

one-half credit - normally the smallest portion of a credit for record-keeping and transcript purposes. You can certainly do a smaller or shorter study of a subject, but it would be better to count it as a *portion* or a *unit* of a "full" study. (¼ credit *might* be acceptable.)

redeem - One of my favorite words, *redeem* means "to take something of little or no value and exchange it, *free of charge,* for something of great value." Many times in my life God has taken something of little or no value and exchanged it for something of *great* value: with my humor, faults, confidence, mistakes, marriage, eating... you name it, He's either already redeemed it, *is* redeeming it, or will begin when "it's time"!

requirements - set up by parents in conjunction with the student, these serve *your* purposes, not vice versa! Once the fear and red tape are taken out of them, *and* they are established by and with loving parents (or other chosen guiders), they can be seen as wisdom-based, loving, acceptable guidelines (for graduation *or* completion of a framework of study) that will help prepare the student for and lead him into "the good works that God has prepared beforehand for him to walk in."

resource - any event, person, circumstance, book, tape, video, or place that will enhance, bring about, provide, or lead to learning. It could be Grampa talking about history, a brochure on gum disease obtained from your dentist's waiting room, a video from the library on how to interview for a job, a tape you got at a science center, the science center itself, a biography, a presentation on a health-related topic at your hospital, a seminar your family attends on "Home & Family Business" (by Gregg Harris), a video on "Why Homeschool Through High School" (by his son, Josh Harris), your vet as she explains how care for a sick pet, a TV special on Martin Luther King Jr., or a town hall meeting on dealing with Y2K in your area. Why, it could even be, in limited doses, a *textbook!*

study - to seek further knowledge and/or experience in an area of interest with more than just a casual level of pursuit. It doesn't mean "cracking the books" positioned at the proverbial desk unless that's how and where the student finds his studying most conducive. For the reading and skill subjects, our kids will more likely be found curled up in a chair, or on (not in) his bed, sitting with Mom on the couch, or snuggled up under a billowing blanket over the heating vent. For more experiential subjects, you might find them helping Dad sheetrock in the new addition, (not just being his "go-for" but actually doing and learning the "stuff"), practicing lines or making a costume for the homeschool group play, working up harmony parts for a trio performance for Home School Presentation Night, watching a community concert harp virtuoso performance, or for P.E. jogging, playing basketball with a group of men, or jumping on our 14-foot diameter trampoline. *(Try it for just three minutes!)* My point in all this is to let God resurrect your concept of "study" and bring it out of the rigid confines of the traditional classroom setting *(not that there's not a place for discipline!)* and into the fresh "universal classroom" of the world around us. Don't discard the "old"; we need to "study to show ourselves approved unto God"; but I think He'd like to that God might want to broaden your concept of study.

T.A.S. - "Task Analysis Sheet." These are described in detail, with many examples given in Section 6-D. They present specific lists of things to be learned, done, practiced, experienced or mastered. For examples, please see the ones for "Carpentry" and "Vehicle Maintenance" under "Home and Life Management Skills" in that section.

teen - (This is a long one!) the collective, generic term for persons having the suffix "teen" at the end of their age, as differentiated from people in their 20's or 60's. In some circles the terms "teen" and "teenager" are somewhat abrasive-sounding, even disdained to the point of not being used. Instead, the terms "youth" and "young adult" are used. ... In my choice of using the word "teen" or teenager, " my intent is not to offend, to be contrary to such perspectives, or to be "trendy." I realize that the heart behind wanting to avoid those "worldly" terms is only to respect and "consecrate" (meaning "to set apart unto the Lord") those young people who might otherwise be unrightfully or incorrectly mislabeled, being lumped in with a group whose values are less than godly. (And I don't see that as being from a haughty heart, in case that might be the assumption.) ... However, I personally believe that Jesus wants to "redeem" many areas of our lives, and take back "territory" that belonged to Him in the first place. I believe that includes the word teen(ager). That term, in reality, refers to nothing more than an age or stage of a maturing person. The term "typical teenager," on the other hand, indicates an element of worldliness that assumes certain negative things about this age of people. "Rebellious," "superficial," "boy-crazy," "irresponsible," etc. So I avoid the "typical teenager" term entirely! (By the way, God can even redeem teenagers who are caught in such elements; read "Changing the Heart of a Rebel.") I do not use the term "youth" to refer to people between the ages of 13 and 19 because it's vague and can just as accurately refer to a 9-year-old. 9-year-olds are not in high school; they are not in their last years of influence at home; they are not thinking about college, careers, or (hopefully or usually) the opposite gender. Teens and teenagers are. ...I personally do not use the term "young adult" either because I don't feel that teenagers *are* "young adults." If anything, they are most accurately "pre-adults" because an adult, by definition, is "full-grown," and, in our society, 21 years or older. So, to be a "young adult" you'd need to be at least 21, 22, or 23. ... When you think of it, even the term "adult" has been "slimed on" in our society. "Adult movies"; "Adult bookstores"; "For Adult Audiences only"... In each one of these the pornographic nature of the content is perfectly understood! So do we stop calling ourselves "adults" because of the world's tainting of the word? I'm not; I'm just helping redeem the word by being what God intended an "adult" to be. ... I just prefer the term "teen" or "teenager" for those persons between thirteen and nineteen. It's quick, it's accurate, and it's getting redeemed by the wonderful kids (woops – sorry; another disliked term by some! See definition earlier in this Glossary.) So now that you know my reasoning behind it, you won't have to cringe every time you pass it in this book!

unit - For our purposes here, a *unit* is an amount of study that doesn't equal (up to) at least ½ credit. (In Marilyn Howshall's treatment of this word, it takes on an entirely different meaning. She delves into the concept of *"unit-of-life"* studies which might be referred to as the "Real McCoy" of the commonly spoken-of term *"unit studies."* The most obvious and exciting difference between the two types of learning is that *unit-of-life* studies are natural, though definitely directed and skill-building, and, as such, *do not burn out the parent!)*

whole lotta good stuff - what I'm hoping you'll feel you have found in this manual! It is with a whole lotta l♥ve that I (and my family) present it to you and your family, which (who) has a whole lotta good stuff ahead of you as you(all)pursue God's very best for you!